Y. Hirabayashi, Y. Igarashi, A.H. Merrill, Jr. (Eds.)

Sphingolipid Biology

Y. Hirabayashi, Y. Igarashi,
A.H. Merrill, Jr. (Eds.)

Sphingolipid Biology

With 118 Figures, Including 1 in Color

 Springer

Yoshio Hirabayashi
Brain Science Institute, The Institute of Physical and Chemical Research (Riken)
2-1 Hirosawa, Wako, Saitama 351-0198, Japan

Yasuyuki Igarashi
Department of Biomembrane and Biofunctional Chemistry, Faculty of Advanced Life
Science and Graduate School of Life Science, Hokkaido University
Kita 12, Nishi 6, Kita-ku, Sapporo 060-0812, Japan

Alfred H. Merrill, Jr.
School of Biology and the Petit Institute for Bioengineering and Bioscience, Georgia
Institute of Technology
Atlanta, GA 30332, USA

ISBN-10 4-431-34198-6 Springer-Verlag Tokyo Berlin Heidelberg New York
ISBN-13 978-4-431-34198-7 Springer-Verlag Tokyo Berlin Heidelberg New York

Library of Congress Control Number: 2006927429

Printed on acid-free paper

Springer is a part of Springer Science+Business Media
springer.com

Typesetting: Camera-ready by the editors and authors
Printing and binding: Shinano Inc., Japan

Foreword

We are now in the new millennium, anticipating the coming of a new age of life sciences. It should not be called the post-genomic age, denoting a mere extension of the current situation, because no one knows what it is and will be.

This book, *Sphingolipid Biology*, was written by leading scientists of sphingolipidomics to introduce recent discoveries in sphingolipid research. The book provides the background of those discoveries and prospects for the future, not only for specialists but also for life science in general, and sheds light on coming developments in association with other research in basic sciences as well as applied life sciences.

The progress made during the past 15 years in sphingolipid research has been remarkable. This fact is readily understood when we open a book with a title similar to the current one, namely, *Sphingolipid Biochemistry*,[1] edited by J.N. Kanfer and S-I Hakomori in 1983, and also a special review issue entitled "Chemistry and Metabolism of Sphingolipids," published in 1970 in the journal *Chemistry and Physics of Lipids*. In the latter publication, one notices the work of Saul Roseman,[2] discoverer of the correct chemical structure of sialic acid, CMP-sialic acid (CMP-Neu5Ac), and the first sialyltransferase. In that issue he first proposed the concept of sphingolipid-mediated cell–cell recognition via molecular mechanism of the ligand-receptor interactions in cell surface membranes, depicting an interesting fishlike model. At the time, sphingomyelin had been regarded as a

---

[1] Kanfer JN, Hakomori S-I (eds) (1983) Sphingolipid Biochemistry. Handbook of Lipid Research, vol 3. Plenum Press, New York & London

[2] Roseman S (1970) The synthesis of complex carbohydrates by multiglycosyl-transferase systems and their potential function in intercellular adhesion. Chem Phys Lipids 5:270–297

most stubborn, mysterious molecule, metabolically and functionally, among not only sphingolipids but phospholipids themselves. No one could have anticipated the remarkable development in sphingomyelin research that has subsequently taken place. In this connection it is worth recalling the work, cited in Part 1, of Johann Ludwig Wilhelm Thudichum, discoverer of sphingosine and cerebroside, and also the founder of neurochemistry.

I should add the words of the late Moto-o Kimura, an eminent molecular evolutionist who proposed the concept of the neutral theory of molecular evolution. He said, "There is a great fissure between genotype (genome) and phonotype (phenome). Nobody has so far been successful in bridging the fissure." This remains our task for the future. The genome may be symbolized by the term "digital" and the phenome by "analog." It is likely that cell membrane lipids, lipid-mediated cell functions, and biosignaling are represented by analog. It is of the utmost importance to elucidate basic mechanisms of coupling between sphingolipids, phenome (analog), and genome (digital) at cellular and molecular levels. Genome (G) versus phenome (P) or phenome versus genome coupling (GVP or PVG coupling) is now proposed as a major issue in the coming post-genomic age.

As I hope is obvious from the foregoing remarks, sphingolipidomics is a subject that today increasingly involves all life scientists, for whom this volume provides a rich source of information.

Yoshitaka Nagai, Ph.D., D.Med.Sc.
Professor Emeritus
University of Tokyo

Contents

Foreword   v
Yoshitaka Nagai

Contributors   xiii

Part 1: Overview

1-1 Sphingolipids Synthesis, Transport and Cellular Signaling   3
Yoshio Hirabayashi, Yasuyuki Igarashi, and
Alfred H. Merrill, Jr.

Part 2: Biosynthesis, Transport of Sphingolipids

2-1 Serine Palmitoyltransferase   25
Jia Wei, Tokumbo Yerokun, Martina Liepelt, Amin Momin,
Elaine Wang, Kentaro Hanada, and Alfred H. Merrill, Jr.

2-2 Ceramide Synthase   49
Irene Pankova-Kholmyansky and Anthony H. Futerman

**2-3 Dihydroceramide: Sphinganine Δ4-Desaturase and
C4-Hydroxylase**   57
Akemi Suzuki, Fumio Omae, and Ayako Enomoto

**2-4 Metabolizing Enzymes Such As Sphingomyelin Synthase
Induce Cell Death by Increasing Ceramide Content**   69
Toshiro Okazaki

2-5 Glucosylceramide and Galactosylceramide Synthase   83
James A. Shayman

**2-6 Synthesis, Metabolism, and Trans-Bilayer Movement of
  Long-Chain Base**   95
Akio Kihara and Yasuyuki Igarashi

**2-7 Molecular Mechanism of Ceramide Trafficking from the
  Endoplasmic Reticulum to the Golgi Apparatus in
  Mammalian Cells**   107
Kentaro Hanada, Miyuki Kawano, and Keigo Kumagai

2-8 Sphingolipid Trafficking   123
Kouichi Funato and Howard Riezman

**2-9 Current Perspectives on *Saccharomyces cerevisiae*
  Sphingolipids**   141
Robert C. Dickson and Robert L. Lester

**Part 3: Generation and Degradation of Sphingolipid
  Signaling Molecules**

**3-1 Generation of Signaling Molecules by De Novo
  Sphingolipid Synthesis**   153
Kazuyuki Kitatani, L. Ashley Cowart, and Yusuf A. Hannun

**3-2 Overview of Acid and Neutral Sphingomyelinases in Cell
  Signaling**   167
Youssef Zeidan, Norma Marchesini, and Yusuf A. Hannun

**3-3 Neutral Ceramidase as an Integral Modulator for the
  Generation of S1P and S1P-Mediated Signaling**   183
Makoto Ito, Motohiro Tani, and Yukihiro Yoshimura

3-4 Activation of Sphingosine Kinase 1   197
Michael Maceyka, Sergio E. Alvarez, Sheldon Milstien, and
Sarah Spiegel

3-5 Ceramide 1-Phosphate   207
Susumu Mitsutake, Tack-Joong Kim, and Yasuyuki Igarashi

3-6 Sphingosine-1-Phosphate Lyase   **219**
Julie D. Saba

Part 4: Membrane Domain and Biological Function

**4-1 Close Interrelationship of Sphingomyelinase and Caveolin
in Triton X-100-Insoluble Membrane Microdomains**   **233**
Keiko Tamiya-Koizumi, Takashi Murate, Katsumi Tanaka,
Yuji Nishizawa, Nobuhiro Morone, Jiro Usukura, and
Yoshio Hirabayashi

**4-2 Roles of Membrane Domains in the Signaling Pathway for
B Cell Survival**   **245**
Miki Yokoyama, Tomoko Kimura, Sachio Tsuchida,
Hiroaki Kaku, Kiyoshi Takatsu, Yoshio Hirabayashi, and
Masaki Yanagishita

4-3 The Role of Lipid Rafts in Axon Growth and Guidance   **253**
Hiroyuki Kamiguchi

4-4 Sphingolipids and Multidrug Resistance of Cancer Cells   **263**
Gerrit van Meer, Maarten Egmond, and David Halter

**Part 5: Membrane Lipid Domain and Human
Pathobiology**

**5-1 A New Pathological Feature of Insulin Resistance and
Type 2 Diabetes: Involvement of Ganglioside GM3 and
Membrane Microdomains**   **273**
Jin-ichi Inokuchi, Kazuya Kabayama, Takashige Sato, and
Yasuyuki Igarashi

5-2 Neuronal Cell Death in Glycosphingolipidoses   **285**
Yaacov Kacher and Anthony H. Futerman

**5-3 Endocytic Trafficking of Glycosphingolipids in
Sphingolipidoses**   **295**
Amit Choudhury, David L. Marks, and Richard E. Pagano

5-4 Ganglioside and Alzheimer's Disease 309
Katsuhiko Yanagisawa

5-5 Modulation of Proteolytic Processing by Glycosphingolipids Generates Amyloid ß-Peptide 319
Irfan Y. Tamboli, Kai Prager, Esther Barth, Micheal Heneka, Konrad Sandhoff, and Jochen Walter

5-6 Hereditary Sensory Neuropathy 329
Simon J. Myers and Garth A. Nicholson

5-7 Ceramide, Ceramide Kinase and Vision Defects: A BLIND Spot for LIPIDS 337
Roser Gonzàlez-Duarte, Miquel Tuson, and Gemma Marfany

5-8 Fumonisin Inhibition of Ceramide Synthase: A Possible Risk Factor for Human Neural Tube Defects 345
Ronald T. Riley, Kenneth A. Voss, Marcy Speer, Victoria L. Stevens, and Janee Gelineau-van Waes

5-9 Sphingolipids and Cancer 363
Eva M. Schmelz and Holly Symolon

Part 6: S1P Signaling and SIPR

6-1 Sphingosine-1-Phosphate and the Regulation of Immune Cell Trafficking 385
Maria Laura Allende and Richard L. Proia

6-2 Sphingolipids and Lung Vascular Barrier Regulation 403
Liliana Moreno, Steven M. Dudek, and Joe G.N. Garcia

6-3 Signaling Mechanisms for Positive and Negative Regulation of Cell Motility by Sphingosine-1-Phosphate Receptors 415
Yoh Takuwa, Naotoshi Sugimoto, Noriko Takuwa, and Yasuyuki Igarashi

6-4 Sphingosine 1-Phosphate-Related Metabolism in the Blood Vessel 427
Yutaka Yatomi, Shinya Aoki, and Yasuyuki Igarashi

Part 7: Advanced Technology in Sphingolipid Biology

7-1 Conventional/Conditional Knockout Mice   **443**
Ichiro Miyoshi, Tadashi Okamura, Noriyuki Kasai,
Tetsuyuki Kitamoto, Shinichi Ichikawa, Soh Osuka, and
Yoshio Hirabayashi

**7-2 Biosynthesis and Function of *Drosophila*
Glycosphingolipids**   **453**
Ayako Koganeya-Kohyama and Yoshio Hirabayashi

**7-3 Characterization of Genes Conferring Resistance Against
ISP-1/Myriocin-Induced Sphingolipid Depletion in Yeast**   **463**
Hiromu Takematsu and Yasunori Kozutsumi

7-4 Lysenin: A New Probe for Sphingomyelin   **475**
Toshihide Kobayashi and Akiko Yamaji-Hasegawa

7-5 Structural Biology of Sphingolipid Synthesis   **483**
Hiroko Ikushiro, Akihiro Okamoto, and Hideyuki Hayashi

**7-6 A Computer Visualization Model for the De Novo
Sphingolipid Biosynthetic Pathway**   **493**
Jin Young Hong, Jeremy C. Allegood, Samuel Kelly,
Elaine Wang, Alfred H. Merrill Jr., and May Dongmei Wang

Appendix

I. Inhibitors of Sphingolipid Biosynthesis   **511**
Jin-ichi Inokuchi

II. The Genes in Yeast and Mammals   **515**
Yasuyuki Igarashi, Akio Kihara, and Yoshio Hirabayashi

III. Structure of Glycosphingolipids and Metabolic Pathway   **519**
Akemi Suzuki

Keyword Index   **529**

Contributors

Jeremy C. Allegood
School of Biology and the Petit Institute of Bioengineering and Bioscience, Georgia Institute of Technology, Atlanta, GA 30332, USA

Maria Laura Allende
Genetics of Development and Disease Branch, National Institutes of Diabetes and Digestive and Kidney Diseases, National Institutes of Health, Bethesda, MD 20892, USA

Sergio E. Alvarez
Department of Biochemistry, Virginia Commonwealth University Medical Center and the Massey Cancer Center, Richmond, Virginia 23298, USA

Shinya Aoki
Department of Laboratory Medicine, Graduate School of Medicine, The University of Tokyo, Tokyo 113-8655, Japan

Esther Barth
Department of Neurology, University of Bonn, Sigmund-Freud-Str. 25, 53127 Bonn, Germany

Amit Choudhury
Department of Biochemistry and Molecular Biology, Thoracic Diseases Research Unit, Mayo Clinic and Foundation, Rochester, Minnesota 55905, USA

L. Ashley Cowart
Department of Biochemistry and Molecular Biology, Medical University of South Carolina, 173 Ashley Avenue, Charleston, South Carolina 29402, USA

Robert C. Dickson
Department of Cellular and Molecular Biochemistry and the Lucille P. Markey Cancer Center, University of Kentucky College of Medicine, Lexington, KY 40536-0298, USA

Steven M. Dudek
Department of Medicine, University of Chicago, 5841 South Maryland Avenue, Chicago, Illinois 60637, USA

Maarten Egmond
Department of Membrane Enzymology, Bijvoet Center and Institute of Biomembranes, Utrecht University, Padualaan 8, 3584 CH Utrecht, The Netherlands

Ayako Enomoto
Sphingolipid Expression Laboratory, RIKEN Frontier Research System, 2-1 Hirosawa, Wako, Saitama 351-0198, Japan

Kouichi Funato
Graduate School of Biosphere Science, Faculty of Applied Biological Science, Hiroshima University, 1-4-4 Kagamiyama, Higashi-Hiroshima 739-8528, Japan

Anthony H. Futerman
Department of Biological Chemistry, Weizmann Institute of Science, Rehovot 76100, Israel

Joe G.N. Garcia
Department of Medicine, University of Chicago, 5841 South Maryland Avenue, Chicago, Illinois 60637, USA

Janee Gelineau-van Waes
Department of Genetics, Cell Biology & Anatomy, University of Nebraska Medical Center, Omaha, Nebraska 68198-5455, USA

Roser Gonzàlez-Duarte
Departament de Genètica, Facultat de Biologia, Universitat de Barcelona, Spain

David Halter
Department of Membrane Enzymology, Bijvoet Center and Institute of Biomembranes, Utrecht University, Padualaan 8, 3584 CH Utrecht, The Netherlands

Kentaro Hanada
Department of Biochemistry and Cell Biology, National Institute of Infectious Diseases, 1-23-1 Toyama, Shinjuku-ku, Tokyo 162-8640, Japan

Yusuf A. Hannun
Department of Biochemistry and Molecular Biology, Medical University of South Carolina, 173 Ashley Avenue, Charleston, South Carolina 29425, USA

Hideyuki Hayashi
Department of Biochemistry, Osaka Medical College, Takatsuki, Osaka 569-8686, Japan

Micheal Heneka
Department of Neurology, University of Bonn, Sigmund-Freud-Str. 25, 53127 Bonn, Germany

Yoshio Hirabayashi
Brain Science Institute, The Institute of Physical and Chemical Research (RIKEN), 2-1 Hirosawa, Wako, Saitama 351-0198, Japan

Jin Young Hong
The Wallace H. Coulter Department of Biomedical Engineering and the Petit Institute of Bioengineering and Bioscience, Georgia Institute of Technology, Atlanta, GA 30332, USA

Shinichi Ichikawa
Department of Applied Life Sciences, Niigata University of Pharmacy and Applied Life Sciences, Niigata 956-8603, Japan

Yasuyuki Igarashi
Department of Biomembrane and Biofunctional Chemistry, Graduate School of Pharmaceutical Sciences, Hokkaido University, Kita 12, Nishi 6, Kita-ku, Sapporo 060-0812, Japan

Hiroko Ikushiro
Department of Biochemistry, Osaka Medical College, Takatsuki, Osaka 569-8686, Japan

Jin-ichi Inokuchi
Department of Biomembrane and Biofunctional Chemistry and CREST, Japan Science and Technology Agency, Graduate School of Pharmaceutical Sciences, Frontier Research Center for Post-Genomic Science and Technology, Hokkaido University, Kita 21, Nishi 11, Kita-ku, Sapporo 001-0021, Japan

Makoto Ito
Department of Bioscience and Biotechnology, Graduate School of Bioresource and Bioenvironmental Sciences, Kyushu University, Hakozaki 6-10-1, Fukuoka 812-8581, Japan
Bio-archtechture Center, Kyushu University, Hakozaki 6-10-1, Fukuoka 812-8581, Japan

Kazuya Kabayama
CREST, Japan Science and Technology Agency, Graduate School of Pharmaceutical Sciences, Frontier Research Center for Post-Genomic Science and Technology, Hokkaido University, Kita 21, Nishi 11, Kita-ku, Sapporo 001-0021, Japan

Yaacov Kacher
Department of Biological Chemistry, Weizmann Institute of Science, Rehovot 76100, Israel

Hiroaki Kaku
Division of Immunology, Department of Microbiology and Immunology, Institute of Medical Science, University of Tokyo, 4-6-1 Shirogane-dai, Minato-ku, Tokyo 108-8639, Japan

Hiroyuki Kamiguchi
Laboratory for Neuronal Growth Mechanisms, RIKEN Brain Science Institute, 2-1 Hirosawa, Wako, Saitama 351-0198, Japan

Noriyuki Kasai
Institute for Animal Experimentation, Tohoku University Graduate School of Medicine, Sendai 980-8575, Japan

Miyuki Kawano
Department of Biochemistry and Cell Biology, National Institute of Infectious Diseases, 1-23-1 Toyama, Shinjuku-ku, Tokyo 162-8640, Japan

Samuel Kelly
School of Biology and the Petit Institute of Bioengineering and Bioscience, Georgia Institute of Technology, Atlanta, GA 30332, USA

Akio Kihara
Department of Biomembrane and Biofunctional Chemistry, Graduate School of Pharmaceutical Sciences, Hokkaido University, Kita 12, Nishi 6, Kita-ku, Sapporo 060-0812, Japan

Tack-Joong Kim
Department of Biomembrane and Biofunctional Chemistry, Graduate School of Pharmaceutical Sciences, Hokkaido University, Kita 12, Nishi 6, Kita-ku, Sapporo 060-0812, Japan

Tomoko Kimura
Department of Hard Tissue Engineering, Biochemistry, Division of Bio-Matrix, Graduate School, Tokyo Medical and Dental University, 1-5-45 Yushima, Bunkyo-ku, Tokyo 113-8549, Japan

Tetsuyuki Kitamoto
Department of Neurological Science, Tohoku University Graduate School of Medicine, Sendai 980-8575, Japan

Kazuyuki Kitatani
Department of Biochemistry and Molecular Biology, Medical University of South Carolina, 173 Ashley Avenue, Charleston, South Carolina 29402, USA

Toshihide Kobayashi
Lipid Biology Laboratory, RIKEN (Institute of Physical and Chemical Research) Discovery Research Institute, 2-1 Hirosawa, Wako, Saitama 351-0198, Japan
Supra-Biomolecular System Research Group, RIKEN Frontier Research System, Saitama 351-0198, Japan
INSERM U585, INSA (Institut National des Sciences Appliquees)-Lyon, 20 Ave A. Einstein, 69621 Villeurbanne, France

Ayako Koganeya-Kohyama
Hirabayashi Laboratory Unit, Brain Science Institute, RIKEN, 2-1 Hirosawa, Wako, Saitama 351-0198, Japan

Yasunori Kozutsumi
Laboratory of Membrane Biochemistry and Biophysics, Graduate School of Biostudies, Kyoto University, Kyoto 606-8501, Japan
CREST, JST, Kawaguchi, Saitama, Japan
Supra-biomolecular System Research Group, Frontier Research System, RIKEN, Saitama 351-0198, Japan

Keigo Kumagai
Department of Biochemistry and Cell Biology, National Institute of Infectious Diseases, 1-23-1 Toyama, Shinjuku-ku, Tokyo 162-8640, Japan

Robert L. Lester
Department of Cellular and Molecular Biochemistry and the Lucille P. Markey Cancer Center, University of Kentucky College of Medicine, Lexington, KY 40536-0298, USA

Martina Liepelt
School of Biology and the Petit Institute for Bioengineering and Bioscience, Georgia Institute of Technology, Atlanta, GA 30332, USA

Michael Maceyka
Department of Biochemistry, Virginia Commonwealth University Medical Center and the Massey Cancer Center, Richmond, Virginia 23298, USA

Norma Marchesini
Department of Biochemistry and Molecular Biology, Medical University of South Carolina, Charleston, SC 29425, USA

Gemma Marfany
Departament de Genètica, Facultat de Biologia, Universitat de Barcelona, Spain

David L. Marks

Department of Biochemistry and Molecular Biology, Thoracic Diseases Research Unit, Mayo Clinic and Foundation, Rochester, Minnesota 55905, USA

Alfred H. Merrill Jr.

School of Biology and the Petit Institute of Bioengineering and Bioscience, Georgia Institute of Technology, Atlanta, GA 30332, USA

Sheldon Milstien

Laboratory of Cellular and Molecular Regulation, National Institute of Mental Health, Bethesda, MD 20892, USA

Susumu Mitsutake

Department of Biomembrane and Biofunctional Chemistry, Graduate School of Pharmaceutical Sciences, Hokkaido University, Kita 12, Nishi 6, Kita-ku, Sapporo 060-0812, Japan

Ichiro Miyoshi

Center for Experimental Animal Science, Nagoya City University Graduate School of Medical Sciences, Nagoya 467-8601, Japan

Amin Momin

School of Biology and the Petit Institute for Bioengineering and Bioscience, Georgia Institute of Technology, Atlanta, GA 30332, USA

Liliana Moreno

Department of Medicine, University of Chicago, 5841 South Maryland Avenue, Chicago, Illinois 60637, USA

Nobuhiro Morone

Nagoya University Graduate School of Medicine, Nagoya 466-8550, Japan

Takashi Murate

Nagoya University Graduate School of Medicine, Nagoya 461-8673, Japan

Simon J. Myers

Northcott Neuroscience Laboratory, ANZAC Research Institute, Hospital Rd, Concord, Sydney, NSW 2139, Australia

Garth A. Nicholson
Northcott Neuroscience Laboratory, ANZAC Research Institute, Hospital Rd, Concord, Sydney, NSW 2139, Australia
Department of Molecular Medicine, Concord Repatriation General Hospital, Hospital Rd, Concord, Sydney, NSW 2139, Australia

Yuji Nishizawa
Nagoya University Graduate School of Medicine, Nagoya 466-8550, Japan

Akihiro Okamoto
Department of Biological Science & Technology, School of High-Technology for Human Welfare, Tokai University, Numazu, Shizuoka 410-0395, Japan

Tadashi Okamura
Department of Infectious Diseases, Research Institute, International Medical Center of Japan, Tokyo 162-8655, Japan

Toshiro Okazaki
Department of Clinical Laboratory Medicine/Hematology, Faculty of Medicine, Tottori University, Yonago, Tottori 683-8504, Japan

Fumio Omae
Sphingolipid Expression Laboratory, RIKEN Frontier Research System, 2-1 Hirosawa, Wako, Saitama 351-0198, Japan

Soh Osuka
CREST and Neuronal Circuit Mechanism Research Group, RIKEN Brain Science Institute, Saitama, Japan

Richard E. Pagano
Department of Biochemistry and Molecular Biology, Thoracic Diseases Research Unit, Mayo Clinic and Foundation, Rochester, Minnesota 55905, USA

Irene Pankova-Kholmyansky
Department of Biological Chemistry, Weizmann Institute of Science, Rehovot 76100, Israel

Kai Prager

Department of Neurology, University of Bonn, Sigmund-Freud-Str. 25, 53127 Bonn, Germany

Richard L. Proia

Genetics of Development and Disease Branch, National Institutes of Diabetes and Digestive and Kidney Diseases, National Institutes of Health, Bethesda, MD 20892, USA

Howard Riezman

Department of Biochemistry, University of Geneva, Sciences II, 30, quai E. Ansermet, CH-1211 Geneva 4, Switzerland

Ronald T. Riley

Toxicology and Mycotoxin Research Unit, United States Department of Agriculture-Agricultural Research Service, PO Box 5677, Athens, Georgia 30604, USA

Julie D. Saba

Children's Hospital Oakland Research Institute, 5700 Martin Luther King Jr. Way, Oakland, CA 94609, USA

Konrad Sandhoff

Kekulé-Institute for Organic Chemistry and Biochemistry, University of Bonn, Gerhard-Domagk-Strasse 1, 53121 Bonn, Germany

Takashige Sato

Department of Biomembrane and Biofunctional Chemistry, Graduate School of Pharmaceutical Sciences, Frontier Research Center for Post-Genomic Science and Technology, Hokkaido University, Kita 21-Nishi 11, Kita-ku, Sapporo 001-0021, Japan

Eva M. Schmelz

Karmanos Cancer Institute, Wayne State University School of Medicine, HWCRC 608, 110 E. Warren Avenue, Detroit, MI 48201, USA. School of Biology and the Petit Institute for Bioengineering and Bioscience, Georgia Institute of Technology, Atlanta, GA 30332, USA

James A. Shayman
Department of Internal Medicine, Division of Nephrology, University of Michigan, 1560 MSRB II, 1150 West Medical Center Drive, Ann Arbor, Michigan 48109-0676, USA

Marcy Speer
Duke Center for Human Genetics, School of Medicine, Duke University, Durham, North Carolina 27710, USA

Sarah Spiegel
Department of Biochemistry, Virginia Commonwealth University Medical Center and the Massey Cancer Center, Richmond, Virginia 23298, USA

Victoria L. Stevens
Epidemiology and Surveillance Research, American Cancer Society, Atlanta, Georgia, USA

Naotoshi Sugimoto
Department of Physiology, Kanazawa University Graduate School of Medicine, 2-6-11 Takara-machi, Kanazawa 920-8640, Japan

Akemi Suzuki
Sphingolipid Expression Laboratory, RIKEN Frontier Research System, 2-1 Hirosawa, Wako, Saitama 351-0198, Japan

Holly Symolon
Karmanos Cancer Institute, Wayne State University School of Medicine, HWCRC 608, 110 E. Warren Avenue, Detroit, MI 48201, USA
School of Biology and the Petit Institute for Bioengineering and Bioscience, Georgia Institute of Technology, Atlanta, GA 30332, USA

Kiyoshi Takatsu
Division of Immunology, Department of Microbiology and Immunology, Institute of Medical Science, University of Tokyo, 4-6-1 Shirogane-dai, Minato-ku, Tokyo 108-8639, Japan

Hiromu Takematsu
Laboratory of Membrane Biochemistry and Biophysics, Graduate School of Biostudies, Kyoto University, Kyoto 606-8501, Japan
CREST, JST, Kawaguchi, Saitama, Japan

Noriko Takuwa
Department of Physiology, Kanazawa University Graduate School of Medicine, 2-6-11 Takara-machi, Kanazawa 920-8640, Japan

Yoh Takuwa
Department of Physiology, Kanazawa University Graduate School of Medicine, 2-6-11 Takara-machi, Kanazawa 920-8640, Japan

Irfan Y. Tamboli
Department of Neurology, University of Bonn, Sigmund-Freud-Str. 25, 53127 Bonn, Germany

Keiko Tamiya-Koizumi
Nagoya University Graduate School of Medicine, Nagoya 461-8673, Japan

Katsumi Tanaka
Nagoya University Graduate School of Medicine, Nagoya 466-8550, Japan

Motohiro Tani
Department of Biomembrane and Biofunctional Chemistry, Graduate School of Pharmaceutical Sciences, Hokkaido University, Kita 12, Nishi 6, Kita-ku, Sapporo 060-0812, Japan

Sachio Tsuchida
Department of Hard Tissue Engineering, Biochemistry, Division of Bio-Matrix, Graduate School, Tokyo Medical and Dental University, 1-5-45 Yushima, Bunkyo-ku, Tokyo 113-8549, Japan

Miquel Tuson
Departament de Genètica, Facultat de Biologia, Universitat de Barcelona, Spain

Jiro Usukura
Nagoya University Graduate School of Medicine, Nagoya 466-8550, Japan

Gerrit van Meer
Department of Membrane Enzymology, Bijvoet Center and Institute of Biomembranes, Utrecht University, Padualaan 8, 3584 CH Utrecht, The Netherlands

Kenneth A. Voss
Toxicology and Mycotoxin Research Unit, United States Department of Agriculture-Agricultural Research Service, PO Box 5677, Athens, Georgia 30604, USA

Jochen Walter
Department of Neurology, University of Bonn, Sigmund-Freud-Str. 25, 53127 Bonn, Germany

Elaine Wang
School of Biology and the Petit Institute of Bioengineering and Bioscience, Georgia Institute of Technology, Atlanta, GA 30332, USA

May Dongmei Wang
The Wallace H. Coulter Department of Biomedical Engineering, and the Petit Institute of Bioengineering and Bioscience, Georgia Institute of Technology, Atlanta, GA 30332, USA

Jia Wei
School of Biology and the Petit Institute for Bioengineering and Bioscience, Georgia Institute of Technology, Atlanta, GA 30332, USA

Akiko Yamaji-Hasegawa
Laboratory of Cellular Biochemistry, RIKEN Discovery Research Institute, Saitama 351-0198, Japan

Katsuhiko Yanagisawa
Department of Alzheimer's Disease Research, National Institute for Longevity Sciences, National Center for Geriatrics and Gerontology, 36-3 Gengo, Morioka, Obu 474-8522, Japan

Masaki Yanagishita
Department of Hard Tissue Engineering, Biochemistry, Division of Bio-Matrix, Graduate School, Tokyo Medical and Dental University, 1-5-45 Yushima, Bunkyo-ku, Tokyo 113-8549, Japan

Yutaka Yatomi
Department of Laboratory Medicine, Graduate School of Medicine, The University of Tokyo, Tokyo 113-8655, Japan

Tokumbo Yerokun
School of Biology and the Petit Institute for Bioengineering and Bioscience, Georgia Institute of Technology, Atlanta, GA 30332, USA

Miki Yokoyama
Department of Hard Tissue Engineering, Biochemistry, Division of Bio-Matrix, Graduate School, Tokyo Medical and Dental University, 1-5-45 Yushima, Bunkyo-ku, Tokyo 113-8549, Japan

Yukihiro Yoshimura
Department of Bioscience and Biotechnology, Graduate School of Bioresource and Bioenvironmental Sciences, Kyushu University, Hakozaki 6-10-1, Fukuoka 812-8581, Japan

Youssef Zeidan
Department of Biochemistry and Molecular Biology, Medical University of South Carolina, Charleston, SC 29425, USA

Acknowledgment

*We are deeply indebted to Bonnie Lee La Madeleine,
Brain Science Institute, RIKEN,
for her excellent editorial assistance.*

Part 1

Overview

1-1 Sphingolipids Synthesis, Transport and Cellular Signaling

Yoshio Hirabayashi[1], Yasuyuki Igarashi[2], and Alfred H. Merrill, Jr.[3]

[1]Brain Science Institute, The Institute of Physical and Chemical Research (Riken), 2-1 Hirosawa, Wako, Saitama 351-0198, Japan; [2]Department of Biomembrane and Biofunctional Chemistry, Graduate School of Pharmaceutical Sciences, Hokkaido University, Kita 12, Nishi 6, Kita-ku, Sapporo 060-0812, Japan; [3]School of Biology and the Petit Institute for Bioengineering and Bioscience, Georgia Institute of Technology, Atlanta, GA 30332, USA

Summary. Sphingolipids are comprised of a sphingoid base backbone (sphingosine, sphinganine or other species) that is synthesized *de novo* from serine and a fatty acyl-coenzyme A then converted into more complex compounds (ceramides, phosphosphingolipids, glycosphingolipids and protein adducts) that are key to the structures of cell membranes, lipoproteins, and the lamellar water barrier of the skin. Many complex sphingolipids as well as simpler sphingoid bases and derivatives are highly bioactive as extra- and intra-cellular regulators of growth, differentiation, migration, survival and numerous cellular responses to stress. The large numbers of functionally significant species requires careful control of sphingolipid biosynthesis, intracellular trafficking and turnover, and the regulatory mechanisms are still being discovered using genetic, biochemical and "sphingolipidomic" approaches. In the approximately 120 years since the compound "sphingosin" was first described, many biological mysteries have been explained by the biophysical and regulatory properties of sphingolipids, and it is certain that clarification of the remaining enigmas of sphingolipidology will contribute profoundly to our understanding of normal and abnormal cell behavior, and new ways to prevent, detect and treat disease.

Keywords.   sphingolipid structure, metabolism, biologic functions, disease

1. Introduction

Sphingolipids are found in essentially all animals, plants, and fungi, as well as some prokaryotic organisms and viruses. They are mostly in membranes, but are also major constituents of lipoproteins, the multilamellar water barrier of skin, and as extracellular signals. This introductory chapter reviews sphingolipid structure and nomenclature, metabolism, trafficking and functions as a general orientation for some readers of this book, and highlights some of the new developments and challenges in the field. For more in-depth descriptions and bibliographies for these topics, please consult the specific chapters of this book.

2. Structures and nomenclature

Sphingolipids can be divided into several major categories: the sphingoid bases and their simple derivatives, ceramides and more complex sphingolipids (Fig. 1). The Lipid MAPS Consortium, after consultation with scientists in Europe and Japan, has recommended a systematic way of depicting sphingolipids (Fahy et al. 2005) that is compatible with the previous recommendations from the IUPAC to facilitate cross-platform communication among scientists in different lipid fields and biological disciplines.

2.1 Sphingoid bases

The name "sphingosin" for a fundamental component of sphingolipids was given by J. L. W. Thudichum in his 1884 treatise *The Chemistry of the Brain* "in commemoration of the many enigmas which it presented to the inquirer." Sphingolipids are now known to be comprised of many backbone "sphingoid bases" (some of which are shown in Fig. 2) and the term sphingosine usually now refers specifically to (2S,3R,4E)-2-aminooctadec-4-ene-1,3-diol, which is also called D-*erythro*-sphingosine and E-sphing-4-enine (nonetheless, one still sometimes encounters studies where the backbone has not been specifically analyzed but is referred to generically as "sphingosine"). A convenient abbreviation for this compound is d18:1 with the first number reflecting the number of carbon atoms, the second the number of double bonds, and "d" referring to the two (<u>di</u>-) hydroxyl groups. Other sphingoid bases vary in alkyl chain length and branching, the number and positions of double bonds, the presence of additional hydroxyl groups, and other features. The structural variation has functional significance; for example, sphingoid bases in skin have an additional hydroxyl at position 4 (with the primary species being 4-hydroxysphinganine, (2S,3R,4R)-2-amino-octadecanetriol, or "phytosphingosine" in the older nomenclature, or in lesser amounts at position 6, and hydroxy-fatty acids in the ceramide, which favor interaction with neighboring molecules, thereby strengthening the permeability barrier of skin.

Sphingoid bases are present in cells primarily as the backbones of more complex sphingolipids, however, as will be discussed later, they also function as intra- and extra-cellular signals and second messengers as free sphingoid bases and sphingoid base 1-phosphates.

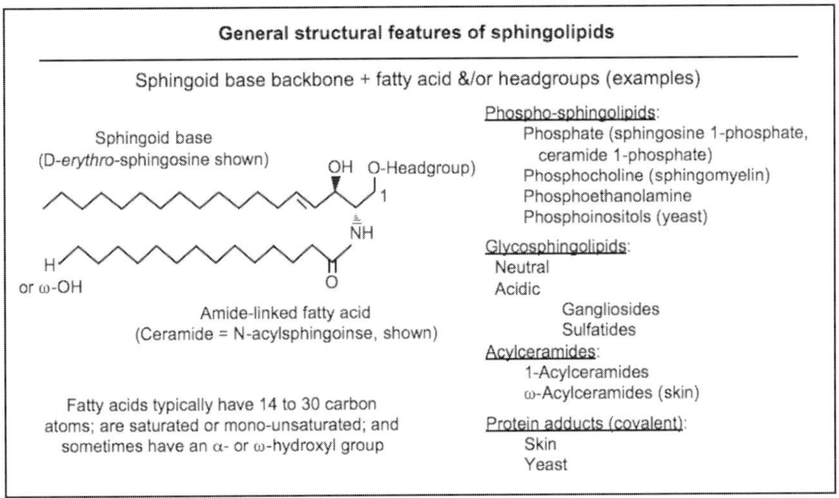

Fig. 1. Overview of sphingolipid structures.

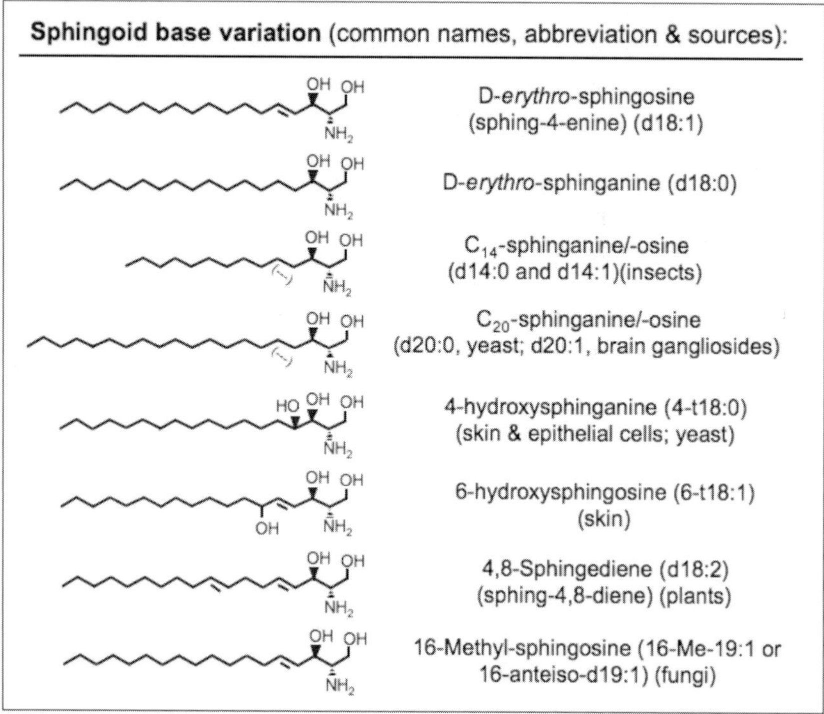

Fig. 2. Structures of the major sphingoid bases of mammals (sphingosine, sphinganine, d20:1 sphingosine, 4-hydroxysphinganine and 6-hydroxysphingosine) and example of sphingoid bases from other species.

2.2 Ceramides

Ceramides are fatty acid derivatives of sphingoid bases (Fig. 1). The fatty acids are typically saturated or mono-unsaturated with chain lengths from 14 to 26 carbon atoms (or even longer in the special case of skin), and sometimes have a hydroxyl group on the ω- or ω-carbon atom (Fig. 3). The predominately saturated alkyl chains of both the sphingoid base and the fatty acid give ceramides high phase transition temperatures (typically >37°) which favors the segregation of ceramides and some complex sphingolipids into specialized regions of the membrane (called "rafts," "microdomains" and, when the protein caveolin is present, "caveolae") that participate in cell signaling, nutrient transport and other functions.

Ceramides also serve as second messengers that regulate cell growth, senescence and programmed cell death (apoptosis). Their biologic activity depends on the type of sphingoid base and fatty acid; for examples, dihydroceramides (i.e., without the 4,5-double bond of the sphingosine back-bone) (Fig. 1) are typically less potent than ceramides as inducers of apoptosis (Futerman and Hannun 2004), whereas, phytoceramides (i.e., with 4-hydroxysphinganine or "phytosphingosine" as the backbone) are sometimes seen to be more potent.

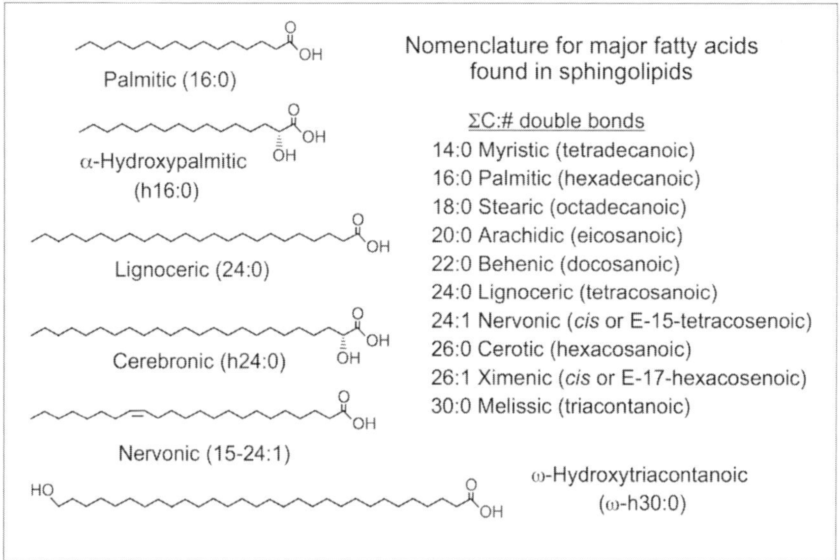

Fig. 3. Major fatty acids found in amide-linkage to sphingoid bases in sphingolipids.

2.3 More complex phospho- and glyco-sphingolipids

The major phosphosphingolipids of mammals are sphingomyelins (cera-mide phosphocholines) (Fig. 1), whereas insects contain mainly ceramide phosphoethanolamines and fungi have phytoceramidephosphoinositols and inositol phosphates. Some aquatic organisms also contain sphingolipids in which the phosphate has been replaced by a phosphono- or arsenate group.

Glycosphingolipids contain one or more carbohydrate groups in—typically--glycosidic linkage ("typically" because this is the usual presumption, however ceramide phosphoinositols can also be categorized glycosphingolipids), as shown in Fig. 4. They are classified several ways, with the most general being the general types of carbohydrates of which they are composed: 1) neutral glycosphingolipids contain one or more uncharged sugars such as glucose (abbreviated Glc, hence, glucosylcera-mide is GlcCer), galactose (Gal), N-acetylglucosamine (GlcNAc),

N-acetylgalactosamine (GalNAc), and fucose (Fuc); and 2) acidic glycosphingolipids contain ionized functional groups (phosphate or sulfate) attached to neutral sugars, or charged sugar residues such as sialic acid (N-acetylneuraminic acid, NeuAc). Among the acidic glycosphingolipids, gangliosides all contain, and the number of sialic acid residues is usually denoted with a subscript letter (i.e., Mono-, Di- or Tri-) plus a number reflecting the subspecies within that category. For a few glycosphingolipids, historically assigned names as antigens and blood group structures are still in common usage (e.g., Lexis x and sialyl Lewis x).

Another subdivision of the glycosphingolipids that is often utilized in naming the neutral species is shown in Fig. 5. Additional information about sphingolipid structures can be obtained in the appendices to this book as well as websites such as http://www.glycoforum.gr.jp/ and www.sphingomap.org (which will soon be posted on www.lipidmap.org), and when the goal is to identify candidate the potential glycosphingolipids from ions that have been obtained by mass spectrometry, tools such as the lipid search engine at http://lipidsearch.jp can be useful.

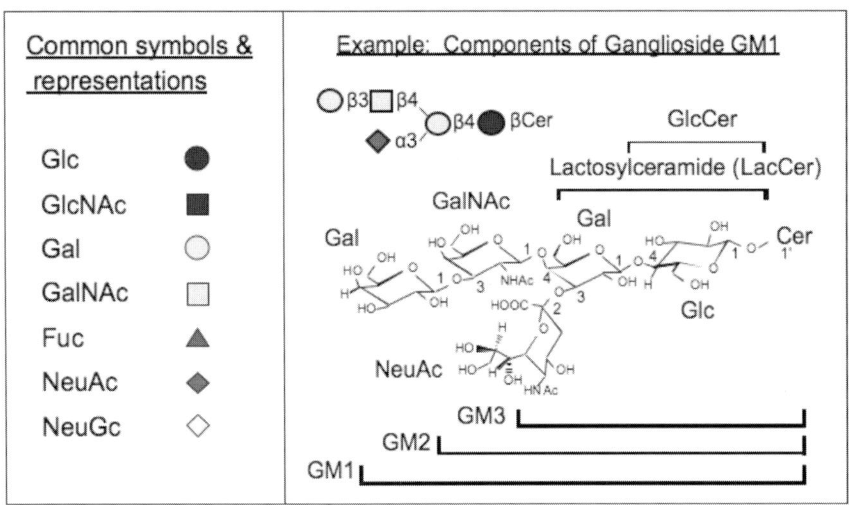

Fig. 4. Ways of depicting glycosphingolipids in two representations (full chemical structures versus symbols for the carbohydrates) using ganglioside GM1 and its components as examples. The abbreviations are glucose (Glc), N-acetylglucosamine (GlcNAc), galactose (Gal), N-acetylgalactosamine (GalNAc), fucose (Fuc), N-acetylneuraminic acid or sialic acid (NeuAc) and N-glycolylneuraminic acid (NeuGc).

Nomenclature for classification of glycosphingolipids

Root name	Carbohydrate in the 'root' structure				Other
(Abbrev)	IV	III	II	I	subcategories

Ganglio (Gg)	Galβ1–3GalNAcβ1–4Galβ1–4Glcβ1–Cer
Lacto (Lc)	Galβ1–3GlcNAcβ1–3Galβ1–4Glcβ1–Cer
Neolacto (nLc)	Galβ1–4GlcNAcβ1–3Galβ1–4Glcβ1–Cer
Globo (Gb)	GalNAcβ1–3Galα1–4Galβ1–4Glcβ1–Cer
Isoglobo (iGb)	GalNAcβ1–3Galα1–3Galβ1–4Glcβ1–Cer
Mollu (Mu)	GalNAcβ1–2Manα1–3Manβ1–4Glcβ1–Cer
Arthro (At)	GalNAcβ1–4GlcNAcβ1–3Manβ1–4Glcβ1–Cer

Ganglioside (Gn) 　(n= # of neuraminic acids; M(ono), D(di-), T(ri-), etc.)	contain N-acetyl (NeuAc) 　or N-glycolyl- (NeuGc) 　neuraminic acid
Sulfatide	contain sulfate

Fig. 5. Nomenclature for classification of the root carbohydrates of glycosphingolipids.

2.4 Protein adducts

Some sphingolipids are covalently attached to protein, for examples: ω-hydroxy- ceramides and -glucosylceramides are attached to surface proteins of skin; and, inositol-phosphoceramides are used as membrane anchors for some fungal proteins, in a manner somewhat analogous to the glycosylphosphatidylinositol (GPI)-anchors that are attached to proteins in other eukaryotes.

3. Biologic functions

The biological functions of sphingolipids are as complex and diverse as their structures, as illustrated in Fig. 6. They range from functions that are attributable to the general biophysical properties of these compounds to subtle and highly specified lipid-protein and lipid-lipid interactions (that have been termed the glycosynapse) to intra- and extra-cellular signaling.

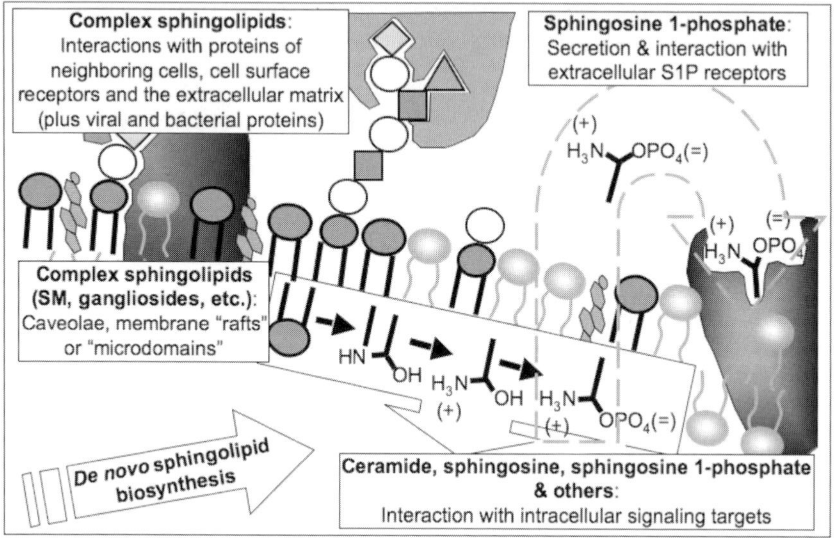

Fig. 6. Illustration of representative biological structures and functions of sphingolipids.

3.1 Biophysical properties of sphingolipids related to cell regulation

The unique biophysical properties of sphingolipids allow them to perform specialized functions while part of membrane bilayers and other structures. Most glycosphingolipids and sphingomyelins tend to cluster dynamically into structures variously called microdomains, rafts and caveolae (when the structure contains caveolin) rather than behave like typical "fluid" membrane lipids due to the typically saturated alkyl sidechains, which allow strong van der Waals interactions, and the ceramide hydroxyls, amide bond and polar headgroups that are capable of hydrogen bonding and dipolar interactions. Although it is often difficult to prove that a particular structure fits these criteria, there is now general consensus that such structures are enriched in growth factor receptors, transporters and other proteins, especially proteins with a glycosylphosphatidylinositiol-lipid anchor. The bilayer "stabilizing" properties of sphingolipids are also important for other types of biological structures, such as the lamellar bodies that maintain the permeability barrier of skin.

In contrast to these functions, which are related to the non-polar properties of sphingolipids, some categories of sphingolipids are relatively water soluble and this undoubtedly has been capitalized upon for their function. Glycosphingolipids with multiple carbohydrates (especially when acidic) are sufficiently polar to partition into the aqueous phase of standard organic solvent extracts, and this is also manifested in the tendency of these compounds to dissociate from cells under certain conditions (such as in association with microvesicles). Sphingosine 1-phosphate is also comparatively water soluble and sphingoid bases in general, although favoring a hydrophobic environment, are sufficiently water soluble to move rapidly between membranes. It is also noteworthy that sphingoid bases are one of the only positively charged lipids of mammalian membranes, and the amino group has a surprisingly low pKa (between 7 and 8) due to the vicinal hydroxyls, hence, can exist in both cationic and neutral forms depending on the environment.

3.2 Biological recognition and regulation

Cellular sphingolipids are located predominantly on the outer leaflet of the plasma membrane, the lumen of intracellular vesicles and organelles (endosomes, Golgi membranes, etc.) as well as in yet undefined locations in mitochondria, nuclei and other intracellular compartments. The complex carbohydrate moieties often interact with complementary ligands, such as extracellular matrix proteins and receptors (Fig. 6), and even other carbohydrates and even other carbohydrates, in what has been referred to as a glycosynapse (Hakomori, 2004). In addition, some sphingolipids interact with proteins on the same cell surface (Fig. 6) to control the location of the protein (for example, in membrane rafts with other signaling proteins) as well as to modify the conformation of the protein and its activity. The latter is exemplified by the binding of gangliosides by several growth factor receptors. Sphingolipids are also recognized by viruses, bacteria and bacterial toxins as a means of both attachment and entry. These types of interactions are clearly both subtle and complex, and—as is often encountered in regulation—highly interconnected with other regulatory pathways so there is not only extensive cross-talk among regulatory systems but also some capacity for redundancy. One of the greatest challenges facing cell biology is the unraveling of the glyco-code.

3.3 Sphingolipids and signal transduction

As already noted, complex sphingolipids participate in cell signaling by affecting the properties of receptors directly (as rafts and allosteric modulators), and—as for other lipid families--the lipid backbones (ceramide, sphingosine, sphingosine 1-phosphate and other species) are also utilized for signal transduction, as noted in Fig. 6. In the "classic" signaling pathways, receptor activation by cytokines (IL1, tumor necrosis factor-ω and many others), growth factors (e.g., platelet derived growth factor) and other agonists induce the hydrolysis of a membrane sphingolipid (sphingomyelin) to ceramide that can serve as a lipid second messenger, or be converted to downstream metabolites (sphingosine, sphingosine 1-phosphate, and the more recently discovered ceramide 1-phosphate), to activate or inhibit downstream targets (protein kinases, phosphoprotein phosphatases and others) that control cell behavior. Because ceramide and sphingosine 1-phosphate often have opposing signaling functions (e.g., induction versus inhibition of apoptosis; inhibition versus stimulation of growth), it has been proposed that cells utilize a ceramide/sphingosine 1-phosphate "rheostat" in deciding between growth arrest/apoptosis versus proliferation/ survival (Chalfant and Spiegel 2005). It is also apparently the case that cells generate multiple mediators to control related steps in a cell activation cascase, as has surfaced in the generation of ceramide 1-phosphate to activate phospholipase A_2 accompanied by induction of COX-2 by sphingosine 1-phosphate. Sphingosine 1-phosphate is additionally released from cells to serve as the agonist for a family of S1P receptors that control multiple cell functions.

In addition to the "classical" signaling pathway, many studies have found that cell regulatory mediators are formed by *de novo* biosynthesis (indeed, this source may prove to be as important as turnover of pre-existing complex sphingolipids). This, however, is also a dangerous game to play because disruption of *de novo* sphingolipid biosynthesis can form toxic intermediates very rapidly, and has been identified to play a role in several diseases.

4. Sphingolipid metabolism

Sphingolipid biosynthesis and turnover has been discussed in depth by several chapters of this book; however, a summary is shown in Fig. 7 to highlight the large number of intermediates that are known to be bioactive as cell regulatory mediators. Hence, it is not surprising that *de novo* sphingolipid biosynthesis as well as turnover has been used to produce (and consume) these compounds for purposes other than just membrane lipid maintenance.

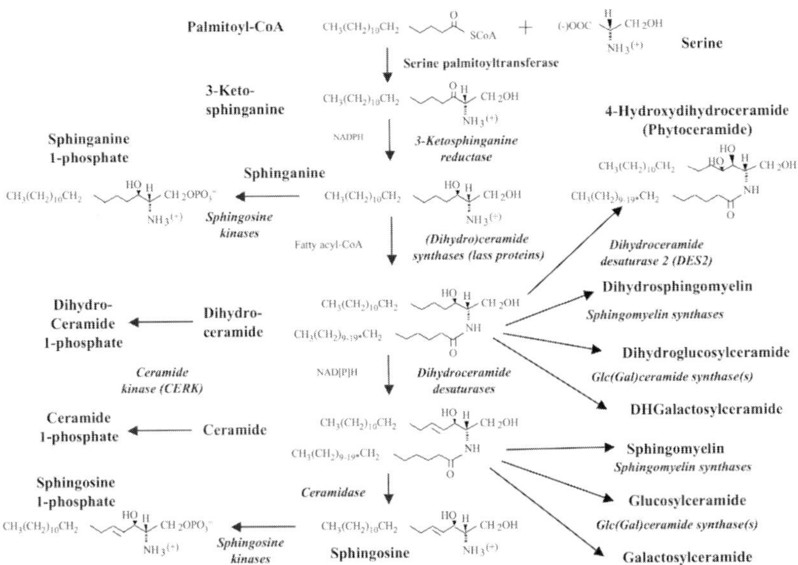

Fig. 7. A summary of the biosynthetic reactions for formation of the lipid backbone of sphingolipids and the initial more complex sphingolipids into which it is incorporated. Not shown are the N-acetyl-sphingoid bases, N-methyl-sphingoid bases, and 1-acylceramides that can be made in some special circumstances. Also not shown are species differences in the pathway, such as that some species use additional fatty acyl-CoA's to produce sphingoid bases of other chain lengths, and that yeast produce 4-hydroxy-sphingoid bases before N-acylation.

One of the major surprises of the last few years has been that sphingoid base acylation is accomplished by ceramide synthases (lass gene family proteins) that are highly specific for the acyl chain lengths. For example, the lass1 gene product essentially utilizes only stearoyl-CoA and produces C18-ceramides (i.e., d18:0/18:0 and d18:1/18:0); furthermore, in the short time since this discovery, it has already been found that genetic and bio-chemical defects in the production of this ceramide subspecies play a role in head and neck cancer (Koybasi et al. 2004). Therefore, despite the complexity of the pathway as it is displayed in Fig. 7, it is actually an oversimplification of the actual pathway, which branches at the sphin-ganine step to form individual dihydroceramides (i.e., d18:0/16:0, d18:0/18:0, d18:0/20:0, etc.) depending on the lass family members that are expressed in that particular cell, after which, each of these species will undergo further partitioning to downstream metabolites (SM, GlcCer, GalCer, etc.)

The complexity of this pathway is more readily appreciated by examination of a pathway model (SphinGOMAP, Fig. 8) that illustrates the currently known downstream intermediates and products of sphingolipid biosynthesis (also see the appendix for this book by Akemi Suzuki). Incredibly, this figure is also a simplification because all of these headgroup variants that are shown have multiple ceramide backbones, and the ceramide subspecies are only illustrated in the box at the upper left of the figure. To see this pathway in a larger version, it can be downloaded from the web site given in the legend. Readers are encouraged to notify the web site operator if thay find errors or omissions so later editions can include these revisions.

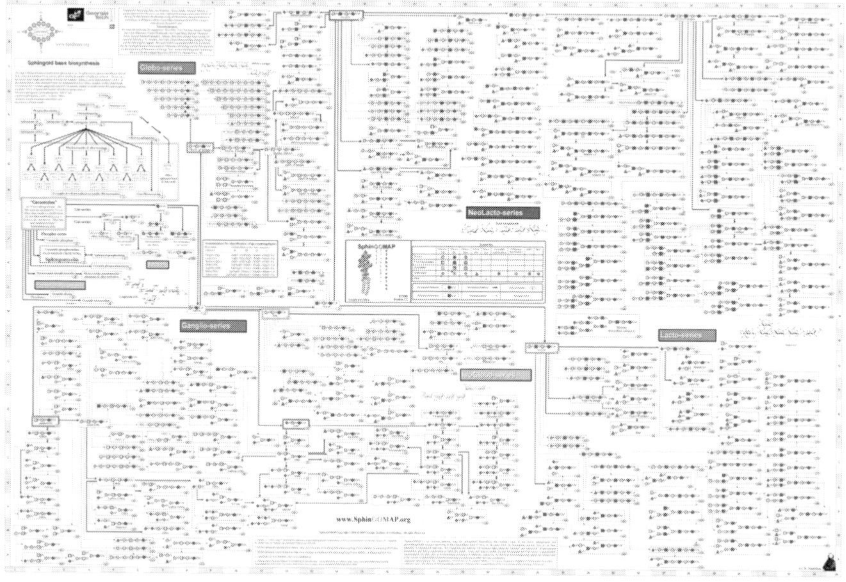

Fig. 8. A working model for sphingolipid biosynthesis that depicts the known sphingolipids as well as likely intermediates despite the lack of report of the observation of these compounds thus far. This figure is available for downloading in .ppt or .pdf formats at www.sphingomap.org.

5. Integration of sphingolipid biosynthesis with membrane trafficking, organelle function and signaling

The underlying biochemical logic of using such complex molecules to control cell behavior is presumably that this allows cells to integrate diverse input "signals" (which may include the availability of pathway precursors) to produce multiple coordinated "outputs". For example, when cells have adequate amounts of the biosynthetic precursors (palmitic and other fatty acids, serine, etc.) this typically elevates the amounts of sphingoid base 1-phosphates (Fig. 9), which are potentially mitogenic and anti-apoptotic; hence, perhaps cells use them as signs that cell cycle progession is "safe"). In contrast, if *de novo* synthesized sphingolipids are not being utilized for membrane biosynthesis (i.e., ceramides and free sphingoid bases are elevated), perhaps this is a signal that cell cycle progression is not desirable (and, indeed, sphingoid bases and ceramides are known to elevate cell cycle inhibitors such as Kip1 and Cip1). In addition, when sphingolipids are made or turned over this not only produces "signaling" metabolites, but alters membrane structure and the behavior of associated receptors and other proteins.

The many roles of ceramides in membrane biogenesis and regulation are probably just beginning to be understood. Ceramide has long been known to influence ER to Golgi trafficking and recent work (Daido et al. 2004) has implicated ceramide in the formation of another important intracellular organelles, the autophagosomes, which are formed continuously by cells to turn over damaged mitochondria, other organelles and cytoplasm, and are induced by starvation to provide essential amino acids and other compounds. It is often noted that the ER comprises the majority of the membrane of cells, and is connected to many of the other organelles: the nuclear membrane, the outer membrane of mitochondria, the plasma membrane (in at least some cell types), *inter alia*, and is constantly changing. Ceramide and other sphingolipids are likely participants—if not key regulators—of these processes.

Sphingolipid metabolism has many facets that reveal the biochemical "logic" of using such complex molecules to control cell behavior. For example, sphingomyelin turnover not only produces "signaling" metabolites, but also alters the structure of membrane domains that depend on the presence of this lipid (furthermore, when ceramide accumulates, its biophysical properties can profoundly affect membrane structure and the behavior of associated receptors and other proteins). As cells change their glycosphingolipid composition, they not only affect the properties of the cell surface but also its interaction with the extracellular matrix and neighboring cells. Thus, sphingolipid metabolism and signaling are almost synonomous as an ensemble of changes in membrane structure and dynamics, the availability of other key precursors, production (and removal) of bioactive metabolites, and the activation and/or inhibition of downstream targets.

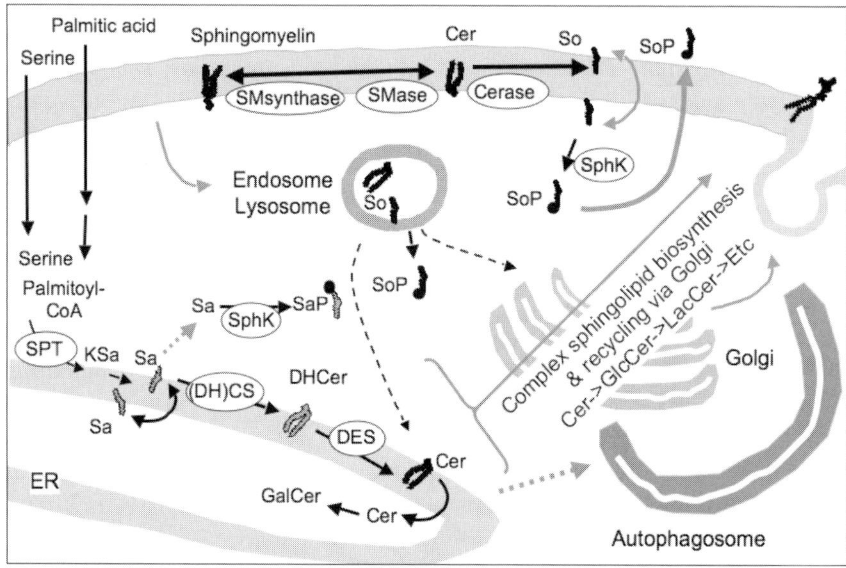

Fig. 9. Some of the interrelationships between sphingolipid biosynthesis, trafficking and menbrane biosynthesis and turnover. The abbreviations are: Cer (ceramide); Cerase (ceramidase); DES (dihydroceramide desaturase); DHCer (dihydroceramide); (DH)CS (ceramide synthase, acting on both sphinganine and sphingosine); ER (endoplasmic reticulum); GalCer (galactosylceramide, made in the lumen of the ER); GlcCer (glucosylceramide); KSa (3-ketosphinganine); LacCer (lactosylceramide); Sa (sphinganine); SaP (sphinganine 1-phosphte); So (sphingosine); SoP (sphingosine 1-phosphate); SM (sphingomyelin); SMase (sphingomyelinase); SMsynthase (SM synthase, which is in the lumen of the Golgi and the plasma membrane); and SphK (sphingosine kinase).

Table 1. Recent animal models with mutations in sphingolipid metabolism*

Enzymatic activities	Animal	Lipid changes	Phenotypes
Serine palmitoyltransferase			
Sptlc2 (subunit 2)	Drosophila**	No sphingolipid or GSL	Lethal (rescued by exogenous sphingolipid)
	mouse		Mice embryonic lethal
Sptlc1 (subunit 1)	human		Hereditary sensory neuropathy type I
Sptlc 1 & Sptlc2 mouse (heterozygote)	Reduction in sphingolipids		None reported
GalCer synthase (CGT	mouse	No GalCer or sulfatide	Mice viable and fertile
			Appearance of GlcCer in myelin
			Defective in axon-myelin interactions
			Loss of insulative function of myelin
GlcCer synthase mouse (GCS/GlcT-1)	No GSL		Embryonic lethal at E7.5
GM3 synthase	mouse	No complex gangliosides	Mice viable and male fertile
			Appearance of asialo-pathway
	human	No GM3 and increase of LacCer in serum	Infantile-onset symptomatic epilepsy
GM2 synthase	mouse	No GM2 ganglioside or complex gangliosides	Mice viable and male in
CERKL (CER kinase?)	human	Not clear	Vision defects (Retinitis pigmentosa)
Neutral SMase (SMPD3)	human	not clear	Defects in osterogenesis and dentinogenesis
Neutral CERase	mouse	Little change in tissue SL	Reduced ceramide hydrolysis in GI tract

*This listing does not include a number of previously existing animal models, such as acidic SMase and CERase.
**Tetradecasphingenine (d14:1) is the main shingosine base in Drosophila, thus, the enzyme should be named serine lauroyltransferase.

6. Emerging areas and perspectives for the future

The chapters of this book provide many examples of exciting new fields of sphingolipidology that are changing our understanding of how biological organisms grow, develop, function, age and die. However, it is also clear that sphingolipid research has only taken the first step of a thousand mile journey when one contemplates that the biologic functions for only a few of the species in Fig. 8 have been found to date, and that one can fairly confidently assume that all of these compounds (and more not shown) still have functions yet-to-be discovered. Nonetheless, the foundations and tools that have been created in our era have well equipped science for this journey, which include genetic models and "sphingolipidomic" technologies.

6.1 Gene targeting technology and new developments in sphingolipid biology

Recent dramatic developments in molecular biology have made it possible to clone most of the genes involved in sphingolipid metabolism. In particular, the genetic information on sphingolipid metabolism and sphingolpid-mediated signaling molecules of *Saccaharomyces cerevisiae*) enabled the identification of potential homologs in higher animals, and once these were cloned, gene-targeting mutant mice have been generated to define the biological roles of genes of interest (Table I).

More importantly, the human genome research has uncovered the sphingolipid-related inherited diseases such as Hereditary sensory neuropathy type I (HSN-1) and Infantile-onset symptomatic epilepsy (Table 1). Interestingly, these diseases were associated with neural malfunctions. This is understandable since sphingo(glyco)lipids are abundant in the central nervous systems and a critical component for development, cell death, and cellular homeostasis. Although much information has been accumulated on the physiological importance, the little is known why and how the defects in sphingo(glyco)lipid metabolism cause deterioration of the neural development and activity. A very recent finding that human Retinitis Pigmentosa is associated with mutation in CERKL, a novel ceramide kinase gene, has provided evidence for a link between retinal neurodegeneration and ceramide-meadiated apoptosis (and, here again, little is known how CERKL is involved in neuronal survival and death). Development of additional knock-out mouse models (especially ones that are organ specific and inducible) as well as introduction of simpler invertebrate systems such as *Drosophila melanogaster* together with RNAi technology, will surely facilitate the understanding of sphingolipid functions and the mechanisms underling development and human diseases.

Biochemical analysis of sphingolipids or related compounds is still important for development of sphingolipid biology. Biochemical analysis of the gene-targeted mice may identify compensatory molecules or uncover the unrecognized components essential for cellular functions. For example, Sandhoff et al. (2005) reported very interesting glycosphingolipids from mice testes, which contained polyunsaturated very-long chain fatty acid in their ceramide moiety. This study indicates that ceramide structure in glycosphingolipids plays an important role in spermatogenesis. Similarly, Nagatsuka et al. have identified a novel glyco-glycerolipid, phosphatidylglucoside, in rodent brains that forms raft-like lipid microdomain. Importantly,it contains solely saturated fatty acids in its diacylglycerol moiety (unpublished results 2006; 2003). Although this lipid is not belonging to sphingolipid, it forms rafts-like lipid microdomain. Interestingly phosphatidylglucoside contained solely unsaturated fatty acids in its diacylglycerol moiety. These studies show that it is quite important to identify glycolipid structures rigorously, including the fatty acyl chain moiety, using methods such as mass spectrometry. Indeed, it is likely that there are yet many unrecognized but biologically important sphingolipids in nature.

6.2 Developments in analytical technology and bioinformatics

The most powerful tool currently available to analyze large numbers of structurally diverse compounds is mass spectrometry, and "sphingolipidomic" approaches are beginning to be attempted (Merrill et al 2005). While these methods are currently only able to analyze a few hundred compounds quantitatively and with a moderate rate of throughput, the growth of new developments in mass spectrometry—including the capacity to perform measurements with tissue slices rather than extracts—allows one to optimistically predict that this and other methods will be able to meet the daunting challenge of analyzing the sphingolipodome. The equally difficult challenge is to develop bioinformatics tools that allow such large data sets to be analyzed and visualized, and the chapter by Hong et al. describe one such system that is under development.

6.3 Bridging discoveries with sphingolipid function and new pharmaceuticals

A desired outcome of basic research is its application to the prevention and treatment of disease. Given the rapid growth in understanding of the functions of sphingosine 1-phosphate in cell regulation, it is perhaps not too surprising that three pharmaceuticals in human clinical trials are related in some way to this pathway (Fig. 10). Safingol (L-threo-sphinganine), which is in Phase I trials for cancer, was initially developed as an inhibitor of protein kinase C, but is also a potent sphingosine kinase inhibitor and may act, at least in part, through that mechanism. Another compound in clinical trials for cancer, phenoxodiol (Fig. 10), has also been reported to inhibit sphngosine kinase, although it is not structurally related to sphingoid bases. Recent analyses using S1P receptor knockout mice and the immunosuppressant FTY720 (Fig. 10), which acts as a ligand for the S1P receptors after phosphorylation, have revealed that S1P plays important functions in the vascular and immune systems.

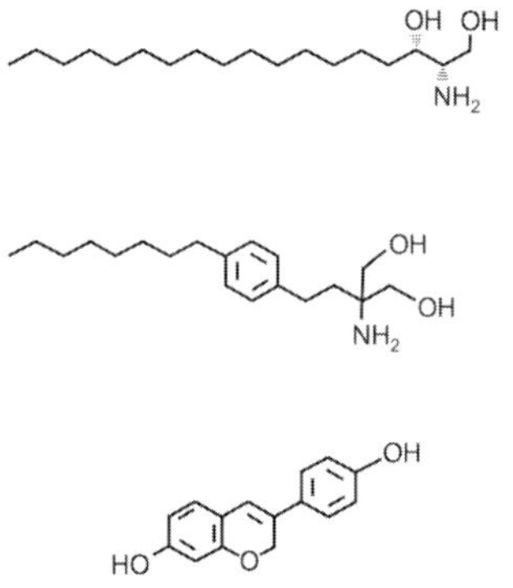

Safingol
L-threo-Dihydrosphingosine
(2S,3S)-2-aminooctadecane-1,3-diol
(2S,3S)-2-amino-1,3-octadecanediol

FTY720 (Fingolimod)
2-amino-2-. [2-(4-octylphenyl)ethyl]-
1,3-propanedio

Phenoxodiol

Fig. 10. Structures of compounds in human clinical trails that inhibit sphingosine kinase.

7. Looking to the future with the wisdom of the past

The field of sphingolipidology—or at least the origin of the root term—may have begun with the publication of *The Chemistry of the Brain* by J. L. W. Thudichum in 1884; however, comments by Thudichum a few years earlier (1876) are in some ways more prescient in telling us what will be needed to fully understand the functions of sphingolipids: "…proceed by a severe process, that of analysis, for nothing less than the results of analysis of work done can establish as proved what many feel as a sentiment…Of science, it is allowed that no part comes out of the human brain alone…Work, work and again work, [are] the three main features of its success."

In this article he respectfully acknowledges his predecessors, and his words too express what today's scientists owe to him: "If we cannot now inscribe their names and likenesses among the stars, and transfer them to an Olympian abobe, yet we can honor them by admiring their works and lessons, by sharing and continuing their works, by, as it were, living their lives with them over again, and thus prolong their memory forward while we prolong our own in the inverse direction…"

8. References

Chalfant CE, Spiegel S. (2005) Sphingosine 1-phosphate and ceramide 1-phosphate: expanding roles in cell signaling. *J Cell Sci,* **118**, 4605-4612.

Daido S, Kanzawa T, Yamamoto A, Takeuchi H, Kondo Y, Kondo S. (2004) Pivotal role of the cell death factor BNIP3 in ceramide-induced autophagic cell death in malignant glioma cells. *Cancer Res,* **64**, 4286-4293.

Fahy E, Subramaniam S, Brown HA, Glass CK, Merrill AH Jr, Murphy RC, Raetz CR, Russell DW, Seyama Y, Shaw W, Shimizu T, Spener F, van Meer G, VanNieuwenhze MS, White SH, Witztum JL, Dennis EA. (2005) A comprehensive classification system for lipids. *J Lipid Res,* **46**, 839-861.

Futerman AH, Hannun YA. (2004) The complex life of simple sphingolipids. *EMBO Rep,* **5**, 777-782.

Hakomori S. (2004) Carbohydrate-to-carbohydrate interaction, through glycosynapse, as a basis of cell recognition and membrane organization. *Glycoconj J,.* **21**, 125-137.

Koybasi S, Senkal CE, Sundararaj K, Spassieva S, Bielawski J, Osta W, Day TA, Jiang JC, Jazwinski SM, Hannun YA, Obeid LM, Ogretmen B. (2004) Defects in cell growth regulation by C18:0-ceramide and longevity assurance gene 1 in human head and neck squamous cell carcinomas. *J Biol Chem,* **279**, 44311-44309.

Merrill AH Jr, Sullards MC, Allegood JC, Kelly S, Wang E. (2005) Sphingolipi-domics: high-throughput, structure-specific, and quantitative analysis of sphingolipids by liquid chromatography tandem mass spectrometry. *Methods*, **36**, 207-224.

Nagatsuka Y, Hara-Yokoyama M, Kasama T, Takekoshi M, Maeda F, Ihara S, Fujiwara S, Ohshima E, Ishii K, Kobayashi T, Shimizu K, Hirabayashi Y. (2003) Carbohydrate-dependent signaling from the phosphatidylgluco-side-based microdomain induces granulocytic differentiation of HL60 cells. *Proc Natl Acad Sci U S A*, **100**, 7454-7459.

Sandhoff R, Geyer R, Jennemann R, Paret C, Kiss E, Yamashita T, Gorgas K, Sijmonsma TP, Iwamori M, Finaz C, Proia RL, Wiegandt H, Grone HJ. (2005) Novel class of glycosphingolipids involved in male fertility. *J Biol Chem*, **280**, 27310-27308.

Thudichum JLW (1876) The character of modern knowledge. *The Popular Science Monthly*, **8**, 724-727.

Part 2

Biosynthesis, Transport of Sphingolipids

2-1 Serine Palmitoyltransferase

Jia Wei[1], Tokumbo Yerokun[1], Martina Liepelt[1], Amin Momin[1],
Elaine Wang[1], Kentaro Hanada[2], and Alfred H. Merrill, Jr.[1]

[1]School of Biology and the Petit Institute for Bioengineering and Bioscience, Georgia Institute of Technology, Atlanta, GA 30332, USA
[2]Department of Biochemistry and Cell Biology, National Institute of Infectious Diseases, 1-23-1 Toyama, Shinjuku-ku, Tokyo 162-8640, Japan

Summary. Serine palmitoyltransferase (SPT) catalyzes the first unique reaction of *de novo* sphingolipid biosynthesis. It is a member of the γ-oxoamine synthase family that utilizes pyridoxal 5'-phosphate as a cofactor. Many factors—such as cytokines, irradiation and stress—alter SPT activity, and in some cases mRNA, and increased ceramide production has been proposed to modulate diverse cell behaviours, including programmed cell death. Normal SPT function is critical because missense mutations in the human *SPTLC1* gene cause hereditary sensory neuropathy type I, and complete knockout of SPT is embryonic lethal; however, the SPT inhibitor myriocin has been found to decrease atherosclerotic lesions in apo-E deficient mice. SPT is clearly involved in critical cell functions and disease, and much remains to be learned about its regulation.

Keywords. Serine palmitoytransferase, sphingolipid biosynthesis, regulation, stress, disease

1. Introduction

Serine palmitoyltransferase (SPT) catalyzes the first unique reaction of *de novo* sphingolipid biosynthesis, the condensation of serine and palmitoyl-CoA to produce 3-ketosphinganine. *De novo* biosynthesis of sphingolipids is essential for many types of cells for membrane structure, cell-cell and cell-matrix interactions, and numerous biological processes that are regulated by bioactive metabolites, such as ceramide and sphingosine 1-phosphate, that are used in cell signaling (for a general introduction to sphingolipids, see Merrill and Sandhoff 2002; for an earlier review on SPT, see Hanada 2003).   This review provides a brief historical background on SPT then highlights some of the information that is currently known about its properties, regulation and roles in cell function.

2. Biochemical and genetic characteristics of SPT

2.1 Early history of SPT

Following over a decade of speculation about how serine and palmitic acid might serve as the precursors for the sphingoid base backbones of sphingolipids, Braun and Snell (1967) and Stoffel et al. (Stoffel et al. 1968) demonstrated that serine and palmitoyl-CoA could be condensed to form 3-ketosphinganine which rapidly reduces to sphinganine (dihydrosphingosine) if NADPH was also present.   This process establishes the first steps of the *de novo* biosynthetic pathways for sphingolipids.   In subsequent decades, some of the properties of SPT were elucidated (Hanada 2003) such as a requirement for pyridoxal 5'-phosphate, a predominate localization of SPT activity in the endoplasmic reticulum (ER), the apparent kinetic properties, and various aspects of its regulation; however, the difficulty of purifying this membrane-associated enzyme until relatively recently (Hanada et al. 2000b) and the lack of knowledge about its gene(s) was a serious impediment to progress.

2.2 SPT genes

In the early 1990's, Lester, Dickson, and their co-workers isolated mutant strains of *Saccharomyces cerevisiae* that require an external supply of phytosphingosine for growth and for the synthesis of complex sphingolipids.   They showed that the mutant strains are defective in SPT activity (Pinto et al. 1992). Mutations that caused complete loss of SPT activity were shown to fall into two genetic complementation groups, *Lcb1* and *Lcb2*, and their wild-type alleles (*LCB1* and *LCB2*) were isolated by functional rescue experiments (Buede et al. 1991; Pinto et al. 1992; Nagiec et al. 1994).   With the availability of the yeast genes for SPT, identification of the mammalian homologs soon followed for mouse (Nagiec et al. 1996), hamster (CHO cells) (Hanada et al. 1997) and human (Weiss and Stoffel

1997). There is ~40% identity between yeast and mammalian SPT proteins, and ~95% identity among the mammalian proteins. As for many genes, the nomenclature for the mammalian SPT genes initially followed that of the microbial genes (e.g., LCB) but was later renamed (i.e., SPTLC1 and SPTLC2 for the human genomic homologs of the yeast LCB1 and LCB2 genes respectively). Nonetheless, there is still some usage of the earlier nomenclature, and examination of the human genome reveals several open reading frames that may represent other isoforms of these proteins. This review will use SPT1 and SPT2 for the SPT subunits encoded by the mammalian (SPTLC1/LCB1 and SPTLC2/LCB2 genes, respectively).

In the human genome, *SPTLC1* comprises 15 exons spanning ~85 kbp in the chromosome 9q21-q22 region, and *SPTLC2* comprises 12 exons spanning ~110 kbp in the chromosome 14q24.3-q31 region. The predicted MW of human SPT1 and SPT2 are 53 kDa and 63 kDa, respectively, and polypeptides of this approximate size are seen on Western blots of human and mouse cells using SPT1 and SPT2 antibodies (Weiss and Stoffel 1997; Hanada et al. 2000b). The subunits have ~20% identity and have a single highly hydrophobic domain in the amino-terminal region, which probably represents a transmembrane domain without cleavable signal sequences (Hanada et al. 1997; Weiss and Stoffel 1997). Indirect immunocytochemical analysis with epitope-tagged SPT1(LCB1) indicated that the N- and C-termini are oriented toward the lumen and cytosolic sides of the ER, respectively (Yasuda et al. 2003), which would place the majority of the active site region on the cytoplasmic side, as earlier established by analysis of the protease-sensitivity of SPT in sealed "right-side out" ER membranes (Mandon et al. 1992).

In addition to the N-terminal hydrophobic sequences, studies of other potential membrane spanning peptides have been conducted with the yeast LCB1 (and to a lesser extent SPTLC1) by insertion of glycosylation and factor Xa cleavage sites at various positions (Han et al. 2004). These studies found that some of the internal polypeptides may also span the membrane, and that the first domain (residues 50 and 84) was not required for the stability, membrane association, interaction with SPT2 protein, or enzyme activity, whereas, the second domain (residues 342-371) and third domain (residues 425-457) were required for protein stability. Therefore, the topology of SPT proteins needs to be further elucidated based on these different observations.

When cells are transfected with epitope-tagged SPT cDNA constructs, the predominant localization of the overexpressed enzyme is in the ER (Yasuda et al. 2003). However, immunohistochemical studies have suggested that SPT or cross-reacting polypeptides are also found in the nucleus, and shift from being predominantly cytosolic in quiescent cells to nuclear in proliferating cells (Carton et al. 2003). We have recently examined the localization of endogenous SPT1 and SPT2 by confocal microscopy using a panel of antibodies (T. Yerokun et al., manuscript in preparation) and also find nuclear immunostaining that appears to be due to a lower MW fragment of as-yet-unknown function.

2.3 Maintenance of SPT2/LCB2 is SPT1/LCB1-dependent and *vice versa*

SPT has been purified as an active form from CHO cells by affinity-peptide chromatography, and this purified enzyme was shown to consist of SPT1 and SPT2 at a 1:1 ratio (Hanada et al. 2000b). Studies using a wide range of experimental models have found that the level of expression of one of the SPT subunits affects the amounts of the other subunit and that this might reflect stabilization of the polypeptides as the dimer, e.g., *i*) LCB2 protein is greatly diminished in *lcb1* mutant yeast cells while *LCB2* mRNA does not change (Gable et al. 2000); *ii*) SPT2 protein (but not mRNA) is far lower in CHO cells deficient in SPT1 compared to wt CHO cells (Hanada et al. 1998), and stable transfection of the SPT1-deficient cells with *SPT1/LCB1* cDNA restored the amount of SPT2 protein to the wild-type level; and *iii*) overexpression of SPT2 protein required co-overexpression of SPT1, however, in this case the amounts of SPT1 could be increased without transfection to overexpress SPT2 (Yasuda et al. 2003). Jiang and coworkers (Hojjati et al. 2005) have also developed heterozygous knockout mice for SPT1 or SPT2 and found that compared with wt mice, the mice that were SPT1(+/-) not only had about half of the normal SPT1 protein mass but also a 53% decrease in SPT2; and SPT2(+/-) mice had a 70% decrease in SPT2 mass plus a 70% decrease in SPT1.

2.4 SPT-associated proteins

Dunn and co-workers have shown that an additional 10-kDa peptide, the product of the *Tsc3* gene, is associated with the SPT1/SPT2 complex in the yeast *S. cerevisiae* (Gable et al. 2000). Tsc3p is not essential for activity, but is required for optimum activity of SPT in yeast. No mammalian homolog of *Tsc3* has been found so far even by computer search of sequence databases nor upon affinity purification of SPT from CHO cells

(Hanada et al. 2000b). Inuzuka et al.(2005) have recentaly described a class of endoplasmic proteins (termed Serinc1 to 5) in yeast and mammalian cells that increase the synthesis of both phosphatidylserine and sphingolipids. Overexpression of Serinc1 in COS cells doubled SPT activity, therefore, Serinc may interact with SPT to facilitate serine utilization.

Using proteomic technologies (tandem-affinity purification and mass spectrometry) to discover protein-protein interactions, a substantial number of proteins have been identified as potential LCB2-associated proteins in *Saccharomyces cerevisiae* (Gavin et al. 2002). These proteins are involved in various biological processes such as vesicle transport, nuclear import and export, among others. A genome-wide yeast two-hybrid analysis in *Drosophila* (Giot et al. 2003) has suggested that SPT2 may interact with 13 proteins, which include a proton transporter, organic cation transporter, hsc-70, and ribosomal proteins, among others.

2.5 Structure and mechanism of the SPT reaction

SPT has significant structural and mechanistic similarities to members of a subfamily of PLP-dependent enzymes that includes 5-aminolevulinic acid synthase, 2-amino-3-ketobutyrate ligase, and 8-amino-7-oxononanoate synthase. These enzymes catalyze condensation of amino acids and the CoA thioesters of carboxylic acids to produce γ-oxoamines, therefore, have been referred to as the PLP-dependent γ-oxoamine synthase (POAS) family. However, the eukaryotic SPT's are the only members of this family that have been identified to date to be membrane bound and heterodimeric--all other members are soluble homodimers. In contrast to eukaryotic SPT, *Sphingomonas paucimobilis* produces a soluble SPT that is ~30% identical to both mammalian SPT1 and SPT2 at the amino acid level, but is a homodimer of a 45-kDa protein (Ikushiro et al. 2001).

POAS members share a conserved motif around the lysine that is responsible for formation of a Schiff's base with pyridoxal 5'-phosphate (PLP)(T[FL][GTS]**K**[SAG][FLV]G). This motif is present in SPT2, but the key lysine is not found in SPT1, therefore, it is likely that only the SPT2 subunit forms this complex. PLP is moderately tightly bound by SPT with an apparent Km of approximately 0.1 µM for the mammalian enzyme (Williams et al. 1984).

It has long been noted (Di Mari et al. 1971) that the involvement of PLP in removing the carboxyl group of serine during the condensation of serine and palmitoyl-CoA allows two mechanisms to be postulated: formation of PLP-stabilized carbanion by decarboxylation of the substrate L-serine, followed by acylation, or formation of a carbanion by removing the α-hydrogen atom of L-serine, followed by acylation and decarboxylation (shown in Fig. 1). Krisnangkura and Sweeley (1976) provided strong evidence for the latter mechanism by demonstrating that the α-hydrogen atom of serine is replaced by a proton from H_2O during the reaction. This reaction has also been found for the condensation reactions catalyzed by 5-aminolevulinic acid synthase (Zaman et al. 1973) and other enzymes of this family. In addition, the structures of several potent SPT inhibitors (*i.e.*, sphingofungins and myriocin) resemble that for a potential transient state intermediate with the carboxyl group of L-serine still present (species IV in Fig. 1), which would be found in latter pathway (as shown in the insert in Fig. 1).

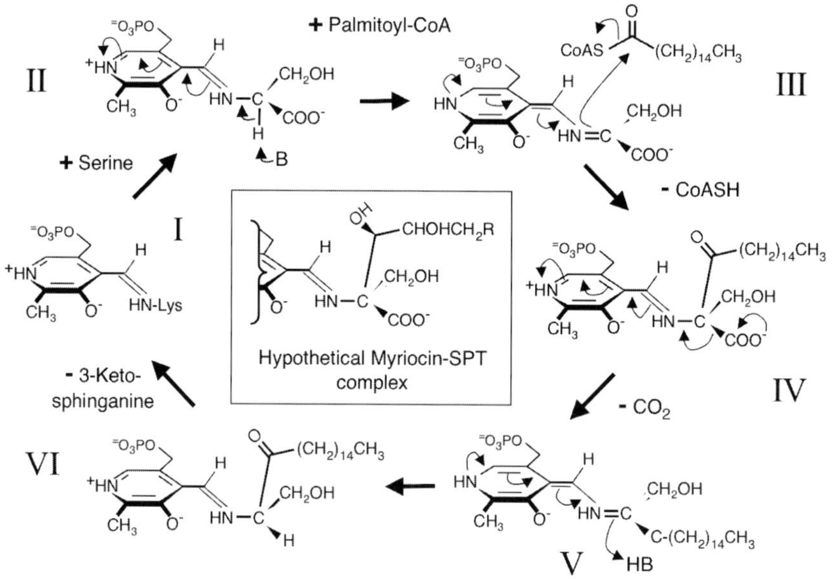

Fig. 1. Suggested catalytic pathway of serine palmitoyltransferase (SPT) and rationale for inhibition by myriocin-like compounds (insert). Starting with the internal aldimine of pyridoxal 5'-phosphate (PLP) with an active site lysine (I), serine is bound by SPT as an external PLP aldimine (II), deprotonated at the α-carbon and condensed with palmitoyl-CoA (III) (possibly through this quinonoid intermediate)(note: it is also possible that the second substrate is bound before the hydrogen is removed), followed by decarboxylation of the doubly β,γ-unsaturated intermediate IV, proton abstraction (V) and release of the produce 3-ketosphinganine (VI). The insert illustrates the similarities between the presumed aldimine of myriocin with SPT and catalytic intermediate IV.

2.5.1 Substrate specificity of SPT

SPT strictly utilizes L-serine as its amino acid substrate, however, D-serine is a competitive inhibitor with an IC_{50} of ~0.3 mM, which is similar to the K_m for L-serine (Hanada et al. 2000b). SPT is also highly selective for the co-substrate fatty acyl-CoA with the mammalian enzyme (Williams et al. 1984; Hanada et al. 2000b) utilizing palmitoyl-CoA (C16:0) > pentade-canoyl- and heptadecanoyl- CoA's (C15:0 and C17:0) >> stearoyl-CoA (C18:1) and essentially no unsaturated species except palmitelaidic acid (t16:1), which has an uncommon *trans-* versus *cis-* double bond. This selectivity, combined with the abundance of palmitoyl-CoA (and scarcity of C15:0 and C17:0 fatty acids in most species except ruminants) probably accounts for the extremely high 18-carbon-chain-length specificity of most sphingoid bases in mammalian sphingolipids. Indeed, in species where other chain lengths are found, such as *Drosophila* which contain 14-carbon sphingoid bases, or *Sphingomonas*, which displays very little sphingoid base specificity, this is reflected in differences in the fatty acyl-CoA selec-tivity of the SPT from these species (Ikushiro et al. 2003; Acharya and Acharya 2005). A question that remains to be answered is why atypically long sphingoid bases (with 20-carbon atoms) are found in human gangli-osides (Merrill 2002). These types of sphingoid bases have mainly been observed in yeast under certain stages of growth and stress, and are thought to have roles in cell signalling (Liu et al. 2005).

These kinetic properties of mammalian SPT, which were measured *in vitro* and are subject to numerous complications inherent in assaying lipid metabolizing enzymes under artificial conditions, have nonetheless been relatively closely replicated in studies of intact LM cells (Merrill et al. 1988) and hepatocytes (Messmer et al. 1989) where the cellular concentra-tions of serine and palmitate were varied. Physiologic concentrations of serine are known to fluctuate above and below the Km for SPT (Merrill et al. 1988), which might influence sphingolipid metabolism and explain why cells have the recently discovered Serinc (for <u>ser</u>ine <u>inc</u>orporating) proteins to facilitate utilization of this precursor (Inuzuka et al. 2005). It has also been noted (Hanada et al. 2000a) that since D-serine is a strong competi-tive inhibitor for L-serine, and substantial quantities of D-serine are present in discrete areas of the brain and some other tissues, D-serine might affect *de novo* sphingolipid synthesis in these cells (also providing, perhaps, one of the rationales for the Serinc proteins). Indeed, serine homeostasis is itself highly complex and part of the one-carbon-metabolism system that includes the biosynthesis and turnover of glycine and methionine, methy-lation of proteins, lipids and nucleic acids, and many other important regulatory pathways. These issues illustrate the fundamental questions that have not yet been answered about how *de novo* sphingoid base biosynthesis is coordinated with other metabolic pathways, which one hopes is at last becoming feasible to address experimentally with new mass spectrometric methods for sphingolipid analysis (Merrill et al. 2005) and conceptually with the types of systems models that have been developed for the metabolic pathway in yeast (Sims et al. 2004; Alvarez-Vasquez et al. 2005; Cowart and Hannun 2005).

2.5.2 Inhibitors of SPT

A fairly high number of naturally occurring inhibitors of SPT have been discovered and many can be rationalized due to their structural similarity to reaction intermediates, such as myriocin shown in the insert of Fig. 1. These include sphingofungins, lipoxamycin (neoenactin M_1), myriocin (ISP-1/thermozymocidin) and sulfamisterin, which are potent (i.e., with IC50 in the nanomolar range) and, as far as has been determined to date, high selective for fungal and mammalian SPT (Zweerink et al. 1992; Mandala et al. 1994; Miyake et al. 1995; Yamaji-Hasegawa et al. 2005). Viridiofungins (Mandala et al. 1997) are also potent inhibitors of mammalian SPT, but additionally inhibit squalene synthase. As expected for an enzyme that utilizes pyridoxal 5'-phosphate, SPT is inhibited by compounds such as L-Cycloserine and γ-chloro-L-alanine, which are sometimes used to inhibit SPT in intact cells, however, this is usually not desirable because these compounds affect too many other metabolic pathways as well.

2.5.3 Kinetic and spectroscopic studies of the mechanism of SPT

Because the SPT from *Sphingomonas paucimobilis* is a soluble homodimer (Ikushiro et al. 2001), it has been much more extensively characterized than the membrane-bound SPT's (Ikushiro et al. 2003; Ikushiro et al. 2004). The purified recombinant SPT has an absorption spectrum with maxima at 426 and 338 nm, which was interpreted as the aldimine (species "I" in Fig. 1) and enolimine tautomer, respectively, of PLP with the active site lysine. Ikushiro et al. also reported that addition of L-serine caused an increase in 426 nm absorption indicative of formation of the external aldimine intermediate (species "II"), however, absorption at ~500 nm was not detected by rapid mixing or static methods, which suggests that the quinonoid intermediate (species "III") does not accumulate. As noted by these authors, this does not necessarily preclude the existence of intermediate "III", but it may not form until addition of palmitoyl-CoA. The absorption and CD spectra of the SPT-myrocin complex was also consistent with the formation of an external aldimine, as hypothesized in Fig. 1. Interestingly, addition of sphinganine or sphingosine also altered the spectrum, although in-depth studies were precluded by the low solubility of these compounds. This may indicate that interaction between SPT and its product (and/or downstream products) might affect activity.

3. Regulation of serine palmitoyltransferase

SPT activity and mRNA have been found in very mammalian tissue and cell type and are especially abundant in organs such as kidney, lung and liver (Merrill et al. 1985; Weiss and Stoffel 1997). In addition, SPT activity depends on the developmental stage of tissues (Longo et al. 1997) and is affected by environmental factors such as diet in several tissues (Rotta et al. 1999; Geelen and Beynen 2000). A growing number of animal studies and investigations with cells in culture are documenting transcriptional and post-transcriptional regulation of SPT activity, however, considering the key role played by this enzyme, surprisingly little is known about its regulation.

3.1 Transcriptional and post-transcriptional regulation in response to extracellular stimuli

Many of the factors that have been reported to increase SPT mRNA and activity are summarized in Table 1. A large fraction of the stimuli are inflammatory, toxic or some other form of stress. Intraperitoneal administration of endotoxin to Syrian hamsters increased SPT activity 2- to 3-fold in the liver, spleen, and kidney with a concomitant increase in *SPT2/LCB2* mRNA (Memon et al. 1998). Similar changes were observed upon administration of interleukin 1ß (IL-1), an inflammatory cytokine. Irradiation of keratinocytes with ultraviolet A (UVA) upregulated SPT activity as well as mRNA (Grether-Beck et al. 2005); ultraviolet light B (UVB) also affects SPT mRNA in epidermal cells (Farrell et al. 1998). SPT activity in epidermal cells appears to be up-regulated in response to barrier requirements of skin (Holleran et al. 1991). These results, along with the observation that inflammation and UVB stimuli also enhance the synthesis of other lipid types, including fatty acids and cholesterol (Memon et al. 1993; Holleran et al. 1997), raise the possibility that a lipogenic signaling pathway(s) is involved in transcriptional control of SPT genes. Consistent with this possibility, both *SPT1* and *SPT2* mRNA are increased when cultured human keratinocytes are treated with nicotinamide, which also enhances fatty acid and cholesterol synthesis (Tanno et al. 2000). The expression of *SPT1* mRNA is also up-regulated in islets of leptin-receptor-deficient obese *fa/fa* rats, compared to the levels in control islets. This up-regulation might be a response to an increase in intracellular fatty acid (Shimabukuro et al. 1998).

Table 1. Stimuli/conditions that up-regulate SPT activity in mammalian cells

Stimulus or condition (tissue or cell types)	Activity	mRNA (protein)		References
		SPT1/ LCB1	SPT2/ LCB2	
UVB (mouse epidermis and cultured human keratinocytes)	1.5	1.5[a]	2-3[a] (2)	(Holleran et al. 1997; Farrell et al. 1998)
UVA (human keratinocytes)	2.5	14	7	(Grether-Beck et al. 2005)
Endotoxin, IL1 (liver, spleen and kidney of Syrian hamster)	2-3	nd	2-4	(Memon et al. 1998; Memon et al. 2001)
Endotoxin, IL1, tumor necrosis factor-γ (human HepG cell line)	2-3	nd	2-3	(Memon et al. 1998)
Leptin receptor mutation (rat pancreas islet)	nd[b]	2-3	nd	(Shimabukuro et al. 1998)
Fatty acids (rat pancreas islet)	nd[b]	1.5-2	nd	(Shimabukuro et al. 1998)
Palmitic acid (rat astrocytes)	1.3	nd	nd (1.4)	(Blazquez et al. 2001)
Nicotineamide (cultured human keratinocytes)	1.2	1.8	1.8	(Tanno et al. 2000)
Etoposide (human leukemia Molt-4 cell line)	2-3	–	–	(Perry et al. 2000)
Retinoic acid (mouse teratocarcinoma PCC7-Mz1 cell line)	3	–	–	(Herget et al. 2000)
Resveratrol (human breast cancer cells)	1.5	–	–	(Scarlatti et al. 2003)
D^9-Tetrahydrocannabinol (a subline of the rat glioma C6 line)	6	1.4 (1.8)	1.1	(Gomez del Pulgar et al. 2002)

Activation of angiotensin II type 2 receptor (a subline of the rat pheochromo-cytoma PC12 cell line)	2	nd	nd	(Lehtonen et al. 1999)
N-(4-hydroxyphenyl)retinamide (human neuroblastoma CHLA-90 cell line)	2	nd	nd	(Wang et al. 2001)
Hexachlorobenzene (rat liver)	1.5-2[a]	nd	nd	(Billi de Catab-bi et al. 2000)
Apolipoprotein E knockout (mouse liver)	2	nd	–	(Jeong et al. 1998)

–, no increase; nd, not determined
[a]Following stimulation, levels initially decreased by ~50%, then increased more than unstimulated control levels.
[b]*De novo* sphingolipid synthesis in intact cells was significantly increased.

Cytotoxic accumulation of palmitate induces apoptosis accompanied by an elevation of the intracellular ceramide level in various cell types, including the hematopoietic LyD9 cell line, *fa/fa* pancreas islets, astrocytes, and CHO cells (Paumen et al. 1997). For most cell types, enhanced *de novo* synthesis of ceramide via SPT is required for palmitate-induced apoptosis (Paumen et al. 1997; Shimabukuro et al. 1998; Blazquez et al. 2001; Listenberger et al. 2001), whereas palmitate-induced apoptosis of CHO cells is suggested to occur through a ceramide-independent, but reactive-oxygen-species-dependent, pathway (Listenberger et al. 2001). Palmitate-induced enhancement of SPT activity in primary astrocytes is prevented by exposure of cells to an activator of AMP-activated protein kinase (AMPK) (Blazquez et al. 2001). The AMPK cascade acts as a metabolic sensor that monitors cellular AMP and ATP levels, and once activated, down-regulates various ATP-consuming anabolic pathways including fatty acid and cholesterol synthesis. The AMPK cascade might also participate in the regulation of sphingolipid synthesis.

Certain types of apoptotic stimuli appear to increase SPT activity with little or no change in mRNA amount, hence, may act post-transcriptionally (Table 1). Upon treatment with retinoic acid, mouse teratocarcinoma PCC7-Mz1 cells undergo apoptosis, but a fraction begins to differentiate in a manner mimicking the early steps of neuronal development. Retinoic acid treatment of PCC7-Mz1 stem cells induces accumulation of ceramide accompanied by an increase in SPT activity without any increases in *SPT1* and *SPT2* mRNA (Herget et al. 2000). Likewise, treatment of Molt-4 human leukemia cells with the chemotherapy agent etoposide elevates both

SPT activity and intracellular ceramide without increasing *SPT* mRNA (Perry et al. 2000).   During cannabinoid-induced apoptosis, both SPT activity and intracellular ceramide increase 4 to 6 fold without major changes in *SPT* mRNA or protein in a subline of the rat glioma C6 cells.   When these cells have been treated also with inhibitors of *de novo* sphingolipid synthesis to block the increase in ceramide, there is usually also a decrease in apoptosis in response to these stimuli (Herget et al. 2000; Perry et al. 2000) which suggests that the elevation in ceramide is a mediator of, rather than merely a consequence of, the pro-apoptotic effects of these agents.

De novo sphingolipid biosynthesis (and SPT activity) is also rapidly up-regulated in Molt-4 cells during heat shock (Jenkins et al. 2002), and heat shock is well known to affect—and to be influenced by--sphingolipid biosynthesis in yeast (Friant et al. 2003).

3.2 Regulation of SPT for homeostasis of cellular sphingolipid amounts

Little is known about what controls the sphingolipid amounts of cells.   A plausible feedback regulator is sphingosine (or perhaps sphinganine) 1-phosphate because addition of sphingosine or an analog that accumulates in the cells as the phosphate metabolite to cultured mouse cerebellar cells significantly decreases SPT activity (which does not appear to be due to direct inhibition of SPT by the sphingoid base because SPT activity is not reduced when microsomes are incubated directly with sphingoid bases) (Mandon et al. 1991; van Echten-Deckert et al. 1997).   Similarly, accumulation of sphingosine-1-phosphate and dihydrosphingosine-1-phosphate by treating primary cultured cerebellar neurons with high concentrations of a dihydroceramide desaturase inhibitor, GT11, reduced SPT activity by about 90% without changing SPT mRNA amount, which could reflect feedback inhibition of SPT (Triola et al. 2004).   Puzzlingly, depression of *de novo* sphingolipid synthesis by exogenous ceramide or its analogs seems not to be accompanied by any inhibition of the activities of anabolic enzymes, including SPT (Ridgway and Merriam 1995; van Echten-Deckert et al. 1998), nor in numerous studies by a co-authors' labs (A. Merrill, unpublished) have we seen upregulation of SPT activity or mRNA when cells are treated with myriocin or fumonisin B1 (Steve Linn, unpublished), which severely decrease total sphingolipid amounts.   It is clear that more studies are required to elucidate the mechanisms underlying the regulation of cell sphingolipid homeostasis via *de novo* synthesis versus recycling of endogenous and exogenous sphingoid bases.

4. Functional significance of SPT in health and disease

4.1 SPT is essential for survival

In addition to being essential for yeast (Pinto et al. 1992), SPT activity (or exogenous addition of sphingoid bases) has been shown to be required for survival of a temperature-sensitive Chinese hamster ovary (CHO) cell mutant (strain SPB-1) with a thermolabile SPT was isolated after selection by *in situ* colony assay (Hanada et al. 1990). Knockout of SPT in fruit flies results in embryonic lethality (Adachi-Yamada et al. 1999), whereas when the SPT deficiency is partial, mutant flies grow into adults with abnormalities which can be rescued by feeding with sphingosine. Attempts to produce homozygous knockout mice were also unsuccessful due to embryonic lethality (Hojjati, Li et al. 2005a), although heterozygots were healthy and had lower SPT mRNA, protein, and activity in some tissues (e.g., liver). In Leishmania, however, targeted deletion of spt2 was not lethal until entry into stationary stage, when the organisms failed to differentiate to infective metacyclic parasites and died (Zhang et al. 2003). Thus, all of the studies to date demonstrate that sphingolipids are essential for survival of the organisms at some stage of development or their life cycle.

4.2 Roles of SPT in disease

4.2.1 Hereditary sensory neuropathy type I (HSN1)

HSN1 is a dominantly inherited disease involving the progressive degeneration of lower limb sensory and autonomic neurons. HSN1 is a genetically heterogenous disease, and at least three gene variants are reported. It has recently been revealed that the genetic defect in the HSN1 families linked to the chromosome 9q22 locus is associated with missense mutations in the human *SPTLC1* gene, which alter a specific amino acid residue (Cys^{133} or Val^{144}) in the SPT1 subunit (Bejaoui et al. 2001, Dawkins et al. 2001).

Expression of *SPT1/LCB1* mutants having the HSN1 mutations has been shown to inhibit SPT activity and sphingolipid synthesis in CHO cells and yeast. The mutated SPT1 proteins remained capable of forming a complex with the SPT2 subunit, but the complex was enzymatically inactive, which suggests that the mutant SPT1 may have a dominant negative-like phenotype (Bejaoui et al. 2002). However, the consequences of the mutations may be more complex because despite the reduction in SPT activity, the rates of sphingolipid biosynthesis, cell proliferation, and death in HSN1 cells were not changed (Dedov et al. 2004).

The amino acid sequence around Cys[133] and Val[144] in SPT1/LCB1 is highly conserved from yeast to mammals. Although the crystallographic analysis of a mammalian SPT has not been conducted, the tertiary structures of other enzymes in this family reveals that the catalytic site is formed at the interface between the subunits, hence, both Cys[133] and Val[144] of SPT1 are predicted to be spatially close to the PLP-binding site of SPT2. Presumably, the amino acid sequence around Cys[133] and Val[144] in SPT1 is involved in the formation of the catalytic site, and the HSN1 mutant types of SPT1 are unable to contribute to the formation of the active catalytic site. It remains unclear why mutations in a protein widely expressed in all tissues trigger pathology that is highly restricted to specific subsets of cells within a tissue.

4.2.2 Roles of SPT in other biological processes and in disease

SPT is the initial enzyme of a complex metabolic pathway that produces thousands, and potentially tens of thousands, of different molecular species of sphingolipids (for a compilation of the major subgroups see the Chapter by Akemi Suzuki in this volume as well as the downloadable pathway map at www.sphingomap.org). While the endproduct complex sphingolipids are known--or in most cases presumed on teleologic grounds--to have cell functions, the turnover products (e.g., ceramide) and biosynthetic intermediates (e.g., sphinganine) have also been shown to be highly bioactive and to participate in both the etiology and sometimes suppression/treatment of disease. Hence, SPT is an interesting candidate as a drug target.

Some existing drugs have already been suggested to act at least in part via increasing SPT activity. N-(4-Hydroxyphenyl)retinamide (4HPR or fenretinide) is a chemotherapy drug undergoing human clinical trials that elevates ceramide in numerous human cancer cell lines (Wang et al. 2003; Reynolds et al. 2004), and the ceramide elevation has been shown to occur by coordinate activation of SPT and ceramide synthase (Wang et al. 2001). The dietary factor resveratrol is antiproliferative and proapoptotic, and its effects correlate with an increase in endogenous ceramide that has been suggested to involve activation of SPT and neutral sphingomylinase (Scarlatti et al. 2003). In Molt-4 human leukemia cells, the chemotherapic agent etoposide increases ceramide due to upregulation of SPT activity (Perry et al. 2000). γ-Tocopherol, the predominant form of vitamin E in diets, induces apoptosis of LNCap cells and addition of myriocin to inhibit SPT protects cells from γ-tocopherol–induced apoptosis (Jiang et al. 2004). Increases in ceramide have also been proposed to mediate the tamoxifen-dependent accumulation of autophagic vacuoles in MCF-7 cells (Scarlatti et al. 2004).

Investigators have for several decades contemplated provocative associations between sphingolipid metabolism and atherosclerosis, however, two recent studies of atherosclerotic lesion development in apoE-deficient mice (Park et al 2004, Hojjati et al 2005b) have provided the strongest link between the possibility that SPT (meaning undoubtedly downstream products) may contribute to this disease. Hojjati et al. have reported that treatment of apoE-deficient mice with myriocin decreases plasma sphingomyelin, ceramide, and sphingosine-1-phosphate and reduces atherosclerotic lesion area (Hojjati et al. 2005b). Therefore, a reduction of SPT activity might be advantageous in decreasing atherogenesis. In addition, Worgall and coworkers (2004) have found that ceramide synthesis correlates with the generation of transcriptionally active SREBP and SRE-mediated gene transcription, perhaps by affecting ER-to-Golgi trafficking. Since SREBP is involved in not only the regulation of cholesterol biosynthesis and lipoprotein utilization but also many other important aspects of lipid regulation, this may be an important link between these pathways.

Manipulation of *de novo* sphingolipid biosynthesis may have benefits for multiple pathologies associated with advanced age because a link is also surfacing between sphingolipids, lipid rafts, and their components (cholesterol) in the processing of γ–amyloid precursor protein (APP). The secretion of soluble APPγ and generation of C-terminal fragment cleaved at γ-site dramatically increased via activation of MAPK/ERK pathway in Chinese hamster ovary cells treated with myriocin and in a mutant, LY-B strain defective in the SPT1 protein (Sawamura et al. 2004)[1]. This indicates modulation of sphingolipid metabolism may able to influence the pathogenesis of Alzheimer's disease by affecting APP cleavage. These examples actually highlight just a few of the promising indications that SPT is a potential pharmaceutical target for prevention and treatment of disease.

---

[1] We have recently found that SPT1 is expressed in LY-B cells but the peptide is mutated (Gly 246 to Arg) and inactive (Martina Leipelt, manuscript in preparation).

5. Future directions

While there has been significant progress in understanding the structure, mechanisms of action, and functional significance of SPT, knowledge about this important enzyme is still frustratingly incomplete. How is the expression of SPT regulated at transcriptional and post-transcriptional levels? Do cells contain only the two SPT peptides characterized to date, or are some of the additional genes with sequence homology to SPT1 and SPT2 (and/or splice variants) also expressed? What types of post-translational modifications (if any) affect SPT structure and function? What is the rate of turnover of SPT proteins, and how is this regulated? How does the cell control the flux through the sphingolipid biosynthetic pathway to minimize accumulation of potentially toxic intermediates? How does the cell regulate the rate of *de novo* sphingoid base biosynthesis versus reutilization of sphingoid bases from membrane lipid turnover and uptake from lipoproteins? How and when are sphingolipids made *de novo* used for cell signaling versus for production of more complex sphingolipids, and how are these coordinated? Fortunately, the intellectual and material resources that have been developed over the past decade now allow these questions to be addressed. Expectations are great for what will be learned about this key enzyme in the decade to come.

6. References

Acharya U and Acharya JK (2005) Enzymes of sphingolipid metabolism in Drosophila melanogaster. *Cell Mol Life Sci*, **62**, 128-142.

Adachi-Yamada T, Gotoh T, Sugimura I, Tateno M, Nishida Y, Onuki T and Date H (1999) De novo synthesis of sphingolipids is required for cell survival by down-regulating c-Jun N-terminal kinase in Drosophila imaginal discs. *Mol Cell Biol*, **19**, 7276-7286.

Alvarez-Vasquez F, Sims KJ, Cowart LA, Okamoto Y, Voit EO and Hannun YA (2005) Simulation and validation of modelled sphingolipid metabolism in Saccharomyces cerevisiae. *Nature*, **433**, 425-430.

Bejaoui K, Uchida Y, Yasuda S, Ho M, Nishijima M, Brown RH Jr, Holleran WM, Hanada K (2002) Hereditary sensory neuropathy type 1 mutations confer dominant negative effects on serine palmitoyltransferase, critical for sphingolipid synthesis. *J Clin Invest*, **110**, 1301-1308.

Bejaoui K, Wu C, Scheffler MD, Haan G, Ashby P, Wu L, de Jong P, Brown RH Jr (2001) SPTLC1 is mutated in hereditary sensory neuropathy, type 1. *Nat Genet*, **27**, 261-262.

Billi de Catabbi SC, Setton-Advruj CP, Sterin-Speziale N, San Martin de Viale LC and Cochon AC (2000) Hexachlorobenzene-induced alterations on neutral and acidic sphingomyelinases and serine palmitoyltransferase activities. A time course study in two strains of rats. *Toxicology*, **149**, 89-100.

Blazquez C, Geelen MJ, Velasco G and Guzman M (2001) The AMP-activated protein kinase prevents ceramide synthesis de novo and apoptosis in astrocytes. *FEBS Lett*, **489**, 149-153.

Braun PE and Snell EE (1967) The biosynthesis of dihydrosphingosine in cell-free preparations of Hansenula ciferri. *Proc Natl Acad Sci U S A*, **58**, 298-303.

Buede R, Rinker-Schaffer C, Pinto WJ, Lester RL and Dickson RC (1991) Cloning and characterization of LCB1, a Saccharomyces gene required for biosynthesis of the long-chain base component of sphingolipids. *J Bacteriol*, **173**, 4325-4332.

Carton JM, Uhlinger DJ, Batheja AD, Derian C, Ho G, Argenteri D and D'Andrea MR (2003) Enhanced serine palmitoyltransferase expression in proliferating fibroblasts, transformed cell lines, and human tumors. *J Histochem Cytochem*, **51**, 715-726.

Cowart LA and Hannun YA (2005) Using genomic and lipidomic strategies to investigate sphingolipid function in the yeast heat-stress response. *Biochem Soc Trans*, **33**, 1166-1169.

Dawkins JL, Hulme DJ, Brahmbhatt SB, Auer-Grumbach M, Nicholson GA (2001) Mutations in SPTLC1, encoding serine palmitoyltransferase, long chain base subunit-1, cause hereditary sensory neuropathy type I. *Nat Genet*, **27**, 309-312.

Dedov VN, Dedova IV, Merrill AH, Jr. and Nicholson GA (2004) Activity of partially inhibited serine palmitoyltransferase is sufficient for normal sphingolipid metabolism and viability of HSN1 patient cells. *Biochim Biophys Acta*, **1688**, 168-175.

Di Mari SJ, Brady RN and Snell EE (1971) Biosynthesis of sphingolipid bases. IV. The biosynthetic origin of sphingosine in Hansenula ciferri. *Arch Biochem Biophys*, **143**, 553-565.

Farrell AM, Uchida Y, Nagiec MM, Harris IR, Dickson RC, Elias PM and Holleran WM (1998) UVB irradiation up-regulates serine palmitoyltransferase in cultured human keratinocytes. *J Lipid Res*, **39**, 2031-2038.

Friant S, Meier KD and Riezman H (2003) Increased ubiquitin-dependent degradation can replace the essential requirement for heat shock protein induction. *Embo J*, **22**, 3783-3791.

Gable K, Slife H, Bacikova D, Monaghan E and Dunn TM (2000) Tsc3p is an 80-amino acid protein associated with serine palmitoyltransferase and required for optimal enzyme activity. *J Biol Chem*, **275**, 7597-7603.

Gavin AC, Bosche M, Krause R, Grandi P, Marzioch M, Bauer A, Schultz J, Rick JM, Michon AM, Cruciat CM, Remor M, Hofert C, Schelder M, Brajenovic M, Ruffner H, Merino A, Klein K, Hudak M, Dickson D, Rudi T, Gnau V, Bauch A, Bastuck S, Huhse B, Leutwein C, Heurtier MA, Copley RR, Edelmann A, Querfurth E, Rybin V, Drewes G, Raida M, Bouwmeester T, Bork P, Seraphin B, Kuster B, Neubauer G and Superti-Furga G (2002) Functional organization of the yeast proteome by systematic analysis of protein complexes. *Nature*, **415**, 141-147.

Geelen MJ and Beynen AC (2000) Consumption of olive oil has opposite effects on plasma total cholesterol and sphingomyelin concentrations in rats. *Br J Nutr*, **83**, 541-547.

Giot L, Bader JS, Brouwer C, Chaudhuri A, Kuang B, Li Y, Hao YL, Ooi CE, Godwin B, Vitols E, Vijayadamodar G, Pochart P, Machineni H, Welsh M, Kong Y, Zerhusen B, Malcolm R, Varrone Z, Collis A, Minto M, Burgess S, McDaniel L, Stimpson E, Spriggs F, Williams J, Neurath K, Ioime N, Agee M, Voss E, Furtak K, Renzulli R, Aanensen N, Carrolla S, Bickelhaupt E, Lazovatsky Y, DaSilva A, Zhong J, Stanyon CA, Finley RL, Jr., White KP, Braverman M, Jarvie T, Gold S, Leach M, Knight J, Shimkets RA, McKenna MP, Chant J and Rothberg JM (2003) A protein interaction map of Drosophila melanogaster. *Science*, **302**, 1727-1736.

Gomez del Pulgar T, Velasco G, Sanchez C, Haro A and Guzman M (2002) De novo-synthesized ceramide is involved in cannabinoid-induced apoptosis. *Biochem J*, **363**, 183-188.

Grether-Beck S, Timmer A, Felsner I, Brenden H, Brammertz D and Krutmann J (2005) Ultraviolet a-induced signaling involves a ceramide-mediated autocrine loop leading to ceramide de novo synthesis. *J Invest Dermatol*, **125**, 545-553.

Han G, Gable K, Yan L, Natarajan M, Krishnamurthy J, Gupta SD, Borovitskaya A, Harmon JM and Dunn TM (2004) The topology of the Lcb1p subunit of yeast serine palmitoyltransferase. *J Biol Chem*, **279**, 53707-53716.

Hanada K (2003) Serine palmitoyltransferase, a key enzyme of sphingolipid metabolism. *Biochim Biophys Acta*, **1632**, 16-30.

Hanada K, Hara T, Fukasawa M, Yamaji A, Umeda M and Nishijima M (1998) Mammalian cell mutants resistant to a sphingomyelin-directed cytolysin. Genetic and biochemical evidence for complex formation of the LCB1 protein with the LCB2 protein for serine palmitoyltransferase. *J Biol Chem*, **273**, 33787-33794.

Hanada K, Hara T and Nishijima M (2000a) D-Serine inhibits serine palmitoyltransferase, the enzyme catalyzing the initial step of sphingolipid biosynthesis. *FEBS Lett*, **474**, 63-65.

Hanada K, Hara T and Nishijima M (2000) Purification of the serine palmitoyltransferase complex responsible for sphingoid base synthesis by using affinity peptide chromatography techniques. *J Biol Chem*, **275**, 8409-8415.

Hanada K, Hara T, Nishijima M, Kuge O, Dickson RC and Nagiec MM (1997) A mammalian homolog of the yeast LCB1 encodes a component of serine palmitoyltransferase, the enzyme catalyzing the first step in sphingolipid synthesis. *J Biol Chem*, **272**, 32108-32114.

Hanada K, Nishijima M and Akamatsu Y (1990) A temperature-sensitive mammalian cell mutant with thermolabile serine palmitoyltransferase for the sphingolipid biosynthesis. *J Biol Chem*, **265**, 22137-22142.

Herget T, Esdar C, Oehrlein SA, Heinrich M, Schutze S, Maelicke A and van Echten-Deckert G (2000) Production of ceramides causes apoptosis during early neural differentiation in vitro. *J Biol Chem*, **275**, 30344-30354.

Hojjati MR, Li Z and Jiang XC (2005a) Serine palmitoyl-CoA transferase (SPT) deficiency and sphingolipid levels in mice. *Biochim Biophys Acta*, **1737**, 44-51.

Hojjati MR, Li Z, Zhou H, Tang S, Huan C, Ooi E, Lu S and Jiang XC (2005b) Effect of myriocin on plasma sphingolipid metabolism and atherosclerosis in apoE-deficient mice. *J Biol Chem*, **280**, 10284-10289.

Holleran WM, Feingold KR, Man MQ, Gao WN, Lee JM and Elias PM (1991) Regulation of epidermal sphingolipid synthesis by permeability barrier function. *J Lipid Res*, **32**, 1151-1158.

Holleran WM, Uchida Y, Halkier-Sorensen L, Haratake A, Hara M, Epstein JH and Elias PM (1997) Structural and biochemical basis for the UVB-induced alterations in epidermal barrier function. *Photodermatol Photoimmunol Photomed*, **13**, 117-128.

Ikushiro H, Hayashi H and Kagamiyama H (2001) A water-soluble homodimeric serine palmitoyltransferase from Sphingomonas paucimobilis EY2395T strain. Purification, characterization, cloning, and overproduction. *J Biol Chem*, **276**, 18249-18256.

Ikushiro H, Hayashi H and Kagamiyama H (2003) Bacterial serine palmitoyltransferase: a water-soluble homodimeric prototype of the eukaryotic enzyme. *Biochim Biophys Acta*, **1647**, 116-120.

Ikushiro H, Hayashi H and Kagamiyama H (2004) Reactions of serine palmitoyltransferase with serine and molecular mechanisms of the actions of serine derivatives as inhibitors. *Biochemistry*, **43**, 1082-1092.

Inuzuka M, Hayakawa M and Ingi T (2005) Serinc, an activity-regulated protein family, incorporates serine into membrane lipid synthesis. *J Biol Chem*, **280**, 35776-83.

Jenkins GM, Cowart LA, Signorelli P, Pettus BJ, Chalfant CE and Hannun YA (2002) Acute activation of de novo sphingolipid biosynthesis upon heat shock causes an accumulation of ceramide and subsequent dephosphorylation of SR proteins. *J Biol Chem*, **277**, 42572-42578.

Jeong T, Schissel SL, Tabas I, Pownall HJ, Tall AR and Jiang X (1998) Increased sphingomyelin content of plasma lipoproteins in apolipoprotein E knockout mice reflects combined production and catabolic defects and enhances reactivity with mammalian sphingomyelinase. *J Clin Invest*, **101**, 905-912.

Jiang Q, Wong J and Ames BN (2004) Gamma-tocopherol induces apoptosis in androgen-responsive LNCaP prostate cancer cells via caspase-dependent and independent mechanisms. *Ann N Y Acad Sci*, **1031**, 399-400.

Krisnangkura K and Sweeley CC (1976) Studies on the mechanism of 3-ketosphinganine synthetase. *J Biol Chem*, **251**, 1597-1602.

Lehtonen JY, Horiuchi M, Daviet L, Akishita M and Dzau VJ (1999) Activation of the de novo biosynthesis of sphingolipids mediates angiotensin II type 2 receptor-induced apoptosis. *J Biol Chem*, **274**, 16901-16906.

Listenberger LL, Ory DS and Schaffer JE (2001) Palmitate-induced apoptosis can occur through a ceramide-independent pathway. *J Biol Chem*, **276**, 14890-14895.

Liu K, Zhang X, Sumanasekera C, Lester RL and Dickson RC (2005) Signalling functions for sphingolipid long-chain bases in Saccharomyces cerevisiae. *Biochem Soc Trans*, **33**, 1170-1173.

Longo CA, Tyler D and Mallampalli RK (1997) Sphingomyelin metabolism is developmentally regulated in rat lung. *Am J Respir Cell Mol Biol*, **16**, 605-612.

Mandala SM, Frommer BR, Thornton RA, Kurtz MB, Young NM, Cabello MA, Genilloud O, Liesch JM, Smith JL and Horn WS (1994) Inhibition of serine palmitoyl-transferase activity by lipoxamycin. *J Antibiot (Tokyo)*, **47**, 376-379.

Mandala SM, Thornton RA, Frommer BR, Dreikorn S and Kurtz MB (1997) Viridiofungins, novel inhibitors of sphingolipid synthesis. *J Antibiot (Tokyo)*, **50**, 339-343.

Mandon EC, Ehses I, Rother J, van Echten G and Sandhoff K (1992) Subcellular localization and membrane topology of serine palmitoyltransferase, 3-dehydrosphinganine reductase, and sphinganine N-acyltransferase in mouse liver. *J Biol Chem*, **267**, 11144-11148.

Mandon EC, van Echten G, Birk R, Schmidt RR and Sandhoff K (1991) Sphingolipid biosynthesis in cultured neurons. Down-regulation of serine palmitoyltransferase by sphingoid bases. *Eur J Biochem*, **198**, 667-674.

Memon RA, Grunfeld C, Moser AH and Feingold KR (1993) Tumor necrosis factor mediates the effects of endotoxin on cholesterol and triglyceride metabolism in mice. *Endocrinology*, **132**, 2246-2253.

Memon RA, Holleran WM, Moser AH, Seki T, Uchida Y, Fuller J, Shigenaga JK, Grunfeld C and Feingold KR (1998) Endotoxin and cytokines increase hepatic sphingolipid biosynthesis and produce lipoproteins enriched in ceramides and sphingomyelin. *Arterioscler Thromb Vasc Biol*, **18**, 1257-1265.

Memon RA, Holleran WM, Uchida Y, Moser AH, Grunfeld C and Feingold KR (2001) Regulation of sphingolipid and glycosphingolipid metabolism in extrahepatic tissues by endotoxin. *J Lipid Res*, **42**, 452-459.

Merrill AH, Jr., Nixon DW and Williams RD (1985) Activities of serine palmitoyltransferase (3-ketosphinganine synthase) in microsomes from different rat tissues. *J Lipid Res*, **26**, 617-622.

Merrill AH, Jr., Sullards MC, Allegood JC, Kelly, S. and Wang, E. (2005) Sphingolipidomics: High-throughput, structure-specific and quantitative analysis of sphingolipids by liquid chromatography tandem mass spectrometry, *Methods*, **36**, 207-224.

Merrill AH, Jr., Wang E and Mullins RE (1988) Kinetics of long-chain (sphingoid) base biosynthesis in intact LM cells: effects of varying the extracellular concentrations of serine and fatty acid precursors of this pathway. *Biochemistry*, **27**, 340-345.

Messmer TO, Wang E, Stevens VL and Merrill AH, Jr. (1989) Sphingolipid biosynthesis by rat liver cells: effects of serine, fatty acids and lipoproteins. *J Nutr*, **119**, 534-538.

Miyake Y, Kozutsumi Y, Nakamura S, Fujita T and Kawasaki T (1995) Serine palmitoyltransferase is the primary target of a sphingosine-like immunosuppressant, ISP-1/myriocin. *Biochem Biophys Res Commun*, **211**, 396-403.

Nagiec MM, Baltisberger JA, Wells GB, Lester RL and Dickson RC (1994) The LCB2 gene of Saccharomyces and the related LCB1 gene encode subunits of serine palmitoyltransferase, the initial enzyme in sphingolipid synthesis. *Proc Natl Acad Sci U S A*, **91**, 7899-7902.

Nagiec MM, Lester RL and Dickson RC (1996) Sphingolipid synthesis: identification and characterization of mammalian cDNAs encoding the Lcb2 subunit of serine palmitoyltransferase. *Gene*, **177**, 237-241.

Park TS, Panek RL, Mueller SB, Hanselman JC, Rosebury WS, Robertson AW, Kindt EK, Homan R, Karathanasis SK, Rekhter MD. (2004) Inhibition of sphingomyelin synthesis reduces atherogenesis in apolipoprotein E-knockout mice. *Circulation*, **110**, 3465-3471.

Paumen MB, Ishida Y, Muramatsu M, Yamamoto M and Honjo T (1997) Inhibition of carnitine palmitoyltransferase I augments sphingolipid synthesis and palmitate-induced apoptosis. *J Biol Chem*, **272**, 3324-3329.

Perry DK, Carton J, Shah AK, Meredith F, Uhlinger DJ and Hannun YA (2000) Serine palmitoyltransferase regulates de novo ceramide generation during etoposide-induced apoptosis. *J Biol Chem*, **275**, 9078-9084.

Pinto WJ, Srinivasan B, Shepherd S, Schmidt A, Dickson RC and Lester RL (1992) Sphingolipid long-chain-base auxotrophs of Saccharomyces cerevisiae: genetics, physiology, and a method for their selection. *J Bacteriol*, **174**, 2565-2574.

Reynolds CP, Maurer BJ and Kolesnick RN (2004) Ceramide synthesis and metabolism as a target for cancer therapy. *Cancer Lett*, **206**, 169-180.

Ridgway ND and Merriam DL (1995) Metabolism of short-chain ceramide and dihydroceramide analogues in Chinese hamster ovary (CHO) cells. *Biochim Biophys Acta*, **1256**, 57-70.

Rotta LN, Da Silva CG, Perry ML and Trindade VM (1999) Undernutrition decreases serine palmitoyltransferase activity in developing rat hypothalamus. *Ann Nutr Metab*, **43**, 152-158.

Sawamura N, Ko M, Yu W, Zou K, Hanada K, Suzuki T, Gong JS, Yanagisawa K and Michikawa M (2004) Modulation of amyloid precursor protein cleavage by cellular sphingolipids. *J Biol Chem*, **279**, 11984-11991.

Scarlatti F, Bauvy C, Ventruti A, Sala G, Cluzeaud F, Vandewalle A, Ghidoni R and Codogno P (2004) Ceramide-mediated macroautophagy involves inhibition of protein kinase B and up-regulation of beclin 1. *J Biol Chem*, **279**, 18384-18391.

Scarlatti F, Sala G, Somenzi G, Signorelli P, Sacchi N and Ghidoni R (2003) Resveratrol induces growth inhibition and apoptosis in metastatic breast cancer cells via de novo ceramide signaling. *Faseb J*, **17**, 2339-2341.

Shimabukuro M, Higa M, Zhou YT, Wang MY, Newgard CB and Unger RH (1998) Lipoapoptosis in beta-cells of obese prediabetic fa/fa rats. Role of serine palmitoyltransferase overexpression. *J Biol Chem*, **273**, 32487-32490.

Sims KJ, Spassieva SD, Voit EO and Obeid LM (2004) Yeast sphingolipid metabolism: clues and connections. *Biochem Cell Biol*, **82**, 45-61.

Stoffel W, LeKim D and Sticht G (1968) Biosynthesis of dihydrosphingosine in vitro. *Hoppe Seylers Z Physiol Chem*, **349**, 664-670.

Tanno O, Ota Y, Kitamura N, Katsube T and Inoue S (2000) Nicotinamide increases biosynthesis of ceramides as well as other stratum corneum lipids to improve the epidermal permeability barrier. *Br J Dermatol*, **143**, 524-531.

Triola G, Fabrias G, Dragusin M, Niederhausen L, Broere R, Llebaria A and van Echten-Deckert G (2004) Specificity of the dihydroceramide desaturase inhibitor N-[(1R,2S)-2-hydroxy-1-hydroxymethyl-2-(2-tridecyl-1-cyclopropenyl)ethyl] o ctanamide (GT11) in primary cultured cerebellar neurons. *Mol Pharmacol*, **66**, 1671-1678.

van Echten-Deckert G, Giannis A, Schwarz A, Futerman AH and Sandhoff K (1998) 1-Methylthiodihydroceramide, a novel analog of dihydroceramide, stimulates sphinganine degradation resulting in decreased de novo sphingolipid biosynthesis. *J Biol Chem*, **273**, 1184-1191.

van Echten-Deckert G, Zschoche A, Bar T, Schmidt RR, Raths A, Heinemann T and Sandhoff K (1997) cis-4-Methylsphingosine decreases sphingolipid biosynthesis by specifically interfering with serine palmitoyltransferase activity in primary cultured neurons. *J Biol Chem*, **272**, 15825-15833.

Wang H, Charles AG, Frankel AJ and Cabot MC (2003) Increasing intracellular ceramide: an approach that enhances the cytotoxic response in prostate cancer cells. *Urology* **61**:1047-52.

Wang H, Maurer BJ, Reynolds CP and Cabot MC (2001) N-(4-hydroxyphenyl)retinamide elevates ceramide in neuroblastoma cell lines by coordinate activation of serine palmitoyltransferase and ceramide synthase. *Cancer Res*, **61**, 5102-5105.

Weiss B and Stoffel W (1997) Human and murine serine-palmitoyl-CoA transferase--cloning, expression and characterization of the key enzyme in sphingolipid synthesis. *Eur J Biochem*, **249**, 239-247.

Williams RD, Wang E and Merrill AH, Jr. (1984) Enzymology of long-chain base synthesis by liver: characterization of serine palmitoyltransferase in rat liver microsomes. *Arch Biochem Biophys*, **228**, 282-291.

Worgall TS, Juliano RA, Seo T, Deckelbaum RJ (2004) Ceramide synthesis correlates with the posttranscriptional regulation of the sterol-regulatory element-binding protein. *Arterioscler Thromb Vasc Biol*, **24**, 943-948.

Yamaji-Hasegawa A, Takahashi A, Tetsuka Y, Senoh Y and Kobayashi T (2005) Fungal metabolite sulfamisterin suppresses sphingolipid synthesis through inhibition of serine palmitoyltransferase. *Biochemistry*, **44**, 268-277.

Yasuda S, Nishijima M and Hanada K (2003) Localization, topology, and function of the LCB1 subunit of serine palmitoyltransferase in mammalian cells. *J Biol Chem*, **278**, 4176-4183.

Zaman Z, Jordan PM and Akhtar M (1973) Mechanism and stereochemistry of the 5-aminolaevulinate synthetase reaction. *Biochem J*, **135**, 257-263.

Zhang K, Showalter M, Revollo J, Hsu FF, Turk J, Beverley SM (2003) Sphingolipids are essential for differentiation but not growth in Leishmania. *EMBO J*, **22**, 6016-6026.

Zweerink MM, Edison AM, Wells GB, Pinto W and Lester RL (1992) Characterization of a novel, potent, and specific inhibitor of serine palmitoyltransferase. *J Biol Chem*, **267**, 25032-25038.

2-2 Ceramide Synthase

Irene Pankova-Kholmyansky and Anthony H. Futerman

Department of Biological Chemistry, Weizmann Institute of Science, Rehovot 76100, Israel

Summary. Ceramide is an important intracellular second messenger, and a key metabolite in the pathway of sphingolipid biosynthesis where it is formed by N-acylation of sphinganine. In this review, we describe the identification and characterization of a gene family consisting of six genes, the LASS genes, that regulate *de novo* ceramide synthesis in mammalian cells. Unexpectedly, each LASS homolog tested to date synthesizes ceramide with a specific fatty acid composition. We recently purified LASS5 to homogeneity and demonstrated that it is a *bona fide* ceramide synthase, displaying the same fatty acid specificity as the membrane-bound enzyme. Moreover, unlike in yeast where an additional subunit, Lip1, is required for ceramide synthase activity, LASS5 does not require any additional subunits, consistent with database analyses showing that mammalian cells do not contain Lip1 homologs. The reasons that mammalian cells contain six LASS genes are not known, but presumably each gene is involved in the synthesis of ceramides containing specific fatty acids for use in the regulation of different biological processes.

Keywords. Dihydroceramide synthase, ceramide, LASS genes, purification.

The first step in the pathway of *de novo* ceramide biosynthesis is condensation of serine and palmitoyl-CoA to form 3-ketosphinganine, which is subsequently reduced by 3-ketosphinganine reductase to sphinganine. The next enzyme in the pathway, dihydroceramide synthase (sphinganine

N-acyl transferase), can *N*-acylate various sphingoid bases, including sphinganine, sphingosine and 4-hydroxysphinganine, can utilize a wide spectrum of fatty acyl-CoAs (Futerman and Hannun, 2004; Merrill, 2002) (Fig. 1), and is inhibited by the mycotoxin, fumonisin B1 (FB1) (Merrill, 2002; Merrill et al., 1993; Wang et al., 1991), a structural analog of sphinganine. Various pieces of evidence had suggested that multiple (dihydro)ceramide synthases exist, including kinetic evidence, the observation that FB1 suppresses the synthesis of most, but not all sphingolipids, and that (dihydro)ceramide synthesis may occur in multiple intracellular locations (Futerman and Hannun, 2004). However, until recently no direct biochemical or molecular data was available to prove the existence, or not, of multiple dihydroceramide synthases.

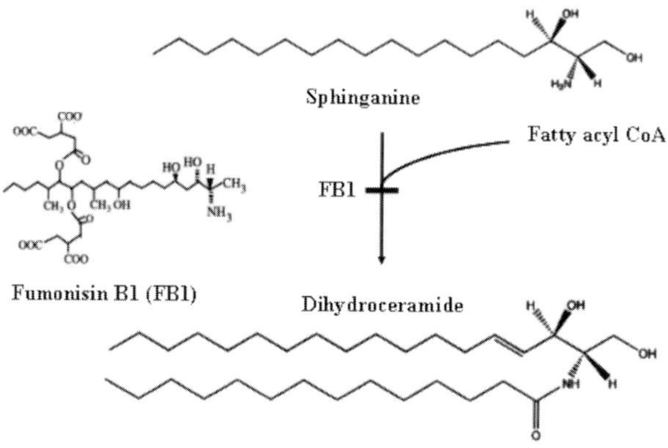

Fig. 1. (Dihydro)ceramide synthase catalyzes the acylation of sphinganine or other sphingoid bases. Fumonisin B1 (FB1) is a specific inhibitor of this reaction.

The first clues to the biochemical identification of ceramide synthases were obtained in *Saccharomyces cerevisiae*, in which two genes, Lag1 and Lac1, were shown to be required for ceramide synthesis using very long chain (C26) fatty acids (Guillas et al., 2001; Schorling et al., 2001). Initially, Lag1p was identified as a regulator of longevity and aging (D'Mello et al., 1994). Lag1p and Lac1p are homologous multi-spanning transmembrane proteins that facilitate ER to Golgi transport of GPI- anchored proteins (Barz and Walter, 1999; Guillas et al., 2001; Schorling et al., 2001; Watanabe et al., 2002). Double mutations of Lag1p and Lac1p resulted in a serious defect in cell growth and a reduction in sphingolipid levels (Barz and Walter, 1999; Guillas et al., 2001; Schorling et al., 2001). A plant homolog, *Alternatia stem cancer locus*-1 (asc-1), was shown to mediate

FB1-resistance in tomato (Brandwagt et al., 2000), implying that ASC1 might encode an FB1-insensitive ceramide synthase.

Subsequently, mammalian homologs of Lag1p were identified and characterized (Mizutani et al., 2005; Riebeling et al., 2003; Venkataraman et al., 2002). Database analysis revealed a family of Lag1p-motif-containing proteins (Jiang et al., 1998; Venkataraman and Futerman, 2002; Winter and Ponting, 2002). Five mammalian proteins, originally characterized as translocating chain-associating membrane (TRAM) protein homologs (TRH), were detected (recently renamed 'LASS' genes (longevity assurance gene homologs) (Fig. 2). TRAM had been identified as a Lag1p homolog, but did not complement the lag1/lac1 deletion mutant (Guillas et al., 2003; Jiang et al., 1998). Another Lag1p/TRAM family member was also discovered, CLN8, a gene defective in a form of neuronal ceroid lipofuscinoses (Kida et al., 2001; Ranta et al., 1999). CLN8, Lag1 and TRAM have five conserved predicted transmembrane α-helices, which are referred to as the TLC (TRAM-Lag1-CLN8) domain (Winter and Ponting, 2002) (Figs. 2, 3). Five of the LASS genes contain a Hox domain, a transcription factor involved in developmental regulation (Fig. 2), suggesting that these LASS proteins could regulate lipid metabolism through transcriptional activation, which would require proteolytic excision of the Hox domain from the membrane-bound protein and its subsequent translocation to the nucleus.

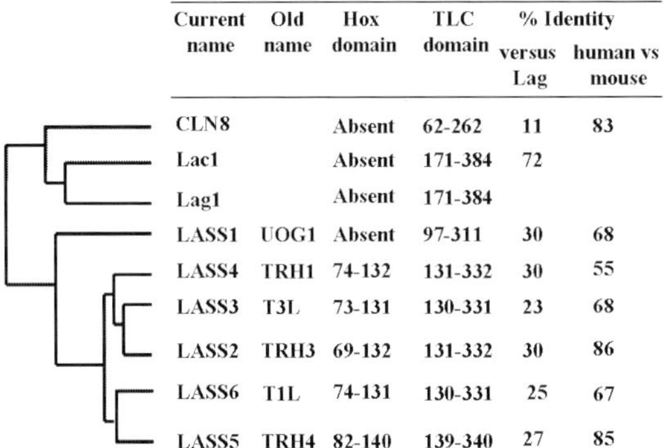

Current name	Old name	Hox domain	TLC domain	% Identity versus Lag	% Identity human vs mouse
CLN8		Absent	62-262	11	83
Lac1		Absent	171-384	72	
Lag1		Absent	171-384		
LASS1	UOG1	Absent	97-311	30	68
LASS4	TRH1	74-132	131-332	30	55
LASS3	T3L	73-131	130-331	23	68
LASS2	TRH3	69-132	131-332	30	86
LASS6	T1L	74-131	130-331	25	67
LASS5	TRH4	82-140	139-340	27	85

Fig. 2. Phenogram showing the relationship between LASS genes. The new LASS terminology is listed next to original terminology. Amino acid residues encompassing the Hox domain are shown for human homologs. The percent amino acid identities of human versus mouse homologs, and the identities of the human homologs versus the yeast Lag gene, are also shown. CLN8 is a member of the same family, but it is not known if this protein is involved in sphingolipid synthesis.

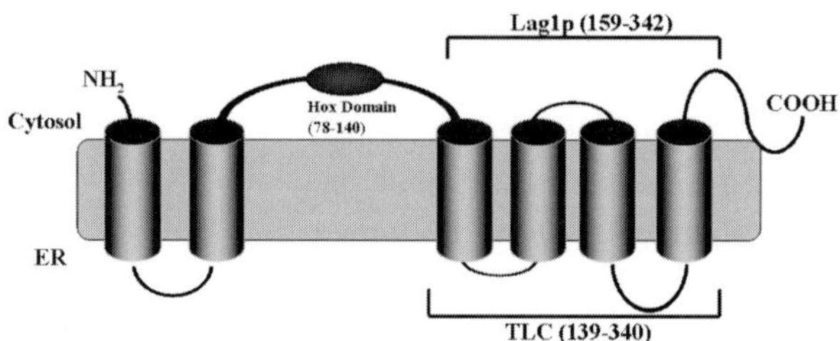

Fig. 3. Putative structure of LASS5. The number of transmembrane domains was determined using SOUSI (http://sosui.proteome.bio.tuat.ac.jp/sosuiframe0. html), and the location of other functional domains using either SMART (http://smart.embl-heidelberg.de) or NCBI-rpsblast (http://www.ncbi.nlm.nih.gov/ BLAST). The Lag1p domain has been postulated to be part of the ceramide synthase complex. The TLC domain is defined as a TRAM-Lag1-CLN8 homology domain.

In our laboratory, we demonstrated that overexpression of a mouse homolog of LAG1 in mammalian cells, upstream of growth and differentiation factor (uog1/LASS1) (Boyer et al., 1993; Jiang et al., 1998), increased the synthesis of ceramide containing one specific fatty acid, namely stearic acid (Venkataraman et al., 2002). This ceramide was then specifically channeled into neutral glycosphingolipids but not into gangliosides. Overexpression of LASS4 (TRH1) and LASS5 (TRH4) in mammalian cells also increased dihydroceramide synthesis, with overexpression of LASS4 resulting in elevated levels of sphingolipids containing stearic and arachidonic acids, and overexpression of LASS5 resulting in synthesis of ceramides containing palmitic acid (Riebeling et al., 2003) (Fig. 4). Another family member, LASS6, was also characterized and demonstrated to produce shorter-acyl chain ceramide species (C14:0- and C16:0-ceramides) (Mizutani et al., 2005). However, even though overexpression of each gene produced ceramides containing specific fatty acids, it was not known at this stage whether these genes modified an endogenous ceramide synthase activity and thereby conferred fatty acid selectivity, or whether the LASS proteins themselves were *bona fide* (dihydro)ceramide synthases.

Ceramide synthase was first purified in an active form from yeast, and a new subunit, Lip1, was identified (Vallee and Riezman, 2005). Lip1 is a single-span membrane protein located in the endoplasmic reticulum that is required for ceramide synthesis *in vivo* and *in vitro* (Vallee and Riezman, 2005); database searches do not reveal any mammalian genes with significant homology to Lip1. The first mammalian ceramide synthase was also recently purified in our laboratory. LASS5 with an HA-tag at the C- terminus was solubilized using digitonin, and purified by immunoprecipitation (Fig. 5a) (Lahiri and Futerman, 2005). Solubilized LASS5-HA demonstrated the same fatty acid specificity as the membrane-bound enzyme. After elution from agarose beads, only one band could be detected by SDS polyacrylamide gel electrophoresis, whose identity was confirmed to be LASS5 by mass spectrometry. Surprisingly, eluted LASS5-HA did not show any dihydroceramide synthase activity unless phospholipids were added exogenously to the eluate. Dioleoylphosphatidylcholine (DOPC) was more efficient at preserving dihydroceramide synthase activity than dioleoylphosphatidylserine (DOPS) (Fig. 5b). Purified LASS5 was highly specific towards palmitoyl-CoA compared to stearoyl CoA (Fig. 5c), and was inhibited by FB1. Unlike in yeast, LASS5 does not require an additional subunit for its activity.

The reason that mammalian cells contain six genes for ceramide synthesis is not known, but is highly suggestive that ceramides containing different fatty acids may play distinct roles in specific cell functions. Ceramide containing C16-fatty acid is thought to be important in apoptosis (Kroesen et al., 2003), and recent studies demonstrated the involvement of C16-ceramide in LNCaP (Eto et al., 2003) and hepatocyte (Osawa et al., 2005) apoptosis. C18-ceramide was selectively down-regulated in head and neck squamous cell carcinomas, and overexpression of LASS1 and generation of C18-ceramide inhibited cell growth and induced apoptosis in these carcinomas (Koybasi et al., 2004).

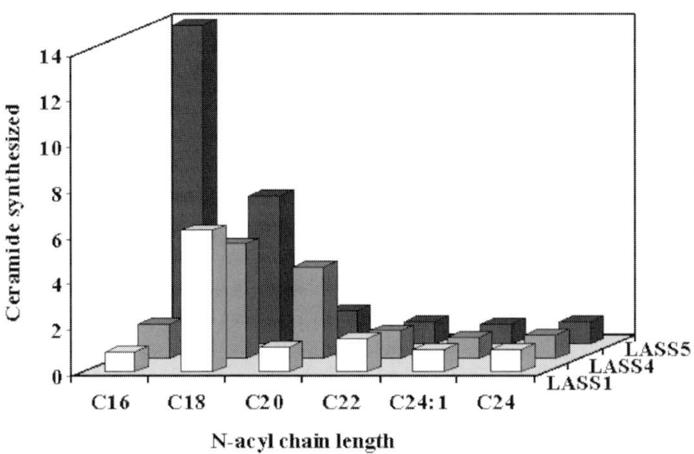

Fig. 4. Fatty acid composition of ceramide in LASS1-, LASS4- and LASS5-overexpressing cells. The fatty acid composition was determined by electrospray tandem mass spectrometry (ESI-MS/MS), and is compared to mock-transfected cells, with a value of 1 assigned to each fatty acid in the latter.

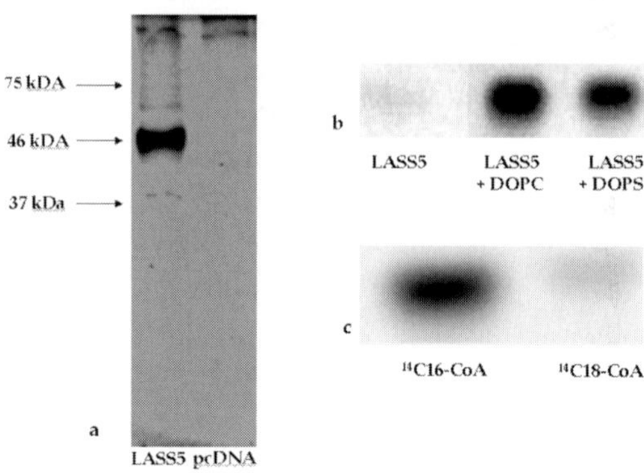

Fig. 5. Purification of LASS5-HA. a. LASS5-HA was solubilized using digitonin, immunoprecipitated, eluted from the beads, and detected by silver staining. The pcDNA lane is taken from an identical experiment in which cells were transfected with pcDNA-HA rather than LASS5-HA. **b.** Dihydroceramide synthase activity of purified LASS5 was analyzed using 1-[^{14}C]-palmitoyl-CoA after no addition or addition of DOPC or DOPS. **c.** Dihydroceramide synthase activity of purified LASS5 was analyzed in the presence of DOPC using 1-[^{14}C]-palmitoyl-CoA (^{14}C16-CoA) or 1-[^{14}C]-stearoyl-CoA (^{14}C18-CoA).

Little is known about how *de novo* sphingolipid synthesis is regulated, and in particular how this pathway is regulated upon consumption of sphingolipids in the signaling pathways in which they are involved. It seems quite likely that the LASS genes will be targets for regulation, and it is surely no coincidence that multiple LASS genes exist; indeed, the dihydroceramide synthases are the only enzymes in the sphingolipid biosynthetic pathway that exist in multiple isoforms (Futerman and Riezman, 2005). Furthermore, LASS genes display different tissue distributions, and the existence of a Hox domain in five of the LASS genes potentially provides an important mode of regulation of ceramide and sphingolipid metabolism.

References

Barz WP, Walter P (1999) Two endoplasmic reticulum (ER) membrane proteins that facilitate ER-to-Golgi transport of glycosylphosphatidylinositol-anchored proteins. *Mol Biol Cell*, **10**, 1043-1059.
Boyer J, Pascolo S, Richard GF, Dujon B (1993) Sequence of a 7.8 kb segment on the left arm of yeast chromosome XI reveals four open reading frames, in-

cluding the CAP1 gene, an intron-containing gene and a gene encoding a homolog to the mammalian UOG-1 gene. *Yeast*, **9**, 279-287.

Brandwagt BF, Mesbah LA, Takken FL, Laurent PL, Kneppers TJ, Hille J, Nijkamp HJ (2000) A longevity assurance gene homolog of tomato mediates resistance to Alternaria alternata f. sp. lycopersici toxins and fumonisin B1. *Proc Natl Acad Sci USA*, **97**, 4961-4966.

D'Mello N P, Childress AM, Franklin DS, Kale SP, Pinswasdi C, Jazwinski SM (1994) Cloning and characterization of LAG1, a longevity-assurance gene in yeast. *J Biol Chem*, **269**, 15451-15459.

Eto M, Bennouna J, Hunter OC, Hershberger PA, Kanto T, Johnson CS, Lotze MT, Amoscato AA (2003) C16 ceramide accumulates following androgen ablation in LNCaP prostate cancer cells. *Prostate*, **57**, 66-79.

Futerman AH, Hannun YA (2004) The complex life of simple sphingolipids. *EMBO Rep*, **5**, 777-782.

Futerman AH, Riezman H (2005) The ins and outs of sphingolipid synthesis. *Trends Cell Biol*, **15**, 312-318.

Guillas I, Jiang JC, Vionnet C, Roubaty C, Uldry D, Chuard R, Wang J, Jazwinski SM, Conzelmann A (2003) Human homologues of LAG1 reconstitute acyl-CoA-dependent ceramide synthesis in yeast. *J Biol Chem*, **278**, 37083-37091.

Guillas I, Kirchman PA, Chuard R, Pfefferli M, Jiang JC, Jazwinski SM, Conzelmann A (2001) C26-CoA-dependent ceramide synthesis of Saccharomyces cerevisiae is operated by Lag1p and Lac1p. Embo J, 20, 2655-2665.

Jiang JC, Kirchman PA, Zagulski M, Hunt J, Jazwinski SM (1998) Homologs of the yeast longevity gene LAG1 in Caenorhabditis elegans and human. *Genome Res*, **8**, 1259-1272.

Kida E, Golabek AA, Wisniewski KE (2001) Cellular pathology and pathogenic aspects of neuronal ceroid lipofuscinoses. *Adv Genet*, **45**, 35-68.

Koybasi S, Senkal CE, Sundararaj K, Spassieva S, Bielawski J, Osta W, Day TA, Jiang JC, Jazwinski SM, Hannun YA, Obeid LM, Ogretmen B (2004) Defects in cell growth regulation by C18:0-ceramide and longevity assurance gene 1 in human head and neck squamous cell carcinomas. *J Biol Chem*, **279**, 44311-44319.

Kroesen BJ, Jacobs S, Pettus BJ, Sietsma H, Kok JW, Hannun YA, de Leij LF (2003) BcR-induced apoptosis involves differential regulation of C16 and C24-ceramide formation and sphingolipid-dependent activation of the proteasome. *J Biol Chem*, **278**, 14723-14731.

Lahiri S, Futerman AH (2005) LASS5 is a bona fide dihydroceramide synthase that selectively utilizes palmitoyl CoA as acyl donor. *J Biol Chem*, **280**, 33735-33738.

Merrill AH, Jr. (2002) De novo sphingolipid biosynthesis: a necessary, but dangerous, pathway. *J Biol Chem*, **277**, 25843-25846.

Merrill AH, Jr., van Echten G, Wang E, Sandhoff K (1993) Fumonisin B1 inhibits sphingosine (sphinganine) N-acyltransferase and de novo sphingolipid biosynthesis in cultured neurons in situ. *J Biol Chem*, **268**, 27299-27306.

Mizutani Y, Kihara A, Igarashi Y (2005) Mammalian longevity assurance homolog LASS6 and its related family members regulate synthesis of specific ceramides. *Biochem J*, **390**, 263-271.

Osawa Y, Uchinami H, Bielawski J, Schwabe RF, Hannun YA, Brenner DA (2005) Roles for C16-ceramide and sphingosine-1-phosphate in regulating hepatocyte apoptosis in response to tumor necrosis factor-alpha. *J Biol Chem*, **280**, 27879-27887.

Ranta S, Zhang Y, Ross B, Lonka L, Takkunen E, Messer A, Sharp J, Wheeler R, Kusumi K, Mole S, Liu W, Soares MB, Bonaldo MF, Hirvasniemi A, de la Chapelle A, Gilliam TC, Lehesjoki AE (1999) The neuronal ceroid lipofuscinoses in human EPMR and mnd mutant mice are associated with mutations in CLN8. *Nat Genet*, **23**, 233-236.

Riebeling C, Allegood JC, Wang E, Merrill AH, Jr., Futerman AH (2003) Two mammalian longevity assurance gene (LAG1) family members, trh1 and trh4, regulate dihydroceramide synthesis using different fatty acyl-CoA donors. *J Biol Chem*, **278**, 43452-43459.

Schorling S, Vallee B, Barz WP, Riezman H, Oesterhelt D (2001) Lag1p and Lac1p are essential for the Acyl-CoA-dependent ceramide synthase reaction in Saccharomyces cerevisae. *Mol Biol Cell*, **12**, 3417-3427.

Vallee B, Riezman H (2005) Lip1p: a novel subunit of acyl-CoA ceramide synthase. *Embo J*, **24**, 730-741.

Venkataraman K, Futerman AH (2002) Do longevity assurance genes containing Hox domains regulate cell development via ceramide synthesis? *FEBS Lett*, **528**, 3-4.

Venkataraman K, Riebeling C, Bodennec J, Riezman H, Allegood JC, Sullards MC, Merrill AH, Jr., Futerman AH (2002) Upstream of growth and differentiation factor 1 (uog1), a mammalian homolog of the yeast longevity assurance gene 1 (LAG1), regulates N-stearoyl-sphinganine (C18-(dihydro)ceramide) synthesis in a fumonisin B1-independent manner in mammalian cells. *J Biol Chem*, **277**, 35642-35649.

Wang E, Norred WP, Bacon CW, Riley RT, Merrill AH, Jr. (1991) Inhibition of sphingolipid biosynthesis by fumonisins. Implications for diseases associated with Fusarium moniliforme. *J Biol Chem*, **266**, 14486-14490.

Watanabe R, Funato K, Venkataraman K, Futerman AH, Riezman H (2002) Sphingolipids are required for the stable membrane association of glycosyl-phosphatidylinositol-anchored proteins in yeast. *J Biol Chem*, **277**, 49538-49544.

Winter E, Ponting CP (2002) TRAM, LAG1 and CLN8: members of a novel family of lipid-sensing domains? *Trends Biochem Sci*, **27**, 381-383.

2-3 Dihydroceramide:Sphinganine Δ4-Desaturase and C4-Hydroxylase

Akemi Suzuki, Fumio Omae, and Ayako Enomoto

Sphingolipid Expression Laboratory, RIKEN Frontier Research System, 2-1 Hirosawa, Wako, Saitama 351-0198, Japan

Summary. *N*-acylsphingenine (ceramide) and *N*-acyl-4-hydroxy- sphinganine (phytoceramide) are intermediates in glycosphingolipid and sphingomyelin biosynthesis. Ceramide has attracted attention because it is a signaling molecule involved in apoptosis. The genes for *N*-acylsphinganine, sphinganine ω4-desaturase and C4-hydroxylase, are responsible for the biosynthesis of ceramide and dihydroceramide. These genes were identified in 2002 through exhaustive search of plant and yeast desaturases, but this review will address primarily the desaturases and hydroxylases of animal origin, particularly the distribution and functions of ceramide and phytoceramide in animal tissues. Ceramide-containing sphingolipids are widely distributed, whereas phytoceramide-containing sphingolipids are limited tissues in the intestine, kidney and skin. Distribution appears to be regulated by the expression of Δ4-desaturase and C4-hydroxylase enzymes. Recent developments concerning these two enzymes are discussed here.

Keywords. Ceramide, phytoceramide, Δ4-desaturase, C4-hydroxylase, intestinal epithelial cells, tissue-specific distribution

1. Introduction

The biosynthetic reactions that produce ceramide (*N*-acyl-Δ4-sphingenine) and phytoceramide (*N*-acyl-4-hydroxysphinganine) are Δ4-desaturation and C4-hydroxylation, respectively. These reactions are commonly discussed together because of similarities in their molecular mechanisms: they both use *N*-acylsphinganine as a substrate and the enzymes that carry out these reactions are homologues. In 1996 Endo *et al.* reported that a mutation of *Des* (degenerative spermatocytes) in Drosophila arrested spermatogenesis at the initiation of meiosis (Endo *et al.* 1996), but the role of *Des* was unknown. A year later, the same group identified a *Des* homologue in the mouse (Endo *et al.* 1997).

Ternes et al described the function of *Des* (Ternes *et al.* 2002) following an extensive database search for genes containing the conserved sequence required for desaturation. *Des* (DES1) and its mouse homologue (DES2) were identified as members of the desaturase family. When yeast *Sur2* mutant, which lacks sphinganine C4- hydroxylase, was transformed with DES1 cDNA, the resulting lipid hydrolysates had Δ4-sphingenine. When DES2 cDNA transformed *Sur2* mutant yeasts, both Δ4-sphingenine and C4-hydroxyl- sphinganine were detected in lipid hydrolysates. They concluded that DES1 was a desaturase responsible for the biosynthesis of Δ4-sphingenine, and that DES2 was a bifunctional desaturase/- hydroxylase responsible for biosynthesis of Δ4-sphingenine and 4-hydroxylsphinganine in mammals.

Despite these significant advances in our understanding of these reactions, their substrates, molecular mechanisms, enzymatic regulation, transcriptional regulation, and product functions remain for future investigations.

2. Δ4-Desaturase

Although studies of desaturase enzymes stretch back to the 1960s, only recent reports are included here. Ceramide synthesis by desaturation of *N*-acylsphinganine (Michel *et al.* 1997) was reported by Michel et al in 1997 based on several observations. First, *N*-acylsphinganine:sphinganine desaturase activity was detected *in vitro* using a rat microsomal fraction and either NADH or NADPH and the apparent K_m values for *N*-acylsphinganine and NADH were 340 and 120 μM respectively. Also, that activity could be influenced by alkyl chains from the sphingosine base and fatty acid. Finally, the enzymatic activity requires molecular oxygen and cyanide, divalent copper, and antibodies against cytochrome b5 inhibited the reaction.

In the same year, Geeraert *et al.* found destaurase activity in intact and permeabilized rat hepatocytes, hepatocyte homogenates, and the microsomal fraction of the homogenates (Geeraert *et al.* 1997). They suggested that desaturation is carried out by a complex composed of flavoprotein, cytochrome b5, and terminal desaturase in the presence of NADPH. They assayed desaturase activity by measuring the tritiated water produced from the substrate *N*-hexanoyl-[4,5-^{3}H]- sphinganine as the result of desaturation. The same group also examined tissue distribution and subcellular localization of desaturase activity and concluded that desaturase activity is distributed throughout various tissues, but highest in the liver where it localizes to the ER (Causeret *et al.* 2000).

Cadena *et al.* cloned MLD as a member of the membrane fatty acid (lipid) desaturase gene family, using a yeast two-hybrid screening with the intracellular kinase domain of the EGF receptor as bait (Cadena *et al.* 1997). It contained two short domains with 40-60% identity to yeast Δ9-desaturase and ω-carotene hydroxylase and three histidine box sequence motifs. These motifs, HX$_3$H, HX$_2$HH, and HX$_2$HHXFP, are conserved in membrane fatty acid desaturases and membrane hydrocarbon hydroxylases (Shanklin *et al.* 1994). MLD transcripts were detected in all 16 organs tested, indicating ubiquitous expression of the enzyme. Analysis of the amino acid sequence of MLD suggested that it contains four membrane-spanning domains, is localized to the ER, and has a cytosolic catalytic site containing three His box sequence motifs. Why the two-hybrid screening picked up MLD is not clear. Cotransfection of MLD with the EGF receptor cDNAs had decreased receptor expression, while overexpression of MLD inhibited biosynthesis of the receptor. This suggests that MLD regulates biosynthesis or processing of the EGF receptor. These authors believed that MLD was a fatty acid desaturase, but it was the human sphingosine Δ4-desaturase DES1 (Ternes *et al.* 2002).

Mikami *et al.* examined the substrate specificity of *N*-acylsphinganine desaturase in homogenates of fetal rat skin. Of the dihydroceramides containing C$_{10}$-C$_{18}$ fatty acids, maximal activity was obtained with C$_{14}$-dihydroceramide (Mikami *et al.* 1998).

As a desaturase, *N*-acylsphinganine appears to be a substrate, its enzymatic activity requires NADH or NADPH, its enzyme is widely distributed in various tissues but localizes in the ER membrane as a membrane-spanning protein.

Ternes *et al.* identified a family of protein sequences from animals, plants, and fungi using a position-specific iterated BLAST search. In this analysis, sequences containing His box sequence motifs were assembled and subsequently grouped into subfamilies according to sequence similarity (Ternes *et al.* 2002). Candidate subfamilies had to consist exclusively of sequences with as-yet-unidentified biochemical functions and contain sequences from animals, plants, and fungi since Δ4-desaturated sphingolipids have been found in all these organisms. Sequences from *S. cerevisiae* were eliminated because this is one of the few eukaryotic organisms that lacks Δ4-desaturated sphingolipids.

As a result of these efforts, Ternes *et al.* successfully cloned DES1 from *H. sapiens*, *M. musculus*, *D. melanogaster*, *L. esculentum*, and *C. albicans*. They identified sphingolipid Δ4-desaturase activity by the presence of sphingenine in the lipid fraction obtained by alkaline hydrolysis of yeasts transformed with the DES1 cDNAs. This study identified the Δ4-desaturase gene as *Des*, which was originally identified as an essential gene for spermatogenesis in *Drosophila* (Endo *et al.* 1996). MLD identified by Cadena *et al,* as described above, became human DES1. In addition, Ternes *et al.* reported that NMR analysis of sphingenine produced by *C. albicans* DES1 confirmed a D-erythro-4-sphingenine conformation, indicating that DES1 produces a stereo-specific product (Ternes *et al.* 2002).

Ceramide is a signaling molecule, a degradation precursor in the biosynthesis of sphingenine and sphingenine-1-phosphate, and the synthetic precursor for glycosphingolipids (Merrill 2002, Spiegel and Milstien 2002). The functions of the ceramide portion of glycosphingolipids remain unclear, but it is important for the formation of microdomains in biological membranes. The microdomains are involved in membrane function and signal transduction. The studies of molecules responsible for the production of ceramide can contribute to the studies on the functions of ceramide and microdomains by manipulating these molecules in living cells.

3. C4-Hydroxylase

C4-hydroxylation has long been considered physiologically important because glycosphingolipids containing C4-hydroxylated sphinganine (phytosphingosine) were restricted to certain tissues in mammals. Early work by Karlsson's group demonstrated that human kidney glycol- sphingolipids and bovine kidney sphingomyelin contained phyto- sphingosine (Karlsson and Martensson 1968, Karlsson and Steen 1968). Several years later, intestinal glycosphingolipids and sphingomyelin were also found to contain phytosphingosine (Breimer *et al.* 1974, Breimer *et al.* 1975, Bouhours and Glickman 1977).

Subsequent reports showed that intestinal epithelial cells and kidney tubular cells express glycosphingolipids and sphingomyelin containing phytosphingosine. Biochemical and immunological analyses of mouse small intestine indicated that high concentrations of phytosphingosine-containing glycosphingolipids are localized at the microvillous membrane of epithelial cells. These results suggested that the function of these glycosphin-golipids was related to active absorption of digested nutrients (Umesaki *et al.* 1981, Suzuki and Yamakawa 1981).

DES2 was the hydroxylase gene responsible for the production of C4-hydroxylated sphinganine identified by Ternes *et al.* (Ternes *et al.* 2002), a homologue of DES1 Δ4-desaturase (Section 1.2). DES2 is a bifunctional enzyme that produces both Δ4-sphingenine and C4-hydroxylated sphinganine. Both Δ4-sphingenine and C4-hydroxylated sphinganine were detected in the alkali-stable lipid fraction prepared from yeasts carrying a defective sphinganine C4-hydroxylase gene by transformation with mouse DES2 cDNA.

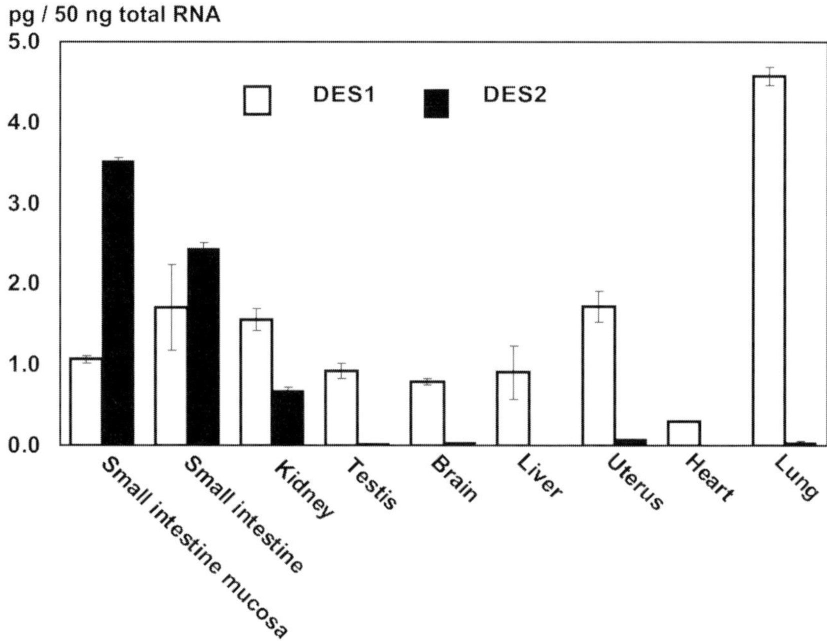

Fig. 1. Tissue distribution of *Des1* and *Des2* mRNAs. Real-time quantitative RT-PCR was performed with 50 ng of total RNA prepared from the tissues indicated. The amounts of *Des1* and *Des2* mRNA were determined using plasmid DNA containing the *Des1* and *Des2* genes as a standard. Data were normalized to the amount of GAPDH mRNA. Data shown are the mean ± SEM; n = 3.

Recently, homogenates of COS-7 cells transfected with mouse DES2 cDNA were used in *in vitro* C4-hydroxylase assays that demonstrated that *N*-acylsphinganine, but not sphinganine, is a substrate for DES2 (Omae *et al.* 2004). When the expression of DES2 mRNA was analyzed in various tissues by a quantitative PCR method, it was found to correlate with the observed tissue distribution of phytosphingosine-containing glyco- sphingolipids, with DES2 mRNA levels being very high in the small intestine, high in the kidney, and low in the liver, brain, heart, and lung. In contrast, DES1 mRNA is rather ubiquitously distributed (Fig. 1). In *in situ* hybridization studies, crypt cells of mouse small intestine exhibited positive signals for DES2 mRNA, indicating that DES2 transcription occurs in crypt cells, not in the further differentiated epithelial cells. These results demonstrate that DES2 is an *N*-acylsphinganine:C4-sphinganine hydroxylase and is responsible for the expression of phyto- sphingosine-containing glycosphingolipids.

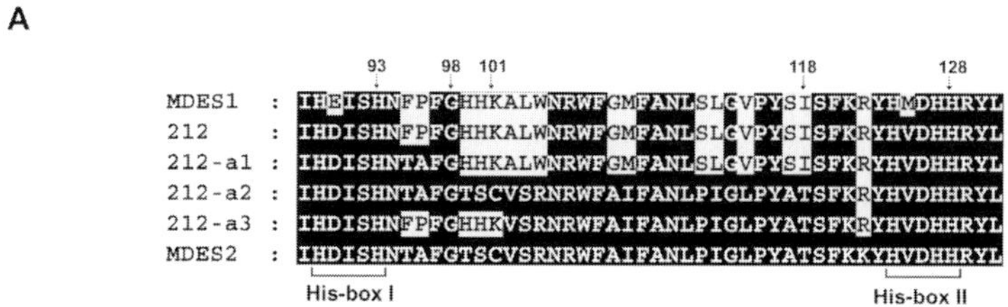

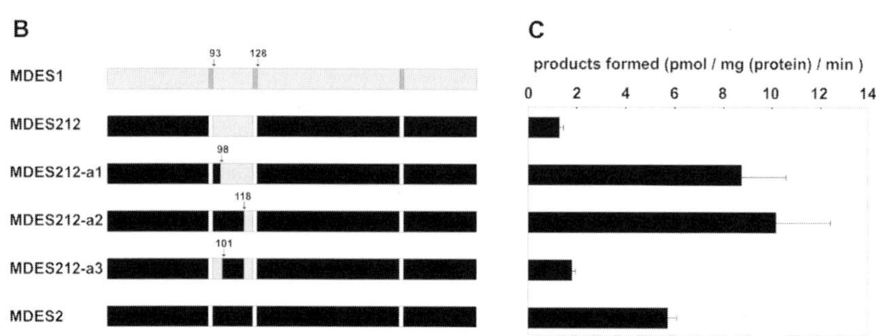

Fig. 2. **The C4-hydroxylase activities of mouse DES2 (MDES2) and its chimeras.** (A) Amino acid sequences of MDES2, mouse DES1 (MDES1), and other chimeras. Black letters on white indicate amino acids identical to those of MDES1; white letters on black indicate those identical to MDES2. (B) The structures of MDES1, MDES2, and the chimeras are shown. MDES1 sequences are gray, and MDES2 sequences are black. Dark grey or white bars indicate His box sequence motifs of MDES1 or MDES2, respectively. (C) C4-hydroxylase-specific activity in homogenates prepared from transfected COS-7 cells.

The bifunctional activity of mouse DES2 was not effectively demonstrated in an *in vitro* assay, suggesting that the assay conditions were suboptimal. For example, the assay may have lacked unidentified essential factors, cytochrome b5, and or NADH or NADPH cytochrome b5 reductase. Further reconstitution experiments using purified proteins will be required to elucidate the hydroxylation reaction mechanism.

Amino acid sequence comparisons of mouse DES1 and DES2 indicate that the sequences are 63% identical and that both sequences share three hydrophobic membrane-spanning domains (Omae *et al.* 2004) and three His box sequence motifs. The sequences required for hydroxylase activity were analyzed by transfection of various hybrid recombinants of mouse DES1 and DES2 cDNAs into COS-7 cells. Although the domain structures were not defined, the [95]XAFGY sequence (X=T or A; Y=T or N) that is C-terminal to the first His box sequence motif of mouse DES2 was shown to be essential for hydroxylase activity (Fig. 2) (Omae *et al.* 2004).

We do not yet understand the biochemical mechanisms of C4-hydroxylation in detail, and further studies are needed. These studies should include gene-targeting experiments and functional analysis of the products of the hydroxylase enzyme.

4. Yeast C4-hydroxylase

In 1997, Haak *et al.* used a yeast transformation experiment to show that yeast SUR (*Sur2p*) encodes sphinganine C4-hydroxylase, (Haak *et al.* 1997). Grilley *et al.* subsequently reported that *SYR2* is identical to *SUR2* and is required for C4-hydroxylation of sphinganine (Grilley *et al.* 1998). C4-hydroxylase activity was detected in *in vitro* assays with sphinganine and dihydroceramide as substrates and the microsomal fraction of SYR2-wild type yeasts as the enzyme source. No activity was detected when SYR2-mutant yeasts were used, indicating that these sphingolipids are good substrates for SYR2. However, when similar assays were performed with homogenates of SYR2 cDNA-transfected COS-7 cells, C4-hydroxylase activity was detected with sphinganine but not with *N*-octanoylsphinganine (Omae *et al.* 2002). Why the results are contradictory remains unknown.

Despite this inconsistency, several conclusions can be drawn. *N*-acylsphinganine, not sphinganine, is a substrate for mouse DES2, sphinganine is a good substrate for SUR2 (SYR2), and the homology between DES2 and DES1 is far greater than that between DES2 and SUR2.

5. Phylogenetic aspects of DES1 and 2

Ternes *et al.* proposed that fatty acid Δ4-desaturase, DES1, and SUR2 (yeast sphinganine C4-hydroxylase) evolved independently in plants, and that DES2 evolved from DES1 after *Drosophila* (Ternes. *et al.* 2002). Figure 3 shows a phylogenetic tree of DES 1 and DES2, constructed from 16 amino acid sequences essential for C4-hydroxylase activity in animals. This tree shows DES2 divergence between *Drosophila* and zebrafish or *Xenopus* (Omae *et al.* 2004).

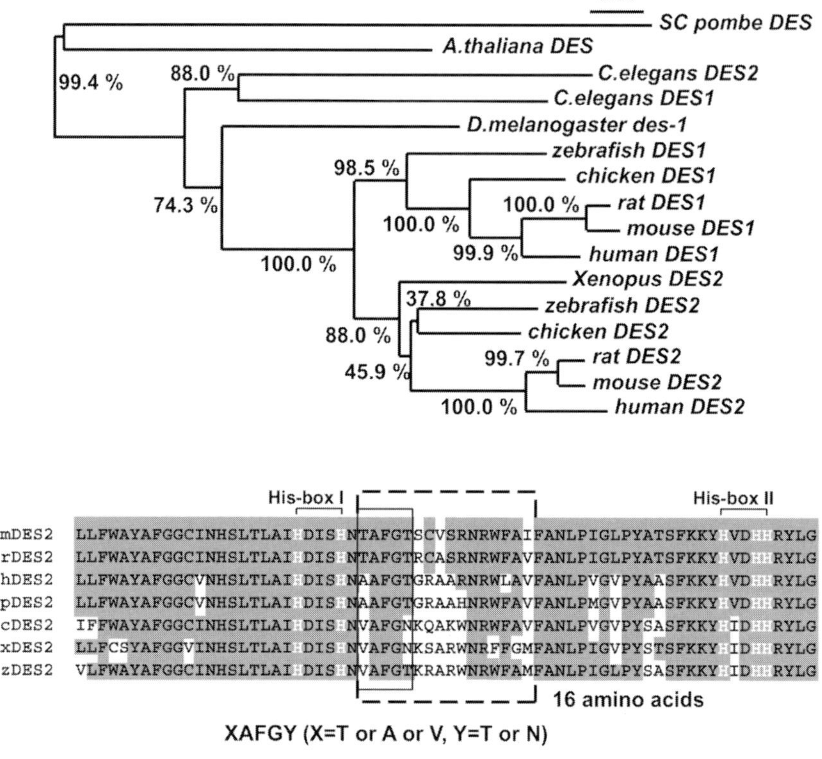

Fig. 3. Phylogenetic tree of DES1 and DES2. The tree was constructed from 16 amino acid sequences using the neighbor-joining method. The branch lengths are proportional to the relative phylogenetic distances between the proteins.

For a better understanding of this issue, a close examination of the particular sequences essential for the catalytic activities of C4-hydroxylase and Δ4-desaturase is required. In addition, X-ray crystallography of DES1 and 2 should yield a better understanding of the critical differences between desaturase and hydroxylase reactions. However, these enzymes are challenging targets for crystallography, since they have three or four transmembrane domains, and these hydrophobic domains are important for substrate interactions. Successful crystallographic experiments will require advances in crystallization and purification methods for membrane proteins.

6. Molecular mechanisms of Δ4-desaturation and C4-hydroxylation

Stereochemistry of C4-hydroxylation of dihydroceramide was first reported by Stoffel and Binczek in 1971 (Stoffel and Binczek 1971). More than 25 years later, Broun *et al.* used fatty acid desaturases and hydroxylases derived from plants to show desaturation for hydroxlyase (Broun *et al.* 1998). Analyses of the amino acid sequences of oleate C12-hydroxylases and oleate Δ9,12-desaturases from different plants revealed that a few conserved desaturase residues were not conserved in the hydroxylases. When seven of these residues were replaced with the corresponding hydroxylase residues, the desaturase was converted into a bifunctional desaturase/hydroxylase. The reciprocal replacement increased the desaturase activity of the hydroxylase.

More recently, Broadwater *et al.* proposed that an activated binuclear iron center formed by His box motifs will hydroxylate an unactivated hydrocarbon substrate by default (Broadwater *et al.* 2002). They proposed that the catalytic function of the binuclear iron center might be changed to desaturation through alteration of the chemical nature of the substrate (effected by intermediate stabilization) or of the position of the substrate in the active site. The large size of the fatty acid substrate and membrane-bound desaturases may allow these enzymes the control, as mediated through extensive protein-substrate interactions, necessary to avoid hydroxylation and instead catalyze desaturation. Presentation of the substrate to the oxidant is a critical factor determining partitioning between hydroxylation or desaturation. Sphinganine and *N*-acylsphinganine substrates may influence catalytic mechanisms differently, but this hypothesis may provide a clue to the molecular mechanisms of DES1 and DES2 catalysis.

Beckmann *et al.* reported that dihydroceramide Δ4-desaturase from *Candida albicans*, cloned and expressed in a SUR-defective yeast strain, eliminated the C-4-H_R and the C-5-H_S in an overall *syn*-elimination of the two vicinal hydrogen atoms (Beckmann *et al.* 2003). They also reported that the enzyme was bifunctional, producing both sphingenine (93%) and 4-hydroxysphinganine (7%). These findings agree with a two-step process involving activation of the dihydroceramide substrate by removal of the C-4-H_R to give a C-central radical or radicaloid followed by either dispro- portionation into an olefin, water, and a reduced diiron complex, or to re- combination of the primary reactive intermediate with an active-bound oxygen to yield a secondary alcohol. The results of preliminary *in vitro* experiments carried out in our laboratory indicate that other molecules may influence DES2 hydroxylase activity. Further studies of the regulation of desaturase and hydroxylase activities of DES1 and 2 are needed.

7. DES1 and DES2 knockout experiments

A conditional knockout mouse study of aryl hydrocarbon receptor nuclear translocator (ARNT) as a transcription factor (Takagi *et al.* 2003) used the *Cre-loxP* system to disrupt the *Arnt* gene in a keratinocyte-specific man- ner. Gene-targeted newborn mice of almost normal appearance died neo- natally of severe dehydration caused by water loss. *Des2* mRNA levels in cultured keratinocytes prepared from the knockout mice were markedly lower than normal, but *Des1* mRNA levels were the same as wild type. The amounts of ω-linoleoyl-*N*-acylsphingenine, *N*-acylsphingenine, and protein-bound ω-hydroxy-*N*-acylsphingenine in the skin of knockout mice were markedly decreased, but *N*-acylsphinganine levels were increased. These apparently contradictory results prompt certain questions. Is sphin- ganine Δ4-desaturation of ω-linoleoyl-*N*-acyl- sphingenine catalyzed by DES2 but not by DES1? Is the desaturation of *N*-acylsphinganine in the skin also catalyzed by DES2 but not by DES1? Conditional *Des1* and *Des2* knockout experiments will provide more precise results. At the same time, the physiological substrates of DES1 and 2 need to be addressed.

8. References

Bouhours JF, Glickman RM (1977) Rat intestinal glycosphingolipids III. Fatty acids and long chain bases of glycosphingolipids from villus and crypt cells. *Biochim Biophys Acta*, **487**, 51-60.

Breimer ME, Karlsson KA, Samuelsson BE (1974) The distribution of molecular species of monoglycosylceramides (cerebrosides) in different parts of bovine digestive tract. *Biochim Biophys Acta*, **348**, 232-240.

Breimer ME, Karlsson KA, Samuelsson BE (1975) Presence of phytosphingosine combined with 2-hydroxy fatty acids in sphingomyelins of bovine kidney and intestinal mucosa. *Lipids*, **10**, 17-19.

Broadwater JA, Whittle E, Shanklin J (2002) Desaturation and hydroxylation: Residues 148 and 324 of *Arabidopsis* FAD2, in addition to substrate chain length, exert a major influence in partitioning of catalytic specificity. *J Biol Chem*, **277**, 15613-15620.

Broun P, Shanklin J, Whittle E, Somerville C (1998) Catalytic plasticity of fatty acid modification enzymes underlying chemical diversity of plant lipids. *Science*, **282**, 1315-1317.

Cadena DL, Kurten RC, Gill GN (1997) The product of the MLD gene is a member of the membrane fatty acid desaturase family: Overexpression of MLD inhibits EGF receptor biosynthesis. *Biochemistry*, **36**, 6960-6967.

Causeret C, Geeraert L, van den Hoeve G, Mannaerts GP, van Veldhoven PP (2000) Further characterization of rat dihydroceramide desaturase: Tissue distribution, subcellular localization, and substrate specificity. *Lipids*, **35**, 1117-1124.

Geeraert L, Mannaerts GP, van Veldhoven PP (1997) Conversion of dihydroceramide into ceramide: involvement of a desaturase. *Biochem J*, **327**, 125-132.

Endo K, Akiyama T, Kobayashi S, Okada M. (1996) Degenerative spermatocyte, a novel gene encoding a transmembrane protein required for the initiation of meiosis in Dorosophila spermatogenesis. *Mol Gen Genet*, **253**, 157-165.

Endo K, Masuda Y, Kobayashi S (1997) Mdes, a mouse homolog of the Drosphila degenerative spermatocyte gene is expressed during mouse spermatogenesis. *Dev Growth Differ*, **39**, 399-403.

Grilley MM, Stock SD, Dickson RC, Lester RL, Takemoto JY (1998) Syringomycin action gene SYR2 is essential for sphingolipid 4-hydrxylation in *Saccharomyces cerevisiae. J Biol Chem*, **273**, 11062-1068.

Haak D, Gable K, Beeler T, Dunn T (1997) Hydroxylation of *Saccharomyces cerevisiae* ceramides requires Sur2p and Scs7p. *J Biol Chem*, **272**, 29704-29710.

Karlsson KA, Martensson E (1968) Studies on sphingosines XIV. On the phytosphingosine content of the major human kidney glycosphingolipids. *Biochim Biophys Acta*, **152**, 230-233.

Karlsson KA, Steen GO (1968) Studies on sphingosones XIII. The existence of phytosphingosine in bovine kidney sphingomyelins. *Biochim Biopys Acta*, **152**, 798-800.

Mikami T, Kashiwagi M, Tsuchihashi K, Akino T, Gasa S (1998) Substrate specificity and some other enzymatic properties of dihydroceramide desaturase (ceramide synthase) in fetal rat skin. *J Biochem*, **123**, 906-911.

Michel C, van Echten-Decker G, Rother J, Sandhoff K, Wang E, Merrill AH (1997) Characterization of ceramide synthesis: A dihydroceramide desaturase introduces the 4,5-trans-double bond of sphingosine at the level of dihydroceramide. *J Biol Chem*, **272**, 22432-22437.

Merrill AH (2002) *De novo* sphingolipid biosynthesis: A necessary, but dangerous, pathway. *J Biol Chem*, **277**, 25843-25846.

Omae F, Miyazaki M, Enomoto A, Suzuki M, Suzuki Y, Suzuki A (2004) DES2 protein is responsible for phytoceramide biosynthesis in the mouse small intestine. *Biochem J*, **379**, 687-695.

Omae F, Miyazaki M, Enomoto A, Suzuki A (2004) Identification of an essential sequence for dihydroceramide C-4 hydroxylase activity of mouse DES2. *FEBS Letters*, **576**, 63-67.

Shanklin J, Whittle E, Fox BG (1994) Eight histidine residues are catalytically essential in a membrane-associated ion enzyme, stearoyl-CoA desaturase, and are conserved in alkane hydroxylase and xylene monooxygenase. *Biochemistry*, **33**, 12787-12794.

Spiegel S, Milstien S (2002) Sphingosine 1-phosphate, a key cell signaling molecule. *J Biol Chem*, **277**, 25851-25854.

Stoffel W, Binczek E (1971) Metabolism of sphingosine bases. XVI. Studies on the stereospecificity of the introduction of the hydroxy group of 4D-hydroxyspinganine (phytosphingosine). *Hoppe-Seyler's Z Physiol Chem*, **352**, 1065-1072.

Suzuki A, Yamakawa T (1981) The different distribution of asialoGM1 and Forssman antigen in the small intestine of mouse demonstrated by immunofluorescence staining. *J Biochem*, **90**, 1541-1544.

Takagi S, Tojo H, Tomita S, Sano S, Itami S, Hara M, Inoue S, Horie K, Kondoh G, Hosokawa K, Gonzalez FJ, Takeda J (2003) Alteration of the 4-sphingenine scaffolds of ceramides in keratinocyte-specific Arnt-deficient mice affects skin barrier function. *J Clin Invest*, **112**, 1372-1382.

Ternes P, Frank S, Zahringer U, Sperling P, Heiz E (2002) Identification and characterization of a sphingolipid ω4-desaturase family. *J Biol Chem*, **277**, 25512-25518.

Umesaki Y, Suzuki A, Kasama T, Tohyama K, Mutai M, Yamakawa T (1981) Presence of asialoGM1 and glucosylceramide in the small intestinal mucosa of mice and induction of fucosylated asialoGM1 by conventionalization of germ-free mice. *J Biochem*, **90**, 1731-1738.

2-4 Metabolizing Enzymes Such As Sphingomyelin Synthase Induce Cell Death by Increasing Ceramide Content

Toshiro Okazaki

Department of Clinical Laboratory Medicine/Hematology, Faculty of Medicine, Tottori University, Yonago, Tottori 683-8504, Japan

Summary. Ceramide, among its other roles, serves as a pro-apoptotic lipid mediator. Ceramide-induced pro-apoptotic signals include caspases, reactive oxygen species (ROS) and c-jun-N-terminal kinase (JNK), and anti-apoptotic signals such as phosphatidylinositides (PI)-3 kinase and protein kinase C are inhibited by ceramide. Sphingosine-1-phosphate (S1P) competes with ceramide-induced cell death by blocking its generation and downstream pathways. This balanced interaction between S1P and ceramide may regulate cell death and survival/growth. Regulating ceramide-metabolizing enzymes, such as sphingomyelin synthase (SMS) and glucosylceramide synthase (GCS), as well as those of ceramide-downstream pathways are crucial to control ceramide signals. It was recently shown that inhibiting SMS and GCS induced apoptotic cell death in human leukemia cell lines. In addition cells possessing less ceramide and have high levels of GCS and SMS were chemoresistant *in vivo*. Membranous SM generated by SMS1 gene, which we cloned, may act as not only the source for generation of ceramide but also as the platform for transmembrane receptors such as Fas antigen. Thus, overwhelming drug- or chemo-resistance in human hematopoietic malignancies by inhibiting ceramide-metabolizing enzymes and enhancing ceramide-downstream signals to intensify ceramide-induced cell death would be one approach to treating these cancers.

Keywords. sphingomyelin, sphingomyelin synthase, ceramide, apoptosis

1. Introduction

Ceramide, originally identified as a structural component of membranes and a source of complex glycosphingolipids, has since been identified as a key lipid mediator for cell death. We found that ceramide increased when exposed to vitamin D3, a differentiation-inducing agent, in human leukemia HL-60 cells (Okazaki et al., 1989) and that synthetic, cell permeable N-acetylsphingosine (C2-ceramide) induced monocytic differentiation and cell growth inhibition (Okazaki et al., 1990). These results suggest that sphingolipid ceramide might induce cell differentiation as well as regulate cell death. 1a, 25-dihydroxyvitamin D3 increased intracellular ceramide levels by activating neutral magnesium-dependent sphingomyelinase (nSMase). Hence, ceramide generated through neutral or acid SMase activation in response to stress may induce cell death while through the *de novo* synthesis pathway for ceramide synthesis from palmitoyl CoA and serine through dihydroceramide leads to a basally generated increase in ceramide. However, the anti-cancer agent doxorubicin which reportedly activates *de novo* synthesis of ceramide, as well as ceramide-metabolizing enzymes such as glucosylceramide synthase (GCS), sphingomyelin synthase (SMS), ceramidase (CDase), sphingosine kinase (SphK) and ceramide kinase (CerK) are other important regulators of ceramide content (Okazaki, 1998).

In addition, ceramide is thought to participate a diverse range of physiological and pathological conditions, including neural and body development, angiogenesis, aging, tumorgenesis, atherosclerosis, infection, allergy and immunosuppression. Here we review the roles of ceramide increase by inhibiting ceramide-metabolizing enzymes such as SMS and GCS to induce *in vitro* and *in vivo* cell death. Then we discuss possible therapeutic avenues to treat hematological malignancies such as leukemia and malignant lymphoma by the mediation of the pro-apoptotic lipid ceramide.

2. Ceramide in cell death and growth/survival

Ceramide was initially found to mediate differentiation and apoptotic cell death in hematopoietic cells. Downstream signals that induce this cell death have been investigated vigorously. However, the primary downstream molecule directing this ceramide induced cell death is unknown. Pro- and anti-apoptotic molecules, such as caspases and protein kinase C, are diverse but appear to response to stressors that induce cell death. The relationship between the generation of reactive oxygen species (ROS) and phophoinositides-3 kinase (PI-3k) regulation within the ceramide signal is discussed here (Okazaki et al., 2002).

Clinically vesnarinone is an effective inotropic agent, but its use is restricted because it can lead to severe agranulocytosis. We investigated the mechanism of agranulocytosis by examining the role of ceramide in vesnarinone-induced ROS accumulation. Vesnarinone induced apoptosis in human leukemia HL-60 cells time- and dose-dependently and increased intracellular content of ceramide. Simultaneously treating cells with vesnarinone and cell permeable, synthetic N-acetylsphingosine (C2-ceramide) enhanced apoptotic cell death. Neither vesnarinone nor C2-ceramide increased ROS generation measured by 2', 7'-dihydrofluorescein (DCFH) method unless cells were not treated with hydroperoxide. In contrast, the effects of vesnarinone and C2-ceramide on lipid peroxidation were assessed by the *cis*-parinaric acid method and nitroblue tetrazolium (NBT) reduction assay. These results suggested that vesnarinone increased lipid peroxidation through ceramide by inhibition of ROS decrease rather than increase of ROS generation. Glutathione, glutathione peroxidase, superoxide dismutase and catalase will eliminate that ROS generation. Catalase is the most effective and active of those molecules against anti-oxidative damage. When HL-60 cells were treated with vesnarinone and C2-ceramide, both induced an increase of catalase both in the activity and protein level. Simultaneous treatment enhanced the increase of catalase function, while the addition of purified catalase into the medium inhibited vesnarinone or C2-ceramide-increased lipid peroxidation and apoptotic HL-60 cell death. These data suggested that vesnarinone and C2-ceramide induced apoptotic cell death by ROS accumulation through suppressing catalase function. HL-60/ves cells, which were resistant to vesnarinone-induced cell death, showed no increase in ceramide content or lipid peroxidation, nor were decreases in catalase protein and activity seen. Further investigation, however, is required to precisely understand how vesnarinone induces only agranulocytosis and not pancytopenia clinically (Kondo et al., 2002b; Iwai et al., 2003).

Insulin-like growth factor-1 (IGF-1) can inhibit stress-induced apoptosis. IGF-1 mediates the anti-apoptotic signal in a PI-3k-dependent manner. Since C2-ceramide-induced apoptosis was inhibited by IGF-1, we investigated the relationship between ceramide increase associated with ROS accumulation and IGF-1-induced PI-3k activation. ROS scavenger catalase was inactivated at the translation and protein levels through ceramide-induced caspase-3. PI-3k inhibited ceramide-increased lipid peroxidation by restoration of caspase-3-inhibited catalase activity. These results suggested that anti-apoptotic IGF-1/PI-3k and pro-apoptotic ceramide are competitive in the induction of cell death through regulation of caspase-3-induced ROS accumulation (Kondo et a., 2002a).

3. SMS in hematological malignancy

When SMS transfers phosphocholine from phosphatidylcholine (PC) to ceramide SM and diaylglycerol (DAG) are generated as ceramide and PC decrease. Since ceramide and DAG are pro-apoptotic and pro-proliferative mediators, respectively, the role of SMS in effectively regulating cell death and survival is crucial. Fas-crosslinking induces apoptosis through ceramide-related enzymes. As acid SMase-deficient fibroblasts taken from acid SMase-/- mice is resistant to Fas-induced cell death, acid SMase may be necessary for Fas-induced lymphocyte apoptosis. Hence, ceramide generation appears to be closely involved in Fas-induced apoptosis.

To test this, apoptosis in Jurkat T cells was induced by Fas-crosslinking. Four-fold and two-fold increases in ceramide content were observed in nuclei and microsome fractions, respectively, when compared to controls. Confocal microscopy using anti-ceramide monoclonal antibody also revealed increases of ceramide in nuclear and plasma membrane fractions after stimulation (Kawase et al., 2002). Ceramide generation in nuclei and microsome fraction is required for Fas-induced lymphocyte apoptosis.

Since neutral SMase activity increased in nuclei and microsome, ceramide was generated by SMase in both fractions. Logical disconnect here. As typical ceramidase activity is barely detectable, we decided to examined changes in ceramide-metabolizing enzymes such as GCS and SMS. Interestingly, Fas-crosslinking inhibited SMS activity in nuclei, but not significantly in microsome. This difference suggests that the "SM cycle", which regulates ceramide action through SMase and SMS function, is active during apoptosis-inducing processes in the nuclei. Inhibiting SMS, further increased of ceramide content in nuclei. This role for nuclear SMS in apoptosis induction was confirmed using D-609, an inhibitor of SMS, to enhance Fas-induced apoptosis by increasing nuclear ceramide content. In addition, a caspase-3 inhibitor, DEVD-CMK inhibited Fas-induced apoptosis by decreasing Fas-activated nuclear caspase-3 activity. DEVD-CMK also inhibited ceramide increases by disrupting the activation of nuclear "SM cycle" (Watanabe et al., 2004). Since caspase-3 is involved in both activation of SMase and inhibition of SMS, caspases may play a comprehensive signally role in regulating of sphingolipids metabolism such as the "SM cycle". The mechanism by which sphingolipids are totally regulated in response to the pro-apoptotic stress should be clarified in the future.

The role of ceramide-content regulation in sustaining cell proliferation was investigated through sphingolipids metabolism by growth factors such as interleukin (IL)-2. Human KHYG-1 cells are natural killer lymphocytes, and grow only in IL-2 supplemental culture condition with 10% fetal calf serum. KHYG-1 cell growth was suppressed 2 days after IL-2 depletion and restored to control levels with additions of IL-2. The similar growth inhibition by IL-2 depletion was found when the cells were treated with exogenous N-acetylceramide (C2-ceramide), which mimics physiological long chain ceramide by activating ceramide-generating enzymes such as SMase. These data suggest that IL-2 depletion inhibited cell growth by increasing ceramide levels. In fact, IL-2 depletion activated acid SMase and inhibited ceramide-metabolizing enzymes while IL-2 supplementation decreased ceramide content by restoring these enzymatic reactions. IL-2 growth signaling is regulated by phosphatidylinositol (PI)-3 kinase, Jak/stat and/or MAP kinase pathway. When cells were treated with specific inhibitors for each of these pathways (LY294002 for PI-3 kinase, AG490 for Jak/stat and PD98059 for MAP kinase) after IL-2 supplementation, only LY294002 inhibited IL-2-resored cell growth. This observation indicates that PI-3 kinase participates in IL-2-induced growth signal in KHYG-1 cells. LY294002 blocked IL-2-induced decrease of ceramide by inhibition of GCS and SMS, and activation of acid SMase. Therefore, PI-3 kinase may work as a regulator of sphingolipid ceramide balance by controlling ceramide-generating and –metabolizing enzymes at the same time. To confirm this, we made cells that overexpressed PI-3 kinase and examined ceramide content and the activities of ceramide-related enzymes. The results showed that overexpression of PI-3k kinase induced higher cell growth and decreased ceramide control by activating GCS and SMS and inhibiting acid SMase. Thus, PI-3 kinase is closely involved in the regulation of ceramide content through enzymatic reaction. LY294002 increased IL-2-induced inhibition of SMase and inhibited IL-2-induced activation of GCS at the protein and mRNA levels. In addition, LY294002 did not affect the breakdown of acid SMase and GCS mRNAs. IL-2-induced transcriptional levels, estimated by run-on assay, increased in acid SMase and decreased in GCS after treatment with LY294002. These results suggest that PI-3 kinase was involved in IL-2-induced regulation of ceramide content at transcriptional level (Taguchi et al., 2004). Taken together, ceramide content through regulation of ceramide-related enzymes seems to be involved in hematopoietic cell death and growth, and may be controlled by unknown comprehensive signaling system related to PI-3 kinase and caspases.

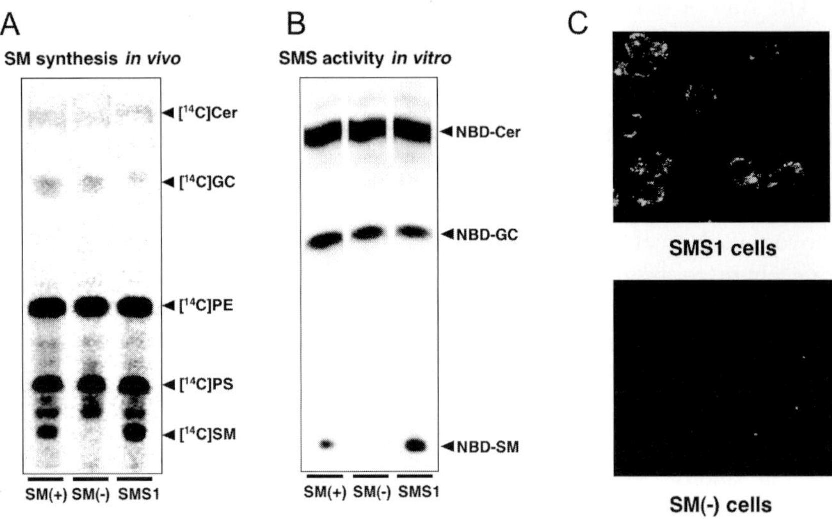

Fig. 1. Restoration of in vivo and in vitro activity of SM synthase in SMS1 cells. A: WR19L/Fas (SM(+)), SMS-deficient murtant WR19L/Fas (SM(-)) and SMS-revertant (SMS1) cells were treated with [14C]serine. [14C]SM was detected in SM(+) and SMS1 cells, but not in SM(-) cells by TLC analysis. B: The protein extracted from SM(+) and SMS1 cells synthesized NBD-SM from NBD-ceramide, but not SM(-) cells. C: Lysenin binding membrane SM was detected by confocal microscopy in SMS1, but not in SM(-) cells.

Chemo-resistance prevents complete remission and cure in many patients with hematological malignancies. Pro-apoptotic molecules such as Bcl-2, myc, cyclin D1, bcr/abl and PI-3 kinase may play a role in chemo-resistance in leukemia and lymphoma. Since ceramide seems to be a pro-apoptotic mediator, the relation of ceramide regulation and drug-resistance was investigated in leukemia cells. In human leukemia HL-60 cells apoptosis was induced by doxorubicin (DOX). This treatment increased ceramide content, whereas in HL-60/ADR cells DOX-induced apoptotic cell death was inhibited and no ceramide increase occurred even after intracellular accumulation of DOX. Drug-sensitivity to DOX in HL-60/ADR cells was restored following treatment with C2-ceramide. As the uptake of C2-ceramide was similar in HL-60/ADR cells as compared to HL-60 cells, the enhancing effect of ceramide on the induction of cell death in HL-60/ADR cells appears unrelated to the ceramide-caused malfunction of MRP-1/MDR-1 pumping system for DOX. When the enzymatic function related to ceramide content was investigated, ceramide-metabolizing enzymatic activities were significantly higher in HL-60/ADR cells than in HL-60 cells. No differences were seen in the activities in neutral and acid SMases, and ceramidases. Overexpressing GCS in HL-60 cells increased resistance to DOX-induced cell death with a decrease of ceramide content. Suppressing ceramide increase by activating ceramide-metabolizing enzymes, GCS and SMS, may be associated with drug-resistance and susceptibility in hematopoietic cells (Itoh et al., 2003).

Ceramide content and the activities of GCS and SMS were investigated in patient samples. The patients were divided into chemo-sensitive and chemo-resistant groups. Chemo-sensitive cases, those patients with acute myeloid and lymphoid leukemia (AML and ALL), were in complete remission (CR) following induction therapy. Chemo-resistant patients with AML, ALL and blast crisis of chronic myeloid leukemia did not respond to various anti-cancer agents and died despite chemotherapy. Ceramide content was significantly lower in chemo-resistant cases than chemo-sensitive cases. Levels of GCS and SMS were higher in chemo-resistant cases than chemo-sensitive cases. Lower ceramide content in chemo-resistant cases was related to Bcl-2 protein levels, but not to MDR1 levels. As ceramide reportedly inhibits Bcl-2 expression, the observed decreased of ceramide content resulted in Bcl-2 activation and thereby produced chemo-resistance (Itoh et al., 2004). Hence, a decrease in ceramide content is involved in chemo-resistance *in vivo* as well as in drug- resistance *in vitro*.

Regulating both GCS and SMS may reduce chemo-resistance by increasing ceramide content. Blocking only one pathway, either GCS or SMS, will not effectively decrease ceramide.

4. Cloning of cDNA for SMS and its implication in Fas-induced cell death

Mouse lymphoid cells WR19L/Fas-SM(-), which show little SMS activity, were isolated from WR19L/Fas cells. WR19L/Fas-SM(-) cells seem to express faint SM in the plasma membrane because a less toxic SM-binding protein lysenin conjugated with MBP showed no fluorescence in the plasma membrane. Toxic lysenin did not induce cell death in WR19L/Fas-SM(-) cells indicating that SM was not expressed on the plasma membrane. When the SM(-) cells were labeled with [^{14}C]serine, no synthesis of [^{14}C]SM was found on thin layer chromatography (TLC) analysis. Enzymatic activity of SMS was measured using NBD-ceramide as a substrate. In SM(-) cells, there was no synthesis of NBD-SM, showing no activity of SMS (Fig. 1). SM binds tightly with cholesterol, which is extruded by methyl-α-cyclodextrin (MCD). When SM(-) cells were treated with MCD, cell death was more prominent, therefore, we transfected human cDNA library into SM(-) cells, and selected the living cells after treatment with MCD. SM-synthesizing transfectants were selected from SM(-) cells. In this way, we established cDNA of SMS (Fig. 2). SMS1 consisting of 413 amino acids has four trans-membrane regions and sterile alpha motif (SAM) domain at N terminus. Following a database search, we found two homologues of SMS1: SMS2 and SMSr. SMS2 has an enzymatic activity *in vitro*, but its overexpression was insufficient for SM in the plasma membrane while those cells overexpressing SMS1 (SMS1 cells) clearly showed an increase of SMS activity and synthesis of [^{14}C]SM in the plasma membrane. PC was a substrate of SMS to transfer phosphocholine to ceramide and synthesize SM, but not CDP-choline. These characteristics seem to be compatible with a previous report for enzyme reaction of SMS. SMSr showed no SMS activity *in vitro* or *in vivo*.

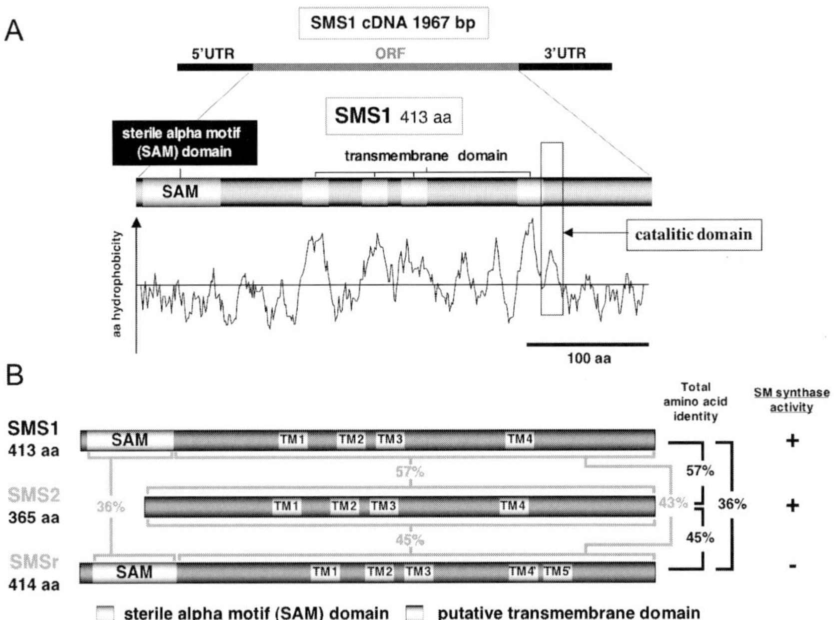

Fig. 2. Structure and hydropathy plot of amino acid sequence of SMS1 and its homologous sequence. A: SMS1 has 413 amino acids with sterile alpha motif (SAM) domain and four transmembrane (TM) domain. Putative catalytic domain is located at the C terminus side of TM4 domain. B: SMS1 has two homologies in the human cDNA library: SMS2 and SMSr. SMS1 and 2 have SMS activity, but not SMSr.

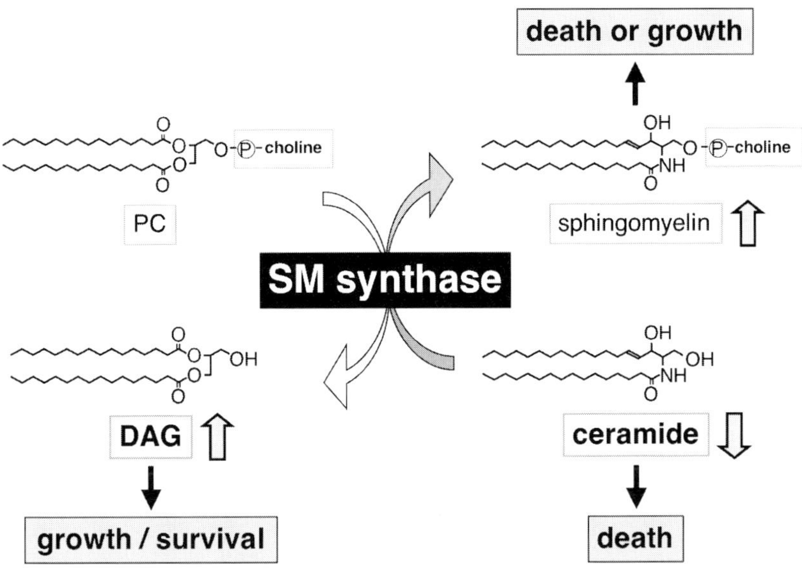

Fig. 3. Role of sphingomyelin synthase (SMS) in cell death and growth. SMS may regulate two critical molecules, diacylglycerol (DAG) and ceramide in cell growth/survival and death.

SMS1 cells grow well in serum-free medium but SM(-) cell growth was less strong unless exogenous SM was added. This difference shows the significance of SM and SMS1 in cell growth (Yamaoka, et al., 2004). The different functions in SMS1 and SMS2, as well as their localizations are being investigated. As specific antibodies for SMS1 or SMS2 and KO animal model are not yet available, it remains unknown whether SMS1 is the only enzyme responsible for mediating cell death and growth by maintaining a balance between SM, ceramide and DAG in mammalian cells (Fig. 3).

Having prepared SMS1 overexpressing cells from WR19l/Fas-SM(-) cells, the role of SM and SMS1 in Fas-induced cell death. SM is rich in the microdomain of plasma membrane. Therefore, SM seems to work not just in ceramide generation but also to give structural support to trans-membrane receptors such as Fas and transferrin receptor. Between SM(-) and SMS1 cells, the levels of expressions in cholesterol judged by polyethylene-derivatized cholesterol, GM1 ganglioside and human Fas antigen were similar to those from FACS marker analysis, but like the parent cells the less-toxic lysenin binding ability was detected in SMS1 cells but not in SM(-) cells. Fas-cross linking induced significantly higher level of apoptosis in SMS1 cells than in SM(-) cells measured by propidium iodide (PI) staining and a loss of mitochondria electron transport. Fas-induced activations of caspase 3 and 8 were inhibited in SM(-) cells, and the formation of Fas-associated death domain (FADD) was less in SM(-) cells than in SMS1 cells. These results suggest that SMS1 plays a role in trans-membrane signaling upstream of caspases and FADD formation. Fas-antigen forms trimer and subsequent multimer of Fas antigen with subsequent cap formation to exert the effective activation of downstream death signal. In SM(-) cells Fas aggregation was impaired because the content of Fas antigen in Triton-x 100-resisitant membrane fraction (DRM) was not increased as compared to SMS1 cells. These results suggest that SM or SMS1 is required to induce the effective formation of Fas-antigen-containing microdomain. When we treated SM(-) cells with long-chain ceramide for 1 h, cells did not undergo apoptosis without Fas-cross-linking. However, after Fas-crosslinking the short treatment with physiological ceramide for 1 h induced a significant increase of apoptosis in SM (-) cells. Ceramide content in DRM was similar at the stable condition between SM (-) and SMS1 cells, but was increased more by Fas-crosslinking in SMS1 cells as compared to SM(-) cells, suggesting that not only SM but also ceramide increase in DRM enhanced Fas-induced trans-membrane signaling required for cell death (Miyaji et al., 2005). Since ceramide is a pro-apoptotic lipid mediator, the search for the direct downstream molecule that mediates ceramide function is on-going. Judging from these findings, it is likely that one of these mechanisms regulates trans-membrane signaling through SM/ceramide-containing platform for cell death-inducing receptors.

5. Involvement of ceramide in angiogenesis in zebrafish

Zebrafish are useful animal models because the primary organs are translucent in embryos and development takes less than a week. Thalidomide, which was withdrawn from the market for causing severe developmental side-effects, has an anti-angiogenetic action that helps in treating multiple myeloma, rectal cancer, and other human malignancies. While the effects of thalidomide have been investigated in *ex vivo* systems, an animal model is not yet established that can explain the drug's action on angiogenesis. Treating zebrafish embryos with thalidomide adversely affected the development of the dorsal artery and posterior cardinal vein. In situ hybridization showed that the expression of VEGF (vascular endothelial growth factor) receptors, such as neuropillin-1 and Flk-1, was inhibited after thalidomide treatment. Ceramide content in zebrafish embryos increases with activation of neutral SMase. Exogenous N-acetylceramide (C2-ceramide) induced similar phenotypes as thalidomide treatment. Thalidomide-induced activation of neutral SMase was blocked by the addition of anti-sense oligodeoxynucleotides (OND) for zebrafish neutral SMase. Subsequently thalidomide-induced inhibition of angiogenesis was completely restored by this anti-sense ODN. These data suggest that thalidomide induced anti-angiogenic effects by increasing ceramide action through SMase inhibition. The direct effect of ceramide was investigated whether C2-ceramide inhibited the growth of HUVECs (human umbilical vein endothelial cells) by affecting the expression of VEGF receptors. As C2 ceramide inhibited VEGF receptor expressions as well as thalidomide, ceramide is suggested to mediate thalidomide-induced anti-angiogenic action in HUVECs. Interestingly these anti-angiogenic effects of thalidomide were blocked by S1P treatment. S1P seems to decrease thalidomide-increased ceramide content by inhibition of SMase. Ceramide can induce apoptotic cell death in malignant cells, but here we suggest that ceramide and S1P together works as a novel regulator of angiogenesis. In summary, sphingolipid balance between ceramide and S1P may control cell growth not only through the induction of apoptosis of tumor cells but also by regulating the environment surrounding the tumor, including mediating nutrient supply through angiogenesis (Yabu et al., 2005).

6. Concluding remarks

PBPP appears to effectively inhibit GCS. When SM(-) cells were treated with PBPP, apoptosis was induced dose-dependently, but SMS1 cells did not show apoptotic cell death even with PBPP. These results clearly showed that simply blocking one ceramide-metabolizing pathway is not enough to induce ceramide increase and cell death. Ceramide-metabolizing enzymes such as ceramidase and ceramide kinases (excluding GCS and SMS) were faintly detected in the primary samples from the patients with hematological malignancy and leukemia cell lines. Therefore, blocking GCS and SMS activities should also induce cell death in drug-resistant leukemia cells. The effectiveness of ceramide increase by inhibiting these two enzymes must also be confirmed in animal models such as zebrafish and mice. At the same time the search for the specific inhibitor for SMS also should be done to innovate the novel ceramide-increasing therapy for refractory hematological malignancy, since nojirimycin was known to be a specific inhibitor for GCS and used in clinical situation. Once an SMS inhibitor has been discovered, a means to induced ceramide increase and cell death in drug-resistance cells becomes feasible. In addition, finding ways to activate ceramide-related cell death signaling that supplement ceramide or its analogues, and intensifies its actions by modulating ceramide-downstream signals would prove beneficial (Fig. 4). In the future, comprehensive therapeutics to increase ceramide pro-apoptotic function through the above strategies may help overcome the chemo-resistance in hematological malignancies.

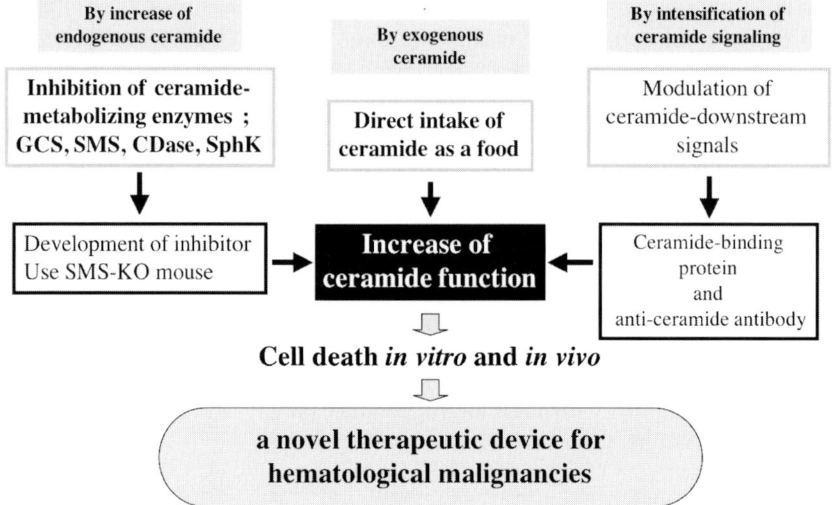

Fig. 4. Molecular targetting of ceramide signal to induce cell death for medical applications. There are three ways to increase ceramide 1) increase of endogenous ceramide, 2) add exogenous ceramide and 3) increase ceramide signaling.

7. References

Itoh M., Kitano T., Watanabe M., Kondo T., Yabu T., Taguchi Y., Iwai K., Matsuda T., Uchiyama T. and Okazaki T. (2003) Possible role of ceramide as an indicator of chemoresistance: decrease of the ceramide content via activation of glucosylceramide synthase and sphingomyelin synthase in chemoresistant leukemia. *Clin. Cancer Res.*, **8**, 415-423.

Iwai K., Kondo T., Watanabe M., Yabu T., Kitano T., Taguchi Y., Umehara H., Takahashi A., Uchiyama T. and Okazaki T. (2003) Ceramide increases oxidative damage due to inhibition of catalase by caspase-3-dependent proteolysis in HL-60 cell apoptosis. *J. Biol. Chem.*, **278**, 9813-9822.

Kawase M., Watanabe M., Kondo K., Yabu T., Taguchi Y., Umehara H., Uchiyama T., Mizuno K. and Okazaki T. (2002) Increase of ceramide in adriamycin-induced HL-60 cell apoptosis: detection by a novel anti-ceramide antibody. *Biochim. Biophys. Acta- Mol. Cell Biol. Lipid*, **1584** (2-3), 104-114.

Kondo T., Iwai K., Kitano T., Watanabe M., Taguchi Y., Yabu T., Umehara H., Domae N., Uchiyama T. and Okazaki T. (2002a) Control of ceramide-induced apoptosis by IGF-1: involvement of PI-3 kinase, caspase-3 and catalase *Cell Death Diff.*, **9**, 682-692.

Kondo T., Suzuki Y., Kitano T., Iwai K., Watanabe M., Umehara H., Daido N., Domae N., Tashima M., Uchiyama T. and Okazaki T. (2002b) Vesnarinone causes oxidative damage by inhibiting catalase function through ceramide action in myeloid cell apoptosis. *Mol. Pharmacol.* **61**, 620-627.

Miyaji M., Xiong J.Z., Yamaoka S., Amakawa R., Fukuhara S., Sato S., Kobayshi T., Domae N., Mimori T., Bloom E., Okazaki T. and Umehara H. (2005) Role of membrane sphingomyelin and ceramide in platform formation for the Fas-mediated apoptosis. *J. Exp. Med.*, **202**, 249-259.

Okazaki T., Bell R. M., and Hannun Y. A. (1989) Sphingomyelin turnover induced by vitamin D3 in HL-60 cells: Role in cell differentiation. *J. Biol. Chem.*, **264**, 19076-19080.

Okazaki T., Bielawaka A., Bell R. M., and Hannun Y. A. (1990) Role of ceramide as a lipid mediator of 1α, 25-dihydroxyvitamin D3-induced HL-60 cell differentiation. *J. Biol. Chem.*, **265**, 15823-15831.

Okazaki T. (1998) Diversity and complexity of ceramide signaling in apoptosis. *Cell Signaling*, **10**, 685-692, 1998

Okazaki T., Kondo T., Watanabe M., Taguchi Y. and Yabu T. (2002) Crosstalk of ceramide with cell survival signaling. in *Cell Signaling* (Futerman A.H., Ed.), pp.91-100, Eurekah.

Taguchi Y., Kondo T., Watanabe M., Miyaji M., Umehara H., Kozutumi Y. and Okazaki T. (2004) Interleukin-2-induced survival of natural killer (NK) cells involving phosphatidylinositol-3 kinase-dependent reduction of ceramide through acid sphingomyelinase, sphingomyelin synthase and glucosylceramide synthase. *Blood*, **104**, 3285-3293.

Watanabe M., Kitano T., Kondo T., Yabu T., Taguchi Y., Tashima M., Umehara H., Domae N., Uchiyama T. and Okazaki T. (2004) Increase of nuclear ceramide through caspase-3-dependent regulation of the 'sphingomyelin (SM) cycle' in Fas-induced apoptosis. *Cancer Res.*, **64**, 1000-1007.

Yabu T., Tomimoto H., Taguchi Y., Yamaoka S., Igarashi Y. and Okazaki T. (2005) Thalidomide-caused vascular defect through ceramide-depleted VEGF receptors, and its restoration by sphingosine-1-phosphate. *Blood*, **106**, 125-134.

Yamaoka S., Miyaji M., Kitano T., Umehara H., and Okazaki T. (2004) Expression cloning of a human cDNA restoring sphingomyelin synthesis and cell growth in sphingomyelin synthase-defective. lymphoid cells. *J. Biol. Chem.*, **279**, 18688-18693.

2-5 Glucosylceramide and Galactosylceramide Synthase

James A. Shayman

Department of Internal Medicine, Division of Nephrology, University of Michigan, 1560 MSRB II, 1150 West Medical Center Drive, Ann Arbor, Michigan 48109-0676, USA

Summary. Most mammalian glycosphingolipid formation begins with the synthesis of cerebrosides, glucosylceramide and galactosylceramide. The synthases catalyzing the formations of glucosylceramide (GlcCer) and galactosylceramide (GalCer) have been characterized at the gene and protein levels. This characterization provides a more detailed understanding of the biochemistry and cellular biology of GlcCer and GalCer as well as their downstream products. Generating mice with targeted deletions of these synthases also led to new insights into the biological importance of glycosphingolipids in general, and cerebrosides in particular. Finally, GlcCer synthase is the target of small molecule inhibitors designed to treat glycolipid storage diseases and proliferative disorders such as cancer.

Keywords. Glucosylceramide, galactosylceramide, PDMP, imino sugars, Gaucher disease

1. Introduction

More than 300 different glycosphingolipids exist within mammalian tissues. With very few exceptions, most of these glycosphingolipids have either GlcCer or GalCer as their base cerebroside. The enzymes that catalyze the formation of GlcCer and GalCer are distinct proteins arising from unique gene products. These two enzymes utilize ceramide as substrate, but they are capable of distinguishing between the nucleotide sugars,

UDP-glucose and UDP-galactose to form their respective products (Figure 1).

Investigators study these enzymes for several reasons. First, they contribute to the regulation of the bioactive compound ceramide at the cellular level. Thus changes in cerebroside synthase expression may attenuate cell susceptibility to proapoptotic and growth inhibitory effects of ceramide. Second, because the cerebroside synthases regulate the activity levels of nearly all glycosphingolipids, experimental manipulations of these activities is insightful for understanding the biological roles of glycosphingolipids in general and of the cerebrosides in particular. Third, cerebroside synthases are targets for pharmacological agents, most notably small molecule inhibitors. Inhibition of GlcCer formation may be a useful strategy for treatment of glycosphingolipidoses, including Gaucher disease, and for other glycosphingolipid mediated diseases such as cancer.

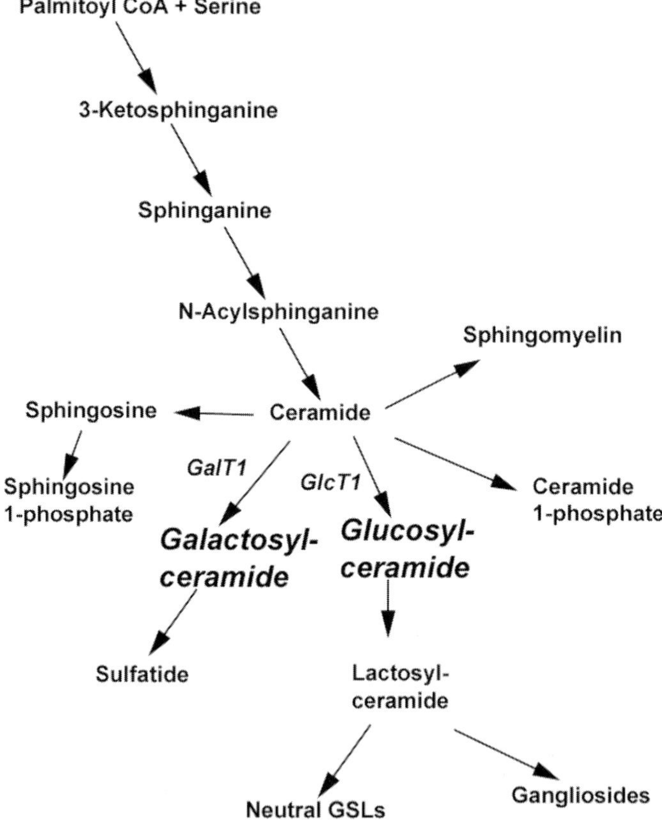

Fig. 1. Pathways of glucosylceramide and galactosylceramide metabolism.

2. Glucosylceramide synthase

2.1 Biochemistry

GlcCer is synthesized from UDP-glucose and ceramide by ceramide gly-cosyltransferase (EC 2.4.1.80, GlcT1). GlcCer synthase catalyzes the transfer of glucose from UDP-glucose to the 1-hydroxyl group of cera-mide. GlcCer synthase was first identified as a distinct enzyme in 1968 (Basu et al., 1968). For many years, the enzyme could only be studied by observing its activity because of the hydrophobic nature of the protein, its low expression level, and its characterization as an integral membrane pro-tein. Hirabayashi and colleagues cloned this protein in 1996, which then accelerated progress in understanding the properties and biological func-tions (Ichikawa et al., 1996b).

Early studies on substrate specificity focused on whether the enzyme displayed a substrate preference for the fatty acyl chain length of the ami-no diols or it preferred the presence of a hydroxyl group in the ceramide (Vunnam and Radin, 1979). These studies demonstrated that the enzyme assay conditions markedly favored the measured activity of the enzyme. Indeed, short chain ceramides were favored under most reaction conditions over the predominantly occurring natural 18 carbon fatty acids. While no preference for fatty acids with hydroxyl groups was observed, different ac-tivities among tissues (liver and brain) for different substrates indicated the presence of t distinct forms of the enzyme. In retrospect, these results probably reflected the extreme hydrophobicity of the enzyme and differ-ences in membrane composition of tissues from which the enzyme was recovered. The latter undoubtedly affected measured activity. To date, there is no evidence that GlcCer arises from more than one gene product.

2.2 Gene and protein structure

GlcCer synthase originally cloned by Ichikawa et al. mapped to 9q31 in humans and 4B3 in mice (Ichikawa et al., 1998a; Ichikawa et al., 1996a). The enzyme has also been cloned from mouse (Ichikawa et al., 1998c), rat (Wu et al., 1999), drosophila (Kohyama-Koganeya et al., 2004), and *C. elegans* (Leipelt et al., 2000). Functional analyses have also led to the identification of GlcCer synthase genes in plants (*Gossypium arboreum* (cotton)) and various fungi (*Magnaporthe grisea, Candida albicans*, and *Pichia pastoris*) (Leipelt et al., 2001). GlcCer synthase is a type III protein but bears limited structural homology to other proteins. The protein contains a type II glycosyltransferase domain. This domain is observed in a diverse family of enzymes, transferring sugar from UDP-glucose, UDP-N-acetyl-galactosamine, GDP-mannose or CDP-abequose, to a range of substrates including cellulose, dolichol phosphate and teichoic acids. The mouse gene, termed *Ugcg*, is 32 kilobases in length with 9 exons and 8 introns.

According to structural and topological studies of the synthase, the enzyme is located on the cytosolic side of the Golgi (Marks et al., 1999). Both the recombinant enzyme and the enzyme isolated from liver Golgi membranes migrate as a 38 kDa protein on SDS-PAGE. However, an anti-GlcCer synthase antibody recognizes a 50 kDa polypeptide when liver Golgi membranes are treated with crosslinking reagents. This is consistent with the presence of a small 15 kDa accessory protein. The characterization and function of this accessory protein is unknown.

A limited amount of work has been performed on the characterization the active site of GlcCer synthase. The amino acid-specific reagent, NEM, appears to inactivate the enzyme. Replacing C207 with alanine reduces the enzyme's activity, and this is consistent NEM targeting of C207. Diethylpyrocarbonate also inactivates the enzyme. This inactivation is blocked by an H193A conversion or by preincubation with excess UDP-glucose. These observations suggest that H193 is near the UDP-glucose binding site. When aligned with other glycosyltransferases, a conserved active site motif consisting of D1, D2, D3, (Q/R)*XX*RW is found in GTF2 α-glycosyltransferases (Marks et al., 2001). These amino acids correspond to D-92, D-144, D-236, R-272, R-275, and W-276 in the rat synthase. Substituting each residue with alanine results in the absence or near total loss of synthase activity. Mutagenesis of H193 of the rat synthase also abolishes the effect of the small molecule inhibitor PDMP (Wu et al., 1999)

2.3 Transcriptional regulation

The initial characterization of the mouse GlcCer synthase gene (*Ugcg*) revealed characteristics that were consistent with that of a housekeeping gene (Ichikawa et al., 1998b). Specifically, there were no TATA and CAAT boxes and the G+C content was high. Analysis of the 5' flanking sequence revealed the presence of several transcription binding sites within the first 1.0 kb. An analysis of this region with GlcCer synthase promoter and luciferase gene constructs suggested that NF-αB/c-Rel and GATA-1 binding sequences at -734 to -688 had negative regulatory effects. By contrast, the promoter activity was highest in the -578 to -103 region. This region contains 4 of 5 Sp1 motifs. Deletion of the second Sp1 motif at region -578 to -537 dramatically decreased promoter activity.

Many researchers suggested that GlcCer synthase regulation in mitigating the effects of ceramide is potentially important. Indeed, when subjected to heat shock, ceramide levels rise in cells (Chang et al., 1995) along with other proapototic stressors including lipopolysaccharide (LPS) and chemotherapeutic agents such as doxorubicin (Lucci et al., 1999). In some systems, such as keratinocytes (Uchida et al., 2002) and endothelial cells (Zhao et al., 2003), higher expressions of GlcCer synthase protects against stress-induced apoptosis. However, this observation has not been universally observed (Tepper et al., 2000).

The importance of the Sp1 sites in the transcriptional regulation of GlcCer synthase was studied in two lines of HL-60 cells that varied in their sensitivity and resistance to doxorubicin. The drug resistant cells upregulate GlcCer synthase in response to doxorubicin as measured by changes in mRNA, protein, and enzyme activity, and the drug sensitive cells do not. Using Sp1 decoy oligodeoxynucleotides to the GC/rich Sp1 region Uchida and colleagues demonstrated that the Sp1 region was necessary for the response to drug resistance (Uchida et al., 2004).

2.4 Biological insights from gene deletion studies

GlcCer synthase has been deleted in two model organisms, mouse (Yamashita et al., 1999) and drosophila (Kohyama-Koganeya et al., 2004). These models helped evaluate the role of glycolipids in embryonic development and in modulating ceramide content. The importance of Glycans, such as the Le^x structure, in embryonic development is well established (Eggens et al., 1989). However, because Le^x structure is present on both glycolipids and glycoproteins, the relative importance of glycolipids in embryonic development was uncertain.

Disrupting mouse *Ugcg* gene led to embryonic lethality. Genotyping was performed to ascertain the time to lethality. Targeted gene disruption in homozygous mice was present in the expected distribution in embryos at day E6.5. Homozygous mouse embryos with the disrupted gene were present at E7.5 and E8.5, but were abnormally small or remnants. By E9.5, no homozygous embryos with the gene deletion could be observed. Embryonic lethality corresponded to the gastrulation stage where massive apoptosis is observed. By injecting syngeneic mice with either wild type ES cells or those with two disrupted alleles, the Proia group demonstrated that the deficient cells could undergo early differentiation into endodermal, mesodermal, and ectodermal derivatives. However, the cells could not develop as wild-type differentiated tissues. These findings are consistent with an essential role for glycosphingolipids in embryonic development. While these findings may be consistent with a secondary elevation of ceramide in the embryonic tissues, no changes in ceramide content were observed in the transfected ES cells.

The drosophila homologue of GlcCer synthase was characterized and shown to be an active GlcCer synthase by transfection in a GlcCer synthase deficient cell line. The drosophila homologue is detectable throughout development and throughout the organism. Of note, at the subcellular level the protein localizes to both the Golgi and the endoplasmic reticulum. Loss of the synthase function by RNA interference enhanced apoptotic cell death. The deficient flies could be partially rescued by the targeted expression of the mammalian GlcCer synthase. The authors suggested that the effects of transfection were due to the downregulation of ceramide in the affected tissues.

2.5 Small molecule inhibitors

GlcCer synthase is also the target of small molecule inhibitor development. Blocking cerebroside synthesis with pharmacological agents, as suggested by Norman Radin in 1978, as a strategy for the treatment of Gaucher disease would also block the synthesis of GlcCer (Radin, 1996). Two general classes of GlcCer synthase inhibitors have been identified. The imino sugars, most notably *N*-butyldeoxynojirimycin, were initially developed as antiviral agents due to their inhibitory effects on α-glucosidase (Ratner et al., 1991). Later, others found that this compound blocked GlcCer in the low micromolar range (Platt et al., 1994). *N*-Butyldeoxynojirimycin also inhibits α-glucocerebrosidase, the enzyme that degrades GlcCer. A recent study suggested that the pharmacological effects of *N*-butyldeoxynojirimycin in Gaucher disease may be due to its role as a chemical chaperone (Alfonso et al., 2005).

A second group of inhibitors was designed to be GlcCer analogues and are typified by D-*threo*-1-phenyl-2-decanoylamino-3-morpholino-propanol (PDMP) (Shayman et al., 2000). Diversification of the three chemical moieties in PDMP, including the fatty acyl, aromatic, and cyclic amine groups, resulted in the discovery of a series of homologues that block GlcCer in the low nanomolar range (Abe et al., 1992; Lee et al., 1999).

Based on their comparatively lower K_is and higher degree of specificity when compared to the imino sugars, PDMP based compounds are favored for studies of the metabolism and function of GlcCer. Collectively, these studies addressed several of the questions of glycobiologists and pharmacologists. Two examples are considered.

First, does inhibition of GlcCer formation block cell growth or induce apoptosis through ceramide accumulation? In early studies with PDMP, a rise in cell ceramide occurred in parallel with a decline in cellular GlcCer content (Rani et al., 1995). These results suggested a substrate accumulation of ceramide when GlcCer synthesis was blocked, however, this interpretation was incorrect as more potent PDMP homologues were developed, such as D-*threo*-1-ethylenedioxyphenyl-2-palmitoylamino-3-pyrrilidino-propanol. This PDMP homologue inhibited GlcCer synthase at an IC50 of 11 nM but only resulted in ceramide accumulation at low micromolar concentrations (Lee et al., 1999). As cationic amphiphile, PDMP compounds appear to accumulate in lysosomes as well as at the Golgi and block the acid ceramidase (Rosenwald and Pagano, 1994). These studies raise questions about using GlcCer synthase blockades as a strategy for inducing apoptosis through cell ceramide accumulation since lysosomal ceramide accumulation as observed in Farber's disease does not result in cell death.

Second, does inhibition of GlcCer synthesis block the formation of GlcCer based glycosphingolipids with consequences for raft formation and cell signaling? Using PDMP and more potent homologues does lower GlcCer based glycolipids. Using single particle tracking, Sheets et al. followed the movement of Thy-1 and compared it to gangliosides GM1 as well as fluorescein Phosphatidylethanolamine (Sheets et al., 1997). They observed that the GPI-linked Thy-1 tracked with the gangliosides in domains of ca. 300 nm in diameter. As predicted, the glycerophospholipid demonstrated significantly less confined diffusion. Of note, when the cells were pretreated with PDMP, the confined domain area decreased 1.5 fold.

3. Galactosylceramide synthase

3.1 Biochemistry

GalCer is synthesized by the transfer of galactose from UDP-galactose to ceramide by the enzyme UDP-galactose:ceramide galactosyltransferase (GalT1, EC 2.4.1.45). The reaction was first characterized by Morrell and Radin (Morell and Radin, 1969). This work clarified the initial belief that GalCer was synthesized by the catalytic addition of a fatty acyl-CoA linked fatty acid directly to psychosine (galactosyl-sphingosine). Because knockout mice targeted for GalCer synthase do not make any glalactosylceramide, only one gene may encode for GalCer synthase. The partially purified enzyme exhibits a 15 fold preference for ceramides containing hydroxyl fatty acids over nonhydroxy fatty acids (Morell and Radin, 1969). This specificity is reproduced in cells transfected with GalCer synthase (Schaeren-Wiemers et al., 1995). The enzyme specificity may therefore explain the presence of hydroxyl-fatty acid containing ceramides in the CNS.

Unlike GlcCer, GalCer serves as the precursor for only a limited number of glycosphingolipids. These include sulfatide (SO_3-3GalCer), galabiaosylceramide (Galα1-4GalCer), and the gangliosides sialo-GalCer (I^3NeuAc-GalCer). In addition, GalCer containing glycosphingolipids are not ubiquitous. Galactose containing glycolipids are major components of the myelin sheath and are present in human urinary and galstrointestinal epithelia. Also in contrast to GlcCer synthase which is localized to the cytosolic side of the Golgi, GalCer synthase is localized to the lumen of the endoplasmic reticulum (Sprong et al., 1998).

3.2 Gene and protein structure

The gene encoding GalCer synthase was first elucidated in rat (Schulte and Stoffel, 1993) and subsequently characterized in mouse (Bosio et al., 1996b; Coetzee et al., 1996) and humans (Bosio et al., 1996a). The exons encoding the respective genes bear a high degree of homology. In addition, the gene is highly homologous with a family of genes encoding UDP-glucuronyltransferase. The human gene has 5 exons and is localized to chromosome 4q26. The encoded protein maintains a canonical KKVK sequence on the cytosolic side of the ER typical of members of this family. In addition, the protein is mannose-rich.

Some work has been performed on the characterization of gene promoter sites. An early analysis of an 8 kb fragment 5' from the ATG start codon revealed a single transcription initiation site without a consensus TATA or CCAAT box. Three positive *cis*-acting regulatory regions were identified at -292 to -256, -747 to -688, and -1325 to -1083. These contain a number of potential binding sites for known transcription factors. In addition, a negative *cis*-acting site at -1595 to -1326 was identified (Tencomnao et al., 2001). A more recent analysis revealed that a GC box at -267 to 259 and a CRE at -697 to -690 were critical for gene expression. By electrophoretic mobility shift assay, nuclear extracts containing Sp1, Sp3, pCREB-1 and ATF-1 were identified (Tencomnao et al., 2004). Differences in the expression levels of these transcription factors, most notably ATF-1 differs between neuroblastoma and oligodendroglioma cell lines (LAN-5 and HOG) that vary in their expression of GalCer synthase.

3.3 Biological insights from knockout studies

GalCer synthase deficient mice were independently created by two groups. These mice displayed hind limb paralysis, tremors, ataxia, and vacuolization of the ventral region of the spinal cord. Notably, the GalCer was replaced by high concentrations of GlcCer and of GlcCer-sulfate. Physiologically, these mice demonstrated electrophysiological defects with histological correlates of impaired oligodendrocyte differentiation, unstable myelin sheaths, and nodal and paranodal structural abnormalities. These studies are consistent with the view that despite the presence of GlcCer and GlcCer sulfatide, GalCer and sulfatide are essential for myelin formation and function.

4. References

Abe, A., Inokuchi, J., Jimbo, M., Shimeno, H., Nagamatsu, A., Shayman, J.A., Shukla, G.S. and Radin, N.S. (1992) Improved inhibitors of glucosylceramide synthase. *J Biochem (Tokyo)*, **111**, 191-196.

Alfonso, P., Pampin, S., Estrada, J., Rodriguez-Rey, J.C., Giraldo, P., Sancho, J. and Pocovi, M. (2005) Miglustat (NB-DNJ) works as a chaperone for mutated acid beta-glucosidase in cells transfected with several Gaucher disease mutations. *Blood Cells Mol Dis.*

Basu, S., Kaufman, B. and Roseman, S. (1968) Enzymatic synthesis of ceramide-glucose and ceramide-lactose by glycosyltransferases from embryonic chicken brain. *J Biol Chem*, **243**, 5802-5804.

Bosio, A., Binczek, E., Le Beau, M.M., Fernald, A.A. and Stoffel, W. (1996a) The human gene CGT encoding the UDP-galactose ceramide galactosyl transferase (cerebroside synthase): cloning, characterization, and assignment to human chromosome 4, band q26. *Genomics*, **34**, 69-75.

Bosio, A., Binczek, E. and Stoffel, W. (1996b) Molecular cloning and characterization of the mouse CGT gene encoding UDP-galactose ceramide-galactosyltransferase (cerebroside synthetase). *Genomics*, **35**, 223-226.

Chang, Y., Abe, A. and Shayman, J.A. (1995) Ceramide formation during heat shock: a potential mediator of alpha B-crystallin transcription. *Proc Natl Acad Sci U S A*, **92**, 12275-12279.

Coetzee, T., Li, X., Fujita, N., Marcus, J., Suzuki, K., Francke, U. and Popko, B. (1996) Molecular cloning, chromosomal mapping, and characterization of the mouse UDP-galactose:ceramide galactosyltransferase gene. *Genomics*, **35**, 215-222.

Eggens, I., Fenderson, B., Toyokuni, T., Dean, B., Stroud, M. and Hakomori, S. (1989) Specific interaction between Lex and Lex determinants. A possible basis for cell recognition in preimplantation embryos and in embryonal carcinoma cells. *J Biol Chem*, **264**, 9476-9484.

Ichikawa, S., Ozawa, K. and Hirabayashi, Y. (1998a) Assignment1 of a UDP-glucose:ceramide glucosyltransferase gene (Ugcg) to mouse chromosome band 4B3 by in situ hybridization. *Cytogenet Cell Genet*, **83**, 14-15.

Ichikawa, S., Ozawa, K. and Hirabayashi, Y. (1998b) Molecular cloning and characterization of the mouse ceramide glucosyltransferase gene. *Biochem Biophys Res Commun*, **253**, 707-711.

Ichikawa, S., Ozawa, K. and Hirabayashi, Y. (1998c) Molecular cloning and expression of mouse ceramide glucosyltransferase. *Biochem Mol Biol Int*, **44**, 1193-1202.

Ichikawa, S., Sakiyama, H., Suzuki, G., Hidari, K.I. and Hirabayashi, Y. (1996a) Expression cloning of a cDNA for human ceramide glucosyltransferase that catalyzes the first glycosylation step of glycosphingolipid synthesis. *Proc Natl Acad Sci U S A*, **93**, 4638-4643.

Ichikawa, S., Sakiyama, H., Suzuki, G., Hidari, K.I. and Hirabayashi, Y. (1996b) Expression cloning of a cDNA for human ceramide glucosyltransferase that catalyzes the first glycosylation step of glycosphingolipid synthesis. *Proc Natl Acad Sci U S A*, **93**, 12654.

Kohyama-Koganeya, A., Sasamura, T., Oshima, E., Suzuki, E., Nishihara, S., Ueda, R. and Hirabayashi, Y. (2004) Drosophila glucosylceramide synthase: a negative regulator of cell death mediated by proapoptotic factors. *J Biol Chem*, **279**, 35995-36002.

Lee, L., Abe, A. and Shayman, J.A. (1999) Improved inhibitors of glucosylceramide synthase. *J Biol Chem*, **274**, 14662-14669.

Leipelt, M., Warnecke, D., Zahringer, U., Ott, C., Muller, F., Hube, B. and Heinz, E. (2001) Glucosylceramide synthases, a gene family responsible for the biosynthesis of glucosphingolipids in animals, plants, and fungi. *J Biol Chem*, **276**, 33621-33629.

Leipelt, M., Warnecke, D.C., Hube, B., Zahringer, U. and Heinz, E. (2000) Characterization of UDP-glucose:ceramide glucosyltransferases from different organisms. *Biochem Soc Trans*, **28**, 751-752.

Lucci, A., Han, T.Y., Liu, Y.Y., Giuliano, A.E. and Cabot, M.C. (1999) Multidrug resistance modulators and doxorubicin synergize to elevate ceramide levels and elicit apoptosis in drug-resistant cancer cells. *Cancer*, **86**, 300-311.

Marks, D.L., Dominguez, M., Wu, K. and Pagano, R.E. (2001) Identification of active site residues in glucosylceramide synthase. A nucleotide-binding catalytic motif conserved with processive beta-glycosyltransferases. *J Biol Chem*, **276**, 26492-26498.

Marks, D.L., Wu, K., Paul, P., Kamisaka, Y., Watanabe, R. and Pagano, R.E. (1999) Oligomerization and topology of the Golgi membrane protein glucosylceramide synthase. *J Biol Chem*, **274**, 451-456.

Morell, P. and Radin, N.S. (1969) Synthesis of cerebroside by brain from uridine diphosphate galactose and ceramide containing hydroxy fatty acid. *Biochemistry*, **8**, 506-512.

Platt, F.M., Neises, G.R., Dwek, R.A. and Butters, T.D. (1994) N-butyldeoxynojirimycin is a novel inhibitor of glycolipid biosynthesis. *J Biol Chem*, **269**, 8362-8365.

Radin, N.S. (1996) Treatment of Gaucher disease with an enzyme inhibitor. *Glycoconj J*, **13**, 153-157.

Rani, C.S., Abe, A., Chang, Y., Rosenzweig, N., Saltiel, A.R., Radin, N.S. and Shayman, J.A. (1995) Cell cycle arrest induced by an inhibitor of glucosylceramide synthase. Correlation with cyclin-dependent kinases. *J Biol Chem*, **270**, 2859-2867.

Ratner, L., vander Heyden, N. and Dedera, D. (1991) Inhibition of HIV and SIV infectivity by blockade of alpha-glucosidase activity. *Virology*, **181**, 180-192.

Rosenwald, A.G. and Pagano, R.E. (1994) Effects of the glucosphingolipid synthesis inhibitor, PDMP, on lysosomes in cultured cells. *J Lipid Res*, **35**, 1232-1240.

Schaeren-Wiemers, N., van der Bijl, P. and Schwab, M.E. (1995) The UDP-galactose:ceramide galactosyltransferase: expression pattern in oligodendrocytes and Schwann cells during myelination and substrate preference for hydroxyceramide. *J Neurochem*, **65**, 2267-2278.

Schulte, S. and Stoffel, W. (1993) Ceramide UDPgalactosyltransferase from myelinating rat brain: purification, cloning, and expression. *Proc Natl Acad Sci U S A*, **90**, 10265-10269.

Shayman, J.A., Lee, L., Abe, A. and Shu, L. (2000) Inhibitors of glucosylceramide synthase. *Methods Enzymol*, **311**, 373-387.

Sheets, E.D., Lee, G.M., Simson, R. and Jacobson, K. (1997) Transient confinement of a glycosylphosphatidylinositol-anchored protein in the plasma membrane. *Biochemistry*, **36**, 12449-12458.

Sprong, H., Kruithof, B., Leijendekker, R., Slot, J.W., van Meer, G. and van der Sluijs, P. (1998) UDP-galactose:ceramide galactosyltransferase is a class I integral membrane protein of the endoplasmic reticulum. *J Biol Chem*, **273**, 25880-25888.

Tencomnao, T., Kapitonov, D., Bieberich, E. and Yu, R.K. (2004) Transcriptional regulation of the human UDP-galactose:ceramide galactosyltransferase (hCGT) gene expression: functional role of GC-box and CRE. *Glycoconj J*, **20**, 339-351.

Tencomnao, T., Yu, R.K. and Kapitonov, D. (2001) Characterization of the human UDP-galactose:ceramide galactosyltransferase gene promoter. *Biochim Biophys Acta*, **1517**, 416-423.

Tepper, A.D., Diks, S.H., van Blitterswijk, W.J. and Borst, J. (2000) Glucosylceramide synthase does not attenuate the ceramide pool accumulating during apoptosis induced by CD95 or anti-cancer regimens. *J Biol Chem*, **275**, 34810-34817.

Uchida, Y., Itoh, M., Taguchi, Y., Yamaoka, S., Umehara, H., Ichikawa, S., Hirabayashi, Y., Holleran, W.M. and Okazaki, T. (2004) Ceramide reduction and transcriptional up-regulation of glucosylceramide synthase through doxorubicin-activated Sp1 in drug-resistant HL-60/ADR cells. *Cancer Res*, **64**, 6271-6279.

Uchida, Y., Murata, S., Schmuth, M., Behne, M.J., Lee, J.D., Ichikawa, S., Elias, P.M., Hirabayashi, Y. and Holleran, W.M. (2002) Glucosylceramide synthesis and synthase expression protect against ceramide-induced stress. *J Lipid Res*, **43**, 1293-1302.

Vunnam, R.R. and Radin, N.S. (1979) Short chain ceramides as substrates for glucocerebroside synthetase. Differences between liver and brain enzymes. *Biochim Biophys Acta*, **573**, 73-82.

Wu, K., Marks, D.L., Watanabe, R., Paul, P., Rajan, N. and Pagano, R.E. (1999) Histidine-193 of rat glucosylceramide synthase resides in a UDP-glucose- and inhibitor (D-threo-1-phenyl-2-decanoylamino-3-morpholinopropan-1-ol)-binding region: a biochemical and mutational study. *Biochem J*, **341 (Pt 2)**, 395-400.

Yamashita, T., Wada, R., Sasaki, T., Deng, C., Bierfreund, U., Sandhoff, K. and Proia, R.L. (1999) A vital role for glycosphingolipid synthesis during development and differentiation. *Proc Natl Acad Sci U S A*, **96**, 9142-9147.

Zhao, H., Miller, M., Pfeiffer, K., Buras, J.A. and Stahl, G.L. (2003) Anoxia and reoxygenation of human endothelial cells decrease ceramide glucosyltransferase expression and activates caspases. *Faseb J*, **17**, 723-724.

2-6 Synthesis, Metabolism, and Trans-Bilayer Movement of Long-Chain Base

Akio Kihara and Yasuyuki Igarashi

Department of Biomembrane and Biofunctional Chemistry, Graduate School of Pharmaceutical Sciences, Hokkaido University, Kita 12, Nishi 6, Kita-ku, Sapporo 060-0812, Japan

Summary. Long-chain bases (LCBs), mainly sphingosine in mammals and phytosphingosine in plants and fungi, act not only as structural constituents of sphingolipids but also as signaling molecules. LCBs can be converted to other bioactive lipid molecules, ceramide and long-chain base 1-phosphates (sphingosine 1-phosphate in mammals). In theory, the balance of these lipids determines cell fate (the sphingolipid rheostat model). Therefore, the regulation of the synthesis and metabolism of LCBs is quite important. To function as a signaling molecule or to become a substrate for certain metabolizing enzymes, the LCB must be localized in a specific leaflet of the lipid bilayer. Thus, regulation of LCB trans-bilayer movement is also important. This review focuses on recent gains in our understanding of the synthesis, metabolism, and trans-bilayer movement of LCBs.

Keywords. Long-chain base, sphingosine, phytosphingosine, flip, flop

1. Introduction

Sphingoid long-chain bases (LCBs), or more simply long-chain bases or sphingoid bases, are the basic building blocks of sphingolipids. LCBs contain two hydroxyl groups at the C1 and C3 positions, and an amino group at the C2 position (Fig. 1). The major LCB in mammals is D-*erythro*-sphingosine (Sph), which contains a double bond between C4 and C5. In fungi and plants, D-*ribo*-phytosphingosine (phyto-Sph) is predominant, and it carries an additional hydroxyl group at C4. D-*erythro*-dihydrosphingosine (dihydro-Sph; sphinganine) is also detectable in most organisms. LCBs vary in chain-length among species, with mammals carrying C18; yeast, C16, C18, and C20 (Ferguson-Yankey et al. 2002); fly, C14 and C16 (Fyrst et al. 2004); and nematode, C17 with an *iso-* or *anteiso*-branched chain (Gerdt et al. 1997).

In addition to functioning as a structural backbone for sphingolipids, LCBs (and their metabolites) are also signaling molecules. In yeast, LCBs (dihydro-Sph and phyto-Sph) participated in endocytosis, cell cycle, and heat stress response. In mammals, Sph and its metabolite Cer induce apoptosis, while its phosphorylated metabolite Sph 1-phosphate (Sph1P) acts in cell proliferation (reviewed in Cuvillier 2002; Pettus et al. 2002; Spiegel and Milstien 2003). In light of these disparate functions, a "sphingolipid rheostat model" has been proposed, in which the cellular balance of these lipids determines cell fate (proliferation or death) (Cuvillier et al. 1996).

2. Production of LCBs

LCBs are produced by one of three pathways: *de novo* sphingolipid biosynthesis, deacylation of (dihydro-)Cer, or dephosphorylation of long-chain base 1-phosphates (LCBPs) (Fig. 1).

In *de novo* sphingolipid biosynthesis, 3-ketodihydrosphingosine (KDS; 3-ketosphinganine), a product of palmitoyl-CoA and L-serine condensation, is reduced at C3 position by KDS reductase, generating dihydro-Sph. The first gene encoding a KDS reductase, *TSC10*, was identified in yeast (Beeler et al. 1998). Subsequently, the mammalian KDS reductase FVT-1 was identified based on its homology to Tsc10 (Kihara and Igarashi 2004a). FVT-1 was found to be expressed ubiquitously among tissues and localized in the endoplasmic reticulum (ER) (Kihara and Igarashi 2004a).

FVT-1 and Tsc10 are members of the short-chain dehydrogenase/reductase (SDR) family. Both FVT-1 and Tsc10 have two transmembrane segments at the C-terminus, but only FVT-1 contains an additional transmembrane segment at its N-terminus. A large hydrophilic domain located N-terminal to the C-terminal transmembrane segments encompasses a Y*XXX*K motif, a conserved motif in SDR members that acts as an active site (Fig. 2). This domain also contains a GXXXGXG segment, a common characteristic of coenzyme binding folds found in

SDR members, which require NAD(H) or NADP(H) as a coenzyme; FVT-1 and Tsc10 both use NADPH.

In yeast, hydroxylation at the C4 position of dihydro-Sph by the hydroxylase Sur2 yields the predominant yeast LCB, phyto-Sph (Haak et al. 1997; Grilley et al. 1998). However, the major mammalian LCB, Sph, cannot be synthesized directly from dihydro-Sph, so the second pathway, deacylation of Cer by ceramidase (CDase) is utilized. The Cer comes from two sources, the degradation of sphingomyelin (by sphingomyelinase) and *de novo* synthesis. In the latter, dihydro-Sph is converted to dihydro-Cer by Cer synthase, then a double bond is introduced between the C4 and C5 by the dihydro-Cer α4-desaturase DES1 (Ternes et al. 2002). Like the yeast C4 hydroxylase Sur2, DES1 is a member of the desaturase/hydroxylase superfamily and an integral membrane protein with several transmembrane segments, though no topology is known.

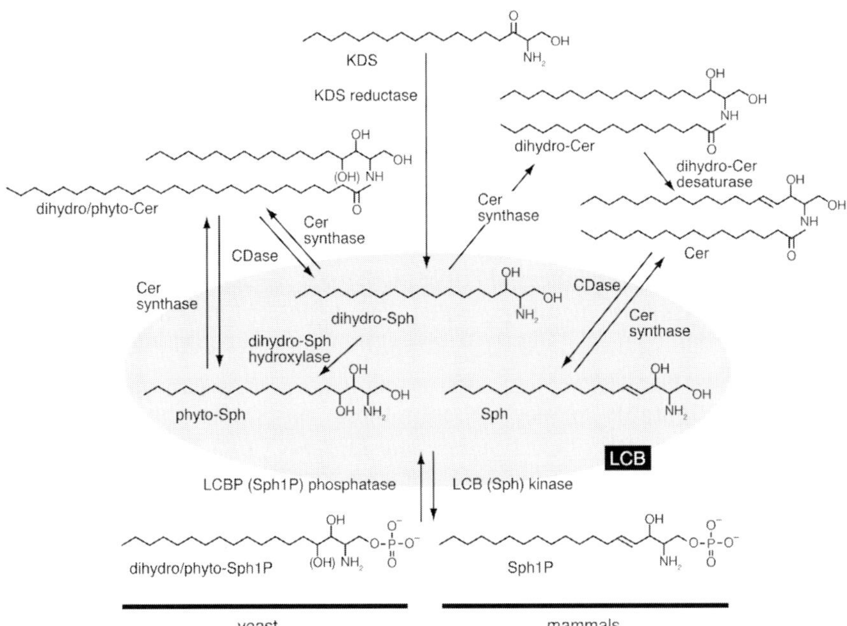

Fig. 1. Synthetic and metabolic pathways of LCBs. Reactions involved in LCB synthesis and metabolism in yeast and mammals are shown.

Both yeast and mammals also utilize CDases. Three types of CDases are known, differing in the optimum pH for activity. Acidic CDase is a soluble protein located in the lumen of the lysosome (Ferlinz et al. 2001). Neutral CDase is a single-span membrane protein localized in the plasma membrane, with a large C-terminal domain that encompasses the active site and is oriented towards the extracellular side (Fig. 2) (Tani et al. 2003). Alkaline CDase is a multi-span membrane protein located in the ER and Golgi (Mao et al. 2001), although the topology of its active site is unclear. Mammals have CDases of all three types, yet yeast express only alkaline CDases. Ypc1 and Ydc1, yeast homologues of mammalian alkaline CDases, preferentially hydrolyze phyto-Cer and dihydro-Cer, respectively (Mao et al. 2000a; Mao et al. 2000b). Additional details of mammalian CDases are available in another section.

LCBs can also be produced by the dephosphorylation of LCBPs. The major LCBP in mammals is Sph1P, whereas yeast have phyto-Sph1P and dihydro-Sph1P. Two lipid phosphatase families, SPP and LPP, are capable of this reaction. *In vitro* assays have demonstrated that the SPP proteins can dephosphorylate LCBPs but not other phosphorylated lipids (Mao et al. 1997; Mandala et al. 2000; Ogawa et al. 2003), while the LPP proteins have broad activities towards substrates that include Sph1P, phosphatidic acid, lyso-phosphatidic acid, Cer 1-phosphate, and diacylglycerol pyrophosphate (Brindley 2004). In yeast (at least), most of the *in vivo* LCBP dephosphorylation activity is attributable to SPP proteins (Kihara et al. 2003), and any involvement of LPP proteins is unclear. Yeast express two SPP family members, Lcb3 and Ysr3, though most activity has been attributed to Lcb3 (Mandala et al. 1998; Kihara et al. 2003). Mammals also have two SPP family members, SPP1 and SPP2 (Mandala et al. 2000; Ogawa et al. 2003), though SPP1 is ubiquitously expressed, whereas SPP2 is rather tissue-specific.

SPP and LPP family members contain three conserved domains: domain 1 (D1), K*XXXXXX*RP; D2, PSGH; and D3, SR*XXXXX*H*XXX*D. These domains, which constitute an active site, are found in a superfamily of phosphatases that includes lipid phosphatases, glucose-6-phosphatases, bacterial nonspecific acid phosphatases, and chloroperoxidase (Stukey and Carman 1997; Sigal et al. 2005). All mammalian LPPs also contain a consensus N-glycosylation site between D1 and D2. Glycosylation studies of this site and reporter analyses suggest that LPPs contain six transmembrane domains and an active site located in the extracytosolic (lumenal) side of the membranes (Barila et al. 1996; Kihara et al. 2003; Sigal et al. 2005). Although all SPP members are known to be localized in the ER, only Lcb3 has been subjected to membrane topology analysis

(Kihara et al. 2003). A C-terminal reporter analysis resulted in a proposed structural model of Lcb3 illustrating 8 membrane-spanning domains and the localization of the highly conserved phosphatase domains within the ER lumen (Fig. 2), a prediction consistent with the sensitivity of Lcb3 to an exogenous proteinase (Kihara et al. 2003). This topology suggests that LCBs are produced in the lumenal leaflet of the ER membrane by the SPP

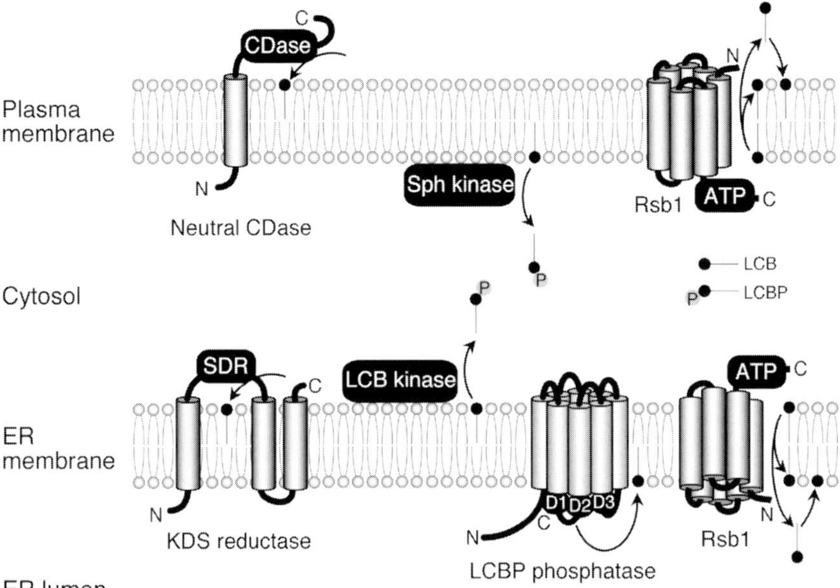

proteins.

Fig. 2. Structures of LCB metabolizing enzymes and the trans-bilayer movement of LCBs. The proposed structure and active site orientations of neutral CDase, LCB (Sph) kinase, Rsb1, KDS reductase (FVT-1), and LCBP phosphatase (Lcb3) are shown. LCBs are produced in the cytosolic leaflet by the KDS reductase but in the extracytosolic leaflet by the neutral CDase and by LCBP phosphatase. Rsb1 translocates LCBs in the cytosolic leaflet to the extracytosolic leaflet. LCB kinase in yeast and Sph kinase in mammals act on the cytosolic surfaces of the ER membrane and the plasma membrane, respectively.

3. LCB metabolism

LCBs can be phosphorylated or metabolized to Cer by a (dihydro)Cer synthase (dihydro-Sph:N-acyltransferase). Cer synthases catalyze an amide linkage between the LCB and a fatty acid. In both yeast and mammals, members of the Lass family have been shown to have Cer synthase activity. Yeast carrying a double deletion of the Lass genes *LAG1* and *LAC1* have drastically reduced sphingolipid levels, and *in vitro* assays using microsomal membranes further demonstrated that both are indispensable for acyl-CoA-dependent Cer synthesis and the residual Cer is actually produced by CDases (Guillas et al. 2001; Schorling et al. 2001). Similarly, mammals express five Lass members (Lass1, Lass2, Lass4, Lass5, and Lass6) that function in mammalian Cer synthesis (Venkataraman et al. 2002; Guillas et al. 2003; Riebeling et al. 2003; Mizutani et al. 2005), each exhibiting a unique tissue distribution pattern.

Each mammalian or yeast Lass protein exhibits its own specific substrate preference for acyl-CoAs that differ in chain-length, although specificity for LCBs (Sph vs. dihydro-Sph vs. phyto-Sph) is not similarly significant. In yeast studies, a novel protein termed Lip1 (Lag1/Lac1 interacting protein) was co-purified with Lag1/Lac1 as an essential component of the Cer synthase (Vallee and Riezman 2005), but no homolog of Lip1 has yet been found in mammals. Both the Lass proteins and Lip1 are localized in the ER, and Lip1, a single-span ER membrane protein, is positioned with its C-terminal large hydrophilic domain facing the ER lumen. Lass members are known to be multi-span membrane proteins, but topology has been analyzed only for mouse Lass6, and that only partially, revealing that its N-terminus and C-terminus are localized in the lumenal and cytosolic sides of the ER, respectively (Mizutani et al. 2005). Furthermore, the active site residues of the Lass proteins have not been determined. Therefore, it remains unclear on which side of the ER membrane Cer is synthesized, although the sensitivity of Cer synthase to an exogenous proteinase or membrane-impermeable chemicals suggest that (dihydro-)Cer is synthesized in the cytosolic leaflet (Mandon et al. 1992; Hirschberg et al. 1993).

LCB kinases, including Sph kinases in mammals, catalyze the phosphorylation of LCBs to form LCBPs. Two mammalian Sph kinases, SPHK1 and SPHK2 are known (Kohama et al. 1998; Liu et al. 2000). SPHK2 exhibits broader substrate specificity and can phosphorylate Sph, dihydro-Sph, phyto-Sph, and FTY720 (a structurally related synthetic immunosuppressant) nearly equally, whereas SPHK1 effectively phosphorylates only Sph and dihydro-Sph (Kohama et al. 1998; Liu et al. 2000; Billich et al. 2003). Both SPHK1 and SPHK2 are expressed ubiquitously, but their amounts vary among tissues. SPHK1 is localized in the cytosol (mainly) and in the plasma membrane, and treatment with phorbol ester or tumor necrosis factor α induces translocation to the plasma membrane (Johnson et al. 2002; Pitson et al. 2003), although the precise mechanism of this remains unclear.

Yeast also have two LCB kinases, Lcb4 and Lcb5. Most cellular activity is attributable to Lcb4 (Nagiec et al. 1998), which is localized mainly in the ER, with some in the plasma membrane. LCB (Sph) kinases have no apparent transmembrane domains or motifs, however there are five highly conserved regions, C1 to C5. The C1 to C3 regions function in Mg^{2+}-ATP binding (Pitson et al. 2002), and C4 is important in binding to the LCB (Yokota et al. 2004). Additionally, the membrane association of Lcb4 is mediated by palmitoylation (Kihara et al. 2005).

4. Trans-bilayer movement of LCB

Lipid molecules traverse between two leaflets of a lipid bilayer. This movement is called flip-flop, with flip meaning movement from the outer (extracytosolic) leaflet to the inner (cytosolic) leaflet. In model membranes, spontaneous flip-flop is slower for lipids with a polar head group than those without, so that Cer moves nearly 10 times faster than sphingomyelin (Bai and Pagano 1997). Although data for the flip-flop of LCBs is not available, it may be even faster than that of Cer.

As described above, LCBs are synthesized both in the cytosolic and extracytosolic leaflets. Dihydro-Sph is produced in the cytosolic leaflet by KDS reductase, and LCBs are generated by LCBP phosphatase or CDases in the extracytosolic (outer) leaflet. Since LCB (Sph) kinase functions on the cytosolic surface of the plasma membrane or the ER, LCBs generated in the extracytosolic leaflet must flip to the cytosolic leaflet to be substrates for the kinase. It is unclear whether this flip reaction is mediated by a protein, a so-called flippase (translocase), or takes place spontaneously, since no such enzyme has been identified.

Such an enzyme has, however, been identified for the flop of LCBs in yeast, Rsb1 (Kihara and Igarashi 2002), though at present no mammalian homolog has been found. Rsb1 is a LCB transporter or translocase, and its overproduction results in an increase in the release of intracellular LCB into the medium. As it is generally difficult to discriminate between a translocase and transporter, it is possible that Rsb1 is a transporter that pumps LCBs from the inner leaflet of the plasma membrane directly to the external medium. Alternatively, Rsb1 may be a translocase that flops LCBs from the inner leaflet to the outer leaflet, from which the LCBs diffuse into the medium. However, even if Rsb1 functions as a transporter, the released LCBs are then re-incorporated into the plasma membrane

(Kihara and Igarashi 2002), so their population in the outer leaflet is increased as well (Fig. 2).

Rsb1 is a membrane protein putatively having seven transmembrane segments. The C-terminal tail of Rsb1 contains an ATP-binding motif (Fig. 2) that is active in the LCB release reaction, which is ATP-dependent. Rsb1 is localized mainly in the ER and in the plasma membrane (Kihara and Igarashi 2004b), so it may also function in the translocation of LCBs from the cytosolic leaflet to the lumenal leaflet of the ER.

In normal growth conditions, expression of Rsb1 is low, but, interestingly, a change in the glycerophospholipid asymmetry in the plasma membrane induces the expression of Rsb1 (Kihara and Igarashi 2004b). In the absence of Rsb1, LCBs may be distributed in both leaflets by spontaneous flip-flop, which is predicted to be fast for these compounds. Induction of Rsb1 must cause an increase in the LCB levels in the extracytosolic (outer) leaflet. Physiologically, this reaction may function in the spatial separation from the cytosol of the LCBs that act as signaling molecules or in maintaining functional lipid asymmetry.

5. Future directions

Most of the biosynthetic and metabolic pathways of LCBs have now been revealed. However, there are many unsolved questions regarding their membrane topology and trans-bilayer movement. For example, the active site topologies of Cer synthase and alkaline CDase are unclear. Additionally, the distribution of LCBs between cytosolic and extracytosolic leaflets in the ER membrane(or in the plasma membrane) has not been investigated. Recent studies have demonstrated that three major groups of enzymes, P-type ATPases, ATP-binding-cassette (ABC) transporters, and scramblases, are active in the trans-bilayer movement of glycerophospholipids, and that proper glycerophospholipid asymmetry is important for several cellular functions (Holthuis and Levine 2005). However, knowledge regarding the trans-bilayer movement of sphingolipids and the factors involved is limited, and the mechanism, spontaneous or protein-mediated, by which LCBs traverse the two leaflets, is almost unknown; thus, future studies are required.

6. References

Bai J, Pagano RE (1997) Measurement of spontaneous transfer and transbilayer movement of BODIPY-labeled lipids in lipid vesicles. *Biochemistry*, **36**, 8840-8848.

Barila D, Plateroti M, Nobili F, Muda AO, Xie Y, Morimoto T, Perozzi G (1996) The *Dri 42* gene, whose expression is up-regulated during epithelial differentiation, encodes a novel endoplasmic reticulum resident transmembrane protein. *J Biol Chem*, **271**, 29928-29936.

Beeler T, Bacikova D, Gable K, Hopkins L, Johnson C, Slife H, Dunn T (1998) The *Saccharomyces cerevisiae TSC10/YBR265w* gene encoding 3-ketosphinganine reductase is identified in a screen for temperature-sensitive suppressors of the Ca^{2+}-sensitive $csg2\alpha$ mutant. *J Biol Chem*, **273**, 30688-30694.

Billich A, Bornancin F, Devay P, Mechtcheriakova D, Urtz N, Baumruker T (2003) Phosphorylation of the immunomodulatory drug FTY720 by sphingosine kinases. *J Biol Chem*, **278**, 47408-47415.

Brindley DN (2004) Lipid phosphate phosphatases and related proteins: signaling functions in development, cell division, and cancer. J Cell Biochem, 92, 900-912.

Cuvillier O, Pirianov G, Kleuser B, Vanek PG, Coso OA, Gutkind S, Spiegel S (1996) Suppression of ceramide-mediated programmed cell death by sphingosine-1-phosphate. *Nature*, **381**, 800-803.

Cuvillier O (2002) Sphingosine in apoptosis signaling. *Biochim Biophys Acta*, **1585**, 153-162.

Ferguson-Yankey SR, Skrzypek MS, Lester RL, Dickson RC (2002) Mutant analysis reveals complex regulation of sphingolipid long chain base phosphates and long chain bases during heat stress in yeast. *Yeast*, **19**, 573-586.

Ferlinz K, Kopal G, Bernardo K, Linke T, Bar J, Breiden B, Neumann U, Lang F, Schuchman EH, Sandhoff K (2001) Human acid ceramidase: processing, glycosylation, and lysosomal targeting. *J Biol Chem*, **276**, 35352-35360.

Fyrst H, Herr DR, Harris GL, Saba JD (2004) Characterization of free endogenous C14 and C16 sphingoid bases from *Drosophila melanogaster*. *J Lipid Res*, **45**, 54-62.

Gerdt S, Lochnit G, Dennis RD, Geyer R (1997) Isolation and structural analysis of three neutral glycosphingolipids from a mixed population of *Caenorhabditis elegans* (Nematoda:Rhabditida). *Glycobiology*, **7**, 265-275.

Grilley MM, Stock SD, Dickson RC, Lester RL, Takemoto JY (1998) Syringomycin action gene *SYR2* is essential for sphingolipid 4-hydroxylation in *Saccharomyces cerevisiae*. *J Biol Chem*, **273**, 11062-11068.

Guillas I, Kirchman PA, Chuard R, Pfefferli M, Jiang JC, Jazwinski SM, Conzelmann A (2001) C26-CoA-dependent ceramide synthesis of *Saccharomyces cerevisiae* is operated by Lag1p and Lac1p. *EMBO J*, **20**, 2655-266.

Guillas I, Jiang JC, Vionnet C, Roubaty C, Uldry D, Chuard R, Wang J, Jazwinski SM, Conzelmann A (2003) Human homologues of *LAG1* reconstitute Acyl-CoA-dependent ceramide synthesis in yeast. *J Biol Chem*, **278**, 37083-37091.

Haak D, Gable K, Beeler T, Dunn T (1997) Hydroxylation of *Saccharomyces cerevisiae* ceramides requires Sur2p and Scs7p. *J Biol Chem*, **272**, 29704-29710.

Hirschberg K, Rodger J, Futerman AH (1993) The long-chain sphingoid base of sphingolipids is acylated at the cytosolic surface of the endoplasmic reticulum in rat liver. *Biochem J*, **290**, 751-757.

Holthuis JC, Levine TP (2005) Lipid traffic: floppy drives and a superhighway. *Nat Rev Mol Cell Biol*, **6**, 209-220.

Johnson KR, Becker KP, Facchinetti MM, Hannun YA, Obeid LM (2002) PKC-dependent activation of sphingosine kinase 1 and translocation to the plasma membrane. Extracellular release of sphingosine-1-phosphate induced by phorbol 12-myristate 13-acetate (PMA). *J Biol Chem*, **277**, 35257-35262.

Kihara A, Igarashi Y (2002) Identification and characterization of a *Saccharomyces cerevisiae* gene, *RSB1*, involved in sphingoid long-chain base release. *J Biol Chem*, **277**, 30048-30054.

Kihara A, Sano T, Iwaki S, Igarashi Y (2003) Transmembrane topology of sphingoid long-chain base-1-phosphate phosphatase, Lcb3p. *Genes Cells*, **8**, 525-535.

Kihara A, Igarashi Y (2004a) FVT-1 is a mammalian 3-ketodihydrosphingosine reductase with an active site that faces the cytosolic side of the endoplasmic reticulum membrane. *J Biol Chem*, **279**, 49243-49250.

Kihara A, Igarashi Y (2004b) Cross talk between sphingolipids and glycerophospholipids in the establishment of plasma membrane asymmetry. *Mol Biol Cell*, **15**, 4949-4959.

Kihara A, Kurotsu F, Sano T, Iwaki S, Igarashi Y (2005) Long-chain base kinase Lcb4 is anchored to the membrane through its palmitoylation by Akr1. *Mol Cell Biol*, in press.

Kohama T, Olivera A, Edsall L, Nagiec MM, Dickson R, Spiegel S (1998) Molecular cloning and functional characterization of murine sphingosine kinase. *J Biol Chem*, **273**, 23722-23728.

Liu H, Sugiura M, Nava VE, Edsall LC, Kono K, Poulton S, Milstien S, Kohama T, Spiegel S (2000) Molecular cloning and functional characterization of a novel mammalian sphingosine kinase type 2 isoform. *J Biol Chem*, **275**, 19513-19520.

Mandala SM, Thornton R, Tu Z, Kurtz MB, Nickels J, Broach J, Menzeleev R, Spiegel S (1998) Sphingoid base 1-phosphate phosphatase: a key regulator of sphingolipid metabolism and stress response. *Proc Natl Acad Sci USA*, **95**, 150-155.

Mandala SM, Thornton R, Galve-Roperh I, Poulton S, Peterson C, Olivera A, Bergstrom J, Kurtz MB, Spiegel S (2000) Molecular cloning and characterization of a lipid phosphohydrolase that degrades sphingosine-1-phosphate and induces cell death. *Proc Natl Acad Sci USA*, **97**, 7859-7864.

Mandon EC, Ehses I, Rother J, van Echten G, Sandhoff K (1992) Subcellular localization and membrane topology of serine palmitoyltransferase, 3-dehydrosphinganine reductase, and sphinganine *N*-acyltransferase in mouse liver. *J Biol Chem*, **267**, 11144-11148.

Mao C, Wadleigh M, Jenkins GM, Hannun YA, Obeid LM (1997) Identification and characterization of *Saccharomyces cerevisiae* dihydrosphingosine-1-phosphate phosphatase. *J Biol Chem*, **272**, 28690-28694.

Mao C, Xu R, Bielawska A, Obeid LM (2000a) Cloning of an alkaline ceramidase from *Saccharomyces cerevisiae*. An enzyme with reverse (CoA-independent) ceramide synthase activity. *J Biol Chem*, **275**, 6876-6884.

Mao C, Xu R, Bielawska A, Szulc ZM, Obeid LM (2000b) Cloning and characterization of a *Saccharomyces cerevisiae* alkaline ceramidase with specificity for dihydroceramide. *J Biol Chem*, **275**, 31369-31378.

Mao C, Xu R, Szulc ZM, Bielawska A, Galadari SH, Obeid LM (2001) Cloning and characterization of a novel human alkaline ceramidase. A mammalian enzyme that hydrolyzes phytoceramide. *J Biol Chem*, **276**, 26577-26588.

Mizutani Y, Kihara A, Igarashi Y (2005) Mammalian Lass6 and its related family members regulate synthesis of specific ceramides. *Biochem J*, in press.

Nagiec MM, Skrzypek M, Nagiec EE, Lester RL, Dickson RC (1998) The *LCB4* (*YOR171c*) and *LCB5* (*YLR260w*) genes of *Saccharomyces* encode sphingoid long chain base kinases. *J Biol Chem*, **273**, 19437-19442.

Ogawa C, Kihara A, Gokoh M, Igarashi Y (2003) Identification and characterization of a novel human sphingosine-1-phosphate phosphohydrolase, hSPP2. *J Biol Chem*, **278**, 1268-1272.

Pettus BJ, Chalfant CE, Hannun YA (2002) Ceramide in apoptosis: an overview and current perspectives. *Biochim Biophys Acta*, **1585**, 114-125.

Pitson SM, Moretti PA, Zebol JR, Zareie R, Derian CK, Darrow AL, Qi J, D'Andrea RJ, Bagley CJ, Vadas MA, Wattenberg BW (2002) The nucleotide-binding site of human sphingosine kinase 1. *J Biol Chem*, **277**, 49545-49553.

Pitson SM, Moretti PA, Zebol JR, Lynn HE, Xia P, Vadas MA, Wattenberg BW (2003) Activation of sphingosine kinase 1 by ERK1/2-mediated phosphorylation. *EMBO J*, **22**, 5491-5500.

Riebeling C, Allegood JC, Wang E, Merrill AH, Jr., Futerman AH (2003) Two mammalian longevity assurance gene (LAG1) family members, *trh1* and *trh4*, regulate dihydroceramide synthesis using different fatty acyl-CoA donors. *J Biol Chem*, **278**, 43452-43459.

Schorling S, Vallee B, Barz WP, Riezman H, Oesterhelt D (2001) Lag1p and Lac1p are essential for the Acyl-CoA-dependent ceramide synthase reaction in *Saccharomyces cerevisae*. *Mol Biol Cell*, **12**, 3417-3427.

Sigal YJ, McDermott MI, Morris AJ (2005) Integral membrane lipid phosphatases/phosphotransferases: common structure and diverse functions. *Biochem J*, **387**, 281-293.

Spiegel S, Milstien S (2003) Sphingosine-1-phosphate: an enigmatic signalling lipid. *Nat Rev Mol Cell Biol*, **4**, 397-407.

Stukey J, Carman GM (1997) Identification of a novel phosphatase sequence motif. *Protein Sci*, **6**, 469-472.

Tani M, Iida H, Ito M (2003) *O*-glycosylation of mucin-like domain retains the neutral ceramidase on the plasma membranes as a type II integral membrane protein. *J Biol Chem*, **278**, 10523-10530.

Ternes P, Franke S, Zahringer U, Sperling P, Heinz E (2002) Identification and characterization of a sphingolipid α4-desaturase family. *J Biol Chem*, **277**, 25512-25518.

Vallee B, Riezman H (2005) Lip1p: a novel subunit of acyl-CoA ceramide synthase. *EMBO J*, **24**, 730-741.

Venkataraman K, Riebeling C, Bodennec J, Riezman H, Allegood JC, Sullards MC, Merrill AH, Jr., Futerman AH (2002) Upstream of growth and differentiation factor 1 (*uog1*), a mammalian homolog of the yeast longevity assurance gene 1 (*LAG1*), regulates *N*-stearoyl-sphinganine (C18-(dihydro)ceramide) synthesis in a fumonisin B_1-independent manner in mammalian cells. *J Biol Chem*, **277**, 35642-35649.

Yokota S, Taniguchi Y, Kihara A, Mitsutake S, Igarashi Y (2004) Asp177 in C4 domain of mouse sphingosine kinase 1a is important for the sphingosine recognition. *FEBS Lett*, **578**, 106-110.

2-7 Molecular Mechanism of Ceramide Trafficking from the Endoplasmic Reticulum to the Golgi Apparatus in Mammalian Cells

Kentaro Hanada, Miyuki Kawano, and Keigo Kumagai

Department of Biochemistry and Cell Biology, National Institute of Infectious Diseases, 1-23-1 Toyama, Shinjuku-ku, Tokyo 162-8640, Japan

Summary. Synthesis and sorting of lipids are essential events for membrane biogenesis and its homeostasis. The endoplasmic reticulum (ER) is the center of the de novo synthesis of various lipid types. Trafficking of various lipids from the ER to other organelles has been suggested to occur by mechanisms different from the vesicle-mediated mechanism for protein trafficking. However, molecular mechanisms underlying intracellular trafficking of lipids remain poorly understood. Ceramide is synthesised at the ER, and translocated to the Golgi compartment for conversion to sphingomyelin (SM). We previously isolated a mammalian cultured cell mutant defective in ceramide trafficking, and have recently identified a key factor (named CERT) for ceramide trafficking in functional rescue experiments and proposed a molecular lipid extraction and transfer model for the non-vesicular mechanism of CERT-mediated trafficking of ceramide.

Keywords. Sphingomyelin, CERT, PH, START, FFAT

1. Introduction

Intracellular transport of lipids from the sites of their synthesis to appropriate destinations must occur, because various steps in lipid biosynthesis occur in different intracellular compartments. The intracellular trafficking of lipids is also important to regulate the lipid composition of organelles. In addition, lipid trafficking might also play a role in signaling by lipid-mediators, because, for a lipid to exert its bio-modulator activity at a site different from that of its synthesis, inter-organelle transport of the lipid must occur.

The trafficking of membrane proteins from the endoplasmic reticulum (ER) to the Golgi apparatus is mediated by transport vesicles, which load the desired set of proteins and deliver them to the correct organelles (Lee et al. 2004; Palmer and Stephens 2004). By contrast, various types of lipids synthesized in the ER are likely sorted to other organelles by nonvesicular mechanisms. For example, transport of de novo synthesized phosphatidylcholine (PC) from the ER to the plasma membranes is not inhibited by energy poisons, which block the ER-to-Golgi transport of proteins (Kaplan and Simoni 1985a). The transport of cholesterol from the ER to the plasma membrane is not affected by brefeldin A (BFA), an inhibitor of anterograde vesicular transport from the ER to the Golgi apparatus (Kaplan and Simoni 1985b). A complex of chaperon proteins and caveolin-1 might mediate the intracellular transport of cholesterol (Smart et al. 2004; Uittenbogaard et al. 1998). After de novo synthesis in the ER, phosphatidylserine (PS) is transported to the matrix of mitochondria to be converted to phosphatidylethanolamine. The ER-to-mitochondrion transport of PS is suggested to occur at a specific subdomain of the ER, where the ER membrane makes contact with the outer membrane of mitochondria (Ardail et al. 1991; Vance 1991; Voelker 1990). Moreover, ceramide also appears to be transported from the ER to the Golgi apparatus in a non-vesicular manner (Collins and Warren 1992; Fukasawa et al. 1999; Funato and Riezman 2001; Kok et al. 1998; Moreau et al. 1993). However, molecular mechanisms underlying the intracellular trafficking of lipids remain poorly understood.

We previously isolated a mammalian cell mutant defective in the intracellular transport of ceramide, and recently identified a molecular factor playing a key role in ceramide trafficking, as summarized in this short review.

2. Biosynthesis and translocation of sphingolipids in mammalian cells

The biosynthesis of sphingolipids in mammalian cells is initiated by the condensation of L-serine and palmitoyl CoA to generate 3-ketodihydrosphingosine (KDS), the reaction catalyzed by serine palmitoyltransferase (SPT) (Hanada 2003; Merrill 2002). KDS is reduced to dihydrosphingosine, which undergoes N-acylation followed by desaturation to generate ceramide (Wang et al. 1991). These reactions to produce ceramide occur at the cytosolic surface of the ER (Mandon et al. 1992). Then, ceramide is delivered to the lumenal side of the Golgi apparatus, and converted to sphingomyelin (SM) by SM synthase catalyzing the transfer of phosphocholine from PC to ceramide (Futerman et al. 1990). The human genome contains genes for two different SM synthases, SMS1 and SMS2 (Huitema et al. 2004). SMS1, which is likely responsible for de novo synthesis of SM (Yamaoka et al. 2004), is localized to the trans Golgi region (Huitema et al. 2004), while SMS2 mainly resides at the plasma membrane (Huitema et al. 2004). SMS2 might catalyze the re-synthesis of SM from ceramide generated at the plasma membrane or de novo synthesis of SM at the Golgi apparatus redundantly with SMS1. Ceramide is also converted to glucosylceramide (GlcCer) by GlcCer synthase catalyzing the transfer of glucose from UDP-glucose to ceramide. Although GlcCer synthase is abundant in the cis Golgi region with a cytosolic orientation of its catalytic site (Futerman and Pagano 1991; Jeckel et al. 1992), this enzyme is also distributed to the ER (Ardail et al. 2003; Futerman and Pagano 1991; Jeckel et al. 1992; Kohyama-Koganeya et al. 2004). Thus, it remains unclear where the primary site for the de novo synthesis of GlcCer is. After being transported to the lumenal side of the Golgi apparatus, GlcCer is further converted to more complex glycosphingolipids (Sandhoff and Kolter 2003).

3. LY-A, a mammalian cell mutant defective in intracellular trafficking of ceramide

Lysenin, a cytolysin derived from the earthworm *Eisenia foetida*, binds to SM with high affinity (Yamaji et al. 1998). To isolate various cell mutants defective in SM metabolism, we selected several types of lysenin-resistant mutants from the parental Chinese hamster ovary-derived CHO-K1 cell line, and obtained two different cell lines deficient in the de novo synthesis of SM (Hanada et al. 1998). One lysenin-resistant mutant cell line, named LY-B, was shown to have a defect in expression of the SPT enzyme (Hanada et al. 1998). Another mutant cell line, named LY-A, was found to have a defect in the transport of ceramide from the site of its synthesis where SM is produced, based on various lines of evidence (Fukasawa et al. 1999): (i) No defect was observed in the synthesis of ceramide and PC (which are precursors of the synthesis of SM) or in the activity of SM synthase. (ii) When cells were treated with BFA to fuse the Golgi apparatus with the ER, the synthesis of SM in LY-A cells was restored to the wild-type level.   (iii) ER-to-Golgi transport of C_5-DMB-ceramide, a fluorescent analogue of ceramide, is far slower in LY-A cells than in wild-type cells.   In addition, we observed that treatment of wild-type cells with energy poisons inhibited the transport of C_5-DMB-ceramide to the Golgi apparatus, while the energy poisons did not affect the behavior of C_5-DMB-ceramide in LY-A cells.   From these results, we concluded that LY-A cells are defective in energy (probably ATP)-dependent trafficking of ceramide from the ER to the Golgi compartment where SM synthase exists (Fukasawa et al. 1999).   There was no appreciable difference in the acquisition rate of endoglycosidase H-resistance of glycoproteins between LY-A and wild-type cells, indicating that ER-to-Golgi trafficking of proteins in LY-A cells is normal (Fukasawa et al. 1999).

4. Reconstitution of ceramide trafficking in semi-intact CHO cells

A reliable in vitro reconstitution system of ER-to-Golgi trafficking of ceramide has been established within semi-intact cells (Funakoshi et al. 2000).   Semi-intact CHO cells, which are prepared by mild perforation of the plasma membrane, retain the ER and Golgi apparatus but not cytosolic soluble proteins (Beckers et al. 1987; Funakoshi et al. 2000).   In the reconstitution system, $[^3H]$ceramide is formed with D-erythro-$[^3H]$sphingosine and palmitoyl CoA by the reaction catalyzed by dihydrosphingosine/sphingosine-N-acyltransferase at the ER in semi-intact cells at 15°C, and the relabelled semi-intact cells are chased in the presence of the cytosolic fraction and an ATP-regeneration system at 37°C to convert $[^3H]$ceramide to $[^3H]$SM (see Fig. 2a for a schematic model of this method).   Importantly, the semi-intact cell system reproduces the phenotype of mutant LY-A cells; the rate of ceramide-to-SM conversion in semi-intact LY-A cells is only ~20% of the wild-type level.   Analysis with the in vitro reconstitution system demonstrated that the ATP-dependent trafficking of ceramide requires cytosol (Funakoshi et al. 2000).   In addition, cytosol-exchange experiments demonstrated that, when semi-intact LY-A cells were chased with the wild-type cytosol, the $[^3H]$ceramide-to-$[^3H]$SM conversion was completely rescued, indicating that the mutant phenotype of LY-A cells results from a deficiency of a cytosolic factor (Funakoshi et al. 2000)(see also Fig. 2b).   This reconstitution system for ceramide trafficking was later shown to be applicable to the rat C6 glioma cell line (Viani et al. 2003).

5. Molecular cloning of CERT, the factor impaired in LY-A cells

To identify the cytosolic factor that is impaired in the LY-A cells, we attempted functional gene rescue experiments. For this approach, we fortunately found that SM-deficient cells such as LY-A cells are hypersensitive to the cholesterol-absorbing agent methy-α-cyclodextrin (MCD) (Fukasawa et al. 2000). This property of SM-deficient mutant cells could be employed for selection of their revertants (Hanada 2003; Yamaoka et al. 2004).

We used retroviral vector-mediated transfection, which can introduce non-viral genes into mitotic cells stably at very high levels of efficiency (Kitamura et al. 2003). Although CHO cells are naturally resistant to infection by ecotropic murine retroviruses, they become the susceptible after expressing the mouse ecotropic retroviral receptor mCAT-1 (Siess et al. 1996). We established an LY-A derivative (named LY-A2) stably expressing the mCAT-1 receptor, and retrovirally transfected LY-A2 cells with a human cDNA library. Then, MCD-tolerant variants were selected, and purified. We could retrieve a cDNA from an MCD-tolerant variant by genomic PCR with vector-derived sequences as primers. When the cloned cDNA was introduced into LY-A and LY-A2 cells, various mutant phenotypes were fully rescued (Hanada et al. 2003). Hence, we named the cDNA product CERT after CERamide Transport (Hanada et al. 2003).

CERT, a hydrophilic 68-kDa protein, consists of three parts (Fig. 1b) (Hanada et al. 2003). The amino terminal ~120 amino acid region forms a pleckstrin homology (PH) domain, which is a phosphoinositide-binding domain (Dowler et al. 2000; Raya et al. 1999). The carboxyl terminal ~230 amino acid region forms a putative lipid-transfer domain, START (Ponting and Aravind 1999). The middle region (MR) is predicted to form no globular domains, but to have a short motif, which is postulated to interact with the ER, as described below.

5.1 START domain of CERT

The spontaneous transfer of long-chain C_{16}-ceramide between phospholipid vesicles is very slow, taking in the order of days (Simon et al. 1999). Analysis with a cell-free assay system of the inter-membrane transfer of ceramide (Fig. 1a) revealed that CERT greatly facilitates the transfer of ceramide, and that the START domain is necessary and sufficient for catalyzing the transfer of ceramide between artificial phospholipid membranes (Fig. 1c) (Hanada et al. 2003).

An analysis of substrate specificity showed that CERT can not transfer sphingosine, SM, PC, or cholesterol (Kumagai et al. 2005). The activity to transfer diacylglycerol, which structurally resembles ceramide, was 5-10% of the activity toward ceramide (Kumagai et al. 2005). Among four stereoisomers of C_{16}-ceramide, CERT specifically recognized the natural D-erythro isomer (Kumagai et al. 2005). CERT efficiently transferred ceramides having C_{14}, C_{16}, C_{18}, and C_{20} chains, but not longer acyl chains, and also mediated the efficient transfer of C_{16}-dihydroceramide and C_{16}-phytoceramide (Kumagai et al. 2005). Binding assays showed that CERT also recognizes short chain fluorescent analogs of ceramide with a stoichiometry of 1:1 (Kumagai et al. 2005). These results indicate that the START domain of CERT catalyzes the intermembrane transfer of various types of natural ceramides and their derivatives.

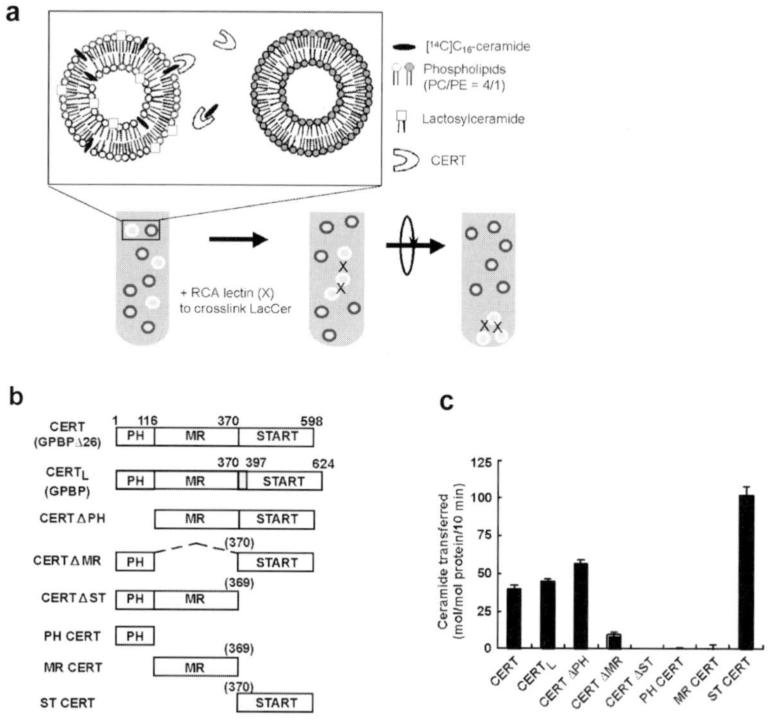

Fig. 1. CERT is a ceramide transfer protein (Hanada et al. 2003; Kumagai et al. 2005). **a** A schematic model of the cell-free assay system for inter-membrane transfer of long-chain ceramide. **b** Domain structure of CERT, its splicing variant $CERT_L$, and deletion mutant constructs. **c** Ceramide transfer assay with various CERT-related constructs.

5.2 PH domain of CERT

The PH domain of the wild-type CERT specifically bound to phosphatidy-linositol-4-monophosphate (PI4P) among various phosphoinositides in protein-lipid overlay assays (Hanada et al. 2003). PI4P-specific PH domains have a Golgi-targeting function, presumably because PI4P is preferentially distributed to the Golgi apparatus in mammalian cells (Balla et al. 2005; Godi et al. 2004; Levine and Munro 2002; Wang et al. 2003; Weixel et al. 2005). Analysis of CERT cDNA isolated from LY-A cells and wild-type CHO cells revealed that the LY-A-derived CERT cDNA has a sole mutation, which changes the 67th amino acid residue from glycine to glutamic acid (this mutation is referred to as G67E) (Hanada et al. 2003). The Gly67, which is located in the PH domain, is conserved among CERT homologs of various multicellular organisms (Hanada et al. 2003). G67E mutation destroyed the PI4P-binding activity of the PH domain. When the intracellular distribution of CERT was examined by a GFP fusion technique, it was found that wild-type CERT is distributed throughout the cytosol, but concentrated in the Golgi region, and that the Golgi association is impaired by the G67E mutation (Hanada et al. 2003). Therefore, CERT efficiently associates with the Golgi apparatus, dependent on its PH domain recognizing PI4P.

5.3 FFAT motif of CERT

Vesicle-associated membrane protein (VAMP)-Associated Protein (VAP) is an ER-resident type II membrane protein. Wyles et al initially showed that oxysterol-binding protein (OSBP) interacts with VAP (Wyles et al. 2002), and, shortly after, Loewen et al showed that conserved short peptide motifs present in OSBP and its yeast relatives are crucial for the interaction of OSBPs and VAP (Loewen et al. 2003). Based on the conserved sequence (EFFDAxE), the motifs are referred to as FFAT motifs (two phenylalanines in an acidic tract) (Loewen et al. 2003). Interestingly, an FFAT motif is present in the middle region of CERT (Fig. 4). We observed that VAP was indeed co-immunoprecipitated with CERT, and that mutations in the FFAT motif of CERT destroyed not only the VAP-CERT interaction but also the CERT-mediated ER-to-Golgi transport of ceramide in cells (M. Kawano, K. Kumagai, M. Nishijima, and K. Hanada, manuscript in preparation).

5.4 CERT-dependent trafficking of ceramide within semi-intact cells

To further examine CERT-dependent trafficking of ceramide, we analyzed ER-to-Golgi transport of ceramide with the semi-intact cell system. The transport of ceramide from the ER to the site where SM is produced in LY-A cells was restored to the wild-type level by adding purified recombinant CERT to the LY-A cytosol (Fig. 2b) (Hanada et al. 2003). Importantly, neither the PH domain-deleted nor G67E mutant CERT rescued the ceramide-to-SM conversion in semi-intact cells (Fig. 2b), although the PH domain was not required for the transfer of ceramide between artificial phospholipid membranes (Fig. 1c) (Hanada et al. 2003). We interpret these results to mean that the PH domain of CERT is crucial for selective targeting to the Golgi when various other organelles coexist like in intact cells.

Sar1[T39N], a dominant negative form of Sar1 protein, inhibits the vesicle-mediated transport of proteins from the ER to the Golgi apparatus (Kuge et al. 1994). However, Sar1[T39N] did not affect the conversion of ceramide to SM (Hanada et al. 2003), suggesting that CERT-dependent trafficking of ceramide is non-vesicular.

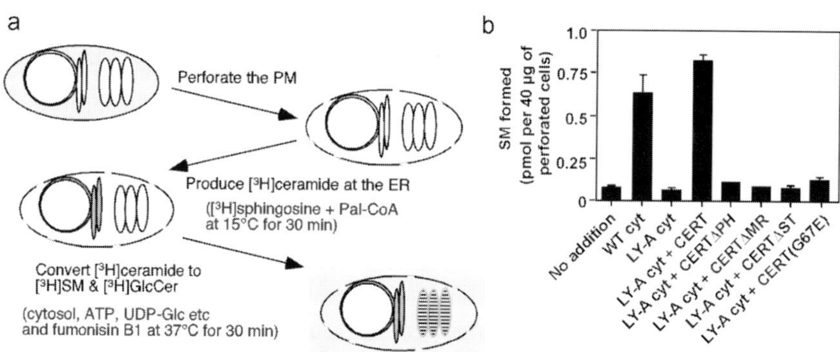

Fig. 2. CERT-mediated trafficking of ceramide within semi-intact cells. a The procedure for the reconstitution of ceramide trafficking is represented schematically (Funakoshi et al. 2000). **b** Radioactive ceramide was formed in perforated LY-A cells (Hanada et al. 2003). The prelabelled semi-intact cells (40 μg) were then incubated in a chase mixture with or without the cytosol (100 μg) of the indicated cell lines and various CERT constructs (10 ng). cyt, cytosol.

6. CERT inhibitor

When examining the effects of various chemically synthesized analogs of ceramides on the metabolic labeling of sphingolipids with [^{14}C]serine in intact CHO cells, we found that a novel compound, N-(3-hydroxy-1-hydroxymethyl-3-phenylpropyl)dodecamide (HPA-12), inhibits de novo SM synthesis severely (Yasuda et al. 2001). It turned out that HPA-12 is a selective, if not specific, inhibitor of ER-to-Golgi transport of ceramide. First, HPA-12 did not affect the enzyme activities of SM synthase and SMases or inhibit the de novo synthesis of PC (the phosphocholine donor for SM synthesis) (Yasuda et al. 2001). Second, ER-to-Golgi transport of C$_5$-DMB-ceramide was inhibited by HPA-12 (Yasuda et al. 2001). In addition, BFA rescued SM synthesis in HPA-12-treated cells. An analysis of the structure-activity relationship showed that only the (1R, 3R) isomer among four stereochemical isomers of HPA-12 is active as an inhibitor (Yasuda et al. 2001) (Fig. 3), and that both the two hydroxyl groups and also middle chain lengths (C11-15) are essential for the bioactivity of HPA compounds (Nakamura et al. 2003). (1R, 3R) HPA-12, but not inactive isomers, inhibited the CERT-mediated inter-membrane transfer of ceramide in a cell-free assay system (Kumagai et al. 2005). (1R, 3R) HPA-12 inhibited de novo SM synthesis in wild-type CHO cells to the level in untreated LY-A cells, while the synthesis in LY-A cells was not affected by the drug (Yasuda et al. 2001). In LY-A cells transfected with wild-type CERT cDNA, (1R, 3R) HPA-12 again inhibited SM synthesis to the level in untreated LY-A cells (Kumagai et al. 2005). Collectively, we concluded that (1R, 3R) HPA-12 is a direct antagonist of CERT.

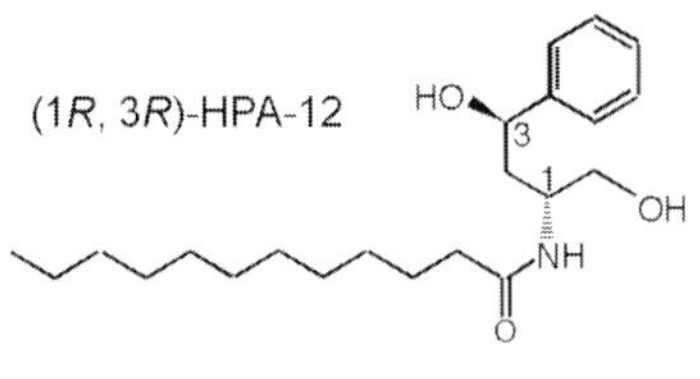

Fig. 3. Structure of a CERT inhibitor.

7. Model of molecular mechanisms of ceramide trafficking

The results described above led us to a non-vesicular model for ER-to-Golgi trafficking of ceramide (Hanada et al. 2003)(Fig. 4). In this model, CERT extracts newly synthesized ceramide from the ER, depending on its START domain, carries the ceramide molecule through the cytosol in a non-vesicular manner, and targets the Golgi apparatus, dependent on its PH domain recognizing PI4P. After the release of ceramide at the Golgi apparatus, CERT might in turn bind diacylglycerol generated during SM synthesis, because CERT can poorly but significantly catalyze the inter-membrane transfer of diacylglycerol in vitro (Kumagai et al. 2005).

Subdomains of the ER are suggested to be spatially very close (~10 nm) to trans Golgi stacks (Levine 2004; Mogelsvang et al. 2004). CERT might quickly shuttle between the two organelle membranes at the sites of contact. If the PH domain and the FFAT motif of CERT can simultaneously associate with the Golgi apparatus and the ER, respectively, the extraction and transfer of ceramide from the ER to the Golgi apparatus might be attained by the 'neck-swinging' movement of the START domain (Hanada et al. 2003; Munro 2003) (Fig. 4, insert).

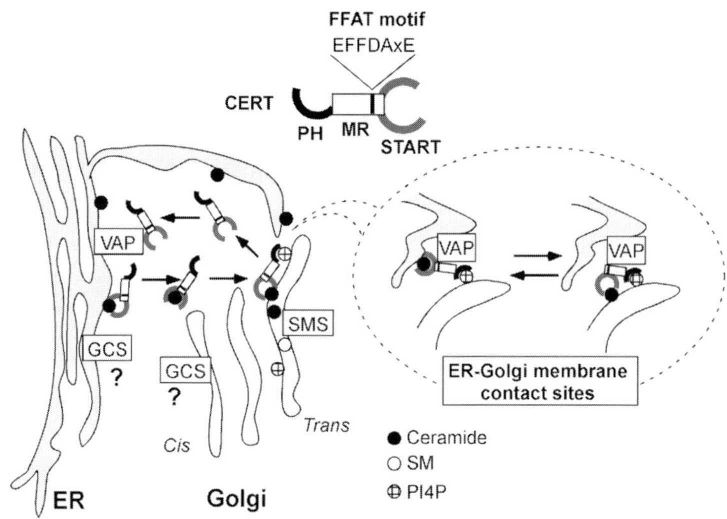

Fig. 4. Molecular extraction and transfer model for CERT-mediated trafficking of ceramide from the ER to the Golgi apparatus (Hanada et al. 2003). CERT-mediated transfer of ceramide from the ER to Golgi apparatus might efficiently occur at the ER-Golgi membrane contact sites by short-distance shuttle. *Insert*, a hypothetical 'neck-swinging' model for CERT-mediated transfer of ceramide at the contact sites.

We have so far failed to reproduce phenotypic differences between LY-A and wild-type cells in a cell-free system with isolated ER and Golgi membranes (Funakoshi, T., Miyuki K., and Hanada K., unpublished results). This might reflect that specific membrane structures such as membrane contact sites play an important role in ceramide trafficking, and that such structures are maintained in semi-intact cells, but not in isolated membranes. Further studies are necessary to elucidate more mechanisms underlying CERT-mediated ceramide trafficking and its regulation.

8. CERT and Goodpasture antigen-binding protein (GPBP)

CERT is identical to GPBPΔ26, a splicing isoform of GPBP (Fig. 1b). Goodpasture disease is an autoimmune disorder described only in humans. In Goodpasture patients, autoantibodies against the non-collagenous carboxyl terminus (NC1) domain of collagen IV α3 cause a rapidly progressive glomerulonephritis. Although the NC1 domain is highly conserved among species, the amino terminal region of the domain is divergent, raising the hypothesis that the divergent region is relevant to the human-specific autoantigenecity of the NC1 domain. Raya et al. screened for human recombinant proteins able to bind the amino terminal peptide of the NC1 domain using the phage display method, and found a 70-kD protein, which they named GPBP (Raya et al. 1999). Raya et al. later found that GPBPΔ26, which has 26 fewer amino acid residues than GPBP, is a more common isoform and is expressed widely in various tissues (Raya et al. 2000). However, both GPBPs were found to be hydrophilic cytoplasmic proteins (Raya et al. 1999; Raya et al. 2000), which conflicts with their proposed role in the interaction with extracellular collagen. As described above, the primary physiological function of CERT/GPBPΔ26 (and CERT$_L$/GPBP) is most likely to mediate intracellular trafficking of ceramide. Therefore, at the present time, we believe it is appropriate to refer to these proteins as CERT and CERT$_L$ in place of GPBPΔ26 and GPBP, respectively, although these proteins might be somehow relevant to Goodpasture autoimmune disease.

Note that, based on the priority of an earlier nomination, COL4A3BP (standing for collagen, type IV, alpha 3-binding protein) is the official nomenclature for the mammalian genes encoding CERTs/GPBPs. In the human genome, the COL4A3BP gene consists of 17 exons encompassing ~130 kbp in the chromosomal 5q13.3 region.

Acknowledgements. This work was supported by Grants-in-aid for Scientific Research on Priority Areas of MEXT of Japan (grant no.12140206 to KH) and for Research on Health Sciences focusing on Drug Innovation of the Japan Health Sciences Foundation (grant no. KH21009 to KH), and by the JSPH Research Fellowship (grant no.16-3830 to MK).

9. References

Ardail D, Lerme F, Louisot P (1991) Involvement of contact sites in phosphatidylserine import into liver mitochondria. *J Biol Chem*, **266**, 7978-7881.

Ardail D, Popa I, Bodennec J, Louisot P, Schmitt D, Portoukalian J (2003) The mitochondria-associated endoplasmic-reticulum subcompartment (MAM fraction) of rat liver contains highly active sphingolipid-specific glycosyltransferases. *Biochem J*, **371**, 1013-1019.

Balla A, Tuymetova G, Tsiomenko A, Varnai P, Balla T (2005) A plasma membrane pool of phosphatidylinositol 4-phosphate is generated by phosphatidylinositol 4-kinase type-III alpha: studies with the PH domains of the oxysterol binding protein and FAPP1. *Mol Biol Cell*, **16**, 1282-1295.

Beckers CJM, Keller DS, Balch WE (1987) Semi-intact cells permeable to macromolecules: use in reconstitution of protein transport from the endoplasmic reticulum to the Golgi complex. *Cell*, **50**, 523-534.

Collins RN, Warren G (1992) Sphingolipid transport in mitotic HeLa cells. *J Biol Chem*, **267**, 24906-24911.

Dowler S, Currie RA, Campbell DG, Deak M, Kular G, Downes CP, Alessi DR (2000) Identification of pleckstrin-homology-domain-containing proteins with novel phosphoinositide-binding specificities. *Biochem J*, **351**, 19-31.

Fukasawa M, Nishijima M, Hanada K (1999) Genetic evidence for ATP-dependent endoplasmic reticulum-to-Golgi apparatus trafficking of ceramide for sphingomyelin synthesis in Chinese hamster ovary cells. *J Cell Biol*, **144**, 673-685.

Fukasawa M, Nishijima M, Itabe H, Takano T, Hanada K (2000) Reduction of sphingomyelin level without accumulation of ceramide in Chinese hamster ovary cells affects detergent-resistant membrane domains and enhances cellular cholesterol efflux to methyl-beta -cyclodextrin. *J Biol Chem*, **275**, 34028-34034.

Funakoshi T, Yasuda S, Fukasawa M, Nishijima M, Hanada K (2000) Reconstitution of ATP- and cytosol-dependent transport of de novo synthesized ceramide to the site of sphingomyelin synthesis in semi-intact cells. *J Biol Chem*, **275**, 29938-29945.

Funato K, Riezman H (2001) Vesicular and nonvesicular transport of ceramide from ER to the Golgi apparatus in yeast. *J Cell Biol*, **155**, 949-959.

Futerman AH, Pagano RE (1991) Determination of the intracellular sites and topology of glucosylceramide synthesis in rat liver. *Biochem J*, **280**, 295-302.

Futerman AH, Stieger B, Hubbard AL, Pagano RE (1990) Sphingomyelin synthesis in rat liver occurs predominantly at the cis and medial cisternae of the Golgi apparatus. *J Biol Chem*, **265**, 8650-8657.

Godi A, Di Campli A, Konstantakopoulos A, Di Tullio G, Alessi DR, Kular GS, Daniele T, Marra P, Lucocq JM, De Matteis MA (2004) FAPPs control Golgi-to-cell-surface membrane traffic by binding to ARF and PtdIns(4)P. *Nat Cell Biol*, **6**, 393-404.

Hanada K (2003) Serine palmitoyltransferase, a key enzyme of sphingolipid metabolism. *Biochim Biophys Acta*, **1632**, 16-30.

Hanada K, Hara T, Fukasawa M, Yamaji A, Umeda M, Nishijima M (1998) Mammalian cell mutants resistant to a sphingomyelin-directed cytolysin. Genetic and biochemical evidence for complex formation of the LCB1 protein with the LCB2 protein for serine palmitoyltransferase. *J Biol Chem*, **273**, 33787-33794.

Hanada K, Kumagai K, Yasuda S, Miura Y, Kawano M, Fukasawa M, Nishijima M (2003) Molecular machinery for non-vesicular trafficking of ceramide. *Nature*, **426**, 803-809.

Huitema K, van den Dikkenberg J, Brouwers JF, Holthuis JC (2004) Identification of a family of animal sphingomyelin synthases. *EMBO J*, **23**, 33-44.

Jeckel D, Karrenbauer A, Burger KNJ, van Meer G, Wieland F (1992) Glucosylceramide is synthesized at the cytosolic surface of various Golgi subfractions. *J Cell Biol*, **17**, 259-267.

Kaplan MR, Simoni RD (1985a) Intracellular transport of phosphatidylcholine to the plasma membrane. *J Cell Biol*, **101**, 441-445.

Kaplan MR, Simoni RD (1985b) Transport of cholesterol from the endoplasmic reticulum to the plasma membrane. *J Cell Biol*, **101**, 446-453.

Kitamura T, Koshino Y, Shibata F, Oki T, Nakajima H, Nosaka T, Kumagai H (2003) Retrovirus-mediated gene transfer and expression cloning: powerful tools in functional genomics. *Exp Hematol*, **31**, 1007-1014.

Kohyama-Koganeya A, Sasamura T, Oshima E, Suzuki E, Nishihara S, Ueda R, Hirabayashi Y (2004) Drosophila glucosylceramide synthase: a negative regulator of cell death mediated by proapoptotic factors. *J Biol. Chem*, **279**, 35995-36002.

Kok JW, Babia T, Klappe K, Egea G, Hoekstra D (1998) Ceramide transport from endoplasmic reticulum to Golgi apparatus is not vesicle-mediated. *Biochem J*, **333**, 779-786.

Kuge O, Dascher C, Orci L, Rowe T, Amherdt M, Plutner H, Ravazzola M, Tanigawa G, Rothman JE, Balch WE (1994) Sar1 promotes vesicle budding from the endoplasmic reticulum but not Golgi compartments. *J Cell Biol*, **125**, 51-65.

Kumagai K, Yasuda S, Okemoto K, Nishijima M, Kobayashi S, Hanada K (2005) CERT mediates intermembrane transfer of various molecular species of ceramides. *J Biol Chem*, **280**, 6488-6495.

Lee MC, Miller EA, Goldberg J, Orci L, Schekman R (2004) Bi-Directional Protein Transport between the ER and Golgi. *Annu Rev Cell Dev Biol*, **20**, 87-123.

Levine T (2004) Short-range intracellular trafficking of small molecules across endoplasmic reticulum junctions. *Trends Cell Biol*, **14**, 483-490.

Levine TP, Munro S (2002) Targeting of Golgi-specific pleckstrin homology domains involves both PtdIns 4-kinase-dependent and -independent components. *Curr Biol*, **12**, 695-704.

Loewen CJ, Roy A, Levine TP (2003) A conserved ER targeting motif in three families of lipid binding proteins and in Opi1p binds VAP. *EMBO J*, **22**, 2025-2035.

Mandon EC, Ehses I, Rother J, van Echten G, Sandhoff K (1992) Subcellular localization and membrane topology of serine palmitoyltransferase, 3-dehydrosphinganine reductase, and sphinganine N-acyltransferase in mouse liver. *J Biol Chem*, **267**, 11144-11148.

Merrill AH, Jr. (2002) De novo sphingolipid biosynthesis: A necessary, but dangerous, pathway. *J Biol Chem*, **277**, 25843-25846.

Mogelsvang S, Marsh BJ, Ladinsky MS, Howell KE (2004) Predicting function from structure: 3D structure studies of the mammalian Golgi complex. *Traffic*, **5**, 338-345.

Munro S (2003) Cell biology: earthworms and lipid couriers. *Nature*, **426**, 775-776.

Nakamura Y, Matsubara R, Kitagawa H, Kobayashi S, Kumagai K, Yasuda S, Hanada K (2003) Stereoselective synthesis and structure-activity relationship of novel ceramide trafficking inhibitors, (1R,3R)-N-(3-hydroxy-1-hydroxymethyl-3-phenylpropyl)dodecanamide (HPA-12) and its analogues. *J Med Chem*, **46**, 3688-3695.

Palmer KJ, Stephens DJ (2004) Biogenesis of ER-to-Golgi transport carriers: complex roles of COPII in ER export. *Trends Cell Biol*, **14**, 57-61.

Ponting CP, Aravind L (1999) START: a lipid-binding domain in StAR, HD-ZIP and signalling proteins. *Trends Biochem Sci*, **24**, 130-132.

Raya A, Revert F, Navarro S, Saus J (1999) Characterization of a novel type of serine/threonine kinase that specifically phosphorylates the human goodpasture antigen. *J Biol Chem*, **274**, 12642-12649.

Raya A, Revert-Ros F, Martinez-Martinez P, Navarro S, Rosello E, Vieites B, Granero F, Forteza J, Saus J (2000) Goodpasture antigen-binding protein, the kinase that phosphorylates the goodpasture antigen, is an alternatively spliced variant implicated in autoimmune pathogenesis. *J Biol Chem*, **275**, 40392-40399.

Sandhoff K, Kolter T (2003) Biosynthesis and degradation of mammalian glycosphingolipids. *Philos Trans R Soc Lond B Biol Sci*, **358**, 847-861.

Siess DC, Kozak SL, Kabat D (1996) Exceptional fusogenicity of Chinese hamster ovary cells with murine retroviruses suggests roles for cellular factor(s) and receptor clusters in the membrane fusion process. *J Virol*, **70**, 3432-3439.

Simon CG, Jr., Holloway PW, Gear AR (1999) Exchange of C(16)-ceramide between phospholipid vesicles. *Biochemistry*, **38**, 14676-14682.

Smart EJ, De Rose RA, Farber SA (2004) Annexin 2-caveolin 1 complex is a target of ezetimibe and regulates intestinal cholesterol transport. *Proc Natl Acad Sci U S A*, **101**, 3450-3455.

Uittenbogaard A, Ying Y, Smart EJ (1998) Characterization of a cytosolic heat-shock protein-caveolin chaperone complex. Involvement in cholesterol trafficking. *J Biol Chem*, **273**, 6525-6532.

Vance JE (1991) Newly made phosphatidylserine and phosphatidylethanolamine are preferentially translocated between rat liver mitochondria and endoplasmic reticulum. *J Biol Chem*, **266**, 89-97.

Viani P, Giussani P, Brioschi L, Bassi R, Anelli V, Tettamanti G, Riboni L (2003) Ceramide in Nitric Oxide Inhibition of Glioma Cell Growth. EVIDENCE FOR THE INVOLVEMENT OF CERAMIDE TRAFFIC. *J Biol Chem*, **278**, 9592-9601.

Voelker DR (1990) Characterization of phosphatidylserine synthesis and translocation in permeabilized animal cells. *J Biol Chem*, **265**, 14340-14346.

Wang E, Norred WP, Bacon CW, Riley RT, Merrill AHJ (1991) Inhibition of sphingolipid biosynthesis by fumonisins: Implications for diseases associated with Fusarium moniliforme. *J Biol Chem*, **266**, 14486-14490.

Wang YJ, Wang J, Sun HQ, Martinez M, Sun YX, Macia E, Kirchhausen T, Albanesi JP, Roth MG, Yin HL (2003) Phosphatidylinositol 4 phosphate regulates targeting of clathrin adaptor AP-1 complexes to the Golgi. *Cell*, **114**, 299-310.

Weixel KM, Blumental-Perry A, Watkins SC, Aridor M, Weisz OA (2005) Distinct golgi populations of phosphatidylinositol 4-phosphate regulated by phosphatidylinositol 4-kinases. *J Biol Chem*, **280**, 10501-10508.

Wyles JP, McMaster CR, Ridgway ND (2002) Vesicle-associated membrane protein-associated protein-A (VAP-A) interacts with the oxysterol-binding protein to modify export from the endoplasmic reticulum. *J Biol Chem*, **277**, 29908-29918.

Yamaji A, Sekizawa Y, Emoto K, Sakuraba H, Inoue K, Kobayashi H, Umeda M (1998) Lysenin, a novel sphingomyelin-specific binding protein. *J Biol Chem*, **273**, 5300-5306.

Yamaoka S, Miyaji M, Kitano T, Umehara H, Okazaki T (2004) Expression cloning of a human cDNA restoring sphingomyelin synthesis and cell growth in sphingomyelin synthase-defective lymphoid cells. *J Biol Chem*, **279**, 18688-18693.

Yasuda S, Kitagawa H, Ueno M, Ishitani H, Fukasawa M, Nishijima M, Kobayashi S, Hanada K (2001) A novel inhibitor of ceramide trafficking from the endoplasmic reticulum to the site of sphingomyelin synthesis. *J Biol Chem*, **276**, 43994-44002.

2-8 Sphingolipid Trafficking

Kouichi Funato[1] and Howard Riezman[2]

[1]Graduate School of Biosphere Science, Faculty of Applied Biological Science, Hiroshima University, 1-4-4 Kagamiyama, Higashi-Hiroshima 739-8528, Japan, [2]Department of Biochemistry, University of Geneva, Sciences II, 30, quai E. Ansermet, CH-1211 Geneva 4, Switzerland

Summary. Sphingolipid synthesis and its localization in different organelles of eukaryotic cells are critical for the biological functions and are highly regulated by selective intracellular trafficking. Transport and sorting of sphingolipids between organelles occurs via vesicular and/or non-vesicular transport mechanisms. Some sphingolipids can also move between the two leaflets of the membrane lipid bilayer and their lateral mobility in the plane of the membrane is probably also be regulated by complex formation with other lipids. This review describes what is known about the pathways and mechanisms of intracellular sphingolipid trafficking.

Keywords. sphingolipid, transport and sorting, translocation, lateral segregation, organelle

1. Introduction

Sphingolipids were so named for their apparent enigmatic, "sphinx-like" nature. Although still an elusive class of lipids, they have recently emerged as bioactive molecules in addition to being critical structural components of cellular membranes. Sphingoid bases, ceramides, and other intermediates of sphingolipid metabolism act as signaling molecules in mediating cell growth, stress responses, and apoptosis. In addition, it has been postulated that sphingolipids can provide a molecular platform for clustering of signal transducing molecules by triggering the formation of lateral lipid assemblies, termed microdomains or rafts. Sphingolipid-enriched microdomains have also been proposed to be involved in protein sorting and transport in both polarized and nonpolarized cells.

The membranes of different intracellular organelles have different protein and lipid compositions that are critical for their proper functions, and therefore, are tightly regulated and controlled by a combination of processes including synthesis, degradation, transport and sorting of proteins and lipids. Over the past few decades numerous studies have brought us detailed knowledge concerning the mechanisms responsible for protein trafficking. However, our insight into how lipids are transported and sorted within cells is far more limited. Lipids are synthesized mainly in the endoplasmic reticulum (ER) of eukaryotic cells and are subsequently transported to their final destination. The transport must occur selectively by specific mechanisms for each lipid, allowing membranes to maintain distinct lipid compositions. Generally, transport of lipids between organelles is thought to occur by one or more of the following possible mechanisms; vesicular transport as for proteins, with budding of vesicles from a donor membrane and targeting to and fusion with the acceptor membranes, transfer of lipid monomers through the cytosol, either via spontaneous diffusion or protein-mediated process, transfer of lipids at specific regions of close apposition between organelles. Lipids can also move between the two leaflets of the membrane bilayer, so-call "flip-flop", and move laterally in the plane of a leaflet, a process termed "lateral diffusion". This chapter will focus on the intracellular trafficking of sphingolipids, and discuss the current understanding of sphingolipid transport and sorting between organelles, as well as their movement within two leaflets of a membrane.

2. Movement of sphingolipids between the two leaflets of ER membrane

The ER is the site for the first steps in sphingolipid biosynthesis that are the condensation of L-serine and palmitoyl-CoA to 3-ketosphinganine (KDS) and its reduction to sphinganine (DHS), followed by further synthesis of dihydroceramide (DH-Cer) in all eukaryotic cells (see the chapter by Hanada and Merrill; the chapter by Futerman). KDS and DHS synthesis occurs on the cytosolic side of the ER membrane (Yasuda et al. 2003; Han et al. 2004; Kihara and Igarashi 2004), but the topology of synthesis of DH-Cer in the ER unclear. Since in mammals, ceramide is utilized for the synthesis of galactosylceramide (GalCer) on the lumenal side of the ER (Sprong et al. 1998), DH-Cer synthesized on the cytosolic surface of the

ER would have to be translocated toward the lumenal side. This may also be the case in yeast, because ceramide is used for the remodeling of glycosylphosphatidylinositol (GPI) lipid moieties, which occurs in the lumen of the ER (Reggiori et al. 1997). Alternatively, if synthesized in the lumen of the ER, DH-Cer would have to be translocated to the cytosolic surface of the ER, where it can be desaturated at C-4,5 by DES1 (see the chapter by Suzuki; Michel and van Echten-Deckert 1997) and probably hydroxylated at C-4 by DES2 (Sur2p in yeast) to generate ceramide and phytoceramide (PH-Cer) (Idkowiak-Baldys et al. 2003), respectively. The external face of the ER is also where ceramide becomes available for delivery to other organelles via non-vesicular transport (see below). Translocation across the ER membrane may not be limited to DH-Cer. In yeast, exogenously added DHS needs to be phosphorylated by Lcb4p and dephosphorylated by Lcb3p before it can be utilized for ceramide synthesis (Funato et al. 2003). The active site of Lcb3p has been proposed to be in the lumen of the ER (Kihara et al. 2003) and therefore, DHS, the precursor of DH-Cer may be generated on the lumenal side of the ER when exogenous sphingoid bases are used for ceramide synthesis. This also implies that DHS-1-phosphate (DHS-1P), which is generated in the cytoplasm, is translocated across the ER membrane.

Recently, Lip1p was identified as an essential component of ceramide synthase (see the chapters by Futerman; Vallee and Riezman 2005). It forms a heteromeric complex with Lac1p and Lag1p, and the functional region of Lip1p is localized in the lumen of the ER or in its transmembrane domain. These findings are consistent with the second model that DH-Cer is synthesized on the lumenal side of the ER, but structural analysis on the polytopic Lag1p and Lac1p membrane proteins will be required to resolve the topology issue.

Recently, an inositol phosphorylceramide phospholipase C, Isc1p, that hydrolyzes yeast inositol sphingolipids into ceramide and the corresponding head group was shown to be predominantly localized in the ER during early growth, whereas it became associated with mitochondria in the late logarithmic growth (Vaena de Avalos et al. 2004), suggesting that its substrates are present in both the ER and mitochondria. As all inositol sphingolipids appear to be synthesized in the lumen of the Golgi in yeast (Levine et al. 2000; Funato et al. 2002; Lisman et al. 2004), which is not connected directly to the mitochondria by vesicular pathways, these sphingolipids might be transported to the mitochondria via a non-vesicular mechanism. A membrane fraction of the ER closely associated to, but not fused with the outer mitochondrial membrane, termed MAM (mitochondria-associated membrane) is thought to be involved in non-vesicular transport of aminophospholipids and other small molecules to mitochondria (Levine 2004). Therefore, if inositol sphingolipids are recycled back to the ER from the Golgi compartment via retrograde vesicular transport, then translocation might occur from the lumenal toward the cytosolic surface of the ER membrane where the sphingolipids could be targeted to the outer mitochondrial membrane. In mammals, neutral sphingomyelinase 1, which cleaves sphingomyelin (SM) to ceramide, has been found in the nuclear matrix (Mizutani et al. 2001). Since SM is synthesized at the lumenal side of the Golgi (Huitema et al. 2004), similar mechanisms could exist to deliver it to the inner leaflet of nuclear envelope membrane.

A sialidase NEU4 is also found in the mitochondria (Yamaguchi et al. 2005). NEU4 hydrolyzes sphingolipids having sialic acid residues, called "gangliosides", and this event could occur in the mitochondria. Indeed, ganglioside GD1b was detected in the mitochondria of malignant hepatoma (Dyatlovitskaya et al. 1976), and GD3 was also found in the mitochondria in response to various apoptotic inducers (Rippo et al. 2000). Interestingly, studies in hepatocytes support the involvement of endosomal vesicle movement in the targeting of GD3 to mitochondria (Garcia-Ruiz et al. 2002). However, the exact route for targeting of GD3 to mitochondria remains unknown (Morales et al. 2004). GD3 is synthesized from the precursor GM3 within the lumen of the Golgi, and resides in the outer leaflet of the plasma membrane (PM) and in the inner leaflet of the endocytic organelles via vesicular transport along the exocytic and endocytic pathways. Thus, the possibility exists that GD3 is translocated across the ER membrane or other membranes like endosomes and Golgi to have access to the cytosolic surface and then targeted to mitochondria. Moreover, a human neutral ceramidase has been found in mitochondria (El Bawab et al. 2000). Likewise, GalCer synthesized in the ER lumen has been proposed to translocate toward the cytosolic surface (Burger et al. 1996; van Meer and Lisman 2002).

3. Transport of sphingolipids between the ER and the Golgi

For the synthesis of complex sphingolipids, ceramides must travel from the ER to the Golgi apparatus. The mechanisms of how ceramides are transported to the Golgi have been studied in mammalian cells and in yeast, *S cerevisiae* (Fukasawa et al. 1999; Funato and Riezman 2001). One of the most important findings is that in both systems, transport of ceramide from the ER to the Golgi can occur through a protein-mediated, non-vesicular pathway. Using a mammalian mutant cell line, LY-A, Hanada and colleagues have identified a protein, CERT that extracts ceramide from the ER and delivers it specifically to the Golgi for conversion into SM (see the chapter by Hanada; Hanada et al. 2003). Furthermore, the fact that SM synthase 1, SMS1 is located in the trans-Golgi compartment (Huitema et al. 2004), which has been observed to be closely associated with the ER has led to the hypothesis that CERT transfers ceramide directly to the trans-Golgi via ER-Golgi membrane contacts (Munro 2003; Riezman and van Meer 2004). In the yeast system, a non-vesicular mechanism also delivers ceramide to the medial-Golgi where it is converted to inositol phosphorylceramide (IPC) (Funato and Riezman 2001). Reconstitution of this pathway in a cell-free system showed that a cytosolic protein is required for this transport. At present, the yeast orthologue of mammalian CERT has not been identified, but the mechanisms analogous to that assumed for CERT-mediated ceramide transport should exist in yeast. Indeed, analysis using an in vitro assay provided a biochemical evidence for an important role of membrane contacts between the ER and the Golgi in the non-vesicular transport of ceramide (Funato and Riezman 2001). As both SM and IPC synthesis occur in the lumen of the Golgi compartments, ceramide delivered to the cytosolic side of the Golgi must be translocated toward the lumenal side once it reaches the Golgi.

In addition to non-vesicular transport, yeast genetic studies have demonstrated the participation of COPII components in ceramide transport to the Golgi site of IPC synthesis (Funato and Riezman 2001), suggesting that some of the ceramide enters into ER-derived COPII coated vesicles and reaches the cis-Golgi compartment either via the cytosolic side or lumenal side of the transport vesicles. Although the mechanism of how ceramide is delivered to the Golgi via vesicular transport remains to be investigated, there may be a selective transport process for ceramides. In yeast, ongoing sphingolipid synthesis has been shown to be specifically required for the ER to Golgi transport of GPI-anchored proteins (Horvath et al. 1994; Watanabe et al. 2002). Further studies demonstrated that GPI-anchored proteins are sorted into distinct vesicles from other secretory proteins upon ER exit (Muniz et al. 2001; Morsomme et al. 2003). Because of these findings and because of the physical properties of GPI-anchored proteins that tend to be associated with sphingolipids rather than glycerolipids (Muniz and Riezman 2000), it is speculated that the two types of vesicles may have different lipid compositions. Therefore, it would be interesting to see if ceramide is co-sorted into GPI-anchored protein containing vesicles.

In most eukaryotic cells (but not *S cerevisiae*) and a few bacteria, ceramide is converted to glucosylceramide (GlcCer) by a UDP-Glc:ceramide glucosyltransferase (GlcCer synthase), which has been identified in many organisms (see the chapter by Shayman). Because the enzymic activity is located mainly at the cytosolic surface of the Golgi compartments (Jeckel et al. 1992), and the conversion of ceramide to GlcCer is most likely energy-independent (Fukasawa et al. 1999), ceramide used for GlcCer synthesis is thought to be transported from the ER to the Golgi by a nonvesicular mechanism (van Meer and Lisman 2002). However, a recent study demonstrated that GlcCer synthase localizes not only to the Golgi, but also to the ER (Kohyama-Koganeya et al. 2004), suggesting that GlcCer can be synthesized in the ER. In this case, ceramide transport is not necessary for GlcCer synthesis. Instead, like GalCer, GlcCer must be transported from the ER to the Golgi where it can be utilized for higher glycosphinglipid synthesis (see below). The process can occur either via the cytosolic side of transport vesicles or monomeric transport through the cytosol, probably mediated by the glycolipid transfer protein GLTP (Lin et al. 2000; Malinina et al. 2004). Both GlcCer and GalCer are then translocated toward the lumenal side of the Golgi by an energy-independent translocator (Lannert H et al. 1994; Burger et al. 1996; Pomorski et al. 2004).

Retrograde transport from the Golgi back to the ER could occur for some sphingolipids. The fact that yeast Isc1p and mammalian sphingomyelinase 1 are localized in the ER and in the nuclear matrix, respectively, may suggest that their substrates, yeast inositol sphingolipids and SM, are recycled back to the ER from the Golgi. Since these sphingolipids are synthesized in the lumenal side of the Golgi, most likely a vesicular mechanism would be used for the retrograde transport to the ER. However, conflicting observations have been made for SM. SM is enriched at the PM and is found in the ER only in low amounts whereas the Golgi has intermediate levels, implying a dynamic sorting of SM along the secretory pathway. Theoretically, the sorting is achieved by either concentration of SM into anterograde transport vesicles or segregation from retrograde transport vesicles. The latter has been experimentally demonstrated by quantitative lipid analysis of COPI-coated vesicles and their parental Golgi membranes (Brugger et al. 2002), providing evidence for an exclusion of SM from COPI vesicles. This may suggest that SM is not recycled back to the ER from the Golgi via COPI dependent vesicular transport. However, there are other possibilities, including other vesicular pathways, to account for the low amounts of SM in the ER.

The best evidence for retrograde transport of sphingolipids could be that cholera toxin associated with ganglioside GM1 travels from the Golgi to the ER where the toxin can be translocated to the cytosol to induce toxicity and it remains bound to GM1 until arrival in the ER (Fujinaga et al. 2003). Also in the absence of toxin, the ER membrane contains GM1 gangliosides that allow toxin binding. These results suggest that there is an endogenous retrograde transport pathway for GM1 that is hijacked by the toxin. Interestingly, the retrograde transport of the toxin-GM1 complex seems to not require COPI-mediated retrograde traffic but involves GM1-rich microdomains or detergent-resistant membranes (DRMs). A similar pathway has been proposed for Shiga toxin that also can bind a ceramide glycolipid, globoside (Sandvig and van Deurs 2002).

4. Intra-Golgi transport of sphingolipids

GlcCer translocated to the Golgi lumen is converted to lactosylceramide (LacCer) by LacCer synthase, and then LacCer is sialosylated to GM3 and GM3 to GD3, by the action of two sialyltransferases (SialT1 and SialT2) (Tettamanti 2004). These glycosphingolipids are further processed by the addition of N-acetylgalactosamine, galactose and sialic acid to form higher order species; LacCer→GA2→GA1→GM1b, GM3→GM2→GM1→ GD1a, GD3→GD2→GD1b→GT1b. N-acetylgalactosaminyltransferase (GalNAcT), galactosyltransferase (GalT2) and sialyltransferase (SiaT4) catalyze the three reactions. Both LacCer and GalCer can also be sulfated to SLacCer and SGalCer or galactosylated to Gb3Cer and Ga2Cer (van Meer 2001). Although early studies suggested that the enzymes for LacCer, GM3 and GD3 synthesis are localized in early Golgi and those synthesizing other higher glycosphingolipids are in late Golgi (Maccioni et al. 1999), more recent studies revealed that LacCer, GM3 and GD3 synthesis resides in the trans-Golgi, whereas GA2, GM2 and GD2 are synthesized in the trans Golgi network (TGN) (Lannert H et al. 1998; Giraudo et al. 1999). This gradient distribution of the enzymes would implicate that LacCer, GM3 and GD3 are transported from the trans-Golgi to the TGN for conversion to GA2, GM2 and GD2, respectively. Because LacCer synthesis and all the stepwise conversions occur in the Golgi lumen (Trinchera et al. 1991; Lannert H et al. 1994; Lannert H et al. 1998), and brefeldin A (BFA), which blocks vesicular transport from the proximal Golgi (cis, medial, and trans) to the TGN inhibits GA2, GM2 and GD2 synthesis (Giraudo et al. 1999), some vesicle traffic seems to be necessary for transport between trans-Golgi and TGN. Interestingly, two TGN-enzymes, GalNAcT and GalT2 have been shown to form a physical and functional complex (Giraudo et al. 2001). In addition, the three trans-Golgi enzymes, LacCer synthase, SialT1 and SialT2 make a multi-protein complex, but fail to interact with GalNAcT and GalT2 (Giraudo et al. 2003). These findings suggest that glycosphingolipid synthesis is organized in distinct complexes, which are concentrated in different sub-Golgi-compartments, supporting a "channeling model" as proposed previously (Roseman 1970).

Is there a similar system for complex inositol sphingolipid biosynthesis in yeast? Inositol sphingolipid biosynthesis begins with the formation of IPC. IPC is synthesized from ceramide and PI by IPC synthase, Aur1p, in the lumen of the medial-Golgi, and then is mannosylated to yield mannosyl-inositol phosphorylceramide (MIPC), which can be modified by Ipt1p to generate the final sphingolipid, mamnosyl-diinositol phosphorylceramide (M(IP)$_2$C) (see the chapter by Dickson). The conversion of IPC to MIPC has been shown to require Csg1p and Csg2p; Csg1p is predicted to be a mannosyltransferase, whereas the function of Csg2p is not obvious. Recent studies have identified a protein, Csh1p that is functionally homologous to Csg1p. Both Csh1p and Csg1p interact with Csg2p (Uemura et al. 2003). Importantly, further studies showed that Csh1p and Csg1p are co-localized with Aur1p to a medial compartment of the Golgi, and the active site of Csg1p, like Aur1p, is most likely located in the lumen of the Golgi (Lisman et al. 2004), indicating that delivery of IPC to other sub-Golgi compartments is not necessary for MIPC synthesis. Still, it is unclear whether Csh1p and Csg1p can form a multi-protein complex with Aur1p. Therefore, the conversion of IPC to MIPC may require a lateral diffusion of IPC to the site of Csh1p/Csg1p/Csg2p complex. Alternatively, the complex may be localized to IPC-enriched microdomains. Whether or not MIPC is transported from the medial-Golgi to a later Golgi compartment for M(IP)$_2$C synthesis is unknown. Further investigation is needed to understand the traffic of inositol sphingolipids in the Golgi complex, and to define the precise localization of Ipt1p, as well as its topology in the Golgi.

5. Transport of sphingolipids from the Golgi complex to the PM

Newly synthesized complex sphingolipids are believed to be transported by vesicular traffic from the Golgi to the PM. This is supported by the fact that most complex sphingolipids are synthesized in the lumen of the Golgi and are exposed on the outer leaflet of the PM. Indeed, it was shown that delivery of newly synthesized SM to the cell surface is blocked by BFA, while the BFA-treatment did not prevent an efficient conversion of endogenous ceramide to SM (van Helvoort et al. 1997), indicating that SM is transported from the Golgi to the PM via vesicular transport. Since SM is synthesized in the trans-Golgi, it is possible that vesicular transport is only required to deliver SM to the TGN, although it is most likely also required from TGN to the PM. A recent study analyzed the transport of GD3 from the TGN to the PM (Crespo et al. 2004). The data revealed that transport is a BFA-insensitive and temperature-dependent process; the former is consistent with the localization of GD3 in the TGN and the latter suggests the involvement of vesicular transport. Very interestingly, a dominant negative form of Rab11, which prevents the exit of vesicular stomatitis virus glycoprotein from the Golgi complex, does not influence the delivery of GD3 to the cell surface, raising the possibility of an alternative vesicle transport route for GD3 from TGN to PM. Furthermore, yeast studies using the secretory mutants *sec1* and *sec6* demonstrated that inositol sphingolipids do not reach the PM of the mutants under nonpermissive conditions (Hechtberger and Daum 1995). Because Sec1 and Sec6 proteins are involved in post-Golgi vesicle trafficking to PM of secretory proteins (Jahn et al 2003), these observations indicate that yeast inositol sphingolipids are transported from the TGN to the PM via the conventional vesicular pathway of secretory proteins.

Polarized cells have different PM domains, which are divided into an apical and a basolateral domain. Each membrane domain has its specific protein and lipid composition, and therefore, newly synthesized proteins and lipids must be delivered to the correct domain. The process occurs via two routes; a direct transport from the Golgi and an indirect endocytic (transcytotic) pathway (see below). Using a fluorescently labeled ceramide analog, C6-NBD-ceramide, it has been demonstrated that in MDCK and Caco-2 epithelial cells, newly synthesized C6-NBD-GlcCer is preferentially delivered to the apical membrane, which results in an enrichment of the GlcCer in the apical PM, while C6-NBD-SM is about equally distributed over apical and basolateral domains (van Meer et al. 1987; van 't Hof and van Meer 1990), suggesting that the short-chain GlcCer and SM are sorted from each other. The sorting is believed to occur in the lumenal leaflet of TGN by lateral segregation, leading to the packaging into distinct vesicles destined for the apical or the basolateral PM domain, and subsequent direct vesicular transport between the TGN and PMs. However, the main site of lateral segregation may be in the proximal Golgi, where GlcCer and SM are synthesized. Similar sorting from the TGN has been suggested in hepatocyte-derived HepG2 cells (Maier and Hoekstra 2003) and in HT26 human colon adenocarcinoma cells (Babia et al. 1994).

6. Endocytosis and sorting of sphingolipids

While some sphingolipids delivered to the cytosolic leaflet of the PM are translocated toward the exoplasmic leaflet, most sphingolipids in the exoplasmic leaflet or supplied exogenously are taken up into the cell by translocation across the PM or via the membrane flux of endocytosis. Yeast as already mentioned and some mammalian cells (Fukasawa et al. 1999; Chigorno et al. 2005) can incorporate exogenous sphingoid bases, as well as sphingoid base phosphates into sphingolipids, so they must be internalized into the cell. This process might require spontaneous or protein-mediated transbilayer movement, and if not must involve endocytosis. Protein-mediated translocation of sphingolipids at the PM has been reported. A member of the ABC transporter CFTR protein, the human cystic fibrosis transmembrane conductance regulator that stimulates uptake of exogenous sphingosine-1-phosphate and the related sphingoid base phosphate, DHS-1P was identified (Boujaoude et al. 2001). In addition, a novel membrane protein, yeast Rsb1p appears to be involved in sphingoid base translocation from the inner to the outer leaflet of PM (see the chapter by Kihara and Igarashi; Kihara A and Igarashi Y 2002). GlcCer delivered to the cytosolic surface of PM seems to be translocated to the outer leaflet by the multidrug transporter MDR1 P-glycoprotein (see the chapter by van Meer; Pomorski et al. 2004).

Recent studies using fluorescent sphingolipid anologues and cholera toxin (bound to GM1) provide evidence that sphingolipids at the PM are internalized by one or more endocytic mechanisms that would determine the subsequent intracellular trafficking (Marks and Pagano 2002). Fluorescent glycosphingolipid analogues such as LacCer, globoside and endogenous GM1 are internalized from the PM almost exclusively by a clathrin-independent, caveolar-related mechanism, whereas an SM analogue is taken up approximately equally via clathrin-dependent and independent pathways (Puri et al. 2001). The LacCer analogue is transported to the early endosomes and recycled back to the PM via the recycling compartment, and most of the LacCer, which is not recycled, is targeted via late endosome to the Golgi apparatus. This Golgi targeting is diverted to lysosomes in presence of elevated cellular cholesterol (or in storage disease cells where cholesterol homeostasis is perturbed; see the chapter by Pagano). In contrast, the excess cholesterol has no effect on the Golgi targeting of the SM analogue that enters via the clathrin-dependent pathway. The difference led to the simple model that sphingolipids internalized by the two pathways are sorted into different vesicles at the PM and do not intermix before arrival in the Golgi. However, further studies showed that the LacCer internalized via the caveolar-related mechanism merges with the clathrin-dependent pathway in early endosomes (Sharma et al. 2003). This finding offers another model, that in the same endocytic compartment, LacCer and SM are segregated into distinct domains, which have different sensitivity to intracellular cholesterol. The internalized SM analogue has also been shown to be recycled back to the PM (Mayor et al. 1993). Sphingolipids destined for degradation would be sorted to the internal vesicles of multivesicular late endosomes or lysosomes (van Genderen et al. 1991; Mobius et al. 1999).

In yeast, IPC is highly enriched in the Golgi and in the vacuole, whereas MIPC and $M(IP)_2C$ accumulate in the PM (Hechtberger et al. 1994). This different distribution implies the existence of a sorting mechanism for inositol sphingolipids at the level of post-Golgi (Funato et al. 2002). A recent study reported evidence that a dominant mutation of yeast NCR1, the orthologue of the Niemann-Pick C (NP-C) gene 1 defective in the NP-C disease, causes an altered distribution of inositol sphingolipids without changes in sterol metabolism (Malathi et al. 2004). Based on these observations, Npc1 protein was proposed to play a role in sorting of inositol sphingolipids; however, no evidence is provided yet for the precise function of Npc1 protein.

Endocytic sorting of sphingolipids occurs in polarized cells. In initial studies in MDCK cells no sorting was observed in the transcytosis of SM and GlcCer analogues (van Genderen and van Meer 1995), but sorting between these lipids was found in HepG2 cells (van IJzendoorn and Hoekstra 1998; Ait Slimane and Hoekstra 2002). SM and GlcCer analogues internalized from apical PM are delivered to a compartment, referred to as subapical compartment (SAC) where endocytic pathways from apical and basolateral PM merge. From here, SM is transported predominantly to the basolateral PM, whereas GlcCer is recycled back to the apical PM. This sorting probably involves lateral segregation in the inner leaflet of SAC and subsequent packaging into distinct vesicles, because it was shown that different vesicles enriched in either SM or GlcCer are recovered from SAC (Maier and Hoekstra 2003).

7. Perspectives

A number of proteins have now been identified that are involved in the synthesis and metabolism of sphingolipids, and overexpression and deletion of these enzymes or yeast genetics techniques has provided important insights into our understanding of the synthetic and metabolic networks and the cellular functions associated with sphingolipids. In addition, the routes and pathways of sphingolipid transport are becoming more clearly defined with the exact localization and topology of these enzymes. However, the molecular mechanisms underlying the transport events are largely unknown. For example, proteins involved in sphingolipid translocation across the ER and the Golgi membrane have not been identified. Other proteins that are required for establishment of specialized organelle-organelle contact sites and the components of the sorting machinery remain to be identified and characterized. Also the quantitative analysis of sphingolipids not only in organelles and transport vesicles but also in microdomains needs to be determined. Another exciting challenge is to develop sensitive and rapid experimental systems designed to follow the real-time movement of individual sphingolipid molecules between the organelles, between the two leaflets within the same membrane, and within one leaflet. Such investigations will contribute to our global knowledge of intracellular trafficking of sphingolipids, and will provide additional clues to their roles in cellular processes and diseases.

Acknowledgements. Supports from the Naito Foundation and ONO Medical Research Foundation (to KF) and the Swiss National Science Foundation (to HR) are gratefully acknowledged.

8. References

Ait Slimane T, Hoekstra D (2002) Sphingolipid trafficking and protein sorting in epithelial cells. *FEBS Lett*, **529**, 54-59.

Babia T, Kok JW, van der Haar M, Kalicharan R, Hoekstra D (1994) Transport of biosynthetic sphingolipids from Golgi to plasma membrane in HT29 cells: involvement of different carrier vesicle populations. *Eur J Cell Biol*, **63**, 172-181.

Boujaoude LC, Bradshaw-Wilder C, Mao C, Cohn J, Ogretmen B, Hannun YA, Obeid LM (2001) Cystic fibrosis transmembrane regulator regulates uptake of sphingoid base phosphates and lysophosphatidic acid: modulation of cellular activity of sphingosine 1-phosphate. *J Biol Chem*, **276**, 35258-35264.

Brugger B, Sandhoff R, Wegehingel S, Gorgas K, Malsam J, Helms JB, Lehmann WD, Nickel W, Wieland FT (2002) Evidence for segregation of sphingomyelin and cholesterol during formation of COPI-coated vesicles. *J Cell Biol*, **151**, 507-518.

Burger KN, van der Bijl P, van Meer G (1996) Topology of sphingolipid galacto-syltransferases in ER and Golgi: transbilayer movement of monohexosyl sphingolipids is required for higher glycosphingolipid biosynthesis. *J Cell Biol*, **133**, 15-28.

Chigorno V, Giannotta C, Ottico E, Sciannamblo M, Mikulak J, Prinetti A, Sonnino S (2005) Sphingolipid uptake by cultured cells: complex aggregates of cell sphingolipids with serum proteins and lipoproteins are rapidly catabolized. *J Biol Chem*, **280**, 2668-2675.

Crespo PM, Iglesias-Bartolome R, Daniotti JL (2004) Ganglioside GD3 traffics from the trans-Golgi network to plasma membrane by a Rab11-independent and brefeldin A-insensitive exocytic pathway. *J Biol Chem*, **279**, 47610-47618.

Dyatlovitskaya EV, Novikov AM, Gorkova NP, Bergelson LD (1976) Gangli-osides of hepatoma 27, normal and regenerating rat liver. *Eur J Biochem*, **63**, 357-364.

El Bawab S, Roddy P, Qian T, Bielawska A, Lemasters JJ, Hannun YA (2000) Molecular cloning and characterization of a human mitochondrial ceramidase. *J Biol Chem*, **275**, 21508-21513.

Fujinaga Y, Wolf AA, Rodighiero C, Wheeler H, Tsai B, Allen L, Jobling MG, Rapoport T, Holmes RK, Lencer WI (2003) Gangliosides that associate with lipid rafts mediate transport of cholera and related toxins from the plasma membrane to endoplasmic reticulm. *Mol Biol Cell*, **14**, 4783-4793.

Fukasawa M, Nishijima M, Hanada K (1999) Genetic evidence for ATP-dependent endoplasmic reticulum-to-Golgi apparatus trafficking of ceramide for sphingomyelin synthesis in Chinese hamster ovary cells. *J Cell Biol*, **144**, 673-685.

Funato K, Riezman H (2001) Vesicular and nonvesicular transport of ceramide from ER to the Golgi apparatus in yeast. *J Cell Biol*, **155**, 949-959.

Funato K, Vallee B, Riezman H (2002) Biosynthesis and trafficking of sphingolip-ids in the yeast Saccharomyces cerevisiae. *Biochemistry*, **41**, 15105-15114.

Funato K, Lombardi R, Vallee B, Riezman H (2003) Lcb4p is a key regulator of ceramide synthesis from exogenous long chain sphingoid base in Saccharo-myces cerevisiae. *J Biol Chem*, **278**, 7325-7334.

Garcia-Ruiz C, Colell A, Morales A, Calvo M, Enrich C, Fernandez-Checa JC (2002) Trafficking of ganglioside GD3 to mitochondria by tumor necrosis fac-tor-alpha. *J Biol Chem*, **277**, 36443-36448.

Giraudo CG, Rosales Fritz VM, Maccioni HJ (1999) GA2/GM2/GD2 synthase lo-calizes to the trans-golgi network of CHO-K1 cells. *Biochem J*, **342**, 633-640.

Giraudo CG, Daniotti JL, Maccioni HJ (2001) Physical and functional association of glycolipid N-acetyl-galactosaminyl and galactosyl transferases in the Golgi apparatus. *Proc Natl Acad Sci U S A*, **98**, 1625-1630.

Giraudo CG, Maccioni HJ (2003) Ganglioside glycosyltransferases organize in distinct multienzyme complexes in CHO-K1 cells. *J Biol Chem*, **278**, 40262-40271.

Han G, Gable K, Yan L, Natarajan M, Krishnamurthy J, Gupta SD, Borovitskaya A, Harmon JM, Dunn TM (2004) The topology of the Lcb1p subunit of yeast serine palmitoyltransferase. *J Biol Chem*, **279**, 53707-53716.

Hanada K, Kumagai K, Yasuda S, Miura Y, Kawano M, Fukasawa M, Nishijima M (2003) Molecular machinery for non-vesicular trafficking of ceramide. *Nature*, **426**, 803-809.

Hechtberger P, Zinser E, Saf R, Hummel K, Paltauf F, Daum G (1994) Characterization, quantification and subcellular localization of inositol-containing sphingolipids of the yeast, Saccharomyces cerevisiae. *Eur J Biochem*, **225**, 641-649.

Hechtberger P, Daum G (1995) Intracellular transport of inositol-containing sphingolipids in the yeast, Saccharomyces cerevisiae. *FEBS Lett*, **367**, 201-204.

Horvath A, Sutterlin C, Manning-Krieg U, Movva NR, Riezman H (1994) Ceramide synthesis enhances transport of GPI-anchored proteins to the Golgi apparatus in yeast. *EMBO J*, **13**, 3687-3695.

Huitema K, van den Dikkenberg J, Brouwers JF, Holthuis JC (2004) Identification of a family of animal sphingomyelin synthases. *EMBO J*, **23**, 33-44.

Idkowiak-Baldys J, Takemoto JY, Grilley MM (2003) Structure-function studies of yeast C-4 sphingolipid long chain base hydroxylase. *Biochim Biophys Acta*, **1618**, 17-24.

Jahn R, Lang T, Sudhof TC (2003) Membrane fusion. *Cell*, **112**, 519-533.

Jeckel D, Karrenbauer A, Burger KN, van Meer G, Wieland F (1992) Glucosylceramide is synthesized at the cytosolic surface of various Golgi subfractions. *J Cell Biol*, **117**, 259-267.

Kihara A, Igarashi Y (2002) Identification and characterization of a Saccharomyces cerevisiae gene, RSB1, involved in sphingoid long-chain base release. *J Biol Chem*, **277**, 30048-30054.

Kihara A, Sano T, Iwaki S, Igarashi Y (2003) Transmembrane topology of sphingoid long-chain base-1-phosphate phosphatase, Lcb3p. *Genes Cells*, **8**, 525-535.

Kihara A, Igarashi Y (2004) FVT-1 is a mammalian 3-ketodihydrosphingosine reductase with an active site that faces the cytosolic side of the endoplasmic reticulum membrane. *J Biol Chem*, **279**, 49243-49250.

Kohyama-Koganeya A, Sasamura T, Oshima E, Suzuki E, Nishihara S, Ueda R, Hirabayashi Y (2004) Drosophila glucosylceramide synthase: a negative regulator of cell death mediated by proapoptotic factors. *J Biol Chem*, **279**, 35995-36002.

Lannert H, Bunning C, Jeckel D, Wieland FT (1994) Lactosylceramide is synthesized in the lumen of the Golgi apparatus. *FEBS Lett*, **342**, 91-96.

Lannert H, Gorgas K, Meissner I, Wieland FT, Jeckel D (1998) Functional organization of the Golgi apparatus in glycosphingolipid biosynthesis. Lactosylceramide and subsequent glycosphingolipids are formed in the lumen of the late Golgi. *J Biol Chem*, **273**, 2939-2946.

Levine TP, Wiggins CA, Munro S (2000) Inositol phosphorylceramide synthase is located in the Golgi apparatus of Saccharomyces cerevisiae. *Mol Biol Cell*, **11**, 2267-2281.

Levine T (2004) Short-range intracellular trafficking of small molecules across endoplasmic reticulum junctions. *Trends Cell Biol*, **14**, 483-490.

Lin X, Mattjus P, Pike HM, Windebank AJ, Brown RE (2000) Cloning and expression of glycolipid transfer protein from bovine and porcine brain. *J Biol Chem*, **275**, 5104-5110.

Lisman Q, Pomorski T, Vogelzangs C, Urli-Stam D, de Cocq van Delwijnen W, Holthuis JC (2004) Protein sorting in the late Golgi of Saccharomyces cerevisiae does not require mannosylated sphingolipids. *J Biol Chem*, **279**, 1020-1029.

Maccioni HJ, Daniotti JL, Martina JA (1999) Organization of ganglioside synthesis in the Golgi apparatus. *Biochim Biophys Acta*, **1437**, 101-118.

Maier O, Hoekstra D (2003) Trans-Golgi network and subapical compartment of HepG2 cells display different properties in sorting and exiting of sphingolipids. *J Biol Chem*, **278**, 164-173.

Malathi K, Higaki K, Tinkelenberg AH, Balderes DA, Almanzar-Paramio D, Wilcox LJ, Erdeniz N, Redican F, Padamsee M, Liu Y, Khan S, Alcantara F, Carstea ED, Morris JA, Sturley SL (2004) Mutagenesis of the putative sterol-sensing domain of yeast Niemann Pick C-related protein reveals a primordial role in subcellular sphingolipid distribution. *J Cell Biol*, **164,** 547-556.

Malinina L, Malakhova ML, Teplov A, Brown RE, Patel DJ (2004) Structural basis for glycosphingolipid transfer specificity. *Nature*, **430**, 1048-1053.

Marks DL, Pagano RE (2002) Endocytosis and sorting of glycosphingolipids in sphingolipid storage disease. *Trends Cell Biol*, **12**, 605-613.

Mayor S, Presley JF, Maxfield FR (1993) Sorting of membrane components from endosomes and subsequent recycling to the cell surface occurs by a bulk flow process. *J Cell Biol*, **121**, 1257-1269.

Michel C, van Echten-Deckert G (1997) Conversion of dihydroceramide to ceramide occurs at the cytosolic face of the endoplasmic reticulum. *FEBS Lett*, **416**, 153-155.

Mizutani Y, Tamiya-Koizumi K, Nakamura N, Kobayashi M, Hirabayashi Y, Yoshida S (2001) Nuclear localization of neutral sphingomyelinase 1: biochemical and immunocytochemical analyses. *J Cell Sci*, **114**, 3727-3736.

Mobius W, Herzog V, Sandhoff K, Schwarzmann G (1999) Intracellular distribution of a biotin-labeled ganglioside, GM1, by immunoelectron microscopy after endocytosis in fibroblasts. *J Histochem Cytochem*, **47**, 1005-1014.

Morales A, Colell A, Mari M, Garcia-Ruiz C, Fernandez-Checa JC (2004) Glycosphingolipids and mitochondria: role in apoptosis and disease. *Glycoconj J*, **20**, 579-588.

Morsomme P, Prescianotto-Baschong C, Riezman H (2003) The ER v-SNAREs are required for GPI-anchored protein sorting from other secretory proteins upon exit from the ER. *J Cell Biol*, **162**, 403-412.

Muniz M, Riezman H (2000) Intracellular transport of GPI-anchored proteins. *EMBO J*, **19**, 10-15.

Muniz M, Morsomme P, Riezman H (2001) Protein sorting upon exit from the endoplasmic reticulum. *Cell*, **104**, 313-320.

Munro S (2003) Earthworms and lipid couriers. *Nature*, **426**, 775-776.

Pomorski T, Holthuis JC, Herrmann A, van Meer G (2004) Tracking down lipid flippases and their biological functions. *J Cell Sci*, **117**, 805-813.

Puri V, Watanabe R, Singh RD, Dominguez M, Brown JC, Wheatley CL, Marks DL, Pagano RE (2001) Clathrin-dependent and -independent internalization of plasma membrane sphingolipids initiates two Golgi targeting pathways. *J Cell Biol*, **154**, 535-547.

Reggiori F, Canivenc-Gansel E, Conzelmann A (1997) Lipid remodeling leads to the introduction and exchange of defined ceramides on GPI proteins in the ER and Golgi of Saccharomyces cerevisiae. *EMBO J*, **16**, 3506-3518.

Riezman H, van Meer G (2004) Lipid pickup and delivery. *Nat Cell Biol*, **6**, 15-16.

Rippo MR, Malisan F, Ravagnan L, Tomassini B, Condo I, Costantini P, Susin SA, Rufini A, Todaro M, Kroemer G, Testi R (2000) GD3 ganglioside directly targets mitochondria in a bcl-2-controlled fashion. *FASEB J*, **14**, 2047-2054.

Roseman S (1970) The synthesis of complex carbohydrates by multiglycosyltransferase systems and their potential function in intercellular adhesion. *Chem Phys Lipids*, **5**, 270-297.

Sandvig K, van Deurs B (2002) Transport of protein toxins into cells: pathways used by ricin, cholera toxin and Shiga toxin. *FEBS Lett*, **529**, 49-53.

Sharma DK, Choudhury A, Singh RD, Wheatley CL, Marks DL, Pagano RE (2003) Glycosphingolipids internalized via caveolar-related endocytosis rapidly merge with the clathrin pathway in early endosomes and form microdomains for recycling. *J Biol Chem*, **278**, 7564-7572.

Sprong H, Kruithof B, Leijendekker R, Slot JW, van Meer G, van der Sluijs P (1998) UDP-galactose:ceramide galactosyltransferase is a class I integral membrane protein of the endoplasmic reticulum. *J Biol Chem*, **273**, 25880-25888.

Tettamanti G (2004) Ganglioside/glycosphingolipid turnover: new concepts. *Glycoconj J*, **20**, 301-317.

Trinchera M, Fabbri M, Ghidoni R (1991) Topography of glycosyltransferases involved in the initial glycosylations of gangliosides. *J Biol Chem*, **266**, 20907-20912.

Uemura S, Kihara A, Inokuchi J, Igarashi Y (2003) Csg1p and newly identified Csh1p function in mannosylinositol phosphorylceramide synthesis by interacting with Csg2p. *J Biol Chem*, **278**, 45049-45055.

Vaena de Avalos S, Okamoto Y, Hannun YA (2004) Activation and localization of inositol phosphosphingolipid phospholipase C, Isc1p, to the mitochondria during growth of Saccharomyces cerevisiae. *J Biol Chem*, **279**, 11537-11545.

Vallee B, Riezman H (2005) Lip1p: a novel subunit of acyl-CoA ceramide synthase. *EMBO J*, **24**, 730-741.

van Genderen IL, van Meer G, Slot JW, Geuze HJ, Voorhout WF (1991) Subcellular localization of Forssman glycolipid in epithelial MDCK cells by immuno-electronmicroscopy after freeze-substitution. *J Cell Biol*, **115**, 1009-1019.

van Genderen I, van Meer G (1995) Differential targeting of glucosylceramide and galactosylceramide analogues after synthesis but not during transcytosis in Madin-Darby canine kidney cells. *J Cell Biol*, **131**, 645-654.

van Helvoort A, Giudici ML, Thielemans M, van Meer G (1997) Transport of sphingomyelin to the cell surface is inhibited by brefeldin A and in mitosis, where C6-NBD-sphingomyelin is translocated across the plasma membrane by a multidrug transporter activity. *J Cell Sci*, **110**, 75-83.

van IJzendoorn SC, Hoekstra D (1998) (Glyco)sphingolipids are sorted in sub-apical compartments in HepG2 cells: a role for non-Golgi-related intracellular sites in the polarized distribution of (glyco)sphingolipids. *J Cell Biol*, **142**, 683-696.

van Meer G, Stelzer EH, Wijnaendts-van-Resandt RW, Simons K (1987) Sorting of sphingolipids in epithelial (Madin-Darby canine kidney) cells. *J Cell Biol*, **105**, 1623-1635.

van Meer G (2001) What sugar next? Dimerization of sphingolipid glycosyltransferases. *Proc Natl Acad Sci U S A*, **98**, 1321-1323.

van Meer G, Lisman Q (2002) Sphingolipid transport: rafts and translocators. *J Biol Chem*, **277**, 25855-25858.

van 't Hof W, van Meer G (1990) Generation of lipid polarity in intestinal epithelial (Caco-2) cells: sphingolipid synthesis in the Golgi complex and sorting before vesicular traffic to the plasma membrane. *J Cell Biol*, **111**, 977-986.

Watanabe R, Funato K, Venkataraman K, Futerman AH, Riezman H (2002) Sphingolipids are required for the stable membrane association of glycosyl-phosphatidylinositol-anchored proteins in yeast. *J Biol Chem*, **277**, 49538-49544.

Yamaguchi K, Hata K, Koseki K, Shiozaki K, Akita H, Wada T, Moriya S, Miyagi T (2005) Evidence for mitochondrial localization of a novel human sialidase (NEU4). *Biochem J*, **390**, 85-93.

Yasuda S, Nishijima M, Hanada K (2003) Localization, topology, and function of the LCB1 subunit of serine palmitoyltransferase in mammalian cells. *J Biol Chem*, **278**, 4176-4183.

2-9 Current Perspectives on *Saccharomyces cerevisiae* Sphingolipids

Robert C. Dickson and Robert L. Lester

Department of Cellular and Molecular Biochemistry and the Lucille P. Markey Cancer Center, University of Kentucky College of Medicine, Lexington, KY 40536-0298, USA

Summary.   This review focuses on recent advances in our understanding of sphingolipid functions in *Saccharomyces cerevisiae*, particularly their role in signal transduction.   The sphingoid long-chain bases, dihydrosphingosine and phytosphingosine, have gained prominence in yeast as regulators of the AGC-type protein kinase Pkh1 and Pkh2, homologs of mammalian phosphoinositide-dependent protein kinase 1 (PDK1).   Pkh1 and Pkh2 activate the downstream kinases Pkc1, Ypk1, Ypk2 and Sch9. In addition, PHS acts downstream of Pkh1 and partially activates Ypk1, Ypk2 and Sch9.   These kinases control a wide range of cellular processes including growth, cell wall integrity, stress resistance, endocytosis and aging.   Our appreciation of long-chain bases as second messengers will grow as we learn more about the processes controlled by AGC kinases as well as other yeast kinases that are likely to be regulated by long-chain bases.

Keywords.   Sphingosine, Pkh1, Pkh2, *Saccharomyces cerevisiae*

1. Introduction

Sphingolipids are abundant structural components of eucaryotic membranes: in baker's yeast, *Saccharomyces cerevisiae*, they comprise nearly thirty percent of the phospholipids or about seven percent of the the plasma membrane mass (Patton and Lester 1991).   Lesser amounts of sphingolipids are found in other yeast membranes (Hechtberger et al. 1994; Schneiter et al. 1999). Sphingolipids also play roles as signaling molecules, particularly ceramide and sphingosine-1-phosphate, and they are a major research focus in mammals and other organisms (Spiegel and Milstien 2003; Ogretmen and Hannun 2004; Goetzl and Rosen 2004) .

In recent years, sphingoid long-chain bases (LCBs), dihydrosphingosine (DHS) and phytosphingosine (PHS), were recognized as signaling molecules in *S. cerevisiae* where they function during heat stress and perhaps under normal, unstressed growth conditions. This review focuses those advances in understanding signal transduction pathways regulated by LCBs. Other functions for sphingolipids and features of sphingolipid metabolism and trafficking in yeast have been reviewed recently and will not be discussed here (Dickson and Lester 2002; Funato et al. 2002; Sims et al. 2004; Cowart and Hannun 2004).

2. Sphingolipid metabolism in *S. cerevisiae*

We present here a brief outline of sphingolipid metabolism in *S. cerevisiae* in order to show how LCBs arise (Fig. 1). Details of metabolism and references to original publications can be found in previous reviews ((Dickson and Lester 2002; Sims et al. 2004)). A long-chain base, usually a linear alkane of 18 or 20 carbons having hydroxyls on C-1 and C-3 and an amino group on C-2, is a defining feature of sphingolipids. The LCBs in *S. cerevisiae* are dihydrosphingosine (DHS, also called sphinganine), and phytosphingosine (PHS), which has an additional hydroxyl on C-4 (see (Dickson and Lester 1999) for structural details). Small amounts of DHS and PHS are present in mammals. Their primary LCB is sphingosine, which is DHS with a 4,5-double bond. The majority of LCBs in yeast are produced by the *de novo* synthesis pathway (Fig. 1) which is in contrast to the situation in mammals where sphingosine is produced primarily through the breakdown of complex sphingolipids, particularly sphingomyelin (Ogretmen and Hannun 2004).

Sphingolipid synthesis begins with the condensation of palmitoyl-CoA and serine to produce 3-ketodihydrosphingosine (3-ketosphinganine, Fig. 1), a reaction catalyzed by serine palmitoyltransferase, a membrane-bound enzyme. In all organisms studied, 3-ketodihydrosphingosine reduces to DHS. In yeast this can be acylated by ceramide synthase on the amide group to form *N*-acylsphinganine (not shown in Fig. 1) or hydroxylated to become PHS, which is then acylated to produce ceramide (sometimes called phytoceramide). A distinguishing feature of sphingolipids in *S.*

cerevisiae is the C_{26}-fattyacyl group. Variations in hydroxylation of the long-chain base and C_{26} fatty acid generate four types of ceramides (not shown) with most hydroxylated forms found in aerobically grown cells (Dickson and Lester 1999; Dickson and Lester 2002). Besides being acylated to form ceramides, DHS and PHS can be phosphorylated by two long-chain base kinases (Fig. 1). Phosphorylated DHS and PHS can be catabolized by Dpl1 lyase or dephosphorylated by Lcb3p or Ysr3p to return to DHS and PHS. All reactions to this point occur in the endoplasmic reticulum (reviewed in Sims et al. 2004).

Ceramides are transported to the Golgi apparatus and sequentially modified by *myo*-inositol phosphate to form inositol-phosphoceramide, IPC (Fig. 1). This is then mannosylated to produce mannose-inositol-phosphoceramide (MIPC) followed by another interaction with *myo*-inositol phosphate to form mannose-(inositol-P)$_2$-ceramide (M(IP)$_2$C).

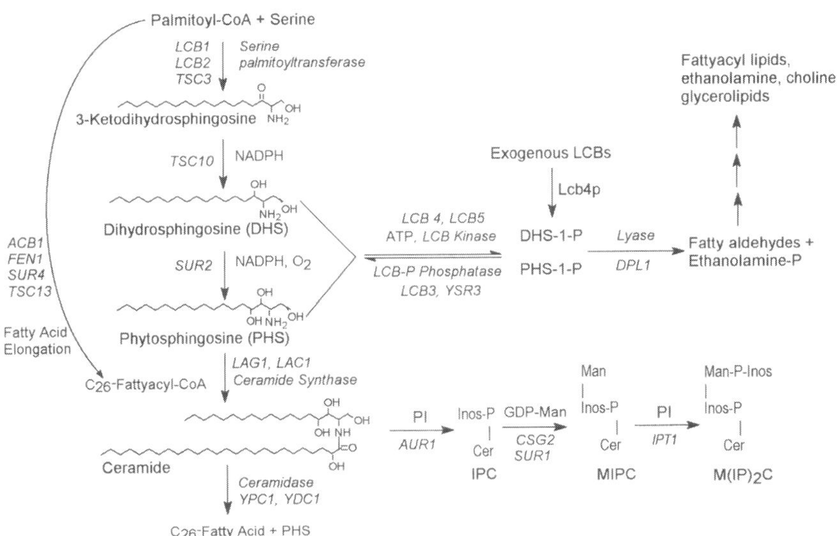

Fig. 1. Diagram of sphingolipid metabolism in *S. cerevisiae*. Gene and enzymes names are shown in italics. The structure of the three complex sphingolipids is incomplete. Reproduced from Dickson and Lester, 2002, with permission.

3. Signaling pathways regulated by LCBs

Sphingolipids were initially implicated in heat stress when growth in strains lacking LCBs but having a compensatory set of glycerol-based lipids that mimicked some sphingolipid function was inhibited by heat, osmotic or low pH stress (Patton et al. 1992). Further evidence for sphingolipid participation in heat stress survival cam with the characterization of the *end8* mutant (now called *lcb1-100*). This mutant, which does not grow at 37°C, was found to be an allele of the *LCB1* gene (Sutterlin et al. 1997), encoding a subunit of serine palmitoyltransferase, the first enzyme in the sphingolipid biosynthetic pathway (Fig. 1). It was not clear from these studies, however, which sphingolipid(s) was required for growth in response to heat stress or whether any sphingolipid was acting as a signaling molecule. Data showing that shifting cells from 25°C to 37°C/39°C caused a 2 to 3-fold increase in C_{18}-DHS and C_{18}-PHS, and a more than 100-fold increase in C_{20}-DHS and C_{20}-PHS implicated LCBs as possible signal molecules. Peak values appeared five to ten minutes after a temperature shift and then returned to normal (Dickson et al. 1997; Jenkins et al. 1997).

Trehalose, a thermoprotectant, accumulates in yeast exposed to heat stress (De Virgilio et al. 1994). This accumulation requires LCBs, although LCB phosphates have not been ruled out (Dickson et al. 1997). Adding DHS to cells also activated transcription of a reporter gene containing carrying multiple copies of stress response elements (STREs) in the promoter. This indicates another potential role for LCBs during heat stress. Yeast cells transiently arrest at the G1 phase of the cell cycle during heat shock. While LCBs are required for that arrest to occur, yet a molecular role in cell cycle arrest remains to be determined (Jenkins and Hannun 2001).

The first evidence for an LCB-regulated signal transduction pathway came from experiments to identify yeast genes that, when present in multiple copies, by-passed growth inhibition by a low concentration of myriocin (Sun et al. 2000). Myriocin inhibits serine palmitoyltransferase and probably inhibits growth by limiting sphingolipid synthesis. One multicopy by-pass gene was *YPK1*, encoding a protein kinase that plays roles in cell wall integrity and actin dynamics (Schmelzle et al. 2002; Roelants et al. 2002), endocytosis (deHart et al. 2002) and translation during nitrogen starvation (Gelperin et al. 2002). Ypk1 and its paralog Ypk2 are related to the mammalian serum- and glucocorticoid-inducible kinase (SGK), a member of the AGC (Protein kinases A, G and C) kinase family. Just prior to these results, Ypk1 was found to be a downstream substrate of the protein kinase Pkh1 (Casamayor et al. 1999). This relationship prompted Sun et al. to determine if multiple copies of *PKH1* also by-passed growth inhibition by myriocin. They did.

Pkh1 is related to mammalian phosphoinositide-dependent protein kinase 1 (PDK1), which is activated by 3-phopshoinositides via binding to a Pleckstrin Homology (PH) domain. Pkh1 has no PH domain and *in vitro* experiments showed that Pkh1 was not activated by phosphoinositides, leaving the upstream signal unknown (Casamayor et al. 1999). Sun et al. (Sun et al. 2000) presented evidence implicating sphingolipids as upstream activation signals. They showed that a phosphorylated species of Ypk1 disappeared when cells were treated with myriocin, presumably inhibiting LCB synthesis, but would reappear *in vivo* even in the presence of myriocin if PHS was present in the culture medium. Whether PHS or some other sphingolipid was the signaling lipid and whether that lipid was directly or indirectly via activation of Pkh1 regulating Ypk1 remain unclear.

Further progress in understanding signaling pathways regulated by LCBs was made searching for multicopy protein kinase genes that restored endocytosis in *lcb1-100* cells at a restrictive temperature (Friant et al. 2000). *PKC1* and *YCK2* were identified as multicopy suppressor genes. Just prior to this publication, others showed that Pkh1 phosphorylates and activates Pkc1 (Inagaki et al. 1999). Another group also suggested this relationship because of the sequence similarities between Ypk1 and Pkc1 and the fact that they are members of the AGC kinase family (Casamayor et al. 1999). Subsequently, multiple copies of *PKH1* or a close relative, *PKH2*, were shown to suppress the endocytosis defect in *lcb1-100* cells. In addition, PHS triggered Pkh1 or Pkh2 phosphorylation of Pkc1 *in vitro* (Friant et al. 2001). Thus, PHS appears stimulate autophosphorylation and activation of Pkh1 or Pkh2, which then phosphorylates and activates Pkc1 (Fig. 2). Pkc1 is also regulated by a protein-activated signaling pathway in which proteins in the plasma membrane sense membrane stretch and transmit a signal to the small G-protein Rho1 that then activates Pkc1 (Heinisch et al. 1999).

Until recently, there has been no direct biochemical demonstration of LCBs stimulation of Pkh1 activity on Ypk1, Ypk2 or Sch9. Using affinity purified protein kinases overproduced in yeast, we successfully demonstrated that PHS stimulates Pkh1 to phosphorylate and activate Ypk1, Ypk2 and Sch9 (Liu et al. 2005). PHS and Pkh1 combined to increase Ypk1 activity 8-fold, and Ypk2 and Sch9 activity 4-fold. Surprisingly, about half of the observed stimulation occurred when Ypk1, Ypk2 or Sch9 were incubated with PHS alone. Hence, PHS appears to activate not just Pkh1 but also partly activate Ypk1, Ypk2 and Sch9 as indicated in Fig. 2. Whether PHS acts directly on Pkc1 remains unclear. The data for Sch9 are the first biochemical demonstrations of an upstream activator for this

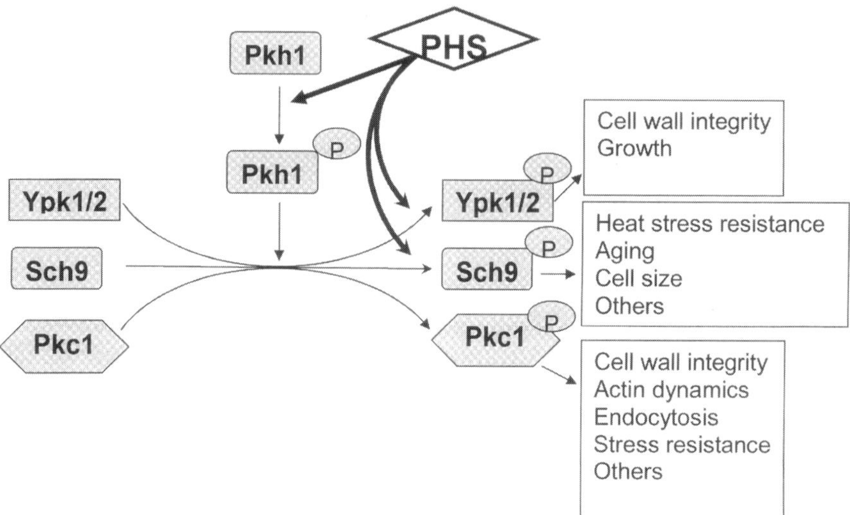

Fig. 2. Regulation of AGC protein kinases by PHS. Available data indicates that PHS stimulates Pkh1, which then activates downstream kinases Ypk1/2, Sch9 and Pkc1. PHS also stimulates autophosphorylation and activation of Ypk1/2 and Sch9. Pkc1 stimulation has not been examined. Phosphorylation is thought to occur in the PDK1 site of the downstream kinases. Reproduced from Liu et al., 2005, with permission.

very interesting kinase. This kinase is related to mammalian Akt/PKBs, which participates in heat stress resistance, aging, Ty1 transposition, cell size, entry into and exit from stationary phases, homologous recombination in ribosomal gene hot spots, and adaptation to changes in nutrients (references in (Liu et al. 2005)).

It is not known what LCBs do to stimulate protein kinase activity, but they probably cause autophosphorylation in the activation loop at the so-called PDK1 site, since phosphorylation of this site is essential for Ypk1, Ypk2 and Sch9 activity (Casamayor et al. 1999; Roelants et al. 2004; Liu et al. 2005). Because the combination of Pkh1 and PHS gives the highest stimulation of Ypk1, Ypk2 and Sch9 activity, they must be phosphorylated at other sites, perhaps the PDK2 site in the C-terminal hydrophobic motif (Casamayor et al. 1999) or the turn motif between the activation loop and the hydrophobic motif (Newton 2002). The kinase responsible for these phosphorylation events remains to be determined.

Pil1 and Lsp1 have also been identified as substrates of Pkh1 and Pkh2, Pil1 and Lsp1. PHS stimulates phosphorylation of Lsp1 by Pkh1 and Pkh2, weakly inhibits phosphorylation of Pil1 by Pkh1, and strongly inhibits phosphorylation of Pil1 by Pkh2 (Zhang et al. 2004). Pil1 and Lsp1 are about 70% identical and act to down-regulate the activity of Pkh1 and Pkh2, at least during heat stress. Other fungi have homologs of Pil1 and Lsp1, but mammals do not. However, mammals may have functional homologs that regulate PDK1 activity.

4. Future directions

The signal transduction pathways and cellular processes regulated by LCBs in yeast are beginning to be understood, and many more functions and signaling pathways await discovery. While AGC protein kinases are the only types known to be regulated by LCBs in yeast, we expect that LCBs will be found to regulate members of other kinase families, since mammalian PAK1, a member of the STE kinase family, is stimulated by sphingosine (King et al. 2000a) as in PAK2 (Roig et al. 2001). Other mammalian protein kinases are also stimulated by LCBs. While mammalian PDK1, the homolog of Pkh1/2 (Fig. 2), is stimulated by 3-phosphoinositides, sphingosine has also been reported to activate it *in vitro* and *in vivo* (King et al. 2000b). Others have found that a cleavage product containing the protein kinase domain of PKCdelta is stimulated by sphingosine to phosphorylate 14-3-3 proteins (Hamaguchi et al. 2003). For all proteins, it will be important to determine where LCBs bind and how they stimulate protein kinase activity.

With the availability of specific mutant strains, pharmacological agents and high-throughput large-scale analyses of various types of special purpose strain sets, yeast should continue to be an excellent model organism to further elucidate the signaling and other functions of LCBs.

Acknowledgements. Our research has been supported by grants GM 41302 and AG 0234377 from the National Institutes of Health. We thank current and past members of our laboratory for their role in understanding the functions of yeast sphingolipids.

5. References

Casamayor A, Torrance PD, Kobayashi T, Thorner J, Alessi DR (1999) Functional counterparts of mammalian protein kinases PDK1 and SGK in budding yeast. *Curr Biol*, **9**, 186-197.

Cowart LA, Hannun YA (2004) Baker's Yeast: a rising foundation for eukaryotic sphingolipid-mediated cell signalling. in *Topics in Current Genetics* (Daum G), Vol 6, pp 383-401, Sppringer-Verlag, Berlin.

De Virgilio C, Hottiger T, Dominguez J, Boller T, Wiemken A (1994) The role of trehalose synthesis for the aquisition of thermotolerance in yeast. I. Genetic evidence that trehalose is a thermoprotectant. *Eur J Biochem,* **219**, 179-186.

deHart AK, Schnell JD, Allen DA, Hicke L (2002) The conserved Pkh-Ypk kinase cascade is required for endocytosis in yeast. *J Cell Biol,* **156**, 241-248.

Dickson RC, Lester RL (1999) Metabolism and selected functions of sphingolipids in the yeast *Saccharomyces cerevisiae. Biochem Biophys Acta,* **1438**, 305-321.

Dickson RC, Lester RL (2002) Sphingolipid functions in *Saccharomyces cerevisiae. Biochem Biophys Acta,* **1583**, 13-25.

Dickson RC, Nagiec EE, Skrzypek M, Tillman P, Wells GB, Lester RL (1997) Sphingolipids are potential heat stress signals in *Saccharomyces. J Biol Chem,* **272**, 30196-30200.

Friant S, Lombardi R, Schmelzle T, Hall MN, Riezman H (2001) Sphingoid base signaling via Pkh kinases is required for endocytosis in yeast. *EMBO J,* **20**, 6783-6792.

Friant S, Zanolari B, Riezman H (2000) Increased protein kinase or decreased PP2A activity bypasses sphingoid base requirement in endocytosis. *EMBO J,* **19**, 2834-2844.

Funato K, Vallee B, Riezman H (2002) Biosynthesis and trafficking of sphingolipids in the yeast Saccharomyces cerevisiae. *Biochemistry,* **41**, 15105-15114.

Gelperin D, Horton L, DeChant A, Hensold J, Lemmon SK (2002) Loss of ypk1 function causes rapamycin sensitivity, inhibition of translation initiation and synthetic lethality in 14-3-3-deficient yeast. *Genetics,* **161**, 1453-1464.

Goetzl EJ, Rosen H (2004) Regulation of immunity by lysosphingolipids and their G protein-coupled receptors. *J Clin Invest,* **114**, 1531-1537.

Hamaguchi A, Suzuki E, Murayama K, Fujimura T, Hikita T, Iwabuchi K, Handa K, Withers DA, Masters SC, Fu H, Hakomori S (2003) Sphingosine-dependent protein kinase-1, directed to 14-3-3, is identified as the kinase domain of protein kinase C delta. *J Biol Chemm,* **278**, 41557-41565.

Hechtberger P, Zinser E, Saf R, Hummel K, Paltauf F, Daum G (1994) Characterization, quantification and subcellular localization of inositol-containing sphingolipids of the yeast, *Saccharomyces cerevisiae. Eur J Biochem,* **225**, 641-649.

Heinisch JJ, Lorberg A, Schmitz HP, Jacoby JJ (1999) The protein kinase C-mediated MAP kinase pathway involved in the maintenance of cellular integrity in *Saccharomyces cerevisiae. Mol Microbiol,* **32**, 671-680.

Inagaki M, Schmelzle T, Yamaguchi K, Irie K, Hall MN, Matsumoto K (1999) PDK1 homologs activate the Pkc1-mitogen-activated protein kinase pathway in yeast. *Mol Cell Biol,* **19**, 8344-8352.

Jenkins GM, Hannun YA (2001) Role for de novo sphingoid base biosynthesis in the heat-induced transient cell cycle arrest of *Saccharomyces cerevisiae. J Biol Chem,* **276**, 8574-8581.

Jenkins GM, Richards A, Wahl T, Mao CG, Obeid L, Hannun Y (1997) Involvement of yeast sphingolipids in the heat stress response of *Saccharomyces cerevisiae. J Biol Chem*, **272**, 32566-32572.

King CC, Gardiner EM, Zenke FT, Bohl BP, Newton AC, Hemmings BA, Bokoch GM (2000a) p21-activated kinase (PAK1) is phosphorylated and activated by 3- phosphoinositide-dependent kinase-1 (PDK1). *J Biol Chem*, **275**, 41201-41209.

King CC, Zenke FT, Dawson PE, Dutil EM, Newton AC, Hemmings BA, Bokoch GM (2000b) Sphingosine is a novel activator of 3-phosphoinositide-dependent kinase 1. *J Biol Chem*, **275**, 18108-18113.

Liu K, Zhang X, Lester RL, Dickson RC (2005) The sphingolipid long-chain base phytosphingosine activates AGC kinases in *Saccharomyces cerevisiae* including Ypk1, Ypk2 and Sch9. *J Biol Chem*, **280**, 22679-22687.

Newton AC (2002) Analyzing protein kinase C activation. *Methods Enzymol*, **345**, 499-506.

Ogretmen B, Hannun YA (2004) Biologically active sphingolipids in cancer pathogenesis and treatment. *Nat Rev Cancer*, **4**, 604-616.

Patton JL, Lester RL (1991) The phosphoinositol sphingolipids of *Saccharomyces cerevisiae* are highly localized in the plasma membrane. *J Bacteriol*, **173**, 3101-3108.

Patton JL, Srinivasan B, Dickson RC, Lester RL (1992) Phenotypes of sphingolipid-dependent strains of *Saccharomyces cerevisiae*. *J Bacteriol*, **174**, 7180-7184.

Roelants FM, Torrance PD, Bezman N, Thorner J (2002) Pkh1 and Pkh2 differentially phosphorylate and activate Ypk1 and Ykr2 and define protein kinase modules required for maintenance of cell wall integrity. *Mol Biol Cell*, **13**, 3005-3028.

Roelants FM, Torrance PD, Thorner J (2004) Differential roles of PDK1- and PDK2-phosphorylation sites in the yeast AGC kinases Ypk1, Pkc1 and Sch9. *Microbiology*, **150**, 3289-3304.

Roig J, Tuazon PT, Traugh JA (2001) Cdc42-independent activation and translocation of the cytostatic p21- activated protein kinase gamma-PAK by sphingosine. *FEBS Lett*, **507**, 195-199.

Schmelzle T, Helliwell SB, Hall MN (2002) Yeast Protein Kinases and the RHO1 Exchange Factor TUS1 Are Novel Components of the Cell Integrity Pathway in Yeast. *Mol Cell Biol*, **22**, 1329-1339.

Schneiter R, Brugger B, Sandhoff R, Zellnig G, Leber A, Lampl M, Athenstaedt K, Hrastnik C, Eder S, Daum G, Paltauf F, Wieland FT, Kohlwein SD (1999) Electrospray ionization tandem mass spectrometry (ESI-MS/MS) analysis of the lipid molecular species composition of yeast subcellular membranes reveals acyl chain-based sorting/remodeling of distinct molecular species en route to the plasma membrane. *J Cell Biol*, **146**, 741-754.

Sims KJ, Spassieva SD, Voit EO, Obeid LM (2004) Yeast sphingolipid metabolism: clues and connections. Biochem Cell Biol, **82**, 45-61.

Spiegel S, Milstien S (2003) Sphingosine-1-phosphate: an enigmatic signalling lipid. *Nat Rev Mol Cell Biol*, **4**, 397-407.

Sun Y, Taniguchi R, Tanoue D, Yamaji T, Takematsu H, Mori K, Fujita T, Kawasaki T, Kozutsumi Y (2000) Sli2 (Ypk1), a homologue of mammalian protein kinase SGK, is a downstream kinase in the sphingolipid-mediated signaling pathway of yeast. *Mol Cell Biol*, **20**, 4411-4419.

Sutterlin C, Doering TL, Schimmoller F, Schroder S, Riezman H (1997) Specific requirements for the ER to Golgi transport of GPI-anchored proteins in yeast. *J Cell Sci*, **110**, 703-2714.

Zhang X, Lester RL, Dickson RC (2004) Pil1p and Lsp1p negatively regulate the 3-phosphoinositide-dependent protein kinase-like kinase Pkh1p and downstream signaling pathways Pkc1p and Ypk1p. *J Biol Chem*, **279**, 22030-22038.

Part 3

Generation and Degradation of Sphingolipid Signaling Molecules

3-1 Generation of Signaling Molecules by De Novo Sphingolipid Synthesis

Kazuyuki Kitatani, L. Ashley Cowart, and Yusuf A. Hannun

Department of Biochemistry and Molecular Biology, Medical University of South Carolina, 173 Ashley Avenue, Charleston, South Carolina, 29425, USA

Summary. Sphingolipids are abundant components of cellular membranes in eukaryotic cells as well as potent signaling molecules. De novo sphingolipid biosynthesis begins with condensation of L-serine and palmitoyl-CoA, generating sphingoid bases, ceramide, and other species through a series of reactions. Ceramide can also be formed from free sphingoid bases, which has been termed the 'salvage' pathway. Sphingolipids from both the de novo and salvage pathways increase with exposure of yeast or mammals to various stimuli such as Fas ligands, chemotherapeutic drugs, tumor necrosis factor-γ and heat stress, and then act as lipid second messengers mediating inflammatory responses, senescence, cell cycle arrest, apoptosis or stress responses. Therefore, generation of signaling molecules by de novo synthesis and/or salvage accounts not only for homeostasis, but also for several disorders resulting from aberrant sphingolipid accumulation or depletion.

Keywords. ceramide, ceramide synthase, de novo sphingolipid synthesis, serine palmitoyltransferase, sphingoid bases

1. Introduction

In recent years sphingolipids, including sphingosine, sphingosine-1-phosphate, ceramide-1-phosphate, and ceramide, have come to the forefront as bioactive lipid mediators playing crucial roles in cell signaling.

These lipids act in cellular pathways regulating cell division, senescence, apoptosis, migration, and responses to stress (reviewed in (Gomez-Munoz, 1998; Kolesnick and Fuks, 2003; Ruvolo, 2003; Spiegel and Milstien, 2003; Ogretmen and Hannun, 2004)). Formation and/or accumulation of these bioactive sphingolipids is stimulated by various extracellular factors including treatment with cytokines (Schutze et al., 1994; Ogretmen and Hannun, 2004), various pharmacological and chemotherapeutic agents (Ogretmen and Hannun, 2004), and free fatty acids (Unger and Orci, 2002), as well as by diverse conditions such as ischemia/reperfusion (Zhang et al., 2001) and heat stress (Jenkins and Hannun, 2001; Jenkins et al., 2002). Several metabolic routes contribute to the observed alterations in sphingolipid metabolism, and the two best studied are cleavage of sphingomyelin to release ceramide, or alternately, synthesis of ceramide de novo. Other pathways include the salvage pathway, hydrolysis of other complex sphingolipids, and regulation of clearance of bioactive sphingolipids. This chapter will focus primarily on the de novo pathway and the related salvage pathway.

2. De novo sphingolipid metabolism

The process of de novo ceramide synthesis begins with condensation of L-serine and palmitoyl-CoA through the action of serine palmitoyltransferase (SPT) to form 3-ketosphinganine, which is then reduced to sphinganine (dihydrosphingosine). Two subunits, LCB1 and LCB2, comprise SPT, which is membrane-bound and localizes to the endoplasmic reticulum (Hanada, 2003). Data from structural studies suggest that sphingoid base synthesis occurs at the interface between the two subunits (Han et al., 2004). In yeast, dihydrosphingosine is converted to phytosphingosine by the hydroxylase Syr2p (Grilley et al., 1998), and either phyto- or dihydrosphingosine can be acylated by ceramide synthase (CS) to produce phyto- or dihydroceramide, respectively (Schorling et al., 2001). A family of CS enzymes exists and can be at least partially distinguished by chain length preferences and sensitivity to the CS inhibitor fumonisin B1 (FB1) (Venkataraman et al., 2002; Riebeling et al., 2003). In mammalian pathways, the introduction of a trans-4,5 double bond by dihydroceramide desaturase converts dihydroceramide to ceramide (Schulze et al., 2000), and production of the sphingoid base sphingosine is achieved through ceramidase-mediated cleavage of ceramide (Slife et al., 1989). Therefore, in yeast, synthesis of bioactive sphingoid bases precedes ceramide synthesis whereas in mammalian cells, sphingosine is produced through the cleavage of ceramide. Importantly, in each case, sphingoid bases can be phosphorylated to the potent signaling molecule sphingosine-1-phosphate and its yeast homologues, phyto- and dihydrosphingosine-1-phosphate (Spiegel and Milstien, 1995; Nagiec et al., 1998).

In addition to de novo sphingolipid synthesis, a related and partially overlapping pathway of sphingolipid turnover modulates sphingolipid levels by recycling (dihydro)ceramide in endosomes/Golgi and salvaging sphingoid bases arising in lysosomes from hydrolysis of sphingolipids. This is termed the "salvage pathway" of sphingolipid synthesis. Interestingly, exogenous short chain ceramide recycles into long chain ceramide via this pathway (Ogretmen et al., 2002). In contrast, sphingolipids in serum that are taken up by the cells directly reach the lysosomes largely to be catabolized, but the majority of catabolic endogenous sphingosine is recycled (Chigorno et al., 2005). A recent study (Becker et al., 2005) demonstrated that phorbol ester-responsive PKC activated the salvage pathway resulting in ceramide elevation(Becker et al., 2005). Further, the resultant ceramide acts as a second messenger for inhibition of classical PKC sequestration. Sphingoid base salvage could represent the predominant pathway of sphingolipid synthesis, but the mechanism(s) for regulation of the salvage pathway remains unclear.

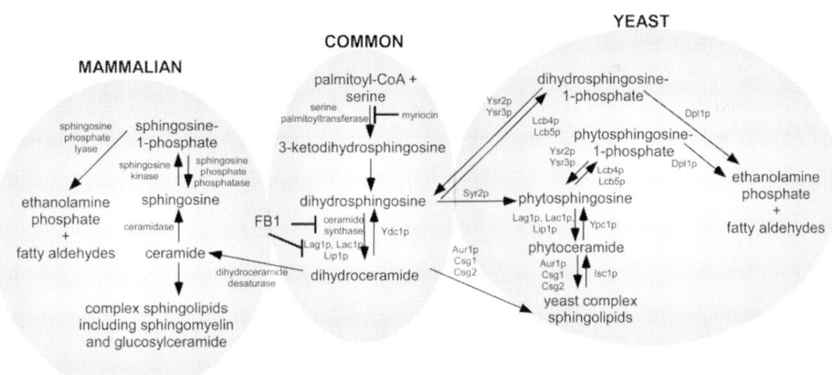

Fig. 1. Scheme of de novo sphingolipid metabolism in yeast and mammals.

3. Distinguishing salvage from de novo sphingolipid synthesis

Several pharmacological inhibitors of enzymes responsible for sphingolipid metabolism have been identified and continue to play a crucial role in addressing the sphingolipid metabolic pathway and sphingolipid dependent events including cell growth, cell death, differentiation and intracellular signal transduction. Two representative inhibitors for ceramide synthesis are myriocin/ISP-1 (Miyake et al., 1995) and FB1 (Wang et al., 1991). Myriocin is a selective inhibitor of SPT, being capable of blocking the de novo pathway. In contrast, FB1 is a potent inhibitor of CS, the enzyme that is responsible for acylation of free long chain sphingoid bases in the de novo synthesis of sphingolipids and the reacylation of sphingosine/sphinganine that is formed upon sphingolipid turnover. Therefore, FB1, but not myriocin, is able to block the salvage pathway, and differential sensitivity of ceramide synthesis to the two inhibitors can distinguish the salvage pathway from de novo biosynthesis. The two pathways can be further distinguished by metabolic labeling, such that activation of the de novo pathway can be detected selectively by pulse labeling with precursors (palmitate or serine) whereas the salvage pathway can be detected selectively by steady state labeling and analysis of turnover. These approaches were illustrated in a recent study (Becker et al., 2005) whereby the release of free sphingosine and ceramide production, that were sensitive to FB1 but not myriocin, was observed upon phorbol ester treatment. Moreover, phorbol esters did not enhance incorporation of precursor palmitate into ceramide but accelerated turnover of complex sphingolipids into ceramide in a process requiring hydrolysis of complex sphingolipids all the way to sphingosine and reacylation of the generated sphingosine to ceramide. The above evidence is able to support the existence of salvage pathway and is a demonstration that it is distinguishable from de novo ceramide synthesis. Thus, these two pharmacological inhibitors provide powerful tools for determining the pathway responsible for ceramide synthesis, but it is important to pay attention to proper use of the inhibitors as well.

4. Mechanisms of regulation of de novo sphingolipid synthesis

Regulation of SPT is key to overall regulation of de novo synthesis because it is the first and rate-limiting enzyme of this pathway (Perry, 2002). Indeed, transcriptional regulation, post-translational regulation, and regulation of substrate availability have each been identified as mechanisms for SPT regulation. Transcriptional regulation has been reported in many systems including keratinocytes (Harris et al., 1997), injured arteries (Uhlinger et al., 2001), and a fibroblast model of wound healing (Carton et al., 2003), and is typically due to up-regulation of transcripts for both subunits, or in a few cases, of the LCB2 (SPTLC2) subunit alone (Farrell et al., 1998). An important observation is that SPT subunit expression tends to be elevated in proliferating tissues (Batheja et al., 2003; Carton et al., 2003).

Post-translational modes of regulation include protein aggregation in response to hypoxia (to inhibit SPT activity)(Dedov et al., 2004) and in yeast, association of the LCB1/2 complex with a third component encoded by TSC3 (Gable et al., 2000). Interestingly, a TSC3 homolog has not been found in any mammalian system to date, and thus, TSC3-mediated SPT activation may represent a yeast-specific evolutionary adaptation. Furthermore, studies performed in pancreatic γ-cells (Shimabukuro et al., 1998) and skeletal muscle cells (Chavez and Summers, 2003) have each described that de novo synthesis can be modulated by increased supply of free fatty acids, especially palmitate.

In addition to their regulation by SPT activity, sphingolipid levels are modulated by relative activities of downstream enzymes as well. An important example of this type of regulation is provided by studies showing that expression of glucosylceramide synthase reduces ceramide levels through incorporation of ceramide into glucosylceramide. This mechanism has been implicated in multidrug resistance in cancer cells (Liu et al., 2001).

Key areas of future research will include the mechanisms for acute regulation of de novo sphingolipid synthesis in response to heat stress and chemotherapeutic agents. Due to the relatively fast response of this pathway to some external factors including heat stress, transcriptional regulation does not seem a likely regulatory mechanism in these instances. Mechanisms for acute regulation such as post-translational modifications and substrate availability are a main focus for future research.

5. Contributions of de novo sphingolipid synthesis to biology

5.1 Sphingolipid function in yeast

In yeast, complex sphingolipids (inositolphosphoryl- and mannosylated derivatives of ceramide) comprise a significant portion of cell membranes. Indeed, strains with deletions in either SPT subunit require sphingoid bases for growth (Pinto et al., 1992), underscoring the importance of de novo sphingolipid synthesis of normal cell function. Furthermore, data suggest that phytosphingosine mediates such important processes as endocytosis (Zanolari et al., 2000), ubiquitination and protein degradation (Chung et al., 2000), and cell cycle control in response to stresses (Jenkins et al., 2002). Additionally, sphingolipid production during heat stress accounts for a major portion of heat induced transcriptional regulation (Cowart et al., 2003).

5.2 Biology of de novo ceramide synthesis in mammalian systems

As referred to in Table 1, several stimuli in cell culture models have been shown to stimulate the de novo ceramide pathway to accumulate ceramide. Particularly, chemotherapeutic agents (Reynolds et al., 2004) are now well-recognized as inducers of this pathway, which appears to mediate, at least partially, cell death through ceramide-dependent events including mitochondrial damage and/or caspase activation (Reynolds et al., 2004). An exposure to pathological or physiological factors such as ischemia/reperfusion (Basnakian et al., 2005), palmitate (Schmitz-Peiffer, 2002), Fas (Chalfant et al., 2001), tumor necrosis factor-γ (Dbaibo et al., 2001), oxidized LDL (Kitatani et al., 2002) and heat stress (Jenkins et al., 2002) not only revealed activation of de novo ceramide synthesis, but also displayed ceramide-dependent cellular responses. These studies in cell models suggest that at least de novo ceramide synthesis controls homeostasis as well as pathological events via the ceramide-dependent signal transduction.

Table. 1. Inducers for de novo ceramide synthesis and associated biological responses in mammalian cells.

Inducer	Responsive targets/events	References
Androgen ablation	Caspases - cell death	(Eto et al., 2003)
Cannabinoids	ERK1/2 - cell death	(Gomez del Pulgar et al., 2002)
Chemotherapeutic drugs	Caspases - cell death	(Reynolds et al., 2004)
Fas	PP1 - SR proteins	(Chalfant et al., 2002)
γ-Tocopherol	Caspases - cell death	(Jiang et al., 2004)
Heat stress	SR proteins	(Jenkins et al., 2002)
Ischemia/reperfusion	Endonuclease - cell death	(Basnakian et al., 2005)
Palmitic acid	Akt/PKB - insulin resistance	(Schmitz-Peiffer, 2002)
Oxidized LDL	Cytosolic PLA_2 - foam cells	(Kitatani et al., 2002)
Tumor necrosis factor-γ	Cell death	(Dbaibo et al., 2001)

5.3 Pathological aspects in mammalian systems

Though sphingolipids are vital for normal function of an organism, aberrant sphingolipid metabolism participates in various pathophysiologies, including type-II diabetes (Schmitz-Peiffer, 2002; Summers and Nelson, 2005), phorbol ester-induced experimental ulcers (Uehara et al., 2003), emphysema caused by cigarette smoking (Petrache et al., 2005), and atherosclerosis (Park et al., 2004). Interestingly, modified LDL, which plays a crucial role in the pathogenesis of atheroscleorosis, has been shown to facilitate de novo ceramide synthesis (Kitatani et al., 2002) and sphingomyelin synthesis (Shiratori et al., 1994) in cell culture models. In an in vivo model (Park et al., 2004), ApoE-knockout mice fed a Western diet containing cholesterol and fat displayed an elevation of plasma sphinganine, ceramide and sphingomyelin, which was significantly suppressed by myriocin. Furthermore, inhibition of de novo ceramide synthesis by myriocin in ApoE-knockout mice lowered levels of plasma neutral lipids, raised HDL cholesterol, and prevented development of atherosclerotic lesions. Additionally, inhibition of de novo sphingolipid synthesis reduced transcription of sterol regulatory element-mediated genes including HMG-CoA synthase by decreasing levels of the transcriptional activator sterol-regulatory element-binding protein (Worgall et al., 2004). Therefore, enzymes responsible for de novo ceramide synthesis could represent a novel class of molecular targets for treatment of dyslipidemia and athero-sclerosis.

On the other hand, inhibition of de novo sphingolipid synthesis in vivo can have toxic effects. Ingestion of FB1, a CS inhibitor, has a wide spectrum of pathophysiologic effects, which include nephropathy, hepatotoxicity and hepatocarcinogenicity (Merrill et al., 1996a; Merrill et al., 1996b; Riley et al., 1996). Inhibition of CS by FB1 in cultured cells or in vivo reduces cellular levels of complex sphingolipids and increases the amount of free sphingoid bases (Schmelz et al., 1998). The free sphingoid bases in particular might contribute to the harmful effects because this class of lipids plays key roles in cell growth and the regulation of apoptosis (Spiegel and Merrill, 1996; Spiegel and Milstien, 2003). The accumulation of the bioactive sphingoid bases as well as the depletion of complex sphingolipids may account for several disorders.

6. Conclusions

Ever-increasing research now focuses on de novo sphingolipid synthesis, and the results continue to reveal the complexity and importance of this metabolic pathway in normal cell function and in disease. Several key areas of research require elucidation of the biochemical and molecular mechanisms by which SPT is regulated, novel signaling functions for de novo sphingolipids in cell signaling, and the effects of aberrations in this pathway on cell function.

Acknowledgements. The authors wish to thank Youssef Zaidan for a careful reading of the manuscript. This work was supported by NIH GM63265.

7. References

Basnakian, A. G., Ueda, N., Hong, X., Galitovsky, V. E., Yin, X., and Shah, S. V. (2005) Ceramide synthase is essential for endonuclease-mediated death of renal tubular epithelial cells induced by hypoxia-reoxygenation. *Am J Physiol Renal Physiol*, **288**, F308-314.
Batheja, A. D., Uhlinger, D. J., Carton, J. M., Ho, G., and D'Andrea, M. R. (2003) Characterization of serine palmitoyltransferase in normal human tissues. *J Histochem Cytochem*, **51**, 687-696.

Becker, K. P., Kitatani, K., Idkowiak-Baldys, J., Bielawski, J., and Hannun, Y. A. (2005) Selective inhibition of juxtanuclear translocation of protein kinase C betaII by a negative feedback mechanism involving ceramide formed from the salvage pathway. *J Biol Chem*, **280**, 2606-2612.

Carton, J. M., Uhlinger, D. J., Batheja, A. D., Derian, C., Ho, G., Argenteri, D., and D'Andrea, M. R. (2003) Enhanced serine palmitoyltransferase expression in proliferating fibroblasts, transformed cell lines, and human tumors. *J Histochem Cytochem*, **51**, 715-726.

Chalfant, C. E., Ogretmen, B., Galadari, S., Kroesen, B. J., Pettus, B. J., and Hannun, Y. A. (2001) FAS activation induces dephosphorylation of SR proteins; dependence on the de novo generation of ceramide and activation of protein phosphatase 1. *J Biol Chem*, **276**, 44848-44855.

Chalfant, C. E., Rathman, K., Pinkerman, R. L., Wood, R. E., Obeid, L. M., Ogretmen, B., and Hannun, Y. A. (2002) De novo ceramide regulates the alternative splicing of caspase 9 and Bcl-x in A549 lung adenocarcinoma cells. Dependence on protein phosphatase-1. *J Biol Chem*, **277**, 12587-12595.

Chavez, J. A., and Summers, S. A. (2003) Characterizing the effects of saturated fatty acids on insulin signaling and ceramide and diacylglycerol accumulation in 3T3-L1 adipocytes and C2C12 myotubes. *Arch Biochem Biophys*, **419**, 101-109.

Chigorno, V., Giannotta, C., Ottico, E., Sciannamblo, M., Mikulak, J., Prinetti, A., and Sonnino, S. (2005) Sphingolipid uptake by cultured cells: complex aggregates of cell sphingolipids with serum proteins and lipoproteins are rapidly catabolized. *J Biol Chem*, **280**, 2668-2675.

Chung, N., Jenkins, G., Hannun, Y. A., Heitman, J., and Obeid, L. M. (2000) Sphingolipids signal heat stress-induced ubiquitin-dependent proteolysis. *J Biol Chem*, **275**, 17229-17232.

Cowart, L. A., Okamoto, Y., Pinto, F. R., Gandy, J. L., Almeida, J. S., and Hannun, Y. A. (2003) Roles for sphingolipid biosynthesis in mediation of specific programs of the heat stress response determined through gene expression profiling. *J Biol Chem*, **278**, 30328-30338.

Dbaibo, G. S., El-Assaad, W., Krikorian, A., Liu, B., Diab, K., Idriss, N. Z., El-Sabban, M., Driscoll, T. A., Perry, D. K., and Hannun, Y. A. (2001) Ceramide generation by two distinct pathways in tumor necrosis factor alpha-induced cell death. *FEBS Lett*, **503**, 7-12.

Dedov, V. N., Dedova, I. V., and Nicholson, G. A. (2004) Hypoxia causes aggregation of serine palmitoyltransferase followed by non-apoptotic death of human lymphocytes. *Cell Cycle*, **3**, 1271-1277.

Eto, M., Bennouna, J., Hunter, O. C., Hershberger, P. A., Kanto, T., Johnson, C. S., Lotze, M. T., and Amoscato, A. A. (2003) C16 ceramide accumulates following androgen ablation in LNCaP prostate cancer cells. *Prostate*, **57**, 66-79.

Farrell, A. M., Uchida, Y., Nagiec, M. M., Harris, I. R., Dickson, R. C., Elias, P. M., and Holleran, W. M. (1998) UVB irradiation up-regulates serine palmitoyltransferase in cultured human keratinocytes. *J Lipid Res*, **39**, 2031-2038.

Gable, K., Slife, H., Bacikova, D., Monaghan, E., and Dunn, T. M. (2000) Tsc3p is an 80-amino acid protein associated with serine palmitoyltransferase and required for optimal enzyme activity. *J Biol Chem*, **275**, 7597-7603.

Gomez del Pulgar, T., Velasco, G., Sanchez, C., Haro, A., and Guzman, M. (2002) De novo-synthesized ceramide is involved in cannabinoid-induced apoptosis. *Biochem J*, **363**, 183-188.

Gomez-Munoz, A. (1998) Modulation of cell signalling by ceramides. *Biochim Biophys Acta*, **1391**, 92-109.

Grilley, M. M., Stock, S. D., Dickson, R. C., Lester, R. L., and Takemoto, J. Y. (1998) Syringomycin action gene SYR2 is essential for sphingolipid 4-hydroxylation in Saccharomyces cerevisiae. *J Biol Chem*, **273**, 11062-11068.

Han, G., Gable, K., Yan, L., Natarajan, M., Krishnamurthy, J., Gupta, S. D., Borovitskaya, A., Harmon, J. M., and Dunn, T. M. (2004) The topology of the Lcb1p subunit of yeast serine palmitoyltransferase. *J Biol Chem*, **279**, 53707-53716.

Hanada, K. (2003) Serine palmitoyltransferase, a key enzyme of sphingolipid metabolism. *Biochim Biophys Acta*, **1632**, 16-30.

Harris, I. R., Farrell, A. M., Grunfeld, C., Holleran, W. M., Elias, P. M., and Feingold, K. R. (1997) Permeability barrier disruption coordinately regulates mRNA levels for key enzymes of cholesterol, fatty acid, and ceramide synthesis in the epidermis. *J Invest Dermatol*, **109**, 783-787.

Jenkins, G. M., Cowart, L. A., Signorelli, P., Pettus, B. J., Chalfant, C. E., and Hannun, Y. A. (2002) Acute activation of de novo sphingolipid biosynthesis upon heat shock causes an accumulation of ceramide and subsequent dephosphorylation of SR proteins. *J Biol Chem*, **277**, 42572-42578.

Jenkins, G. M., and Hannun, Y. A. (2001) Role for de novo sphingoid base biosynthesis in the heat-induced transient cell cycle arrest of Saccharomyces cerevisiae. *J Biol Chem*, **276**, 8574-8581.

Jiang, Q., Wong, J., Fyrst, H., Saba, J. D., and Ames, B. N. (2004) gamma-Tocopherol or combinations of vitamin E forms induce cell death in human prostate cancer cells by interrupting sphingolipid synthesis. *Proc Natl Acad Sci U S A*, **101**, 17825-17830.

Kitatani, K., Nemoto, M., Akiba, S., and Sato, T. (2002) Stimulation by de novo-synthesized ceramide of phospholipase A(2)-dependent cholesterol esterification promoted by the uptake of oxidized low-density lipoprotein in macrophages. *Cell Signal*, **14**, 695-701.

Kolesnick, R., and Fuks, Z. (2003) Radiation and ceramide-induced apoptosis. *Oncogene*, **22**, 5897-5906.

Liu, Y. Y., Han, T. Y., Giuliano, A. E., and Cabot, M. C. (2001) Ceramide glycosylation potentiates cellular multidrug resistance. *Faseb J*, **15**, 719-730.

Merrill, A. H., Jr., Liotta, D. C., and Riley, R. T. (1996a) Fumonisins: fungal toxins that shed light on sphingolipid function. *Trends Cell Biol*, **6**, 218-223.

Merrill, A. H., Jr., Wang, E., Vales, T. R., Smith, E. R., Schroeder, J. J., Menaldino, D. S., Alexander, C., Crane, H. M., Xia, J., Liotta, D. C., Meredith, F. I., and Riley, R. T. (1996b) Fumonisin toxicity and sphingolipid biosynthesis. *Adv Exp Med Biol*, **392**, 297-306.

Miyake, Y., Kozutsumi, Y., Nakamura, S., Fujita, T., and Kawasaki, T. (1995) Serine palmitoyltransferase is the primary target of a sphingosine-like immunosuppressant, ISP-1/myriocin. *Biochem Biophys Res Commun*, **211**, 396-403.

Nagiec, M. M., Skrzypek, M., Nagiec, E. E., Lester, R. L., and Dickson, R. C. (1998) The LCB4 (YOR171c) and LCB5 (YLR260w) genes of Saccharomyces encode sphingoid long chain base kinases. *J Biol Chem*, **273**, 19437-19442.

Ogretmen, B., and Hannun, Y. A. (2004) Biologically active sphingolipids in cancer pathogenesis and treatment. *Nat Rev Cancer*, **4**, 604-616.

Ogretmen, B., Pettus, B. J., Rossi, M. J., Wood, R., Usta, J., Szulc, Z., Bielawska, A., Obeid, L. M., and Hannun, Y. A. (2002) Biochemical mechanisms of the generation of endogenous long chain ceramide in response to exogenous short chain ceramide in the A549 human lung adenocarcinoma cell line. Role for endogenous ceramide in mediating the action of exogenous ceramide. *J Biol Chem*, **277**, 12960-12969.

Park, T. S., Panek, R. L., Mueller, S. B., Hanselman, J. C., Rosebury, W. S., Robertson, A. W., Kindt, E. K., Homan, R., Karathanasis, S. K., and Rekhter, M. D. (2004) Inhibition of sphingomyelin synthesis reduces atherogenesis in apolipoprotein E-knockout mice. *Circulation*, **110**, 3465-3471.

Perry, D. K. (2002) Serine palmitoyltransferase: role in apoptotic de novo ceramide synthesis and other stress responses. *Biochim Biophys Acta*, **1585**, 146-152.

Petrache, I., Natarajan, V., Zhen, L., Medler, T. R., Richter, A. T., Cho, C., Hubbard, W. C., Berdyshev, E. V., and Tuder, R. M. (2005) Ceramide upregulation causes pulmonary cell apoptosis and emphysema-like disease in mice. *Nat Med*, **11**, 491-498.

Pinto, W. J., Srinivasan, B., Shepherd, S., Schmidt, A., Dickson, R. C., and Lester, R. L. (1992) Sphingolipid long-chain-base auxotrophs of Saccharomyces cerevisiae: genetics, physiology, and a method for their selection. *J Bacteriol*, **174**, 2565-2574.

Reynolds, C. P., Maurer, B. J., and Kolesnick, R. N. (2004) Ceramide synthesis and metabolism as a target for cancer therapy. *Cancer Lett*, **206**, 169-180.

Riebeling, C., Allegood, J. C., Wang, E., Merrill, A. H., Jr., and Futerman, A. H. (2003) Two mammalian longevity assurance gene (LAG1) family members, trh1 and trh4, regulate dihydroceramide synthesis using different fatty acyl-CoA donors. *J Biol Chem*, **278**, 43452-43459.

Riley, R. T., Wang, E., Schroeder, J. J., Smith, E. R., Plattner, R. D., Abbas, H., Yoo, H. S., and Merrill, A. H., Jr. (1996) Evidence for disruption of sphingolipid metabolism as a contributing factor in the toxicity and carcinogenicity of fumonisins. *Nat Toxins*, **4**, 3-15.

Ruvolo, P. P. (2003) Intracellular signal transduction pathways activated by ceramide and its metabolites. *Pharmacol Res*, **47**, 383-392.

Schmelz, E. M., Dombrink-Kurtzman, M. A., Roberts, P. C., Kozutsumi, Y., Kawasaki, T., and Merrill, A. H., Jr. (1998) Induction of apoptosis by fumonisin

B1 in HT29 cells is mediated by the accumulation of endogenous free sphingoid bases. *Toxicol Appl Pharmacol*, **148**, 252-260.

Schmitz-Peiffer, C. (2002) Protein kinase C and lipid-induced insulin resistance in skeletal muscle. *Ann N Y Acad Sci*, **967**, 146-157.

Schorling, S., Vallee, B., Barz, W. P., Riezman, H., and Oesterhelt, D. (2001) Lag1p and Lac1p are essential for the Acyl-CoA-dependent ceramide synthase reaction in Saccharomyces cerevisae. *Mol Biol Cell*, **12**, 3417-3427.

Schulze, H., Michel, C., and van Echten-Deckert, G. (2000) Dihydroceramide desaturase. *Methods Enzymol*, **311**, 22-30.

Schutze, S., Machleidt, T., and Kronke, M. (1994) The role of diacylglycerol and ceramide in tumor necrosis factor and interleukin-1 signal transduction. *J Leukoc Biol*, **56**, 533-541.

Shimabukuro, M., Zhou, Y. T., Levi, M., and Unger, R. H. (1998) Fatty acid-induced beta cell apoptosis: a link between obesity and diabetes. *Proc Natl Acad Sci U S A*, **95**, 2498-2502.

Shiratori, Y., Okwu, A. K., and Tabas, I. (1994) Free cholesterol loading of macrophages stimulates phosphatidylcholine biosynthesis and up-regulation of CTP: phosphocholine cytidylyltransferase. *J Biol Chem*, **269**, 11337-11348.

Slife, C. W., Wang, E., Hunter, R., Wang, S., Burgess, C., Liotta, D. C., and Merrill, A. H., Jr. (1989) Free sphingosine formation from endogenous substrates by a liver plasma membrane system with a divalent cation dependence and a neutral pH optimum. *J Biol Chem*, **264**, 10371-10377.

Spiegel, S., and Merrill, A. H., Jr. (1996) Sphingolipid metabolism and cell growth regulation. *Faseb J*, **10**, 1388-1397.

Spiegel, S., and Milstien, S. (1995) Sphingolipid metabolites: members of a new class of lipid second messengers. *J Membr Biol*, **146**, 225-237.

Spiegel, S., and Milstien, S. (2003) Sphingosine-1-phosphate: an enigmatic signalling lipid. *Nat Rev Mol Cell Biol*, **4**, 397-407.

Summers, S. A., and Nelson, D. H. (2005) A role for sphingolipids in producing the common features of type 2 diabetes, metabolic syndrome X, and Cushing's syndrome. *Diabetes*, **54**, 591-602.

Uehara, K., Miura, S., Takeuchi, T., Taki, T., Nakashita, M., Adachi, M., Inamura, T., Ogawa, T., Akiba, Y., Suzuki, H., Nagata, H., and Ishii, H. (2003) Significant role of ceramide pathway in experimental gastric ulcer formation in rats. *J Pharmacol Exp Ther*, **305**, 232-239.

Uhlinger, D. J., Carton, J. M., Argentieri, D. C., Damiano, B. P., and Dandrea, M. R. (2001) Increased expression of serine palmitoyltransferase (SPT) in balloon-injured rat carotid artery. *Thromb Haemost*, **86**, 1320-1326.

Unger, R. H., and Orci, L. (2002) Lipoapoptosis: its mechanism and its diseases. *Biochim Biophys Acta*, **1585**, 202-212.

Venkataraman, K., Riebeling, C., Bodennec, J., Riezman, H., Allegood, J. C., Sullards, M. C., Merrill, A. H., Jr., and Futerman, A. H. (2002) Upstream of growth and differentiation factor 1 (uog1), a mammalian homolog of the yeast longevity assurance gene 1 (LAG1), regulates N-stearoyl-sphinganine (C18-(dihydro)ceramide) synthesis in a fumonisin B1-independent manner in mammalian cells. *J Biol Chem*, **277**, 35642-35649.

Wang, E., Norred, W. P., Bacon, C. W., Riley, R. T., and Merrill, A. H., Jr. (1991) Inhibition of sphingolipid biosynthesis by fumonisins. Implications for diseases associated with Fusarium moniliforme. *J Biol Chem*, **266**, 14486-14490.

Worgall, T. S., Juliano, R. A., Seo, T., and Deckelbaum, R. J. (2004) Ceramide synthesis correlates with the posttranscriptional regulation of the sterol-regulatory element-binding protein. *Arterioscler Thromb Vasc Biol*, **24**, 943-948.

Zanolari, B., Friant, S., Funato, K., Sutterlin, C., Stevenson, B. J., and Riezman, H. (2000) Sphingoid base synthesis requirement for endocytosis in Saccharomyces cerevisiae. *Embo J*, **19**, 2824-2833.

Zhang, D. X., Fryer, R. M., Hsu, A. K., Zou, A. P., Gross, G. J., Campbell, W. B., and Li, P. L. (2001) Production and metabolism of ceramide in normal and ischemic-reperfused myocardium of rats. *Basic Res Cardiol*, **96**, 267-274.

3-2 Overview of Acid and Neutral Sphingomyelinases in Cell Signaling

Youssef Zeidan, Norma Marchesini, and Yusuf A. Hannun

Department of Biochemistry and Molecular Biology, Medical University of South Carolina, Charleston, SC 29425, USA

Summary. Ceramide, which forms through the activation of sphingomyelinases (SMases), is as a bioactive lipid that mediates cell growth, differentiation, stress responses, and programmed cell death (apoptosis). Molecular and biochemical examinations to determine the role of these enzymes in ceramide-mediated cell signaling is possible with the recent availability of the cloned acid SMase (A-SMase) and neutral SMase (N-SMase). Here we review the recent experimental data and discuss its relevance for understanding the biochemical and molecular properties, regulation, mechanisms and roles of A-SMase and N-SMase.

Keywords. sphingomyelinase, ceramide, apoptosis, Niemann Pick disease (NPD), FAN (factor associated with N-SMase activation)

1. Introduction

To understand the regulation of ceramide generation it is crucial to study the enzymes catalyzing the hydrolysis of sphingomyelin (SM) into phosphocholine and ceramide, SMases. Exogenous stimuli activate N- and A-SMases and contribute to stimulus-induced increases of ceramide levels. Therefore, N- and A- SMases are likely the principal pathways for ceramide production in cell regulation (Chatterjee 1999; Goni and Alonso 2002; Gulbins and Kolesnick 2002; Levade and Jaffrezou 1999). Existing controversies on the roles of these enzymes focused special attention on data from studies using A-SMase and N-SMase deficient mice and cell lines either overexpressing or down-regulating A-SMase and N-SMase.

2. Acid sphingomyelinase

2.1 Gene, message, and protein

A single gene, SMPD1 which spans 5 Kb on the short arm of chromosome 11(11p15.1-11p15.4), encodes human A-SMase (da Veiga Pereira et al. 1991). Of particular interest are events regulating transcription of SMPD1. Indeed, studies of the promoter regions of murine and human SMPD1 showed that transcription factors SP-1 and AP-2 activate those regions (Langmann et al. 1999; Newrzella and Stoffel 1992).

A-SMase undergoes extensive posttranslational modifications. Glyco-sylation determines the stability, catalytic activity and cellular trafficking of the enzyme. Studies of murine and human A-SMases identified six as-paragine residues that can be glycosylated (Ferlinz et al. 1997; Newrzella and Stoffel 1996). Of major importance are the Golgi transferases con-ferring mannose-6-phosphate (M6P) residues to the glycoprotein. Similar to other lysosomal proteins. the recognition of these M6P residues by their trans Golgi receptors is crucial for lysosomal trafficking. Interestingly, a secretory isoform of A-SMase (S-SMase) lacking M6P residues, charac-terized by Tabas group (Schissel et al. 1996), may play a role in inflam-matory responses and atherosclerosis (Wong et al. 2000).

2.2 Receptor and non-receptor mediated regulation of A-SMase

Researchers observed activation of A-SMase in response to several cellular stress agents, including cytokines (TNFβ, IL-1, IL-2), chemotherapeutic agents (fenretinide, cisplatin, doxorubicin), radiation (UV and $\beta\beta$radiation), reactive oxygen species, and bacterial (*Staphylococcus aureus*, *Pseudo-monas aeroginosa*) and viral (*Rhinovirus*) infections.

Several studies also addressed A-SMase regulation in response to TNFβ. When TNF interacted with its receptor, adaptor proteins TRADD and FADD, as well as caspase 8, are recruited to TNF-R55 and the death inducing signaling complex, or DISC, is formed (Schwandner et al. 1998). Schutze and co-workers showed that the internalization of TNF-R55 is crucial for DISC formation and suggested that further downstream signal-ing, including activation of A-SMase, occurs after caspase 8 signaling

(Schneider-Brachert et al. 2004; Schutze et al. 1999). Earlier reports, however, suggested that overexpression of this protease did not enhance A-SMase activation by TNF even though pancaspase inhibition (ZVAD and crmA overexpression) attenuated that A-SMase activation (Schwandner et al. 1998). On the other hand, ligation of a closely related death receptor (Fas/CD 95) activates A-SMase in a caspase 8 dependent manner (Rotolo et al. 2005). DISC orientation interferes with the direct interaction of A-SMase with caspase 8, or other members of DISC. This is because DISC faces the cytoplasmic side of the TNF receptosome whereas A-SMase resides in either the lumen of the endolysosomes or on the external leaflet of the plasma membrane (Kronke 1999).

With UV radiation, the enzyme will translocate to the exoplasmic leaflet, where most cellular SM is located (Charruyer et al. 2005; Rotolo et al. 2005). Unlike CD95 translocation process was independent of caspase 8, but the mechanisms regulating this translocation have not been examined. To date, the molecular mechanisms for the activation of A-SMase, in response to either cytokines or other stress agents, remain undetermined.

2.3 Role of A-SMase in raft remodeling

Cell surface receptor clustering is an early step in many signal transduction pathways that many different classes of receptors share, including receptor tyrosine kinases (RTKs) such as EGFR and PDGFR and death receptors such as TNFR and CD95. Ceramide generated within lipid rafts might play a key role in raft formation. Gulbins et al. first suggested that receptor ligand binding induces rapid translocation of A-SMase to membrane rafts (Gulbins and Grassme 2002) and that subsequent SM hydrolysis and ceramide generation result in the fusion of raft microdomains and form polar signaling platforms. Although this model was originally proposed to describe CD95 mediated cytotoxicity, later it was used to describe the processes other signaling cascades including CD14, CD40 and EGFR signaling (Grassme et al. 2002; Huang et al. 2005). Interestingly, recent reports indicate that platforms can form in response to cytotoxic agents such as cisplatin, suggesting that non-ligands may also stimulate raft clustering (Lacour et al. 2004). Although this model is plausible, questions remain unanswered. Do these mega platforms exist physiologically and is ceramide formation necessary? How much ceramide is required for this process? Which form of A-SMase (lysosomal-SMase or secretory-SMase) translocates and by what mechanism?

2.4 Downstream targets of A-SMase derived ceramide

Although several events downstream of A-SMase activation have been described, only a few proteins interact directly with ceramide. In fact, Schutze and coworkers suggested that the acidic aspartate protease, cathepsin D, serves as a direct target for endolysosomal ceramide (Heinrich et al. 1999). After its activation by ceramide, cathepsin D matures to its 32 KDa active form before released into the cytosol. Consequently, cytosolic cathepsin D triggers the mitochondrial death pathway by cleaving Bid, which promotes cytochrome c release (Heinrich et al. 2004). Another recent study showed that ceramide, generated by activation of A-SMase in response to UV light, caused Bax oligomerization and subsequent cytochrome c release (Kashkar et al. 2005). They found that A-SMase deficient lymphoblasts were resistant to UV light unless they were reconstituted with A-SMase. Interestingly, Bax activation was detected in intact cells as well as in isolated mitochondria after treatment with ceramide, suggesting a direct lipid-protein interaction. Therefore, the mitochondrion and its associated proteins may represent key targets for A-SMase derived ceramide. Yet, it is not clear how lysosomal ceramide reaches the mitochondria.

2.5 Functions of A-SMase in cell regulation

Several studies addressing the role of A-SMase in cytokine and stress response initially relied on desipramine (or other related compounds) as an inhibitor. However, desiprmaine is not a specific inhibitor of A-Smase, it indirectly induces proteolytic cleavage of the enzyme, making these studies somewhat suspect.

The A-SMase knock-out mouse can now provide an in vivo model to study the enzyme. This knock-out mouse recapitulates the phenotype of Niemman-Pick disease (NPD), a neurodegenerative lysosomal storage disorder resulting from A-SMase deficiency (Horinouchi et al. 1995;Stoffel 1999). In vivo resistance to CD95 mediated apoptosis was reported in different cell types of this knock out mouse, including splenocytes, hepatocytes and lymphocytes (Kirschnek et al. 2000). Moreover, grafted fibrosarcomas and melanomas had faster rates of growth in A-SMase[-/-] mice when compared to wild type mice. In addition, tumor endothelial cells failed to undergo radiation-induced apoptosis in the A-SMase knock out mouse (Garcia-Barros et al. 2004). Taken together, these results suggest a role for A-SMase in tumor suppression and as a positive response to radiotherapy. However, other groups reported a normal apoptotic response to

CD95 in cell lines derived from NPD patients (Segui et al. 2000). Moreover, an unexpected apoptotic response to CD95 activation was observed in cells derived from the knock-out mouse generated by Stoffel et al. In this strain, A-SMase$^{-/-}$ splenocytes pretreated with anti-CD3 demonstrated higher sensitivity to CD95 cytotoxicity than its wild type (Nix and Stoffel 2000). An explanation for this discrepancy between the NPD cell culture models and the A-SMase null mouse models has not yet been provided (Lozano et al. 2001).

In the case of TNF, Kronke's team proposed a central role for A-SMase in NF-kB activation (Schutze et al. 1992). Yet mouse embryonic fibroblasts (MEFs) from A-SMase deficient mice respond normally to TNF stimulation (Zumbansen and Stoffel 1997). So, the authors concluded that A-SMase is not involved in TNF signaling. More studies are required to accurately determine the importance of A-SMase in TNF mediated responses.

A pivotal role for A-SMase during UV mediated cellular injury was recently reported. As noted above, UV radiation can activate A-SMase which leads to the formation of ceramide enriched signaling platforms, a process requiring intact rafts and microtubule and cytoskeletal networks. Mechanistically, the report demonstrated that A-SMase activation is necessary for activation of Bax and the induction of the mitochondrial death pathway by UV (Kashkar et al. 2005).

3. Neutral Mg^{2+}-dependent sphingomyelinases

Neutral Mg^{2+}-dependent sphingomyelinases are integral membrane proteins in mammals and soluble proteins in bacteria. Many groups have attempted to purify membrane-bound neutral SMase from different sources (Goni and Alonso 2002; Marchesini and Hannun 2004; Sawai and Hannun 1999). Two mammalian membrane-associated N-SMases (designated nSMase1 and nSMase2) were recently identified based on their sequence similarities with *Bacillus cereus*.

3.1 Neutral Mg^{2+}-dependent sphingomyelinase 1

nSMase1 encodes an ubiquitously expressed 47 kDa protein (Tomiuk et al. 1998). Sawai et al. demonstrated that, although nSMase1 exhibited sphingomyelinase activity in vitro, cells overexpressing the protein show no sphingomyelin metabolic changes (Sawai et al. 1999). Rather, it was the putative N-SMase protein that acted as a lyso-PAF phospholipase C in vitro and in cells. This has been confirmed by other studies.

3.2 Neutral Mg^{2+}-dependent sphingomyelinase 2

Hofmann et al (2000) reported the cloning of a 71 kDa mammalian brain-specific, Mg^{2+}-dependent N-SMase, nSMase2 (Hofmann et al. 2000). We (Marchesini et al. 2003) expressed and characterized the mammalian enzyme in yeast and in MCF7 breast cancer cells. Our results showed that nSMase2 does indeed act as a neutral sphingomyelinase in mammalian cells. Cells overexpressing nSMase2 had lower sphingomyelin levels and higher ceramide levels than corresponding cells transfected with a vector only. nSMase2 is activated by phosphatidylserine, a requirement for anionic phospholipid that is shared with the related yeast enzyme ISC1, that possesses an inositol phosphosphingolipid phospholipase C activity (Sawai et al. 2000) and with the N-SMase purified from rat brain (Liu et al. 1998). In addition nSMase2 was inhibited by GW4869, a newly developed inhibitor of N-SMase (Luberto et al. 2002).

Several reports suggest nSMase2 involvement in SM/ceramide signaling. nSMase2 is activated in MCF7 after TNF-β stimulation (Marchesini et al. 2003) and in hepatocytes after IL-1β stimulation (Karakashian et al. 2004). Immunocytochemistry studies by Stoffel and co-workers (Hofmann et al. 2000; Stoffel et al. 2005) used Golgi marker and observed endogenous and overexpressed nSMase2 co-localization in several cell lines. However, others studies showed that the enzyme displays intracellular as well as plasma membrane localization (Karakashian et al. 2004; Marchesini et al. 2004).

A specific role for nSMase2 in cell growth regulation and confluence-induced cell cycle arrest has been suggested (Marchesini et al. 2004). Confluency induces the enzyme's messenger RNA, and the gene was originally cloned as cell arrest associated protein-1 (CCA-1), independently isolated from rat 3Y1 cells (Hayashi et al. 1997). Overexpression of nSMase2 resulted in a premature cell cycle arrest, whereas specific down regulation of the enzyme by siRNA delayed confluence-induced cell cycle arrest. Moreover, while in subconfluent cells nSMase2 was primarily intracellular, the enzyme became markedly enriched at the plasma membrane upon cell-cell contact. These results suggest that nSMase2 localization may be a dynamic process that is subjected to regulatory mechanisms (Marchesini et al. 2004). Interestingly, nSMase2 is localized to chromosome 16q22.1, near the cadherin cluster. Cadherins are involved in contact inhibition, adhesion migration and play a role in the wingless signal. The close chromosomal localization of cadherins and nSMase2 and their conserved order in human and mouse chromosomes might indicate that they share genetic control elements. Further studies should clarify this point and may reveal still other functions of this interesting enzyme.

3.3 Neutral sphingomyelinase: role in the stress response

Several cellular stresses including various members of the tumor necrosis factor (TNF) receptor family, anticancer drugs, oxidants, hypoxia and amyloid-β peptide (Aβ) have been shown to activate N-SMase (Chatterjee 1999; Kronke 1999; Levade and Jaffrezou 1999). Interestingly, Niemann-Pick cells could generate ceramide and undergo apoptosis after CD95 (Fas) stimulation, exposure to anti-class I antibodies, or TNF-β stimulation. This implies an action of other sphingomyelinases (and/or other pathways of ceramide generation) (Levade et al. 2002). Several N-SMase inhibitors have been used in vivo to implicate N-SMase action in specific responses. Scyphostatin has been employed to elucidate the role of N-SMase in eNOS activation (Barsacchi et al. 2003) and to prevent N-SMase induced mechanotransduction in vascular endothelium (Czarny et al. 2003). The N-SMase inhibitor GW4869 was developed following a high throughput screen and exhibited significant and specific inhibitory activity on N-SMase in vitro and in cells (Luberto et al. 2002). This inhibitor implicated activation of N-SMase in the regulation of apoptosis induced by TNF-β (Luberto et al. 2002). In addition, Chatterjee et al. (Kolmakova et al. 2004) showed that GW4869 abrogated apoC-I-induced ceramide generation and apoptosis in smooth muscle cells. Also the new inhibitor, C11AG, blocks LPS-stimulated sphingomyelin degradation and NF kappa B activation in macrophages (Amtmann et al. 2003). In addition, in neuronal cells the role of neutral sphingomyelinase in lipopolysaccharide (LPS) and beta-amyloid-induced cytotoxity was evidenced by the N-SMase inhibitor 3-O-methyl sphingomyelin (Ayasolla et al. 2004; Won et al. 2004). Indeed, genetic knockdown of nSMase2 using siRNA has clearly implicated this enzyme in the regulation of apoptosis in response to amyloid β-peptide (Aβ) in neuronal cells (Jana and Pahan 2004; Lee et al. 2004; Yang et al. 2004).

3.4 N-SMase mechanisms of activation

Several factors appear to modulate, either positively or negatively, the activity of N-SMase (Chatterjee et al. 1999; Marchesini and Hannun 2004)). Two of the better studied mechanisms are discussed here.

3.4.1 Regulation by oxidative stress

Several reports have shown that exogenous oxidants or conditions known to promote elevated cellular levels of reactive oxygen species (ROS) lead to activation of N-SMase. Tripeptide glutathione (GSH) plays a major role in redox homeostasis, and N-SMase can be regulated by GSH in vivo. Liu and Hannun (1997) were the first to demonstrate that in the human acute lymphoblastic leukemic Molt-4 cells, the depletion of intracellular GSH activated N-Smase while the addition of GSH inhibited the enzyme. In addition, exogenous GSH inhibited the TNFβ induced activation of N-SMase in MCF-7 cells (Liu et al. 1998) as well as the formation of ceramide in response to hypoxia in neuronal cells (Yoshimura et al. 1999). N-acetylcysteine (NAC), a well characterized thiol reducing agent that raises the intracellular GSH pool, provided further evidence that GSH may regulate N-SMase under oxidative stress conditions induced by cytokines as well as Aß. Indeed, pretreatment with NAC, in daunorubicin- (Mansat-de Mas et al. 1999), 1-h-D-arabinofuranosylcytosine- (Ara-C) (Bezombes et al. 2001), doxorubicin-, or diamide-treated cells (Gouaze V et al. 2001) inhibited N-SMase activation. Recently, various reports demonstrated that Aß induces cell death by activation the N-SMase-ceramide pathway via an oxidative mechanism (Alessenko et al. 2004; Lee et al. 2004). Indeed, it was shown that NADPH oxidase-mediated superoxide production in neurons may be responsible for A   activation of N-SMase (Jana and Pahan 2004).

One interesting observation suggested that N-SMase is regulated by plasma membrane antioxidants, such that ceramide production and cell death induced by serum withdrawal in leukemic cells was prevented by ubiquinone (coenzyme Q) (Fernandez-Ayala et al. 2000; Lopez-Lluch et al. 1999).

3.4.2 Regulation by FAN

Adam et al (1996) identified a novel adaptor protein FAN (factor associated with N-SMase activation) that specifically binds to the TNF receptor. It may also couple TNF receptor to N-SMase activation (Adam-Klages et al. 1996) In addition, using deletion mutants of the p55 TNF receptor

(TNF-R55), they identified a domain (NSD) adjacent to the death domain that was specifically required for activation of N-SMase (Adam et al. 1996). The role of FAN in enhanced N-SMase activity in TNF signaling was demonstrated by overexpressing full-length FAN and was later confirmed in cells derived from FAN-deficient mice (Kreder et al. 1999). These and others studies (Segui et al. 1999; Segui et al. 2001) define a specific pathway leading to the receptor mediated activation of N-SMase; however the mechanisms coupling FAN to N-SMase are unknown. Recently, the receptor for activated C-kinase 1 (Tcherkasowa et al. 2002) and caveolin-1 (Veldman et al. 2001) was shown to interact with N-SMase, and regulate its activity. These results suggest that the mechanism of regulation of N-SMase may involve further members of a putative multicomplex protein.

3.5 nSMase1 and/or nSMase2 knockout mouse

Knock-out mice for nSMase1 and nSMase2 have been generated (Stoffel et al. 2005; Zumbansen and Stoffel 2002). In agreement with the biochemical properties of nSMase1 and its proposed role as a lysoPAF phospholipase and not a sphigomyelinase, the nSMase1 mouse demonstrated no perturbations in SM metabolism (Zumbansen and Stoffel 2002). A nSMase2 knock-out and a double nSMase1-nSMase2 knock-out have also been generated. Interestingly, the double mutant mice completely lacked N-SMase activity in vitro (under basal conditions). However, it did not develop SM storage disease or impaired apoptosis. Stoffel et al. (2005), with the those two strains revealed a pivotal function of nSMase2 in the control of the hypothalamus–pituitary growth axis. They proposed a role for nSMase2 as a regulator of postnatal development.

4. Future directions

Although the precise cellular role of each of the above SMases in SM hydrolysis and in specific cell functions remains unclear, recent advances in research on A-SMase and N-SMase are beginning to elucidate characteristics of the enzyme and mechanisms of regulation in vitro and in cells. The molecular cloning of mammalian SMases and the increasing availability of specific tools such as siRNA, specific pharmacologic inhibitors, antibodies, as well as the recent generation of knockout mice promise accelerated insight into mechanisms of regulation and the identification of the sphingomyelinases involved in specific signaling pathways and their respective biological significance for cellular processes. The proposed relationship among SMases and various factors are summarized in Fig. 1.

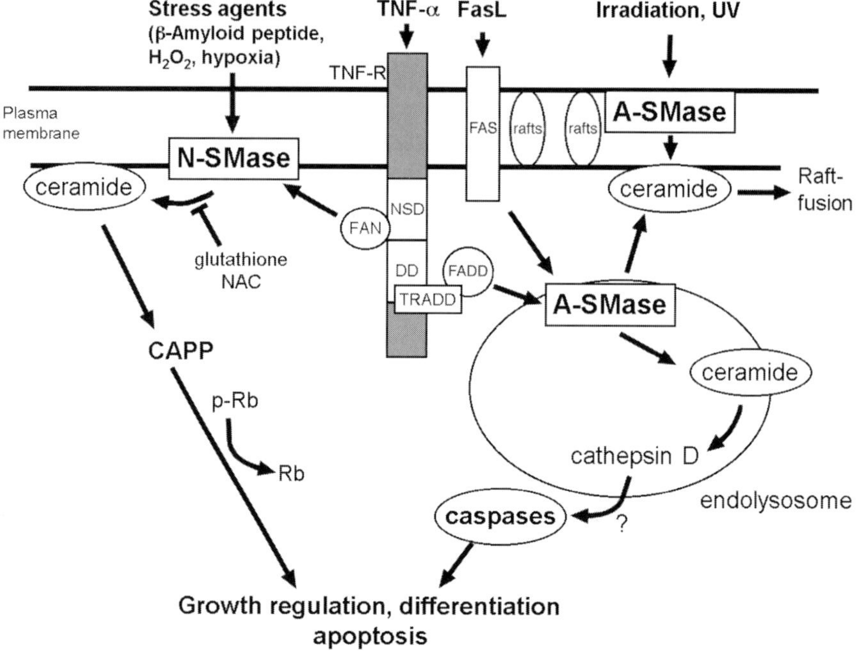

Fig. 1. Neutral and acid SMase activation in cell signaling. Abbreviations: CAPP, ceramide activated protein phosphatase; DD, death domain; FasL, Fas ligand; NAC, N-acetylcysteine; NSD; neutral SMase activation domain; Rb, Retinoblastoma, SM, sphingomyelin.

5. References

Adam-Klages S, Adam D, Wiegmann K, Struve S, Kolanus W, Schneider-Mergener J, Krönke M (1996) FAN, a novel WD-repeat protein, couples the p55 TNF-receptor to neutral sphingomyelinase. *Cell*, **86**, 937-947.

Adam D, Wiegmann K, Adam-Klages S, Ruff A, Krönke M (1996) A novel cytoplasmic domain of the p55 tumor necrosis factor receptor initiates the neutral sphingomyelinase pathway. *J Biol Chem*, **271**, 14617-14622.

Alessenko AV, Bugrova AE, Dudnik LB (2004) Connection of lipid peroxide oxidation with the sphingomyelin pathway in the development of Alzheimer's disease. *Biochem Soc Trans*, **32**, 144-146.

Amtmann E, Baader W, Zoller M (2003) Neutral sphingomyelinase inhibitor C11AG prevents lipopolysaccharide-induced macrophage activation. *Drugs Exp Clin Res*, **29**, 5-13.

Ayasolla K, Khan M, Singh AK, Singh I (2004) Inflammatory mediator and beta-amyloid (25-35)-induced ceramide generation and iNOS expression are inhibited by vitamin E. *Free Radic Biol Med*, **37**, 325-338.

Barsacchi R, Perrotta C, Bulotta S, Moncada S, Borgese N, Clementi E (2003) Activation of endothelial nitric-oxide synthase by tumor necrosis factor-alpha: a novel pathway involving sequential activation of neutral sphingomyelinase, phosphatidylinositol-3' kinase, and Akt. *Mol Pharmacol*, **63**, 886-895.

Bezombes C, Plo I, Mansat-De Mas V, Quillet-Mary A, Negre-Salvayre A, Laurent G, Jaffrezou JP (2001) Oxidative stress-induced activation of Lyn recruits sphingomyelinase and is requisite for its stimulation by Ara-C. *Faseb J*, **15**, 1583-1585.

Charruyer A, Grazide S, Bezombes C, Muller S, Laurent G, Jaffrezou JP (2005) UV-C light induces raft-associated acid sphingomyelinase and JNK activation and translocation independently on a nuclear signal. *J Biol Chem*, **280**, 19196-19204.

Chatterjee S (1999) Neutral sphingomyelinase: past, present and future. *Chem Phys Lipids*, **102**, 79-96.

Chatterjee S, Han H, Rollins S, Cleveland T (1999) Molecular cloning, characterization, and expression of a novel human neutral sphingomyelinase. *J Biol Chem*, **274**, 37407-37412.

Czarny M, Liu J, Oh P, Schnitzer JE (2003) Transient mechanoactivation of neutral sphingomyelinase in caveolae to generate ceramide. *J Biol Chem*, **278**, 4424-4430.

da Veiga Pereira L, Desnick RJ, Adler DA, Disteche CM, Schuchman EH (1991) Regional assignment of the human acid sphingomyelinase gene (SMPD1) by PCR analysis of somatic cell hybrids and in situ hybridization to 11p15.1----p15.4. *Genomics*, **9**, 229-234.

Ferlinz K, Hurwitz R, Moczall H, Lansmann S, Schuchman EH, Sandhoff K (1997) Functional characterization of the N-glycosylation sites of human acid sphingomyelinase by site-directed mutagenesis. *Eur J Biochem*, **243**, 511-517.

Fernandez-Ayala DJ, Martin SF, Barroso MP, Gomez-Diaz C, Villalba JM, Rodriguez-Aguilera JC, Lopez-Lluch G, Navas P (2000) Coenzyme Q protects cells against serum withdrawal-induced apoptosis by inhibition of ceramide release and caspase-3 activation. *Antioxid Redox Signal*, **2**, 263-275.

Garcia-Barros M, Lacorazza D, Petrie H, Haimovitz-Friedman A, Cardon-Cardo C, Nimer S, Fuks Z, Kolesnick R (2004) Host acid sphingomyelinase regulates microvascular function not tumor immunity. *Cancer Res*, **64**, 8285-8291.

Goni FM, Alonso A (2002) Sphingomyelinases: enzymology and membrane activity. *FEBS Lett*, **531**, 38-46.

Grassme H, Bock J, Kun J, Gulbins E (2002) Clustering of CD40 ligand is required to form a functional contact with CD40. *J Biol Chem*, **277**, 30289-30299.

Gulbins E, Grassme H (2002) Ceramide and cell death receptor clustering. *Biochim Biophys Acta*, **1585**, 139-145.

Gulbins E, Kolesnick R (2002) Acid sphingomyelinase-derived ceramide signaling in apoptosis. *Subcell Biochem*, **36**, 229-244.

Hayashi Y, Kiyono T, Fujita M, Ishibashi M (1997) cca1 is required for formation of growth-arrested confluent monolayer of rat 3Y1 cells. *J Biol Chem*, **272**, 18082-18086.

Heinrich M, Neumeyer J, Jakob M, Hallas C, Tchikov V, Winoto-Morbach S, Wickel M, Schneider-Brachert W, Trauzold A, Hethke A, Schutze S (2004) Cathepsin D links TNF-induced acid sphingomyelinase to Bid-mediated caspase-9 and -3 activation. *Cell Death Differ*, **11**, 550-563.

Heinrich M, Wickel M, Schneider-Brachert W, Sandberg C, Gahr J, Schwandner R, Weber T, Saftig P, Peters C, Brunner J, Kronke M, Schutze S (1999) Cathepsin D targeted by acid sphingomyelinase-derived ceramide. *Embo J*, **18**, 5252-5263.

Hofmann K, Tomiuk S, Wolff G, Stoffel W (2000) Cloning and characterization of the mammalian brain-specific, Mg2+-dependent neutral sphingomyelinase. *Proc Natl Acad Sci U S A*, **97**, 5895-5900.

Huang Y, Yang J, Shen J, Chen FF, Yu Y (2005) Sphingolipids are involved in N-methyl-N'-nitro-N-nitrosoguanidine-induced epidermal growth factor receptor clustering. *Biochem Biophys Res Commun*, **330**, 430-438.

Jana A, Pahan K (2004) Fibrillar amyloid-beta peptides kill human primary neurons via NADPH oxidase-mediated activation of neutral sphingomyelinase. Implications for Alzheimer's disease. *J Biol Chem*, **279**, 51451-51459.

Karakashian AA, Giltiay NV, Smith GM, Nikolova-Karakashian MN (2004) Expression of neutral sphingomyelinase-2 (NSMase-2) in primary rat hepatocytes modulates IL-beta-induced JNK activation. *Faseb J*, **18**, 968-970.

Kashkar H, Wiegmann K, Yazdanpanah B, Haubert D, Kronke M (2005) Acid Sphingomyelinase Is Indispensable for UV Light-induced Bax Conformational Change at the Mitochondrial Membrane. *J Biol Chem*, **280**, 20804-20813.

Kirschnek S, Paris F, Weller M, Grassme H, Ferlinz K, Riehle A, Fuks Z, Kolesnick R, Gulbins E (2000) CD95-mediated apoptosis in vivo involves acid sphingomyelinase. *J Biol Chem*, **275**, 27316-27323.

Kolmakova A, Kwiterovich P, Virgil D, Alaupovic P, Knight-Gibson C, Martin SF, Chatterjee S (2004) Apolipoprotein C-I induces apoptosis in human aortic smooth muscle cells via recruiting neutral sphingomyelinase. *Arterioscler Thromb Vasc Biol*, **24**, 264-269.

Kreder D, Krut O, Adam-Klages S, Wiegmann K, Scherer G, Plitz T, Jensen JM, Proksch E, Steinmann J, Pfeffer K, Kronke M (1999) Impaired neutral sphingomyelinase activation and cutaneous barrier repair in FAN-deficient mice. *Embo J*, **18**, 2472-2479.

Kronke M (1999) Involvement of sphingomyelinases in TNF signaling pathways. *Chem Phys Lipids*, **102**, 157-166.

Lacour S, Hammann A, Grazide S, Lagadic-Gossmann D, Athias A, Sergent O, Laurent G, Gambert P, Solary E, Dimanche-Boitrel MT (2004) Cisplatin-induced CD95 redistribution into membrane lipid rafts of HT29 human colon cancer cells. *Cancer Res*, **64**, 3593-3598.

Langmann T, Buechler C, Ries S, Schaeffler A, Aslanidis C, Schuierer M, Weiler M, Sandhoff K, de Jong PJ, Schmitz G (1999) Transcription factors Sp1 and AP-2 mediate induction of acid sphingomyelinase during monocytic differentiation. *J Lipid Res*, **40**, 870-880.

Lee JT, Xu J, Lee JM, Ku G, Han X, Yang DI, Chen S, Hsu CY (2004) Amyloid-beta peptide induces oligodendrocyte death by activating the neutral sphingomyelinase-ceramide pathway. *J Cell Biol*, **164**, 123-131.

Levade T, Jaffrezou JP (1999) Signalling sphingomyelinases: which, where, how and why? *Biochim Biophys Acta*, **1438**, 1-17.

Liu B, Andrieu-Abadie N, Levade T, Zhang P, Obeid LM, Hannun YA (1998) Glutathione regulation of neutral sphingomyelinase in tumor necrosis factor-alpha-induced cell death. *J Biol Chem*, **273**, 11313-11320.

Lopez-Lluch G, Barroso MP, Martin SF, Fernandez-Ayala DJ, Gomez-Diaz C, Villalba JM, Navas P (1999) Role of plasma membrane coenzyme Q on the regulation of apoptosis. *Biofactors*, **9**, 171-177.

Lozano J, Morales A, Cremesti A, Fuks Z, Tilly JL, Schuchman E, Gulbins E, Kolesnick R (2001) Niemann-Pick Disease versus acid sphingomyelinase deficiency. *Cell Death Differ*, **8**, 100-103.

Luberto C, Hassler DF, Signorelli P, Okamoto Y, Sawai H, Boros E, Hazen-Martin DJ, Obeid LM, Hannun YA, Smith GK (2002) Inhibition of tumor necrosis factor-induced cell death in MCF7 by a novel inhibitor of neutral sphingomyelinase. *J Biol Chem*, **277**, 41128-41139.

Mansat-de Mas V, Bezombes C, Quillet-Mary A, Bettaieb A, D'Orgeix A D, Laurent G, Jaffrezou JP (1999) Implication of radical oxygen species in ceramide generation, c-Jun N-terminal kinase activation and apoptosis induced by daunorubicin. *Mol Pharmacol*, **56**, 867-874.

Marchesini N, Hannun YA (2004) Acid and neutral sphingomyelinases: roles and mechanisms of regulation. *Biochem Cell Biol*, **82**, 27-44.

Marchesini N, Luberto C, Hannun YA (2003) Biochemical properties of mammalian neutral sphingomyelinase 2 and its role in sphingolipid metabolism. *J Biol Chem*, **278**, 13775-13783.

Marchesini N, Osta W, Bielawski J, Luberto C, Obeid LM, Hannun YA (2004) Role for mammalian neutral sphingomyelinase 2 in confluence-induced growth arrest of MCF7 cells. *J Biol Chem*, **279**, 25101-25111.

Newrzella D, Stoffel W (1992) Molecular cloning of the acid sphingomyelinase of the mouse and the organization and complete nucleotide sequence of the gene. *Biol Chem Hoppe Seyler*, **373**, 1233-1238.

Newrzella D, Stoffel W (1996) Functional analysis of the glycosylation of murine acid sphingomyelinase. *J Biol Chem*, **271**, 32089-32095.

Nix M, Stoffel W (2000) Perturbation of membrane microdomains reduces mitogenic signaling and increases susceptibility to apoptosis after T cell receptor stimulation. *Cell Death Differ*, **7**, 413-424.

Rotolo JA, Zhang J, Donepudi M, Lee H, Fuks Z, Kolesnick R (2005) Caspase-dependent and-independent activation of acid sphingomyelinase signaling. *J Biol Chem*, **280**, 26425-26434.

Sawai H, Domae N, Nagan N, Hannun YA (1999) Function of the cloned putative neutral sphingomyelinase as lyso-platelet activating factor-phospholipase C. *J Biol Chem*, **274**, 38131-38139.

Sawai H, Hannun YA (1999) Ceramide and sphingomyelinases in the regulation of stress responses. *Chem Phys Lipids*, **102**, 141-147.

Sawai H, Okamoto Y, Luberto C, Mao C, Bielawska A, Domae N, Hannun YA (2000) Identification of ISC1 (YER019w) as inositol phosphosphingolipid phospholipase C in Saccharomyces cerevisiae. *J Biol Chem*, **275**, 39793-39798.

Schissel SL, Schuchman EH, Williams KJ, Tabas I (1996) Zn2+-stimulated sphingomyelinase is secreted by many cell types and is a product of the acid sphingomyelinase gene. *J Biol Chem*, **271**, 18431-18436.

Schneider-Brachert W, Tchikov V, Neumeyer J, Jakob M, Winoto-Morbach S, Held-Feindt J, Heinrich M, Merkel O, Ehrenschwender M, Adam D, Mentlein R, Kabelitz D, Schutze S (2004) Compartmentalization of TNF receptor 1 signaling: internalized TNF receptosomes as death signaling vesicles. *Immunity*, **21**, 415-428.

Schutze S, Machleidt T, Adam D, Schwandner R, Wiegmann K, Kruse ML, Heinrich M, Wickel M, Kronke M (1999) Inhibition of receptor internalization by monodansylcadaverine selectively blocks p55 tumor necrosis factor receptor death domain signaling. *J Biol Chem*, **274**, 10203-10212.

Schutze S, Potthoff K, Machleidt T, Berkovic D, Wiegmann K, Kronke M (1992) TNF activates NF-kappa B by phosphatidylcholine-specific phospholipase C-induced "acidic" sphingomyelin breakdown. *Cell*, **71**, 765-776.

Schwandner R, Wiegmann K, Bernardo K, Kreder D, Kronke M (1998) TNF receptor death domain-associated proteins TRADD and FADD signal activation of acid sphingomyelinase. *J Biol Chem*, **273**, 5916-5922.

Segui B, Andrieu-Abadie N, Adam-Klages S, Meilhac O, Kreder D, Garcia V, Bruno AP, Jaffrezou JP, Salvayre R, Kronke M, Levade T (1999) CD40 signals apoptosis through FAN-regulated activation of the sphingomyelin-ceramide pathway. *J Biol Chem*, **274**, 37251-37258.

Segui B, Bezombes C, Uro-Coste E, Medin JA, Andrieu-Abadie N, Auge N, Brouchet A, Laurent G, Salvayre R, Jaffrezou JP, Levade T (2000) Stress-induced apoptosis is not mediated by endolysosomal ceramide. *Faseb J*, **14**, 36-47.

Segui B, Cuvillier O, Adam-Klages S, Garcia V, Malagarie-Cazenave S, Leveque S, Caspar-Bauguil S, Coudert J, Salvayre R, Kronke M, Levade T (2001) Involvement of FAN in TNF-induced apoptosis. *J Clin Invest*, **108**, 143-151.

Stoffel W, Jenke B, Block B, Zumbansen M, Koebke J (2005) Neutral sphingomyelinase 2 (smpd3) in the control of postnatal growth and development. *Proc Natl Acad Sci U S A*, **102**, 4554-4559.

Tcherkasowa AE, Adam-Klages S, Kruse ML, Wiegmann K, Mathieu S, Kolanus W, Kronke M, Adam D (2002) Interaction with factor associated with neutral sphingomyelinase activation, a WD motif-containing protein, identifies receptor for activated C-kinase 1 as a novel component of the signaling pathways of the p55 TNF receptor. *J Immunol*, **169**, 5161-5170.

Tomiuk S, Hofmann K, Nix M, Zumbansen M, Stoffel W (1998) Cloned mammalian neutral sphingomyelinase: functions in sphingolipid signaling? *Proc Natl Acad Sci U S A*, **95**, 3638-3643.

Veldman RJ, Maestre N, Aduib OM, Medin JA, Salvayre R, Levade T (2001) A neutral sphingomyelinase resides in sphingolipid-enriched microdomains and is inhibited by the caveolin-scaffolding domain: potential implications in tumour necrosis factor signalling. *Biochem J*, **355**, 859-868.

Won JS, Im YB, Khan M, Singh AK, Singh I (2004) The role of neutral sphingomyelinase produced ceramide in lipopolysaccharide-mediated expression of inducible nitric oxide synthase. *J Neurochem*, **88**, 583-593.

Wong ML, Xie B, Beatini N, Phu P, Marathe S, Johns A, Gold PW, Hirsch E, Williams KJ, Licinio J, Tabas I (2000) Acute systemic inflammation up-regulates secretory sphingomyelinase in vivo: a possible link between inflammatory cytokines and atherogenesis. *Proc Natl Acad Sci U S A*, **97**, 8681-8686.

Yang DI, Yeh CH, Chen S, Xu J, Hsu CY (2004) Neutral sphingomyelinase activation in endothelial and glial cell death induced by amyloid beta-peptide. *Neurobiol Dis*, **17**, 99-107.

Yoshimura S, Banno Y, Nakashima S, Hayashi K, Yamakawa H, Sawada M, Sakai N, Nozawa Y (1999) Inhibition of neutral sphingomyelinase activation and ceramide formation by glutathione in hypoxic PC12 cell death. *J Neurochem*, **73**, 675-683.

Zumbansen M, Stoffel W (1997) Tumor necrosis factor alpha activates NF-kappaB in acid sphingomyelinase-deficient mouse embryonic fibroblasts. *J Biol Chem*, **272**, 10904-10909.

Zumbansen M, Stoffel W (2002) Neutral sphingomyelinase 1 deficiency in the mouse causes no lipid storage disease. *Mol Cell Biol*, **22**, 3633-3638.

3-3 Neutral Ceramidase as an Integral Modulator for the Generation of S1P and S1P-Mediated Signaling

Makoto Ito[1,2] , Motohiro Tani[1,3,4], and Yukihiro Yoshimura[1]

[1]Department of Bioscience and Biotechnology, Graduate School of Bioresource and Bioenvironmental Sciences, Kyushu University, Hakozaki 6-10-1, Fukuoka 812-8581, Japan, [2]Bio-archtechture Center, Kyushu University, Hakozaki 6-10-1, Fukuoka 812-8581, Japan, [3]Department of Biomembrane and Biofunctional Chemistry, Graduate School of Pharmaceutical Sciences, Hokkaido University, Kita 12, Nishi 6, Kita-ku, Sapporo 060-0812, Japan, [4]Present address: Department of Biochemistry and Molecular Biology, Medical University of South Carolina, Charleston, South Carolina 29425, USA

Summary.   Neutral ceramidases (CDases) are widely distributed from bacteria to humans.   Enzymes from bacteria and drosophila are exclusively secretory while those in zebrafish and mammals are typical type II integral membrane proteins that occasionally detach from cells after their NH_2-terminal anchor is processed.   We found that the difference in enzymatic localization is entirely due to the presence of a mucin-like domain located downstream of the NH_2-terminal signal/anchor sequence of vertebrate enzymes.   This sequence is not found in enzymes from bacteria and invertebrates.   Overexpression of mouse neutral CDase on either the cell-surface or in the extracellular milieu significantly increased the intracellular concentration of sphingosine.   Sphingosine 1-phosphate (S1P) also increased when the amount of cell-surface ceramide increased following cell treatment with bacterial sphingomyelinase.   Increases were not seen in the ER. In contrast, knockdown of CDase reduced the amounts of these sphingolipid metabolites.   Hence, this enzyme seems to participate in vascular S1P production where cell- and lipoprotein-bound sphingomyelin are likely to be a source of ceramide.   Knockdown of neutral

CDase also impaired S1P-mediated angiogenesis and heart development during zebrafish embryogenesis thereby impairing blood circulation. The CDase enzyme participates in ceramide metabolism at plasma membranes and in the extracellular milieu to regulate S1P generation and S1P-mediated signaling through the production of sphingosine.

Keywords. angiogenesis, neutral ceramidase, ceramide, sphigosine 1-phosphate (S1P), S1P receptor, zebrafish early development

1. Introduction

Ceramidase (CDase) was first identified in rat brain by Simon Gatt in 1963 (1963). In 1995, Sandhoff's team purified a lysosomal CDase from human urine (Bernardo et al. 1995) and cloned its DNA (Koch et al. 1996). The enzyme is crucial for the catabolism of ceramide (Cer) in lysosomes and its genetic mutation causes Farber disease, a condition that stems from Cer accumulation in lysosomes. Schuchman's group cloned the mouse homologue (Li et al. 1998) and generated knockout mice which were embryonically lethal (Li et al. 2002). On the other hand, several lines of evidence indicated the presence of the non-lysosomal CDases showing the neutral to alkaline pH optimum (tentatively designated neutral CDase) in mammalian tissues (Nilson 1969, Slife 1989). Unexpectedly, however, neutral CDase was first purified and cloned from an opportunistic pathogen, *Pseudomonas aeruginosa* (Okino et al. 1998, 1999). Then mouse and rat homologues were purified and cloned by Tani et al. (2000a, 2000b) and Mitsutake et al. (2001), respectively. Hannun and his team also purified the rat homologue (El Bawb et al. 1999) and cloned the human homologue (El Bawab et al. 2000). Neutral CDase homologues have since been cloned and characterized in the fruit fly (Yoshimura et al. 2002), slime mold (Monjusho et al. 2003) and zebrafish (Yoshimura et al. 2004). These neutral CDase homologues have a pH optimum of 6.5~8.5 except the slime mold homologue which has an extremely acidic pH optimum, 3.5 (Monjusho et al. 2003). Interestingly, however, the neutral CDase homologues differ completely from the lysosomal acid CDase in primary structure. On the other hand, Obeid and her associates cloned a different class of CDase having an extremely alkaline pH optimum from yeast, mouse, and human (tentatively designated alkaline CDase), which also differed extensively from the acid and neutral CDases in primary structure (Mao et al. 2000a, 2000b, 2001, 2003). Collectively, CDases are classified into three groups, lysosomal acid, neutral, and alkaline enzymes according to optimum pH and primary structure (Ito et al. 2003). Interestingly, the genomic information on neutral CDase is highly conserved from bacteria to humans, while it is still unclear whether acid and alkaline enzymes are present in prokaryotes.

In contrast to the lysosomal acid CDase, the biological significance of neutral and alkaline CDases is little understood. However, accumulating evidence suggests that neutral CDases regulate the intracellular concentration of Cer and thereby Cer-mediated signaling (Franzen et al. 2001, 2002; Achaya et al. 2003, 2004). In general, CDase is crucial for not only the regulation of Cer content but also the generation of sphingosine (Sph) in cells, since Sph is considered not to be generated by *de novo* synthesis (Michel et al. 1997) because of the specificity of delta 4 desatulase which introduces a double bond into dihydroCer to form Cer but not into dihydroSph (Omae et al. 2004). Sph is the sole precursor for Sph 1-phosphate (S1P) which is a ligand for 7-membrane spanning, G-protein coupled receptors, $S1P_{1-5}$ (Hla 2003). Intracellular S1P is generated through phosphorylation of Sph by Sph kinase which is mainly located in the cytosol (Spiegel and Kolesnick 2002). S1P-mediated signaling has been implicated in the regulation of a variety of cellular processes, including cell proliferation, differentiation, migration and apoptosis (Spiegel and Kolesnick 2002). However, how or where Sph is generated and supplied to Sph kinases has yet to be clarified. In the vascular system, platelets are thought to be a source of S1P because they contain a very large amount of S1P due to the strong activity of the Sph kinase and a lack of S1P lysase. Platelets also release S1P when stimulated (Yatomi et al. 1997). Interestingly, neutral CDase is present in platelets and possibly participates the generation of S1P (Tani et al. 2005a). Alternatively, it was reported that intracellular S1P (Edsall et al. 2001) and Sph kinase (Ancellin et al. 2002) were released from cells through unknown mechanisms.

In this chapter, we describe the role of neutral CDase in the generation of S1P and S1P-mediated signaling at the plasma membrane and in the extracellular milieu.

2. Intracellular distribution and topology of neutral CDase

Neutral CDase localizes at apical membranes of rat kidney where it is concentrated in lipid microdomains with cholesterol and glycosphingolipids (Mitsutake, 2001). The enzyme is also co-localized with caveolin-1 (Romiti et al. 2001) and actively released from murine endothelial cells (Romiti et al. 2000). To clarify the intracellular distribution and topology, a double-tagged rat CDase, with a *FLAG*-tag at the NH$_2$-terminus and *Myc*-tag at the COOH-terminus, was constructed (Fig. 1A) and expressed in HEK293 cells (Tani et al. 2003). Under permeable conditions in the presence of Triton X-100, both *FLAG* and *Myc* signals were observed in ER/Golgi compartments as well as on plasma membranes (Fig. 1B, a, c). However, only the *Myc*-signal was observed under impermeable conditions in the absence of the detergent (Fig. 1B, b, d). These results indicate that the COOH-terminus of the CDase resides on the extracellular side of the plasma membrane whereas the NH$_2$-terminus is on the cytoplasmic side, indicating the enzyme is a typical type II integral membrane protein when expressed in HEK293 cells (Fig. 1C).

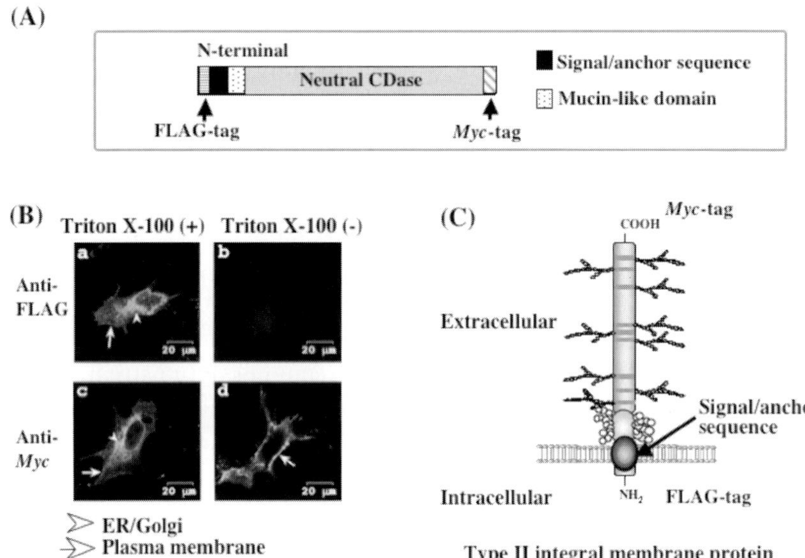

Fig. 1. Topology of rat neutral CDase. (A) Construct of double-tagged rat neutral CDase. (B) Expression of rat neural CDase in HEK293 cells. (a) stained with anti-*FLAG* antibody after permeablization with Triton X-100, (b) stained with anti-*FLAG* antibody without treatment of Triton X-100, (c) stained with anti-*Myc* antibody after permeablization with Triton X-100, (d) stained with anti-*Myc* antibody without treatment of Triton X-100. (C) Putative topology of rat neutral CDase expressed in HEK293 cells. Details are described in (Tani et al. 2003).

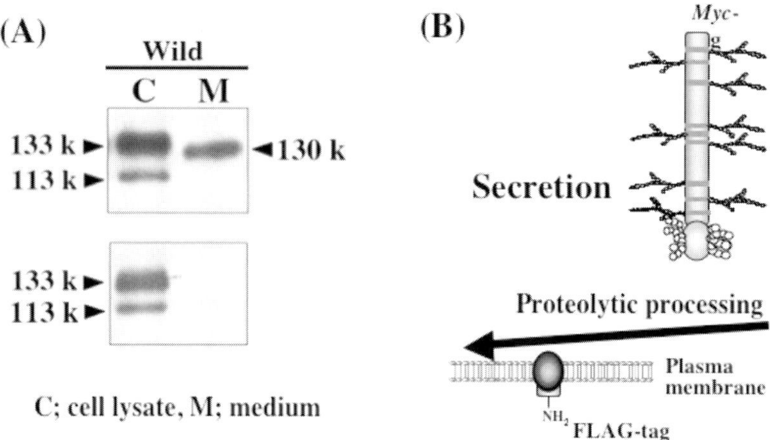

Fig. 2. Release of neutral CDase from the plasma membrane of HEK293 cells. (A) Western blot of neutral CDase expressed in HEK293 cells. Upper panel, stained with anti-*Myc* antibody. Lower panel, stained with anti-*FLAG* antibody. (B) Scheme for detachment of neutral CDase from the plasma membrane of HEK293 cells. Details are described in (Tani et al. 2003).

Interestingly, some neutral CDase was released into the medium when expressed in HEK 293 or CHOP cells.   To clarify how neutral CDase detaches from the cell, Western blotting was performed using anti-*Myc* antibody which recognizes the COOH-terminal end and anti-*FLAG* antibody which targets the NH₂-terminal end.   A 113-kDa and a 130-kDa *Myc*-positive band were detected in cell lysates, while only the 130-kDa band was detected in the medium (Fig. 2A).   The 113-kDa and 133-kDa proteins were glycosylated with high-mannose type *N*-glycans and high mannose/complex/hybrid type *N*-glycans, respectively, suggesting that they are immature in the ER but mature in the Golgi apparatus (Mitsutake 2001).   On the other hand, two *FLAG*-positive bands corresponding to the *Myc*-positive bands were detected in the cell lysate but not in the culture medium (Fig. 2A).   These results indicate that the NH₂-terminal signal/anchor sequence remains intact before secretion and the enzyme is secreted into the medium possibly after cleavage.   In conclusion, the neutral CDase is expressed on plasma membranes as a type II integral membrane protein anchoring to the membranes with an internal NH₂-terminal signal/anchor sequence (Fig. 1C), occasionally detaching from the cells after cleavage of the NH₂-terminal sequence (Fig. 2B).   This conclusion was also confirmed using a full-length human neutral CDase expressed in HEK293 cells (Hwang et al. 2005).   Although immnunohistochemical analysis clearly showed that the neutral CDase was mainly localized at the apical membrane of proximal tubules, distal tubules, and collecting ducts in rat kidney. The enzyme was distributed with endosome-like organelles in hepatocytes in rat liver (Mitsutake et al. 2001). Interestingly, liver enzyme could be extracted simply by the freeze-thawing without use of detergents while kidney enzyme could not be so obtained (Tani et al. 2000a).

3. A function of the evolutionary-acquired mucin-like domain of neutral CDase

Vertebrate neutral CDases have a Ser/Thr/Pro-rich domain (tentatively designated the mucin-like domain) downstream of the NH_2-terminal hydrophobic region (Fig. 3). Mucin-like domains are observed in CDases from vertebrates, but not from bacteria and invertebrates, suggesting that CDases evolved into an enzyme. Interestingly, a mutated rat CDase deleted of this mucin-like domain was actively released into the medium and not expressed on the cell-surface. This observation was consistent with the finding that the neutral CDase lacking mucin-like domain from bacteria and invertebrates is a secretory protein. The secretion of both wild-type and mucin like domain-deleted mutant rat CDases from HEK293 cells was strongly inhibited by brefeldin A and treatment at 5°C but not by cytochalasin D, suggesting that the two CDases were processed and secreted by a pathway via ER/Golgi compartments (Tani et al. 2003).

We found that the domain was glycosylated with *O*-glycans using a core type I-specific lectin, peanut agglutinin. The *O*-glycosylation of the domain may mediate cell-surface expression of the enzyme because an Ala-mutant Cdase, in which all Ser and Thr residues in the domain are replaced with Ala, was continuously released into the medium. Consequently, its cell-surface expression was markedly reduced when compared to the expression activity of wild-type CDase. Green fluorescence protein (GFP), a soluble protein, is expressed on the cell-surface when the signal/anchor sequence is fused to GFP via the mucin-like domain, indicating that the domain functions as a potential signal to express proteins on the cell surface as a type II integral protein (Tani et al. 2003).

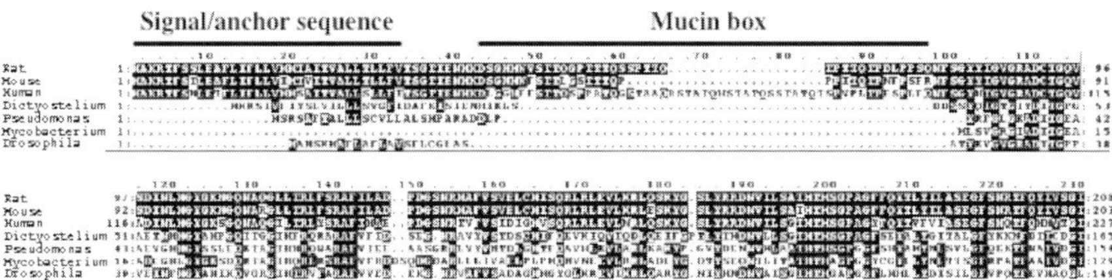

Fig. 3. Alignment of N-terminal regions of neutral CDases from vertebrates with those from invertebrates and bacteria.

4. Involvement of a plasma membrane-bound neutral CDase and a secreted neutral CDase in the generation of Sph and S1P

Since the catalytic domain of neutral CDase faces the extracellular space as shown in Fig. 2B, cell-surface Cer is expected to be hydrolyzed by the enzyme. However, no significant increase in Sph was observed when mouse neutral CDase was overexpressed in CHOP cells. The lack of increase may be due to the limited amount of free Cer in the outer leaflet of plasma membranes. Once the amount of cell-surface ceramide was increased by the treatment of cells with bacterial sphingomyelinase (SMase), a transient increase in the amounts of cellular Sph and S1P peaking at 0.5-1 h was detected in CDase transfectants as well as mock transfectants (Fig. 4A). As expected, the increase in these metabolites was much greater in the CDase transfectants than mock transfectants.

Overexpressed CDase and endogenous CDase appear to be involved in the generation of Sph and S1P, since the production of these metabolites in B16 cells was significantly reduced by CDase siRNA, but not SCR siRNA (negative control), during SMase treatment (Fig. 4B). It is noteworthy that about 50% of the activity of endogenous neutral CDase was reduced by the CDase siRNA. Collectively, neutral CDase may be responsible for the generation of Sph as well as S1P by hydrolysis of cell-surface Cer.

To disclose the effects of the subcellular distribution of neutral CDase on the generation of Sph and S1P, two constructs were generated; a secretable mutant CDase (SecCD) and an ER-retainable mutant CDase (ERCD). SecCD was exclusively released into the medium and no signals were detected on the plasma membranes under impermeable conditions. In contrast, ERCD was localized to the ER and not present on the cell surface and showed strong activity in the cell lysate, but not in the medium. A significant increase in the amounts of Sph and S1P was observed in the cells transfected with wild-type CDase and SecCD, but not ERCD, compared to mock transfectants after the treatment of cells with bacterial SMase.   However, without the SMase treatment, no significant increase in Sph and S1P was observed.   These results indicate that plasma membrane-bound and secreted neutral CDases, but not the ER-retainable enzyme, are involved in the generation of Sph by hydrolysis of Cer at the outer leaflets of plasma membranes leading to an increase in the cellular S1P level (Tani et al. 2005b).

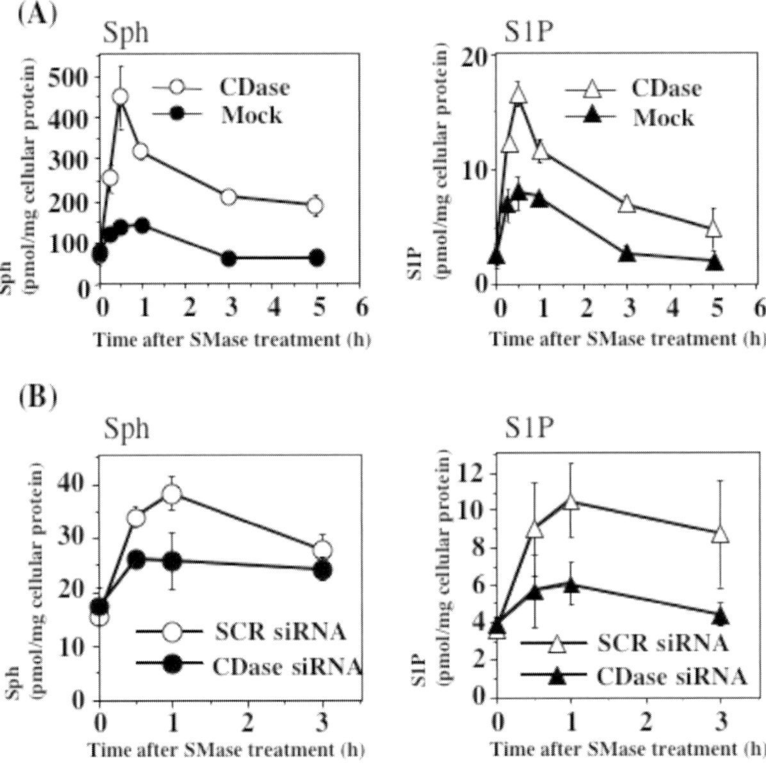

Fig. 4. Generation of Sph and S1P in neutral CDase-overexpressing cells (A) and CDase–knocking down cells (B). (A) CHOP cells were transfected with the CDase cDNA or empty vector (Mock).   (B) Mouse B16 melanoma cells were co-transfected with siRNA specific to mouse neutral CDase (CDse siRNA) or control siRNA (SCR siRNA).   For both (A) and (B), the generation of Sph and S1P was measured by HPLC after treatment of cells with bacterial SMase for the period indicated (Tani et al 2005b).

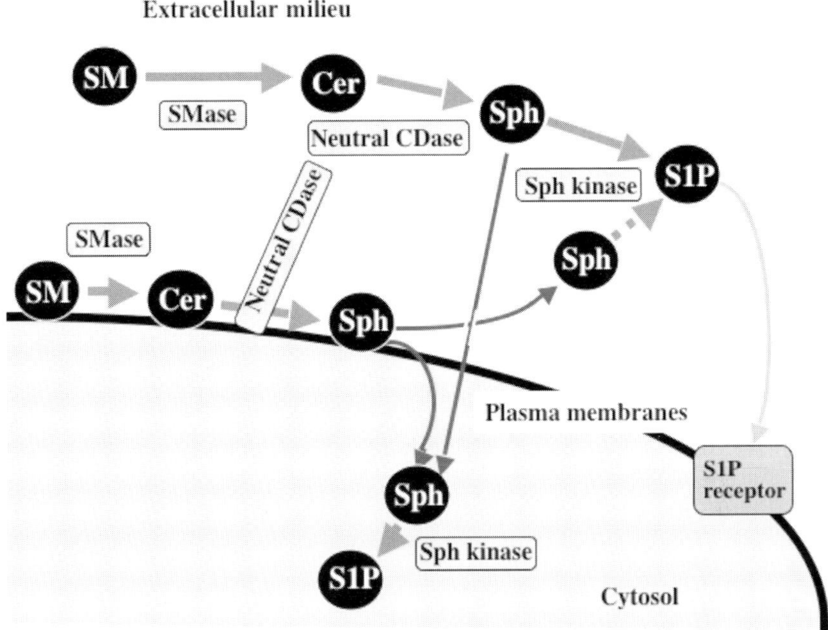

Fig. 5. Scheme for metabolic pathway from SM to S1P involving neutral CDase at the outer leaflet of the plasma membrane and in the extracellular milieu.

Although the generation of S1P is closely linked to that of Sph by neutral CDase, the sites where these two metabolites are generated seem to be different. The Sph produced by the CDase was easily withdrawn by BSA whereas S1P was not, suggesting that Sph is generated at the cell surface whereas S1P is generated inside of the cell. Also, the production of S1P was much greater in CHO-SPHK1 cells than in parental CHO cells following bacterial SMase treatment even though the amounts of S1P generated were similar. The former cells are stable transfectants expressing cytoplasmic Sph-kinase type I (SPHK1). These results strongly suggested that Sph was generated at the outer leaflet of plasma membranes by neutral CDase, then promptly incorporated into the cells after which some of it was phosphorylated by the cytosolic Sph kinase (Tani et al. 2005b).

It is assumed that cell-bound and secreted neutral CDases participate in the metabolism of serum-derived SM and Cer in the vascular system because both sphingolipids are associated with low density lipoproteins (Merrill Jr AH et al 1995, Subbaiah RV et al 1999, Boyanovsky et al 2003). Significant amounts of Sph as well as S1P were generated in the fetal bovine serum (FBS)-containing conditioned medium of SecCD and

wild-type CDase transfectants, compared to mock transfectants, after incubation with bacterial SMase.   The increase in the amount of both sphingolipid metabolites was much greater in SecCD transfectants than in
wild-type CDase transfectants, and depended on the amount of CDase in
the conditioned medium.   It is worth noting that no increase in these metabolites was observed if FBS or SMase was omitted from the conditioned
medium, suggesting that the major source of Cer is the FBS-derived SM
and both Sph kinase and ATP are probably present in FBS.   We detected
Sph kinase activity in the FBS used in this experiment (0.82 ± 0.03 nU/ml,
n=3).   The S1P generated in the serum by the combined actions of CDase
and Sph kinase seems to be functional because it triggers the internalization of $S1P_1$ from the cell surface to intracellular vesicles (Tani et al.
2005b).   In summary, neutral CDase is involved in the metabolic pathway
from SM to S1P at the plasma membranes and in the extracellular region,
regulating the generation of S1P and S1P-mediated signaling through the
production of Sph (Fig. 5).   It is also noteworthy that a part of Sph generated outside of cells could be incorporated into cells and then converted to
S1P by cytosolic Sph kinase (Fig. 5).

5. Involvement of neutral CDase in S1P-mediated angiogenesis during zebrafish embryogenesis

Previously, we reported that the knockdown of neutral CDase during zebrafish embryogenesis caused a lack of circulation of blood cells (Yoshimura et al. 2004).   Neutral CDase was shown to be involved in the pathway for the generation of S1P leading to the S1P-mediated activation of
$S1P_1$ (Tani et al. 2005b), and thus further investigation was undertaken to
reveal whether the enzyme is responsible for S1P-mediated angiogenesis
using zebrafish.

The translation of neutral CDase mRNA was specifically inhibited by
the injection of the anti-sense morpholino oligonucleotide (AMO) into one
to four-cell embryos.   The AMO was designed using the
5'-untranslational region of zebrafish neutral CDase (znCD-AMO) (Yoshimura et al. 2004).   Knocking down the enzyme with znCD-AMO increased the number of embryos with severe morphological and cellular
abnormalities such as pericardiac edema and a lack of circulation of blood
cells (attached movie in Yoshimura et al. 2004).   The lack of circulation
of blood cells seems to be caused by an impairment of both angiogenesis
in segmental vessels and the development of the heart (Yoshimura et al.
unpublished results).   We found that the knockdown of $S1P_1$ by a specific

AMO (zS1P$_1$-AMO) resulted in almost the same phenotype caused by znCD-AMO. Interestingly, typical phenotypes such as the lack of circulation of blood cells and pericardiac edema caused by znCD-AMO were partly rescued by injection of S1P into embryos (blood cell-circulation positive cells; control 96.5±2.6%, znCD AMO 34.9±15.1%, znCD AMO+S1P 71.2±15.6%, average of 4 independent experiments with SD). However, the same phenotype caused by zS1P$_1$-AMO was not rescued by the injection of S1P (zS1P$_1$ AMO 13.1±9.7%, zS1P$_1$ AMO+S1P 7.6±8.1%, average of 4 independent experiments with SD). These results may indicate that neutral CDase is integral to the generation of S1P and S1P-mediated angiogenesis during embryogenesis in the zebrafish.

6. References

Acharya U, Patel S, Koundakjian E, Nagashima K, Han X, Acharya JK (2003) Modulating sphingolipid biosynthetic pathway rescues photoreceptor degeneration. *Science*, **299**, 1740-1743.

Acharya U, Mowen BM., Nagashima K, Acharya JK (2004) Ceramidase expression facilitates membrane turnover and endocytosis of rhodopsin in photoreceptors. *Proc Natl Acad Sci U S A*, **101**, 1922-1926.

Ancellin N, Colmont C, Su J, Li Q, Mittereder N, Chae S, Stefansson S, Liau G, Hla T (2002) Extracellular export of sphingosine kinase-1 enzyme. Sphingosine 1-phosphate generation and the induction of angiogenic vascular maturation. *J Biol Chem*, **277**, 6667-6675.

Bernardo K, Hurwitz R, Zenk T, Desnick RJ, Ferlinz K, Schuchman EH, Sandhoff K (1995) Purification, characterization, and biosynthesis of human acid ceramidase. *J Biol Chem*, **270**, 11098-11102.

Boyanovsky B, Karakashian A, King K, Gilitiay N, Nikolova-Karakashian M (2003) Uptake and metabolism of low density lipoproteins with elevated ceramide content by human microvascular endothelial cells. *J Biol Chem*, **278**, 26992-26999.

Edsall LC, Cuvillier O, Twitty S, Spiegel S, Milstien S (2001) Sphingosine kinase expression regulates apoptosis and caspase activation in PC12 cells. *J Neurochem*, **76**, 1573-1584.

El Bawab S, Bielawska A, and Hannun YA (1999) Purification and characterization of a membrane-bound nonlysosomal ceramidase from rat brain. *J Biol Chem*, **274**, 27948-27955.

El Bawab S, Roddy P, Qian T, Bielawska A, Lemasters JJ, Hannun YA (2000) Molecular cloning and characterization of a human mitochondrial ceramidase. *J Biol Chem*, **275**, 21508-21513.

Franzen R, Pautz A, Brautigam L, Geisslinger G, Pfeilschifter J, Huwiler A (2001) Interleukin-1 induces chronic activation and *de novo* synthesis of neutral ceramidase in renal mesangial cells. *J Biol Chem*, **276**, 35382-35389.

Franzen R, Fabbro D, Aschrafi A, Pfeilschifter J, Huwiler A (2002) Nitric oxide Induces degradation of the neutral ceramidase in rat renal mesangial cells and is counterregulated by protein kinase C. *J Biol Chem*, **277**, 46184-46190.

Gatt S (1963) Enzymatic hydrolysis and synthesis of ceramidase. *J Biol Chem*, **238**, 3131-3133.

Hla T (2003) Signaling and biological actions of sphingosine 1-phosphate. *Pharmacol Res*, **47**, 401-407.

Hwang YH, Tani M, Nakagawa T, Okino N, Ito M (2005) Subcellular localization of human neutral ceramidase expressed in HEK293 cells. *Biochem Biophys Res Commun*, **331**, 37-42.

Ito M, Okino N, Tani M, Mitsutake S, Kita K (2003) Molecular evolution of neutral ceramidase : From bacteria to mammals. *In 'Ceramide signaling'* (Anthony, H. Futerman, ed.), Landes Bioscience, ISBN:0-306-47442-5, pp 41-48

Koch J, Gartner S, Li CM, Quintern LE, Bernardo K, Levran O, Schnabel D, Desnick RJ, Schuchman EH, Sandhoff K (1996) Molecular Cloning and Characterization of a full-length complementary DNA encoding human acid ceramidase. *J Biol Chem*, **271**, 33110-33115.

Li CM, Hong SB, Kopal G, He X, Linke T, Hou WS, Koch J, Gatt S, Sandhoff K, Schuchman EH (1998) Cloning and characterization of the full-length cDNA and genomic sequences encoding murine acid ceramidase. *Genomics*, **50**, 267-274.

Mao C, Xu R, Bielawska A, Szulc ZM, Obeid LM (2000a) Cloning and characterization of a *Saccharomyces cerevisiae* alkaline ceramidase with specificity for dihydroceramide. *J Biol Chem*, **275**, 31369-31378.

Mao C, Xu R, Bielawska A, Obeid LM (2000b) Cloning of an alkaline ceramidase from *Saccharomyces cerevisiae*. *J Biol Chem*, **275**, 6876-6884.

Mao C, Xu R, Szule ZM, Bielawska A, Galadari SH, Obeid LM (2001) Cloning and characterization of a novel human alkaline ceramidase. *J Biol Chem*, **276**, 26577-26588.

Mao C, Xu R, Szulc ZM, Bielawski J, Becker KP, Bielawska A, Galadari SH, Hu W, Obeid LM (2003) Cloning and characterization of a mouse endoplasmic reticulum alkaline ceramidase. *J Biol Chem*, **278**, 31184-31191.

Merrill Jr AH, Lingrell S, Wang E, Nikolova-Karakashian M, Vales TR, Vance DE (1995) Sphingolipid biosynthesis de novo by rat hepatocytes in culture. *J Biol Chem*, **270**, 13834-13841.

Michel C, van Echten-Deckert G, Rother J, Sandhoff K, Wang E, Merrill AHJr (1997) Characterization of ceramide synthsis. A dihydroceramide desaturase introduces the 4,5-trans-double bond of sphingosine at the level of dihydroceramide. *J Biol Chem*, **272**, 22432-22437.

Mitsutake S, Tani M, Okino N, Mori K, Ichinose S, Omori A, Iida H, Nakamura T, Ito M (2001) Purification, characterization, molecular cloning, and subcellular distribution of neutral ceramidase of rat kidney. *J Biol Chem*, **276**, 26249-26259.

Monjusho H, Okino N, Tani M, Maeda M, Yoshida M, Ito M. (2003) A neutral ceramidase homologue from *Dictyosterium discoideum* exhibits an acidic pH optimum. *Biochem J*, **376**, 473-479.

Nilsson A (1969) The presence of sphingomyelin- and ceramide-cleaving enzymes in the small intestinal tract. *Biochim Biophys Acta*, **176**, 339-347.

Okino N, Tani M, Imayama S, Ito M (1998) Purification and characterization of a novel ceramidase from *Pseudomonas aeruginosa*. *J Biol Chem*, **273**, 14368-14373.

Okino N, Ichinose S, Omori A, Imayama S, Nakamura T and Ito M (1999) Molecular cloning, sequencing, and expression of the gene encoding alkaline ceramidase from *Pseudomonas aeruginosa*. *J Biol Chem*, **274**, 36616-36622.

Omae F. Miyazaki M, Enomoto A, Suzuki M, Suzuki Y, Suzuki A (2004) DES2 protein is responsible for phytoceramide biosynthesis in the mouse small intestine. *Biochem J*, **379**, 687-695.

Romiti E, Meacci E, Tani M, Nuti F, Farnararo M, Ito M, Bruni P (2000) Neutral/alkaline and acid ceramidases activities are actively released by murine endotherial cells. *Biochem Biophys Res Commun*, **275**, 746-751.

Romiti E, Meacci E, Tanzi, G, Becciolini L, Mitsutake S, Farnararo M, Ito M, Bruni P (2001) Localization of neutral ceramidase in caveolin-enriched light membranes of murine endothelial cells. *FEBS Lett*, **506**, 163-168.

Romiti E, Meacci E., Donati C., Formigli L., Zeccchi-Orlandini S,Farnararo M, Ito M, Bruni P (2003) Neutral ceramidase secreted by endotherial cells is released in part associated with caveolin-1. *Arch Biochem Biophys*, **417**, 27-33.

Slife CW, Wang E, Hunter R, Wang C, Burgess C, Liotta DC, Merrill Jr AH (1989) Free sphingosine formation from endogenous substrates by a liver plasma membrane system with a divalent cation dependence and a neutral pH optimum. *J Biol Chem*, **264**, 10371-10377.

Spiegel S, Kolesnick R (2002) Sphingosine 1-phosphate as a therapeutic agent. *Leukemia*, **16**, 1596-1602.

Subbaiah, PV, Subramanian VS, Wang K (1999) Novel physiological function of sphingomyelin in plasma. *J Biol Chem*, **274**, 36409-36414.

Tani M, Okino N, Mitsutake S, Tanigawa T, Izu H, Ito M (2000a) Purification and characterization of a neutral ceramidase from mouse liver. A single protein catalyzes the reversible reaction in which ceramide is both hydrolyzed and synthesized. *J Biol Chem*, **275**, 3462-3468.

Tani M, Okino N, Mori K, Tanigawa T, Izu H, Ito M (2000b) Molecular cloning of the full-length cDNA encoding mouse neutral ceramidase. *J Biol Chem*, **275**, 11229-11234.

Tani M, Iida H, Ito M (2003) *O*-Glycosylation of Mucin-like Domain Retains the Neutral Ceramidase on the Plasma Membranes as a Type II Integral Membrane Protein. *J Biol Chem*, **278**, 10523-10530.

Tani M, Ito M, Igarashi Y (2005a) Mechanism of sphingosine and sphingosine 1-phosaphate generation in human platelets. *J Lipid Res*, **46**, 2458-2467.

Tani M, M, Igarashi Y, Ito M (2005b) Involvement of neural ceramidase in ceramide metabolism at the plasma membrane and in extracellular milieu. *J Biol Chem*, **280**, 36592-36600.

Yatomi Y, Yamamura S, Ruan F, Igarashi Y (1997) Sphingosine 1-phosphate induces pltelet activation through an extracellular action and shares a platelet surface receptor with lysophosphatidic acid. *J Biol Chem,* **272**, 5291-5297.
Yoshimura Y, Okino N, Tani M, Ito M (2002) Molecular cloning and characterization of a secretory neutral ceramidase of *Drosophila melanogaster. J Biochem,* **132**, 229-236.
Yoshimura Y, Tani M, Okino N, Iida H, Ito M (2004) Molecular cloning and functional analysis of zebrafish neutral ceramidase. *J Biol Chem,* **279**, 44012-44022.

3-4 Activation of Sphingosine Kinase 1

Michael Maceyka[1], Sergio E. Alvarez[1], Sheldon Milstien[2], and Sarah Spiegel[1]

[1]Department of Biochemistry, Virginia Commonwealth University Medical Center and the Massey Cancer Center, Richmond, Virginia 23298, USA, [2]Laboratory of Cellular and Molecular Regulation, National Institute of Mental Health, Bethesda, MD 20892, USA

Summary. Sphingosine-1-phosphate (S1P) is a potent bioactive lipid that regulates many important biological processes as a ligand for cell surface receptors and as an intracellular second messenger. S1P levels inside cells are regulated in a temporal and spatial manner by the balance between its formation, catalyzed by two sphingosine kinase isoenzymes, SphK1 and SphK2, and degradation by S1P lyase and S1P-specific phosphatases. Activation of SphK1 is a key step in the agonist-induced regulation of S1P levels. SphK1 is a cytosolic enzyme and a growing body of evidence suggests that S1P production is also regulated by the translocation of SphK1 to membranes, where its substrate sphingosine is formed. SphK1 translocation to membranes can be mediated by interactions with lipids and proteins. Here we review the current knowledge of SphK1 activation and translocation.

Keywords. Sphingosine, sphingosine-1-phosphate, sphingosine kinase

1. Functions of S1P

Ceramide, sphingosine (Sph) and S1P are potent signaling molecules that are involved in numerous cell functions including proliferation, apoptosis and cell migration. The three molecules are inter-convertible (Maceyka et al., 2002). This makes the enzymes responsible for converting ceramide to Sph to S1P and back key elements regulating cell fate. While ceramide and

Sph typically promote apoptosis and inhibit cell growth, S1P inhibits apoptosis and promotes cell growth (Cuvillier et al., 1996). This review will focus only on the role of S1P and the enzymes that produce it.

S1P has a primary role in cell migration, proliferation, survival, angiogenesis, vascular maturation, immunity and allergic responses. It has been detected in plants, yeast, slime molds, worms, flies and mammals (Spiegel and Milstien, 2003). The biological effects of S1P are mediated by a family of five S1P receptors (S1PRs), termed $S1P_{1-5}$. These are G protein-coupled receptors that couple differentially to heterotrimeric G proteins and regulate multiple signaling pathways. Although S1P may act in a paracrine fashion, evidence suggests that S1P can act in an autocrine fashion, a process also called transactivation. Thus, external stimuli can activate cytosolic SphK, which in turn releases S1P from the cell to activate cell surface S1PRs. For example, PDGF-stimulated motility of fibroblasts depends both on activation of SphK and $S1P_1$ (Hobson et al., 2001). Although no direct molecular target has been identified, substantial evidence points to a role for intracellular S1P that is independent of cell surface receptors. This includes the use of caged S1P which demonstrated that intracellular S1P caused Ca^{2+} release from intracellular stores that was not mimicked by addition of extracellular S1P (Meyer zu Heringdorf et al., 2003). Perhaps the most convincing evidence comes from cells from knockout mice lacking functional S1PRs. Again, intracellular S1P, but not exogenously added S1P, promoted cell growth and survival (Olivera et al., 2003). Moreover, in plants which do not express S1PRs, intracellular S1P regulates stomatal aperture (Coursol et al., 2005).

2. The kinases that produce S1P

To better understand how cells regulate S1P levels, much effort has focused on sphingosine kinases (SphKs), the enzymes that catalyze the phosphorylation of the 1-OH of Sph. SphKs from diverse organisms have been cloned, including mammals, yeast, protozoan, plants, and flies (Liu et al., 2002). All members of the SphK family have five conserved domains thought to be involved in substrate binding (Liu et al., 2002). SphK catalytic activity can be eliminated by mutating the consensus ATP binding domain of SphK1. Then mutated SphK1 can be used as a "dominant-negative" to reveal the biological roles of SphK1 (Pitson et al., 2000). Two isozymes of SphK have been cloned from mammals, SphK1 and SphK2, which differ in their tissue distribution and susceptibility to inhibition by detergents and high salt concentrations (Liu et al., 2000).

Accumulating evidence indicates that SphK1 is required for the biological effects of S1P. SphK1 can be activated by a long list of agonists including growth factors, cytokines, ligands for GPCRs, crosslinking of immunoglobulin receptors, and lipopolysaccharide. SphK1 must be active for for the promotion of cell growth, cell migration and inhibition of apoptosis (Spiegel and Milstien, 2003). Although SphK2 is less well studied, it sometimes seems to function in an opposite manner to SphK1. For example, while overexpression of SphK2 increases cellular Sph phosphorylating activity to levels similar to those of cells overexpressing SphK1, its expression promotes apoptosis, in part through its function as a BH3-only protein (Liu et al., 2003). Moreover, in some cell types, SphK2 either resides in the nucleus or is translocated there, where it acts to retard cell growth (Igarashi et al., 2003). However, it appears that endogenous SphK2 may have redundant or overlapping functions to SphK1 in cell motility (Hait et al., 2005).

3. Regulation of SphK1 by post-translational modifications

Translocation of SphK1 from the cytosol to the plasma membrane brings it to the cellular location where its Sph substrate resides. This also enables SphK1 to produce S1P locally, which may account for the specificity of its actions through one or a subset of nearby and thus specifically activated S1PRs. Altering the subcellular location of SphK1 can have marked effects on cell function because S1P activates different effectors depending on its subcellular location. Indeed, targeting SphK1 to the plasma membrane by adding an amino-terminal myristylation sequence decreased cell proliferation (Safadi-Chamberlain et al., 2005).

Since SphK1 does have putative phosphorylation sites for PKA, PKC, and casein kinase II, SphK1 activity may be regulated by phosphorylation (Kohama et al., 1998). Indeed, PKC-dependent activation of SphK1 and translocation to the plasma membrane increases secretion of S1P, allowing for autocrine/paracrine signaling (Johnson et al., 2002). Recently, it has been demonstrated that ERK1 phosphorylates SphK1 on serine 225 (Pitson et al., 2003). Moreover, this phosphorylation is required for subsequent translocation to the plasma membrane and is critical for its oncogenic properties (Pitson et al., 2005). Although it is likely that phosphorylation induces a conformational change to facilitate activation, sequence analysis failed to identify a recognizable activation loop like those present in other kinases that might be altered by phosphorylation of serine 225 (Pitson et al., 2003). Interestingly, this work did reveal that SphK1 activation and translocation can be uncoupled. Constitutively targeting a non-phosphorylatable Ser225A-SphK1 mutant to the plasma membrane rescued the effects of the mutation on the ability of SphK1 to increase cell proliferation and reduce apoptosis while having no effect on SphK1 activity (Pitson et al., 2003). Moreover, targeting both wild-type and Ser225A-SphK1 to the plasma membrane increased S1P levels, again suggesting that SphK1 activity is regulated in part by sequestration from its lipid substrate.

Numerous studies have shown that cross-linking of the high affinity receptor for IgE, FcγRI, activates SphK1 and induces its translocation to the plasma membrane (Jolly et al., 2004; Melendez and Khaw, 2002). Moreover, this activation of SphK1 and increased formation of S1P is required for normal mast cell degranulation and may be involved in movement of mast cells to sites of inflammation (Jolly et al., 2004). Recently, it was shown that SphK1 interacts directly with the tyrosine kinase Lyn and that this interaction leads to its recruitment to FcγRI. Furthermore, interaction of SphK1 with Lyn enhanced their enzymatic activities (Urtz et al., 2004). These findings position the activation of SphK1 as a FcγRI proximal event.

4. Regulation of SphK1 expression

Given the importance of SphK1-produced S1P, it is not surprising that cells regulate the ability of SphK1 to make S1P at many levels. Several recent reports demonstrate that SphK1 is also regulated at the level of transcription. For example, the anti-apoptotic effects of IGFBP-3 were correlated with increased SphK1 mRNA and inhibition of SphK1 activity reversed the anti-apoptotic effects of this ligand (Granata et al., 2004). Similarly, SphK1 mRNA, protein, and activity are increased by TGF-γ, and this was required for its ability to upregulate TIMP-1 (Yamanaka et al., 2004). Interestingly, PKC-induced increases in SphK1 mRNA have been linked to the transcription factors AP2 and Sp1 and their binding to the SphK1 5'-promoter (Nakade et al., 2003). The significance of alterations in SphK1 mRNA is suggested by the observation that SphK1 mRNA levels are elevated in a variety of solid tumors, compared with normal tissue from the same patient (French et al., 2003; Johnson et al., 2005).

SphK1 activity also depends on the turnover of the protein. Recently, it was demonstrated that SphK1-mediated S1P production can be regulated by proteolysis of the protein. DNA damaging chemotherapeutic agents induced p53-dependent degradation of SphK1 through cathepsin B and perhaps several effector caspases (Taha et al., 2004). Moreover, cathepsin B has been implicated in SphK1 degradation induced by TNF-γ (Taha et al., 2005). These results suggest that downregulation of SphK1 by TNF may be dependent on the "lysosomal pathway" of apoptosis.

5. Regulation of SphK1 by interaction with lipids

As mentioned above, many signaling pathways involve the translocation of SphK1 from the cytosol to cellular membranes, where its substrate Sph is generated. For translocation to occur, SphK1 must interact with component(s) of these membranes. The lipids of the bilayer are obvious candidates. It has long been known that acidic phospholipids stimulate SphK activity (Olivera et al., 1996), and it was originally suggested that phosphatidic acid (PA) might regulate SphK in vivo. Indeed, it has recently been reported that SphK1 is a PA effector (Delon et al., 2004). This conclusion was supported by several lines of evidence. First, SphK1 binds to PA covalently coupled to beads. Second, SphK1 co-localizes with PLD1, the enzyme that makes PA, to a juxta-nuclear compartment, suggesting that SphK1 translocates to areas where PA is produced. Other studies also demonstrate that SphK1 is downstream of PLD1 activity in the FcγRI-triggered signal transduction pathways leading to calcium increases and mast cell degranulation (Melendez and Khaw, 2002). These studies add SphK1 to the growing list of proteins shown to directly interact with and be regulated by membrane lipids.

6. Regulation of SphK1 by interaction with proteins

Many studies have shown that SphK1 can also interact with other proteins to alter its localization or activity (Table 1). The first demonstration that SphK1 can interact with other proteins came from our purification of SphK1 to homogeneity which relied in part on affinity chromatography with immobilized Ca^{2+}/calmodulin (Olivera et al., 1998). It was later shown that Ca^{2+}/calmodulin was required for translocation of SphK1 to the PM (Young et al., 2003). Interestingly, although a Ca^{2+}/calmodulin inhibitor reduced translocation and agonist-induced S1P levels, it had little effect on SphK1 activity. This suggests that translocation to membranes is more important for the production of bioactive S1P.

TNF receptor-associated factor 2 (TRAF2), an adaptor protein downstream of TNF-γ, has been identified as another protein that interacts with SphK1. A TRAF2-binding motif on SphK1 was identified that mediated the interaction between TRAF2 and SphK1. This interaction activates SphK1, and was also required for TRAF2-mediated activation of NF-γ B and the anti-apoptotic effect of TNF-γ (Xia et al., 2002).

Utilizing yeast two-hybrid screens, a number of other proteins that interact with SphK1 have been identified (Table 1). SphK1 binds a protein of unknown function, named SKIP, present in a human brain cDNA library (Lacana et al., 2002). SKIP has homology to PKA anchoring proteins, suggesting that it acts as a signaling scaffold. Overexpression of SKIP inhibited SphK1 functions, although the physiological significance of this interaction remains to be elucidated. SphK1 also interacts with RPK118, a protein with a phosphatidylinositol-3-phosphate-binding PX domain that targets SphK1 to early endosomes (Hayashi et al., 2002). SphK1 is also targeted to endosomes as well as the PM through interaction with aminoacylase 1 (Acy1) (Maceyka et al., 2004). Though Acy1 expression slightly decreased SphK1 activity, it enhanced SphK1-induced protection from apoptosis and promotion of cell growth. SphK1 also co-localizes to endosomal-like structures with γ-catenin, also known as NPRAP (Fujita et al., 2004). γ-Catenin is a PDZ domain containing protein thought to be involved in cell-cell interactions. Overexpression of γ-catenin increased SphK1 activity, and γ-catenin also increased the activity of SphK1 in in vitro assays. SphK1 has also been shown to interact with another protein involved in cell-cell interactions, platelet endothelial cell adhesion molecule-1 (PECAM-1) (Fukuda et al., 2004), a cell surface protein that causes aggregation of endothelial cells. PECAM-1 and SphK1 co-localized to the PM, particularly at the sites of cell-cell contact. Overexpressing PECAM-1 reduced SphK1 activity, and activation and phosphorylation of specific tyrosine residues on PECAM-1 reduced its association with SphK1 (Fukuda et al., 2004). These results suggest that PECAM-1 functions to sequester and attenuate SphK1 activity in unstimulated cells.

We have found additional proteins that interact with SphK1 through a two-hybrid screen (Table 1), but further work is necessary to confirm that these are bona fide SphK1 interacting proteins. In summary, although a plethora of proteins are capable of interacting with SphK1 and changing its activity or localization, the physiological significance of these interactions is still unclear and will unquestionably be an area of immense future interest.

Acknowledgements. We thank Dr. Victor Nava for the initial cloning of SphK1 interacting proteins. This work was supported by NIH grants R37GM043880, R01CA61774, and RO1AI50094 to SS, and a NRSA-Kirschstein postdoctoral fellowship to MM.

Table 1: SphK1 interacting proteins.

Protein	Method	Confirmed	Function	Ref
Ca^{2+}/calmodulin	Affinity column	Binding, signaling	Targets SphK1 to PM	(Kohama et al., 1998)
TRAF2	IP	IP, signaling	Required for SphK1 stimulation by TNF-γ	(Xia et al., 2002)
AKAP	Two-hybrid	IP, signaling	Reduce SphK1 activity	(Lacana et al., 2002)
RPK118	Two-hybrid	IP, signaling, colocalization	Translocates SphK1 to early endosomes	(Hayashi et al., 2002)
PECAM-1	Two-hybrid	IP, signaling, colocalization	Reduce SphK1 activity, translocate to PM	(Fukuda et al., 2004)
γ-catenin/NPRAP	Two-hybrid	IP, activity, co-localization	Activate SphK1; membrane translocation	(Fujita et al., 2004)
Lyn	Antibody array	IP, activity, co-localization, signaling	Increase SphK1 activity, translocates SphK1 to FcγR1 signaling complex	(Urtz et al., 2004)
Aminoacylase 1	Two-hybrid	IP, co-localization, signaling	Translocates SphK1 to membranes; reduces SphK1 activity	(Maceyka et al., 2004)
Succinic semial-dehyde reductase	Two-hybrid	ND	Synthesizes γ-hydroxybutyrate	This report
Guanylate kinase 1	Two-hybrid	ND	Involved in GDP, cGMP metabolism	This report
Peroxin 12	Two-hybrid	ND	Part of the peroxisomal targeting signal-1 import complex	This report
HYPK	Two-hybrid	ND	WW protein, binds huntingtin	This report

List of proteins reported to physically interact with SphK1. (IP, immunoprecipitation; ND, not done)

7. References

Coursol, S., Le Stunff, H., Lynch, D. V., Gilroy, S., Assmann, S. M., and Spiegel, S. (2005). Arabidopsis sphingosine kinase and the effects of phytosphingosine-1-phosphate on stomatal aperture. *Plant Physiol,* **137**, 724-737.

Cuvillier, O., Pirianov, G., Kleuser, B., Vanek, P. G., Coso, O. A., Gutkind, S., and Spiegel, S. (1996). Suppression of ceramide-mediated programmed cell death by sphingosine-1-phosphate. *Nature,* **381**, 800-803.

Delon, C., Manifava, M., Wood, E., Thompson, D., Krugmann, S., Pyne, S., and Ktistakis, N. T. (2004). Sphingosine kinase 1 is an intracellular effector of phosphatidic acid. *J Biol Chem,* **279**, 44763-44774.

French, K. J., Schrecengost, R. S., Lee, B. D., Zhuang, Y., Smith, S. N., Eberly, J. L., Yun, J. K., and Smith, C. D. (2003). Discovery and evaluation of inhibitors of human sphingosine kinase. *Cancer Res,* **63**, 5962-5969.

Fujita, T., Okada, T., Hayashi, S., Jahangeer, S., Miwa, N., and Nakamura, S. (2004). Delta-catenin/NPRAP (neural plakophilin-related armadillo repeat protein) interacts with and activates sphingosine kinase 1. *Biochem J,* **382**, 717-723.

Fukuda, Y., Aoyama, Y., Wada, A., and Igarashi, Y. (2004). Identification of PECAM-1 association with sphingosine kinase 1 and its regulation by agonist-induced phosphorylation. *Biochim Biophys Acta,* **1636**, 12-21.

Granata, R., Trovato, L., Garbarino, G., Taliano, M., Ponti, R., Sala, G., Ghidoni, R., and Ghigo, E. (2004). Dual effects of IGFBP-3 on endothelial cell apoptosis and survival: involvement of the sphingolipid signaling pathways. *FASEB J,* **18**, 1456-1458.

Hait, N. C., Sarkar, S., Le Stunff, H., Mikami, A., Maceyka, M., Milstien, S., and Spiegel, S. (2005). Role of sphingosine kinase 2 in cell migration towards epidermal growth factor. *J Biol Chem,* **280**, 29462-29469.

Hayashi, S., Okada, T., Igarashi, N., Fujita, T., Jahangeer, S., and Nakamura, S. (2002). Identification and characterization of RPK118, a novel sphingosine kinase-1-binding protein. *J Biol Chem,* **277**, 33319-33324.

Hobson, J. P., Rosenfeldt, H. M., Barak, L. S., Olivera, A., Poulton, S., Caron, M. G., Milstien, S., and Spiegel, S. (2001). Role of the sphingosine-1-phosphate receptor EDG-1 in PDGF-induced cell motility. *Science,* **291**, 1800-1803.

Igarashi, N., Okada, T., Hayashi, S., Fujita, T., Jahangeer, S., and Nakamura, S. I. (2003). Sphingosine kinase 2 is a nuclear protein and inhibits DNA synthesis. *J Biol Chem,* **278**, 46832-46839.

Johnson, K. R., Becker, K. P., Facchinetti, M. M., Hannun, Y. A., and Obeid, L. M. (2002). PKC-dependent activation of sphingosine kinase 1 and translocation to the plasma membrane. Extracellular release of sphingosine-1-phosphate induced by phorbol 12-myristate 13-acetate (PMA). *J Biol Chem,* **277**, 35257-35262.

Johnson, K. R., Johnson, K. Y., Crellin, H. G., Ogretmen, B., Boylan, A. M., Harley, R. A., and Obeid, L. M. (2005). Immunohistochemical distribution of sphingosine kinase 1 in normal and tumor lung tissue. *J Histochem Cytochem*, **53**, 1159-1166.

Jolly, P. S., Bektas, M., Olivera, A., Gonzalez-Espinosa, C., Proia, R. L., Rivera, J., Milstien, S., and Spiegel, S. (2004). Transactivation of sphingosine-1-phosphate receptors by Fc{epsilon}RI triggering is required for normal mast cell degranulation and chemotaxis. *J Exp Med*, **199**, 959-970.

Kohama, T., Olivera, A., Edsall, L., Nagiec, M. M., Dickson, R., and Spiegel, S. (1998). Molecular cloning and functional characterization of murine sphingosine kinase. *J Biol Chem*, **273**, 23722-23728.

Lacana, E., Maceyka, M., Milstien, S., and Spiegel, S. (2002). Cloning and characterization of a protein kinase A anchoring protein (AKAP)-related protein that interacts with and regulates sphingosine kinase 1 activity. *J Biol Chem*, **277**, 32947-32953.

Liu, H., Chakravarty, D., Maceyka, M., Milstien, S., and Spiegel, S. (2002). Sphingosine kinases: a novel family of lipid kinases. *Prog Nucl Acid Res*, **71**, 493-511.

Liu, H., Sugiura, M., Nava, V. E., Edsall, L. C., Kono, K., Poulton, S., Milstien, S., Kohama, T., and Spiegel, S. (2000). Molecular cloning and functional characterization of a novel mammalian sphingosine kinase type 2 isoform. *J Biol Chem*, **275**, 19513-19520.

Liu, H., Toman, R. E., Goparaju, S., Maceyka, M., Nava, V. E., Sankala, H., Payne, S. G., Bektas, M., Ishii, I., Chun, J., *et al.* (2003). Sphingosine kinase type 2 is a putative BH3-Only protein that induces apoptosis. *J Biol Chem*, **278**, 40330-40336.

Maceyka, M., Nava, V. E., Milstien, S., and Spiegel, S. (2004). Aminoacylase 1 is a sphingosine kinase 1-interacting protein. *FEBS Lett*, **568**, 30-34.

Maceyka, M., Payne, S. G., Milstien, S., and Spiegel, S. (2002). Sphingosine kinase, sphingosine-1-phosphate, and apoptosis. *Biochim Biophys Acta*, **1585**, 193-201.

Melendez, A. J., and Khaw, A. K. (2002). Dichotomy of Ca^{2+} signals triggered by different phospholipid pathways in antigen stimulation of human mast cells. *J Biol Chem*, **277**, 17255-17262.

Meyer zu Heringdorf, D., Liliom, K., Schaefer, M., Danneberg, K., Jaggar, J. H., Tigyi, G., and Jakobs, K. H. (2003). Photolysis of intracellular caged sphingosine-1-phosphate causes Ca^{2+} mobilization independently of G-protein-coupled receptors. *FEBS Lett*, **554**, 443-449.

Nakade, Y., Banno, Y., K, T. K., Hagiwara, K., Sobue, S., Koda, M., Suzuki, M., Kojima, T., Takagi, A., Asano, H., *et al.* (2003). Regulation of sphingosine kinase 1 gene expression by protein kinase C in a human leukemia cell line, MEG-O1. *Biochim Biophys Acta*, **1635**, 104-116.

Olivera, A., Kohama, T., Tu, Z., Milstien, S., and Spiegel, S. (1998). Purification and characterization of rat kidney sphingosine kinase. *J Biol Chem*, **273**, 12576-12583.

Olivera, A., Rosenfeldt, H. M., Bektas, M., Wang, F., Ishii, I., Chun, J., Milstien, S., and Spiegel, S. (2003). Sphingosine kinase type 1 induces

G12/13-mediated stress fiber formation yet promotes growth and survival independent of G protein coupled receptors. *J Biol Chem*, **278**, 46452-46460.

Olivera, A., Rosenthal, J., and Spiegel, S. (1996). Effect of acidic phospholipids on sphingosine kinase. *J Cell Biochem*, **60**, 529-537.

Pitson, S. M., Moretti, P. A., Zebol, J. R., Lynn, H. E., Xia, P., Vadas, M. A., and Wattenberg, B. W. (2003). Activation of sphingosine kinase 1 by ERK1/2-mediated phosphorylation. *EMBO J*, **22**, 5491-5500.

Pitson, S. M., Moretti, P. A., Zebol, J. R., Xia, P., Gamble, J. R., Vadas, M. A., D'Andrea, R. J., and Wattenberg, B. W. (2000). Expression of a catalytically inactive sphingosine kinase mutant blocks agonist-induced sphingosine kinase activation: a dominant negative sphingosine kinase. *J Biol Chem*, **275**, 33945-33950.

Pitson, S. M., Xia, P., Leclercq, T. M., Moretti, P. A., Zebol, J. R., Lynn, H. E., Wattenberg, B. W., and Vadas, M. A. (2005). Phosphorylation-dependent translocation of sphingosine kinase to the plasma membrane drives its oncogenic signalling. *J Exp Med*, **201**, 49-54.

Safadi-Chamberlain, F., Wang, L. P., Payne, S. G., Lim, C. U., Stratford, S., Chavez, J. A., Fox, M. H., Spiegel, S., and Summers, S. A. (2005). Effect of a membrane-targeted sphingosine kinase 1 on cell proliferation and survival. *Biochem J*, **88**, 827-834.

Spiegel, S., and Milstien, S. (2003). Sphingosine-1-phosphate: an enigmatic signalling lipid. *Nature Rev Mol Cell Biol*, **4**, 397-407.

Taha, T. A., Kitatani, K., Bielawski, J., Cho, W., Hannun, Y. A., and Obeid, L. M. (2005). Tumor necrosis factor induces the loss of sphingosine kinase-1 by a cathepsin B-dependent mechanism. *J Biol Chem*, **280**, 17196-17202.

Taha, T. A., Osta, W., Kozhaya, L., Bielawski, J., Johnson, K. R., Gillanders, W. E., Dbaibo, G. S., Hannun, Y. A., and Obeid, L. M. (2004). Down-regulation of sphingosine kinase-1 by DNA damage: dependence on proteases and p53. *J Biol Chem*, **279**, 20546-20554.

Urtz, N., Olivera, A., Bofill-Cardona, E., Csonga, R., Billich, A., Mechtcheriakova, D., Bornancin, F., Woisetschlager, M., Rivera, J., and Baumruker, T. (2004). Early activation of sphingosine kinase in mast cells and recruitment to FcepsilonRI are mediated by its interaction with Lyn kinase. *Mol Cell Biol*, **24**, 8765-8777.

Xia, P., Wang, L., Moretti, P. A., Albanese, N., Chai, F., Pitson, S. M., D'Andrea, R. J., Gamble, J. R., and Vadas, M. A. (2002). Sphingosine kinase interacts with TRAF2 and dissects tumor necrosis factor-alpha signaling. *J Biol Chem*, **277**, 7996-8003.

Yamanaka, M., Shegogue, D., Pei, H., Bu, S., Bielawska, A., Bielawski, J., Pettus, B., Hannun, Y. A., Obeid, L., and Trojanowska, M. (2004). Sphingosine kinase (SPHK1) is induced by TGF-beta and mediates TIMP-1 upregulation. *J Biol Chem*, **279**, 53994-54001.

Young, K. W., Willets, J. M., Parkinson, M. J., Bartlett, P., Spiegel, S., Nahorski, S. R., and Challiss, R. A. (2003). Ca^{2+}/calmodulin-dependent translocation of sphingosine kinase: role in plasma membrane relocation but not activation. *Cell Calcium*, **33**, 119-128.

3-5 Ceramide 1-Phosphate

Susumu Mitsutake, Tack-Joong Kim, and Yasuyuki Igarashi

Department of Biomembrane and Biofunctional Chemistry, Graduate
School of Pharmaceutical Sciences, Hokkaido University, Kita 12, Nishi 6,
Kita-ku, Sapporo 060-0812, Japan

Summary. Over the past couple of decades, ceramide (Cer) has emerged
as a lipid mediator of cell signaling in a variety of events, including apoptosis and cell differentiation, and its intracellular levels are tightly controlled. Several enzymes are known to be regulators of Cer- levels. One of
these, ceramide kinase (CERK), catalyzes the conversion of ceramide
(Cer) to ceramide 1-phosphate (C1P). Although the activity of this enzyme
was reported in 1989, the CERK- gene was only recently identified.
CERK, which is activated by Ca^{2+}, possesses a typical diacylglycerol
kinase catalytic domain, a pleckstrin homology domain, and a Ca^{2+}/CaM
binding domain. The groups that examined CERK/C1P functions suggested that CERK/C1P is involved in many processes including: membrane
fusion, inflammation, DNA-synthesis, intracellular Ca^{2+} increase, and other
processes. These CERK/C1P functions and the investigations over of the
past decade that revealed them are reviewed here.

Keywords. ceramide kinase, ceramide 1-phosphate, pleckstrin homology
domain

1. Introduction

Cer is widely recognized as a lipid second messenger in various cell types,
and participates in a variety of cellular events such as apoptosis and differentiation (Perry and Hannun 1998; Obeid el al. 1993; Okazaki et al.
1990). Cer is generated by *de novo* synthesis or sphingomyelin (SM) hydrolysis, and is converted to SM, glucosylceramide (GlcCer), or sphingos-

ine (Sph), which itself can act as a signaling molecules, as can its derivative sphingosine 1-phosphate (S1P) (Igarashi 1997; Spiegel and Merrill 1996).

Cabot's group showed that a Cer- reducing enzyme, Glucosyltransferase (GlcT), was crucial for the multidrug resistance of certain cancer cell types (Liu Y.Y. et al. 2001). Their work suggests that Cer- production and Cer-reduction are important to regulate cellular Cer levels and whether Cer induces apoptosis. To date, four enzymes are known to reduce cellular Cer levels, GlcT, SM-synthase, ceramidase (CDase) and ceramide kinase (CERK). .All of these enzymes are now cloned. In particular, the recent identification by Sugiura of the gene encoding CERK (Sugiura et al. 2002) has provided insight into the previously studied activities of CERK and enabled us to investigate a new Cer- metabolic pathway and its related metabolite, C1P.

2. Ceramide kinase

2.1 History of CERK

In 1989, Bajjalieh and co-workers identified new enzyme in rat brain, synaptosomes, that phosphorylated ceramide (Bajjalieh et al. 1989; Bajjalieh and Batchelor 2000). During the purification step, CERK activity was successfully separated from diacylglycerol kinase (DGK), indicating that the enzyme was a Cer-specific lipid kinase, which was activated by calcium ions. Kolesnick's group subsequently reported CERK activity in HL60 cells (Kolesnick and Hemer 1990). Additionally, they labeled HL60 cell with $[^{32}P]$orthophosphoric acid, and confirmed the *in situ* formation of $[^{32}P]$C1P (Dressler and Kolesnick 1990). Following these reports, several groups described similar CERK activity as well as the existence of C1P, identifying functions for both (Bajjalieh and Batchelor 2000, Gómez-Muñoz 2004a). In 2002, Sugiura and co-workers eventually cloned the CERK gene, based on sequence similarity to sphingosine kinase type 1 (Sugiura et al. 2002). Their successful cloning and the characterization of the enzyme facilitated investigation into the physiological functions of CERK/C1P.

2.2 Primary structure and general properties of CERK

Human CERK is composed of 538 amino acid with a catalytic region that shares a high degree of similarity with the DGK catalytic domain. hCERK also exhibits high similarity to sphingosine kinase 1 (SPHK1), especially in the DGK catalytic region (57.1% similarity) (Fig. 1). Additionally, hCERK has a pleckstrin homology (PH) domain on its NH_2 terminus (Fig. 1), which is also myristoylated (Carré et al. 2004).

PH domains are known to mediate protein-protein and protein-membrane association and to act through high affinity binding to phosphatidylinositol phosphate (PIP) (Lemmon et al. 1996). Our studies indicate that the PH domain of CERK interacts mainly with $PI(4,5)P_2$, and only slightly with $PI(3,4)P_2$, $PI(3,5)P_2$ and $PI(3,4,5)P_3$ (Kim et al. paper in preparation). On the other hand, the PH domain is also important for CERK enzymatic activity since its deletion diminishes the activity significantly (Carré et al. 2004). Moreover, a mutation from Leu to Ala at the 10th amino acid from the NH_2 terminus dramatically reduces enzymatic activity, although, interestingly, this mutation dose not affect the affinity of the substrate. Thus, the PH domain of CERK is not only important for PIP-binding it might also act as a regulator of these enzymes (Kim et al. 2005). PIPs may also affect CERK. CERK is translocated to the plasma membrane in response to osmotic stress (Carré et al. 2004). Since osmotic stress is known to alter the intracellular content of PIPs (Sbrissa et al. 2002), the CERK translocation might be due to this alteration.

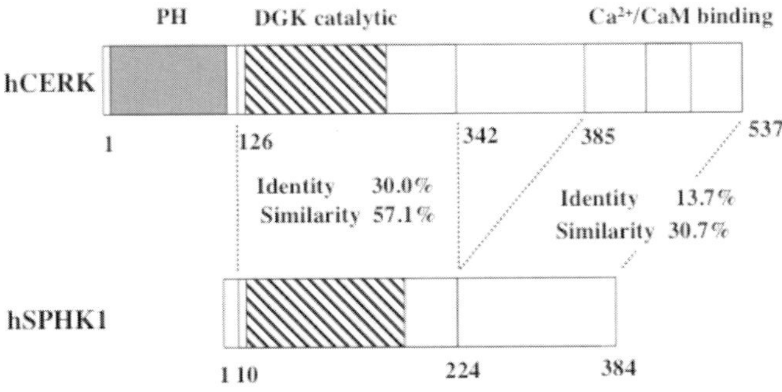

Fig. 1. Primary structures of hCERK and hSPHK1. hCERK was cloned, based on sequence similarity with hSPHK1. It has a typical pleckstrin homology (PH) domain (grey box), diacylglycerol kinase (DGK) catalytic domain (shaded box) and a Ca^{2+}/calmoduline binding domain (CaM; open box).

While CERK is activated by Ca^{2+}, there are no apparent Ca^{2+} binding domains in its primary structure, such as an EF hand or C2 domain. A Ca^{2+} regulatory mechanism was not easily explained. Our studies found that Ca^{2+}/calmodulin (CaM) is involved in C1P formation. Residue 422-435 of this enzyme contains the 1-8-14B type CaM binding motif. In response to intracellular Ca^{2+} elevation, Ca^{2+}/CaM enhances the Ca^{2+}-sensitiviy of CERK (Mitsutake et al. paper in preparation).

2.3 Tissue and intracellular distribution of CERK

The mRNA expression of hCERK is high in the brain, heart, skeletal muscle, kidney, and liver. During mammalian embryogenesis, CERK exhibits especially high expression at embryonic day 7 (Sugiura et al. 2002). Similarly, in Drosophila, the hCERK homologue CG16708 is expressed in the hindgut and the midgut by stage 15 (Renault et al. 2002).

As CERK activity was initially identified as a synaptic vesicle-associated protein (Bajjalieh et al. 1989), it was believed to be associated with some cellular component. In HEK293 cells overexpressing hCERK, activity was recovered in membrane fractions (Sugiura et al. 2002). However, in our immunohistochemical studies using a polyclonal antibody raised against mCERK, areas of partial colocalization with the endoplasmic reticulum were observed in RBL-2H3 cells. Little colocalization with the Golgi apparatus or plasma membrane was observed, although we did find strong CERK expression in the plasma membrane of some cells. CERK was predominantly localized in the cytosol along with more than 70% of the CERK activity (Mitsutake et al. 2004). In contrast, when Carré and co-workers examined the CERK distribution in COS1 cells using GFP-tagged CERK, they found perinuclear and punctuate cytoplasmic localization (Carré et al. 2004). According to database searches, no distinct signal sequence or organelle retention signals have been found in the amino acid sequence of CERK. Thus, CERK might initially be expressed in the cytosolic region, with the distribution being altered by cell-type or cell-condition. Further investigation, including additional analysis of the PH domain, might resolve this question.

3. Physiological role of CERK and C1P

3.1 C1P formation and membrane fusion

An elegant study by Shayman and co-workers demonstrated that in neutrophils CERK plays an important role in phagolysosome formation. In these cells, CERK activation occurs in a time-dependent fashion, peaking 10 min after formyl peptide stimulation and challenge with antibody-coated erythrocyte (Hinkovska-Galcheva et al. 1998). That laboratory also showed that adding exogenous C1P can promote liposome fusion in cell-free systems. They confirmed that overexpression of CERK enhances phagocytosis (Hinkovska-Galcheva et al. 2005).

Additionally, we have shown that C1P formation is involved in calcium-dependent degranulation of mast cells, and that C1P formation is enhanced during activation induced by IgE/antigen or by the Ca^{2+}-ionophore A23187. In fact, exogenous introduction of CERK into permeabilized RBL-2H3 cells is sufficient to cause degranulation (Mitsutake et al. 2004). Moreover, our newly developed synthetic inhibitor of CERK efficiently inhibits mast cell degranulation (Kim et al. paper in preparation).

Both phagocytosis and degranulation include a membrane-fusion step, and CERK was originally found in the synaptosome. Together, these facts imply that CERK/C1P are involved in calcium dependent membrane fusion, however, the exact nature of this involvement is still in question. The metabolism of Cer is crucial for membrane turnover and endocytosis of rhodopsin (Acharya et al. 2003; Rohrbough et al. 2004), so Cer metabolism might also be important for membrane dynamics.

Studies of phosphatidic acid (PA), which is a lipid structurally similar to C1P, might provide us with some insight. Bader provides one hypothetical role for PA (Badder et al. 2004). During membrane fusion, PA is produced by the action of phospholipase D (PLD) from phosphatidylcholine (PC). PA is a cone-shaped lipid, and one predicted effect of its generation at the granule docking site is to promote the negative curvature of the cytoplasmic membrane leaflet, thereby facilitating the formation of the hemi-fusion intermediates required for the fusion of the two membranes. Additional study is required to clarify the CERK/C1P role in membrane fusion.

3.2 CERK/C1P and inflammation

CERK has also been implicated in the inflammatory response (Baumruker et al. 2005). Petuus demonstrated that C1P directly enhances the activity of phospholipase A2, and a siRNA against CERK blocked both the arachidonic acid release and the subsequent PGE2 production after stimulation (Pettus et al. 2003, 2004). Moreover, direct binding of C1P and PLA2 has been confirmed. C1P act as allosteric regulator of PLA2 (Subramanian et al. 2005). In plants, mutation of CERK induces enhanced disease symptoms during pathogen attack (Liang et al. 2003), this might indicate C1P involvement in the host defense systems of plants.

3.3 The effect of exogenously added C1P

In fibroblasts, both exogenously added short chain C2 or C8-C1P or natural long-chain C1P can stimulate the incorporation of [^{3}H]thymidine into DNA, without conversion of C1P to S1P (Gómez-Muñoz et al. 2004a). Additionally, in bone marrow-derived macrophages, exogenously added C1P induces cell survival via the inhibition of SMase activity (Gómez-Muñoz et al. 2004b). C2-C1P reportedly causes intracellular increases in Ca^{2+} in calf pulmonary artery endothelial cells (Tornquist et al. 2004) and thyroid FRTL-5 cells (Hogback et al. 2003), however Riley determined that the formation of C8-C1P is not involved in the Ca^{2+} increase or activation of neutrophils (Riley et al. 2003).

4. CERK/C1P related enzymes

4.1 CERK-like protein

Single-nucleotide polymorphism (SNP) homozygosity mapping revealed that a mutation in CERK-like (CERKL) protein causes autosomal recessive retinitis pigmentosa (RP) (Tuson et al. 2004). CERKL was originally identified as a CERK homologue, but in characterization studies, CERKL didn't phosphorylate Cer in vitro, and [^{32}P]C1P formation was not detected in [^{32}P] labeled COS-1 cells transiently expressing CERKL (Bornancin et al. 2004). GFP-tagged CERKL is distributed in many cell compartments, including a specific association with nucleoli, indicating the presence of a nuclear localizing signal (NLS). Interestingly, the CERKL mutant (R257X) linked to the pathology of RP accumulated in nucleus but not in associated nucleoli. The detailed functions of CERKL, however, have yet to be discovered.

4.2 C1P-phosphatase

C1P is produced from Cer generated via both the hydrolysis of SM or the recycling pathway of Sph (Riboni et al. 2002). For a C1P metabolic pathway, one could imagine two pathways (Fig. 2). Several groups reportedly found a C1P-phosphatase. Futerman's group distinguished C1P-phosphatase activity from that of PA- phosphatase (Boudker and Futerman 1993) and showed that exogenously added C1P was rapidly hydrolyzed by C1P- phosphatase. In the same year, Shinghal found C1P-phosphatase activity in rat brain synaptosomes (Shinghal et al. 1993).

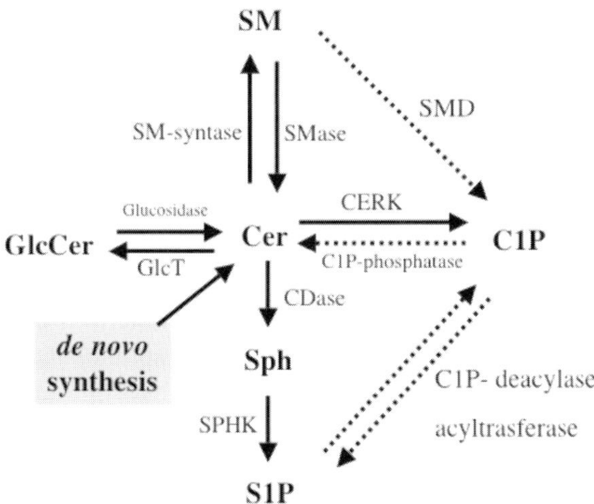

Fig. 2. Generation and possible metabolic pathways of C1P. Sphingolipids are metabolized and various lipid mediators are generated in known and propose pathways. SMD, C1P- deacylase and acyltransferase (dotted arrows) have not been identified yet in mammalian cells. Abbreviations used are: SM, sphingomyelin; SMase, sphingomyelinase; SMD, sphingomyelinase D; Cer, ceramide; CERK, ceramide kinase; C1P, ceramide 1-phosphate; CDase, ceramidase; Sph, sphingosine; S1P, sphingosine 1-phosphate; SPHK, sphingosine kinase; GlcCer, Glucosylceramide; GlcT, ceramide glucosyltransferase.

Lipid phosphate phosphatases (LPPs) can hydrolyze a variety of lipid phosphates including lysophosphatidic acid, PA, S1P and C1P *in vitro* (Brindley et al. 2000, Kai et al. 1997). The catalytic domain of the LPP family is thought to be located in the outer leaflet of the plasma membrane or the luminal face of the intracellular organelle (Zhang et al. 2000), so, these enzymes might might hydrolyzed extracellular substrates. However, as noted above, C1P production might mainly occur in the cytosol or the cytosolic face of organelles. So far, whether C1P can be hydrolyzed by LPPs *in vivo* has not been determined. There might be an alternate phosphatase that is able to hydrolyze C1P in the cytosolic side of the membrane.

4.3 Sphingomyelinase D and C1P-deacylase

The brown recluse spider (*Loxosceles reclusa*) and the bacterium *Corynebacterium pseudotuberculosis* possess sphingomyelinase D (SMD), which hydrolyses SM to produce C1P (Bernheimer et al. 1985, Songer et al. 1990). If this type of enzyme exists in mammalian cells, C1P could be produced independently from CERK action. As shown in Figure 2, N-deacylation of C1P could produce S1P independently from sphingosine kinase. We examined whether neutral CDase (Mitsutake et al. 2001) could hydrolyze C1P, but found it could not. At present, neither SMD nor C1P-deacylase have been found in mammalian cells.

4.4 Future aspects of CERK study

Cer is the precursor for all sphingolipids and functions as a second messenger in a variety of cellular events including apoptosis and cell differentiation. Its intracellular levels are tightly controlled. No direct target of Cer has yet been identified, however protein kinase C (PKC), phosphatidylinositol 3-kinase (PI3K) and PLD, each of which have important role in biological processes, are all regulated by Cer (Gómez-Muñoz et al. 1994, Huwiler et al. 1998, Kondo et al. 2002). These targets are all active in the cytosolic face of cellular organelles. Of the enzymes known to control Cer levels (CDase, GlcT, SM-synthase and CERK), only CERK appears able to metabolize Cer in cytosolic region. Considering these facts, CERK might act as a key enzyme in Cer signaling. The CERK- gene has only recently been identified, so the details of CERK/C1P functions will likely be revealed in the future. This may facilitate our understanding of Cer- signaling.

5. References

Acharya U, Patel S, Koundakjian E, Nagashima K, Han X and Acharya J.K. (2003) Modulating sphingolipid biosynthetic pathway rescues photoreceptor degeneration. *Science*, **299**, 1740-1743.

Bader MF, Doussau F, Chasserot-Golaz S, Vitale N and Gasman S (2004) Coupling actin and membrane dynamics during calcium-regulated exocytosis: a role for Rho and ARF GTPases. *Biochim Biophys Acta*, **1742**, 37-49.

Bajjalieh S, and Batchelor R (2000) Ceramide kinase. *Methods Enzymol*, **311**, 207-215.

Bajjalieh SM, Martin TF and Floor E (1989) Synaptic vesicle ceramide kinase. A calcium-stimulated lipid kinase that co-purifies with brain synaptic vesicles. *J Biol Chem*, **264**, 14354-14360.

Baumruker T, Bornancin F and Billich A (2005) The role of sphingosine and ceramide kinases in inflammatory responses. *Immunol Lett*, **96**, 175-185.

Bernheimer AW, Campbell BJ and Forrester LJ (1985) Comparative toxinology of Loxosceles reclusa and Corynebacterium pseudotuberculosis. *Science*, **228**, 590.

Bornancin F, Mechtcheriakova D., Stora S, Graf C, Wlachos A, Devay P, Urtz N, Baumruker and T, Billich A (2005) Characterization of a ceramide kinase-like protein. *Biochim Biophys Acta*, **1687**, 31-43.

Boudker O. and Futerman AH (1993) Detection and characterization of ceramide-1-phosphate phosphatase activity in rat liver plasma membrane. *J Biol Chem*, **268**, 22150-22155.

Brindley DN, Xu J, Jasinska R and Waggoner DW (2000) Analysis of ceramide 1-phosphate and sphingosine-1-phosphate phosphatase activities. *Methods Enzymol*, **311**, 233-244.

Carré A, Graf C, Stora S, Mechtcheriakova D, Csonga R, Urtz N, Billich A, Baumruker T and Bornancin F (2004) Ceramide kinase targeting and activity determined by its N-terminal pleckstrin homology domain. *Biochem Biophys Res Commun*, **324**, 1215-1219.

Dressler KA and Kolesnick RN (1990) Ceramide 1-phosphate, a novel phospholipid in human leukemia (HL-60) cells. Synthesis via ceramide from sphingomyelin. *J Biol Chem*, **265**, 14917-14921.

[a]Gómez-Muñoz A (2004) Ceramide-1-phosphate: a novel regulator of cell activation. *FEBS Lett*, **562**, 5-10.

[b]Gómez-Muñoz A, Kong JY, Salh B and Steinbrecher UP (2004) Ceramide-1-phosphate blocks apoptosis through inhibition of acid sphingomyelinase in macrophage. *J Lipid Res*, **45**, 99-105.

Gómez-Muñoz A, Martin A, O'Brien L and Brindley DN (1994) Cell-permeable ceramides inhibit the stimulation of DNA synthesis and phospholipase D activity by phosphatidate and lysophosphatidate in rat fibroblasts. *J Biol Chem*, **269**, 8937-8943.

Hinkovska-Galcheva V, Boxer LA, Kindzelskii A, Hiraoka M, Abe A, Goparju S, Spiegel S, Petty HR and Shayman JA (2005) Ceramide 1-phosphate, a mediator of phagocytosis. *J Biol Chem*, **280**, 26612-26621.

Hinkovska-Galcheva VT, Boxer LA, Mansfield PJ, Harsh D, Blackwood A, and Shayman JA (1998) The formation of ceramide-1-phosphate during neutrophil phagocytosis and its role in liposome fusion. *J Biol Chem*, **273**, 33203-33209.

Hinkovska-Galcheva VT, Boxer LA, Kindzelskii A, Hiraoka M, Abe A, Goparju S, Spiegel S, Petty HR and Shayman JA (2005) Ceramide 1-phosphate, a mediator of phagocytosis. *J Biol Chem*, **280**, 26612-26621.

Hogback SLP, Rudnas B, Bjorklund S, Slotte JP, and Tornquist K (2003) Ceramide 1-phosphate increases intracellular free calcium concentrations in thyroid FRTL-5 cells: evidence for an effect mediated by inositol 1,4,5-trisphosphate and intracellular sphingosine 1-phosphate. *Biochem J*, **370**, 111-119.

Huwiler A, Fabbro D and Pfeilschifter J (1998) Selective ceramide binding to protein kinase C-alpha and -delta isoenzymes in renal mesangial cells. *Biochemistry*, **37**, 14556-14562.

Igarashi Y (1997) Functional roles of sphingosine, sphingosine 1-phosphate, and methylsphingosines: in regard to membrane sphingolipid signaling pathways. *J Biochem (Tokyo)*, **122**, 1080-1087.

Kai M, Wada I, Imai S, Sakane F and Kanoh H (1997) Cloning and characterization of two human isozymes of Mg^{2+}-independent phosphatidic acid phosphatase. *J Biol Chem*, **272**, 24572-24578.

Kim TJ, Mitsutake S, Kato M and Igarashi Y (2005) The leucine 10 residue in pleckstrin homology domain of ceramide kinase is crucial for its catalytic activity. *FEBS lett* in press

Kolesnick RN and Hemer MR (1990) Characterization of a ceramide kinase activity from human leukemia (HL-60) cells. Separation from diacylglycerol kinase activity. *J Biol Chem*, **265**, 18803-18808.

Kondo T, Kitano T, Iwai K, Watanabe M, Taguchi Y, Yabu T, Umehara H, Domae N, Uchiyama T and Okazaki T (2002) Control of ceramide-induced apoptosis by IGF-1: involvement of PI-3 kinase, caspase-3 and catalase. *Cell Death Differ*, **9**, 682-692.

Lemmon MA, Ferguson KM and Schlessinger J (1996) PH domains: diverse sequences with a common fold recruit signaling molecules to the cell surface. *Cell*, **85**, 621-624.

Liang H, Yao N, Song JT, Luo S, Lu H and Greenberg JT (2003) Ceramides modulate programmed cell death in plants. *Genes Dev*, **17**, 2636-2641.

Liu YY, Han HY, Giuliano AE, and Cabot MC (2001) Ceramide glycosylation potentiates cellular multidrug resistance. *FASEB J*, **15**, 719-730.

Mitsutake S, Kim TJ, Inagaki Y, Kato M, Yamashita T, and Igarashi Y (2004) Ceramide kinase is a mediator of calcium-dependent degranulation in mast cells. *J Biol Chem*, **279**, 17570-17577.

Mitsutake S, Tani M, Okino N, Mori K, Ichinose S, Omori A, Iida H, Nakamura T and Ito M (2001) Purification, characterization, molecular cloning, and subcellular distribution of neutral ceramidase of rat kidney. *J Biol Chem*, **276**, 26249-26259.

Obeid LM, Linardic CM, Karolak LA and Hannun YA (1993) Programmed cell death induced by ceramide. *Science*, **259**, 1769-1771.

Okazaki T, Bell RM and Hannun YA (1989) Sphingomyelin turnover induced by vitamin D3 in HL-60 cells. Role in cell differentiation. *J Biol Chem*, **264**, 19076-19080.

Perry DK and Hannun YA (1998) The role of ceramide in cell signaling. *Biochim Biophys Acta*, **1436**, 233-243.

Pettus BJ, Bielawska A, Spiegel S, Roddy P, Hannun YA, and Chalfant CE. (2003) Ceramide kinase mediates cytokine- and calcium ionophore-induced arachidonic acid release. *J Biol Chem*, **278**, 38206-38213.

Pettus BJ, Bielawska A., Subramanian P, Wijesinghe DS, Maceyka M, Leslie CC, Evans JH, Freiberg J, Roddy P, Hannun YA, and Chalfant CE. (2004) Ceramide 1-phosphate is a direct activator of cytosolic phospholipase A2. *J Biol Chem*, **279**, 11320-11326.

Renault AD, Starz-Gaiano M and Lehmann R (2002) Metabolism of sphingosine 1-phosphate and lysophosphatidic acid: a genome wide analysis of gene expression in Drosophila. *Gene Expression Patterns*, **2**, 337-345.

Riboni L, Bassi R, Anelli V and Viani P. (2002) Metabolic formation of ceramide-1-phosphate in cerebellar granule cells: evidence for the phosphorylation of ceramide by different metabolic pathways. *Neurochem Res*, **27**, 711-716.

Rile G, Yatomi Y, Takafuta T and Ozaki Y (2003) Ceramide 1-phosphate formation in neutrophils. *Acta Haematol*, **109**, 76-83.

Rohrbough J, Rushton E, Palanker L, Woodruff E, Matthies HJ, Acharya U, Acharya JK and Broadie K (2004) Ceramidase regulates synaptic vesicle exocytosis and trafficking. *J Neurosci*, **24**, 7789-7803.

Sbrissa D, Ikonomov OC, Deeb R and Shisheva A (2002) Phosphatidylinositol 5-phosphate biosynthesis is linked to PIKfyve and is involved in osmotic response pathway in mammalian cells. *J Biol Chem*, **277**, 47276-47284.

Shinghal R, Scheller RH and Bajjalieh SM (1993) Ceramide 1-phosphate phosphatase activity in brain. *J Neurochem*, **61**, 2279-2285.

Songer JG, Libby SJ, Iandolo JJ and Cuevas WA (1990) Cloning and expression of the phospholipase D gene from Corynebacterium pseudotuberculosis in Escherichia coli. *Infect Immun*, **58**, 131-136.

Spiegel S and Merrill AH (1996) Sphingolipid metabolism and cell growth regulation. *FASEB J*, **10**, 1388-1397.

Subramanian P, Stahelin RV, Szulc Z, Bielawska A, Cho W, and Chalfant CE. (2005) Ceramide 1-phosphate acts as a positive allosteric activator of group IVA cytosolic phospholipase A2alpha and enhances the interaction of the enzyme with phosphatidylcholine. *J Biol Chem*, **280**, 17601-17607.

Sugiura M, Kono K, Liu H, Shimizugawa T, Minekura H, Spiegel S, and Kohama T (2002) Ceramide kinase, a novel lipid kinase. Molecular cloning and functional characterization. *J Biol Chem*, **277**, 23294-23300.

Tornquist K, Blom T., Shariatmadari R, and Pasternack M (2004) Ceramide 1-phosphate enhances calcium entry through voltage-operated calcium channels by a protein kinase C-dependent mechanism in GH4C1 rat pituitary cells. *Biochem J*, **380**, 661-668.

Tuson M, Marfany G and Gonzalez-Duarte R (2004) Mutation of CERKL, a novel human ceramide kinase gene, causes autosomal recessive retinitis pigmentosa (RP26). *Am J Hum Genet*, **74**, 128-138.

Zhang QX, Pilquil CS, Dewald J, Berthiaume LG and Brindley DN (2000) Identification of structurally important domains of lipid phosphate phosphatase-1: implications for its sites of action. *Biochem J*, **345**, 181-184.

3-6 Sphingosine-1-Phosphate Lyase

Julie D. Saba

Children's Hospital Oakland Research Institute, 5700 Martin Luther King Jr. Way, Oakland, CA 94609, USA

Summary. Sphingosine-1-phosphate lyase is responsible for the ultimate step in sphingolipid degradation. Sphingosine-1-phosphate and other phosphorylated long chain bases generated through the actions of sphingosine kinase may be dephosphorylated by lipid phosphatases or cleaved at the C_{2-3} carbon-carbon bond by the pyridoxal 5'-phosphate-dependent enzyme, sphingosine-1-phosphate lyase. By regulating intracellular levels of phosphorylated long chain bases, sphingosine-1-phosphate lyase may control intracellular and extracellular signaling events mediated by these bioactive molecules. Through additional effects mediated by its substrate and products on the biosynthesis of sphingolipids, their conversion into phospholipids, and the synthesis of sterols and fatty acids, sphingosine-1-phosphate lyase may regulate the flow of lipid intermediates through several metabolic pathways. Subsequent to the cloning of the first sphingosine-1-phosphate lyase gene from *Saccharomyces cerevisiae*, putative homologs have been identified in various plant and animal species, indicating that the enzyme is highly conserved. Sphingosine-1-phosphate lyase mutant phenotypes in fungi, simple metazoans, and mammalian cells suggest that the enzyme plays a role in the regulation of apoptosis, stress responses, reproduction, development, and tissue integrity and repair. Analysis of sphingosine-1-phosphate lyase gene expression has shed light on how the enzyme may be regulated under physiological conditions. Identification of specific inhibitors and generation of constitutive and conditional mouse knockout models should facilitate the dissection of sphingosine-1-phosphate lyase's role within the context of mammalian development, physiology and the pathophysiology of various disease states.

Keywords. sphingosine-1-phosphate, S1P lyase (SPL), pyridoxal 5'-phosphate, *Sgpl1*, long chain base

1. Introduction

Sphingosine-1-phosphate lyase (SPL) is responsible for the ultimate step in sphingolipid degradation. Long chain bases (LCBs) generated either through de novo synthesis or metabolism of higher order sphingolipids are subject to phosphorylation by the actions of sphingosine kinases. These phosphorylated long chain bases (LCBPs) may then be dephosphorylated by lipid phosphatases, or cleaved at the C_{2-3} carbon-carbon bond by the pyridoxal 5'-phosphate-dependent enzyme, SPL (Van Veldhoven, 2000). The latter reaction, which is irreversible, depletes intracellular stores of the potent mitogen and motogen sphingosine-1-phosphate (S1P) and results in the formation of two products, ethanolamine phosphate and a long chain aldehyde two carbons shorter than the substrate molecule (Van Veldhoven, 2000).

By regulating intracellular levels of S1P and other LCBPs, SPL may control intracellular signaling events mediated by these bioactive molecules. SPL may also regulate the amount of S1P available for cellular export, thus impacting autocrine and/or paracrine signaling through extracellular S1P receptors. Inhibition of SPL expression also results in accumulation of sphingosine and other LCBs, which may have additional biological import (Cuvillier, 2002; Suzuki et al., 2004). Through both its substrate and products, SPL may regulate the flow of lipid intermediates through several metabolic pathways. For example, there is evidence that LCBPs inhibit the enzyme serine palmitoyltransferase, which catalyzes the rate-limiting step in sphingolipid biosynthesis (van Echten-Deckert et al., 1997). In human cells, SPL overexpression leads to a modest increase in sphingolipid biosynthesis, suggesting that the lyase may participate in a feedback loop that promotes de novo synthesis of sphingolipids (Reiss et al., 2004). The fatty aldehyde product of the reaction can be converted to the corresponding fatty acid by fatty aldehyde dehydrogenase, then covalently linked to Coenzyme A through the actions of acyl-CoA synthetase and incorporated into phospholipids. The ethanolamine phosphate product can also be incorporated into phospholipids by the actions of phosphoethanolamine cytidylyltransferase. The products of the SPL reaction may themselves influence cell proliferation and survival, as indicated by recent studies demonstrating that SPL stimulates mitogenesis through a mechanism that is independent of S1P (Kariya et al., 2005). Inhibition of SPL expression in *Drosophila melanogaster* leads to dysregulated processing of the transcription factor sterol regulatory element binding protein (SREBP). SPL may indirectly influence SREBP processing by altering levels of membrane phosphatidylethanolamine, which is likely to function as a sensor for sterol gene regulation equivalent to the role of cholesterol in mammalian cells (Dobrosotskaya et al., 2002). SPL can, thus, be viewed as an enzyme uniquely situated at a crossroads between several lipid metabolic pathways, with the potential to influence the degradation and biosynthesis of sphingolipids, their conversion into phospholipids, and the synthesis of sterols and fatty acids through its effects on SREBP (Figure 1). Although the influence of SPL on S1P-independent aspects of lipid homeostasis has not been thoroughly examined, this gatekeeper function may ultimately prove to play a significant role in mediating the full effect of SPL on cell biology, physiology and development.

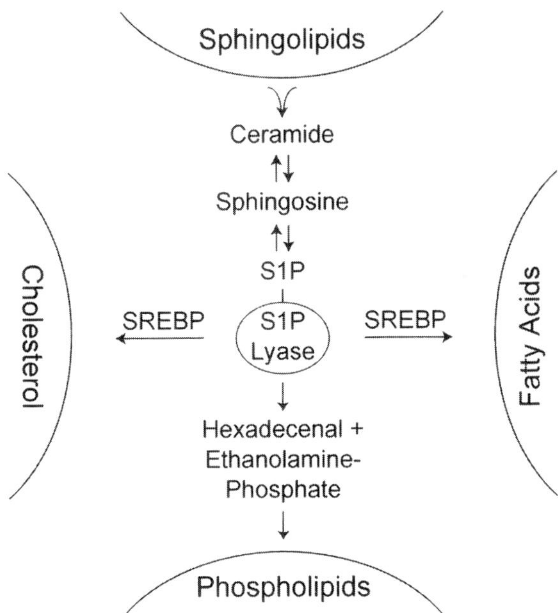

Fig. 1. Sphingosine phosphate lyase and lipid homeostasis.

2. Enzyme properties, subcellular localization and topology

Studies performed by Stoffel and colleagues in the 1960s utilizing D,L-*erythro*-dihydrosphingosine-1-phosphate (DHS1P) substrate and rat liver as an enzyme source found the primary intracellular sites of SPL activity to be endoplasmic reticulum (ER) and inner mitochondrial membranes (Stoffel et al., 1969). This is consistent with findings that SPL activity is enriched in the cell membrane fractions of yeast and other organisms, further corroborated by confocal microscopy studies of an SPL-green fluorescence protein fusion protein in human embryonic kidney cells (Reiss et al., 2004). Recent studies performed using epitope-tagged human SPL proteins demonstrated equal recovery from rough and smooth ER fractions, and confirmed that SPL is an integral membrane protein (Ikeda et al., 2004).

With DHS1P as substrate, the reaction catalyzed by SPL results in the direct formation of one molecule of palmitaldehyde representing the C_{3-18} portion of the substrate, and ethanolamine phosphate resulting from the C_{1-2} portion. The reaction is enhanced up to six-fold with the addition of the cofactor pyridoxal 5'-phosphate. Phosphatases that rapidly hydrolyze both cofactor and substrate are inhibited by sodium fluoride, which has no effect on SPL at low concentrations, thus allowing SPL-mediated catalysis to be measured in isolation. The standard assay utilizes a D-*erythro*-dihydrosphingosine [4,5-^{3}H] 1-phosphate substrate and follows the formation of radiolabeled aldehyde product (Van Veldhoven and Mannaerts, 1991). The naturally occurring D(+)*erythro* isoforms of DHS1P and S1P are recognized and cleaved by the enzyme. DHS1P of various carbon chain lengths from C_7 to C_{20} are adequate substrates for SPL, as are some S1P analogs containing methyl groups at C_4 or C_5. Accumulation of phytosphingosine-1-phosphate in yeast SPL mutants and correction of the defect by expression of murine or human SPL cDNA provides indirect evidence that phytosphingosine-1-phosphate also serves as a substrate for most SPL enzymes. Development of a more flexible assay system is needed to address the relative activity of SPL enzymes toward unique LCBP substrates endogenous to other species, such as the C17 iso and anteiso sphingosines of *C. elegans*, the phytosphingosines of *S. cerevisiae* and the sphingenines, hydroxysphingenines and sphingadienines of plants (Chitwood et al., 1995; Dickson and Lester, 1999; Dunn et al., 2004; Lynch and Dunn, 2004).

The enzyme is inhibited by sulfhydryl reagents, the pyridoxal 5'-phosphate analog deoxypyridoxine, and carbonyl-reactive compounds that complex with the cofactor. Recently, more specific inhibitors have been identified, including sphinganine phosphonate and the 2D,3L optical isomer of DHS1P which both serve as competitive inhibitors, as well as the ceramide analog GT11, which inhibits dihydroceramide desaturase at low concentrations but affects SPL activity at higher concentrations (Triola et al., 2004; Van Veldhoven, 2000).

We recently explored interactions between SPL and the sphingosine analog and immune modulator FTY720, which functions as an S1P receptor antagonist. FTY720 was shown to inhibit SPL activity in vitro and in vivo, leading to S1P accumulation in tissues of treated animals concomitant with drug-induced lymphopenia (P. Bandhuvula, Y. Tam, B. Oskouian and J.D. Saba, unpublished observations, 2005). These findings raise the possibility that SPL inhibition may contribute to the immunomodulatory effects of the drug and suggest that LCB analogs designed as S1P receptor agonists/antagonists may exert biological effects through interactions with enzymes of S1P metabolism.

3. SPL-encoding genes

The first known SPL gene, *DPL1*, was cloned in 1997 by virtue of its ability to confer sphingosine resistance when overexpressed in *S. cerevisiae* (Saba et al., 1997). At this time, functional homologs of *DPL1* have been confirmed experimentally in *Homo sapiens*, *Mus musculus*, *C. elegans*, *D. melanogaster* and *A. thaliana*, demonstrating that this enzyme is highly conserved throughout evolution (Herr et al., 2003; Mendel et al., 2003; Van Veldhoven et al., 2000; Zhou and Saba, 1998)(P. Bandhuvula, D. Worrall and J.D. Saba, unpublished observations, 2005). To date, *DPL1* and its homologs are the only genes known to encode SPL enzymes. In *S. cerevisiae* and *C. elegans*, the *DPL1* homolog is the sole or predominant SPL, since deletion or knock-down of the gene results in loss of detectible enzyme activity. Discrepancies between enzyme activity and *Sgpl1* gene expression in different mouse tissues suggest the possibility that additional SPL enzymes may exist. In support of this notion, murine F9 embryonal carcinoma cells disrupted for *Sgpl1* contain residual SPL activity and can metabolize exogenously added dihydrosphingosine (Ikeda et al., 2005).

Certain structural features are retained in all Dpl1p homologs. A hydrophobic N-terminal domain demonstrating low sequence similarity between SPLs of different species serves to anchor the protein to the outer membrane of the ER but is not critical for enzyme activity (Van Veldhoven et al., 2000). Two conserved cysteines (218 and 317) are necessary for enzymatic activity in human SPL. A highly conserved sequence surrounding human SPL K353 between 344-359 is presumed to be required for cofactor binding, since this sequence is also found in glutamate decarboxylase and other pyridoxal 5'-phosphate-requiring enzymes, and mutation of this residue destroys enzyme activity. The enzyme is oriented with the cofactor and substrate-binding central hydrophilic domain facing the cytosol (Ikeda et al., 2004). Despite the information that has been gained since the cloning of *DPL1*, many questions remain, including whether the enzyme acts as a monomer or multimer, if different isoforms or SPL genes exist, if the enzyme is regulated physiologically through posttranslational modifications or other mechanisms, and whether physiologically relevant polymorphisms exist in human *Sgpl1*.

4. Expression patterns and gene regulation

In *D. melanogaster* and *C. elegans*, SPL gene expression is developmentally regulated and spatially restricted to the embryonic and adult gut. Gut-specific expression in simple metazoans may reflect a role for SPL in digestion of sphingolipids found in the diet. However, loss of SPL expression in these organisms leads to striking phenotypes in tissues other than the alimentary tract (see below). Sphingolipid metabolism may be centralized in these organisms, with the gut acting as a lipid sink and/or generating lipid intermediates that diffuse or are transported to other sites or affect other tissues indirectly through membrane or signaling functions.

In mammals, variable levels of SPL activity have been demonstrated in most tissues, with highest levels in intestine, liver, kidney and thymus. Only platelets have no detectible SPL activity. During murine development, *Sgpl1* gene expression has been demonstrated as early as embryonic day 4.5, with highest expression levels attained on days 6.5-9.5 (Ikeda et al., 2004). Murine SPL protein expression is upregulated in F9 embryonal carcinoma cells stimulated with retinoic acid to undergo differentiation to primitive endoderm, providing the first indication that SPL expression may be dynamically regulated in physiologically relevant contexts (Kihara et al., 2003).

A conserved mechanism involving the GATA family of transcription factors was found to regulate SPL expression (Oskouian et al., 2005). A combination of promoter deletion analysis, site-specific mutagenesis, RNA interference and transgenic reporter assays were used to show that the nematode GATA factor ELT-2 is required for tissue-specific expression of SPL in the developing worm. Human SPL gene expression also appears to be regulated by GATA factors, as transient expression of GATA-4 or -6 led to an increase in expression of a human SPL luciferase reporter as well as endogenous protein expression and activity.

We have recently observed induction of SPL reporter gene expression in response to serum deprivation, hypoxia, loss of cell anchorage and treatment with DNA damaging agents (B. Oskouian and J.D. Saba, unpublished observations, 2005). While these observations remain preliminary, they suggest that additional regulatory mechanisms controlling SPL expression remain to be identified.

5. SPL in biology and disease

SPL mutant phenotypes suggest that the enzyme plays a role in the normal physiological processes of a wide variety of organisms including fungi, simple metazoans and mammals (Acharya and Acharya, 2005; Oskouian and Saba, 2004; Saba, 2004). In *S. cerevisiae*, the haploid *dpl1* mutant demonstrates changes in calcium homeostasis, marked resistance to heat stress and high culture saturation density (Birchwood et al., 2001; Gottlieb et al., 1999; Skrzypek et al., 1999). The latter two phenotypes have been linked to alterations in cell cycle control (Sims et al., 2004). In addition, *Dpl1* overexpression suppresses the lethality of some endocytosis mutants (Grote et al., 2000). Interactions between Dpl1p and VAMP proteins involved in exocytosis were implicated in this process, suggesting the possibility that Dpl1p may play a role in regulation of membrane trafficking.

In *D. discoideum*, loss of function mutations in the SPL gene confer resistance to the alkylating agent cisplatin and the related compound carboplatin (Li et al., 2000). The effect appears to be related to regulation of S1P levels, since SK overexpression also promoted drug resistance (Min et al., 2004; Min et al., 2005a). The SPL mutant also demonstrated defects throughout the developmental program of the organism, including abnormal fruiting body formation, depleted spore mass, decreased survival during stationary phase, and poor slug migration, probably due to an absence of filopodia and atypical actin organization (Li et al., 2001). Many of the defects were reproduced in wild-type cells by addition of S1P. Thus, SPL and regulation of S1P levels are necessary for normal *Dictyostelium* development, as well as for determining viability in response to cytotoxic stress.

The *D. melanogaster Sply* mutant illustrates the influence of SPL on animal development and tissue integrity (Herr et al., 2003). The mutant is homozygous for a transposon-insertion null allele at the *Drosophila* SPL locus. Phenotypically, *Sply* mutants exhibit semilethality, diminished egg laying, extra spermathecae and gross pattern abnormalities in dorsal longitudinal flight muscles that power the wings, leading to flightlessness. Timed experiments on consecutive days after eclosion revealed a progressive loss of individual muscle fibers. Further, dysregulated apoptosis was observed in mutant embryos, especially in the region of the developing genital disc. These defects were corrected either by restoring *Sply* expression or by introducing a mutant allele of the *lace* gene, which encodes a serine palmitoyltransferase subunit, leading to diminished sphingolipid synthesis and a reduction in the accumulation of sphingolipid intermediates characteristic of these mutants. These studies established the importance of precise regulation of sphingolipid metabolism and *Sply* expression for development, regulation of apoptosis, reproduction, and tissue integrity.

Inhibition of S1P lyase expression by RNA interference in *C. elegans* hermaphrodites leads to poor feeding, delayed growth, reproductive abnormalities and intestinal damage in the F1 and F2 progeny (Mendel et al., 2003). Intestinal abnormalities observed in treated worms include areas of prominent intestinal constriction and vacuolization. Reproductive phenotypes include poor egg laying, asynchronous egg development, ovulation defects and withering of the reproductive tract. In *C. elegans*, the intestinal cells produce yolk proteins and provide nurse cell-like functions that support egg production and development, which may potentially explain the prominent reproductive tract abnormalities observed in association with depletion of SPL, which is a gut-specific protein in nematodes.

As mentioned above, SPL expression is upregulated when F9 embryonal carcinoma cells are induced to differentiate to primitive endoderm (Kihara et al., 2003). Targeted disruption of SPL in F9 cells led to the accumulation of S1P and accelerated the differentiation to primitive endoderm, but loss of SPL expression had no effect on further differentiation to parietal endoderm. The effect on early endodermal differentiation was emulated by overexpression of sphingosine kinase, but addition of exogenous S1P had no effect. These findings suggest that SPL expression may control the rate of endodermal differentiation in the early embryo and that this effect may be through an intracellular rather than extracellular S1P signaling mechanism.

In mammalian cells, SPL overexpression results in S1P depletion, ceramide accumulation and enhanced apoptosis in response to stressful conditions, including DNA damage and serum deprivation (Min et al., 2005b; Reiss et al., 2004). Inhibition of SPL expression through small interfering RNA techniques diminishes apoptosis under the same conditions, portending physiological relevance of this phenomenon (B. Oskouian, P. Sooriyakumaran and J.D. Saba, unpublished observations, 2005). SPL appears to stimulate apoptosis through p53- and p38-dependent mechanisms. The potential for SPL to regulate cell fate decisions in physiological contexts is supported by several recent findings. The high level of SPL expression in tissues marked by rapid cell turnover such as the intestinal villi and olfactory epithelium as well as in degenerating intrahepatic bile duct epithelial cells in primary biliary cirrhosis indicates that SPL may be involved in facilitating the clearance of injured cells in damaged or regenerating tissues (Genter et al., 2003; Tanaka et al., 2001). In addition, downregulation of SPL expression in metastatic compared to primary malignant tissues suggests that SPL loss could play a role in cancer progression by allowing cells to evade normal apoptotic mechanisms (Ramaswamy et al., 2003). The recent identification of murine *Sgpl1* as a transcriptional target of platelet-derived growth factor indicates that SPL may play a role in mediating platelet-derived growth factor signaling and suggests that the full impact of SPL on biology has not yet been determined (Chen et al., 2004).

6. Conclusions and future directions

Studies performed in a variety of organisms and cell systems have established that SPL expression and function play a role in development, tissue integrity and stress responses. SPL activity influences the level of phosphorylated and unphosphorylated LCBs and ceramides. However, SPL may exert S1P-independent and potentially catalysis-independent effects. SPL protein purification, crystallization and molecular modeling should help to elucidate how SPL interacts with substrate, cofactor and membrane environment and carries out its enzymatic functions. Identification of mechanisms involved in the regulation of SPL expression may provide additional clues regarding SPL function. The development of simpler assay systems, knockout mouse models, and more potent and specific biochemical inhibitors are critical next goals. These tools should greatly facilitate dissection of SPL-mediated functions in development, immunity and other physiological processes and may substantiate its relevance as a pharmacological target in the treatment of human disease.

7. References

Acharya U, Acharya J (2005) Enzymes of sphingolipid metabolism in *Drosophila melanogaster. Cell and Mol Life Sci*, **62**, 128-142.

Birchwood CJ, Saba JD, Dickson RC, Cunningham KW (2001) Calcium influx and signaling in yeast stimulated by intracellular sphingosine-1-phosphate accumulation. *J Biol Chem*, **276**, 11712-11718.

Chen W, Delrow J, Corrin P, Frazier J, Soriano P (2004) Identification and validation of PDGF transcriptional targets by microarray-coupled gene-trap mutagenesis. *Nature Genetics*, **36**, 304-312.

Chitwood DJ, Lusby WR, Thompson MJ, Kochansky JP, Howarth OW (1995) The glycosylceramides of the nematode *Caenorhabditis elegans* contain an unusual, branched-chain sphingoid base. *Lipids*, **30**, 567-573.

Cuvillier O (2002) Sphingosine in apoptosis signaling. *Biochim Biophys Acta*, 1585, 153-162

Dickson RC, Lester RL (1999) Yeast sphingolipids. *Biochim Biophys Acta*, **1426**, 347-357.

Dobrosotskaya I, Seegmiller A, Brown M, Goldstein J, Rawson R (2002) Regulation of SREBP processing and membrane lipid production by phospholipids in Drosophila. *Science*, **296**, 879-883.

Dunn T, Lynch D, Michaelson L, Napier J (2004) A post-genomic approach to understanding sphingolipid metabolism in *Arabidopsis thaliana. Annals of Botany*, **93**, 483-497.

Genter MB, Van Veldhoven PP, Jegga AG, Sakthivel B, Kong S, Stanley K, Witte DP, Ebert CL, Aronow BJ (2003) Microarray-based discovery of highly expressed olfactory mucosal genes: potential roles in the various functions of the olfactory system. *Physiol Genomics*, **16**, 67-81.

Gottlieb D, Heideman W, Zhou J, Oskouian B, Saba J (1999) The *DPL1* gene is involved in mediating the response to nutrient deprivation in *Saccharomyces cerevisiae. Mol Cell Biol Res Comm*, **1**, 66-71.

Grote E, Vlacich G, Pypaert M, Novick PJ (2000) A snc1 endocytosis mutant: phenotypic analysis and suppression by overproduction of dihydrosphingosine phosphate lyase. *Mol Biol Cell*, **11**, 4051-4065.

Herr DR, Fyrst H, Phan V, Heinecke K, Georges R, Harris GL, Saba JD (2003) Sply regulation of sphingolipid signaling molecules is essential for Drosophila development. *Development*, **130**, 2443-2453.

Ikeda M, Kihara A, Igarashi Y (2004) Sphingosine-1-phosphate lyase SPL is an endoplasmic reticulum-resident, integral membrane protein with the pyridoxal 5'-phosphate binding domain exposed to the cytosol. *Biochem Biophys Res Commun*, **325**, 338-343.

Ikeda M, Kihara A, Kariya Y, Lee YM, Igarashi Y (2005) Sphingolipid-to-glycerophospholipid conversion in SPL-null cells implies the existence of an alternative isozyme. *Biochem Biophys Res Commun*, **329**, 474-479.

Kariya Y, Kihara A, Ikeda M, Kikuchi F, Nakamura S, Hashimoto S, Choi C, Lee Y, Igarashi Y (2005) Products by the sphingosine kinase/sphingosine 1-phosphate (S1P) lyase pathway but not S1P stimulate mitogenesis. *Genes Cells*, **10**, 605-615.

Kihara A, Ikeda M, Kariya Y, Lee E, Lee Y, Igarashi Y (2003) Sphingosine-1-phosphate lyase is involved in the differentiation of F9 embryonal carcinoma cells to primitive endoderm. *J Biol Chem*, **278**, 14578-14585.

Li G, Alexander H, Schneider N, Alexander S (2000) Molecular basis for resistance to the anticancer drug cisplatin in Dictyostelium. *Microbiology*, **146 (Pt 9)**, 2219-2227.

Li G, Foote C, Alexander S, Alexander H (2001) Sphingosine-1-phosphate lyase has a central role in the development of Dictyostelium discoideum. *Development*, **128**, 3473-3483.

Lynch D, Dunn T (2004) An introduction to plant sphingolipids and a review of recent advances in understanding their metabolism and function. *New Phytologist*, **161**, 677-702.

Mendel J, Heinecke K, Fyrst H, Saba JD (2003) Sphingosine phosphate lyase expression is essential for normal development in *Caenorhabditis elegans*. *J Biol Chem*, **278**, 22341-22349.

Min J, Stegner A, Alexander H, Alexander S (2004) Overexpression of sphingosine-1-phosphate lyase or inhibition of sphingosine kinase in Dictyostelium discoideum results in a selective increase in sensitivity to platinum-based chemotherapy drugs. *Eukaryotic Cell*, **3**, 795-805.

Min J, Traynor D, Stegner AL, Zhang L, Hanigan MH, Alexander H, Alexander S (2005a) Sphingosine kinase regulates the sensitivity of Dictyostelium discoideum cells to the anticancer drug cisplatin. *Eukaryot Cell*, **4**, 178-189.

Min J, Van Veldhoven P, Zhang L, Hanigan M, Alexander H, Alexander S (2005b) Sphingosine-1-phosphate lyase regulates sensitivity of human cells to select chemotherapy drugs in a p38-dependent manner. *Mol Cancer Res*, **3**, 287-296.

Oskouian B, Saba J (2004) Death and taxis: what non-mammalian models can tell us about sphingosine 1-phosphate. *Cell and Developmental Biology*, **15**, 529-540.

Oskouian B, Mendel J, Shocron E, Lee MA Jr, Fyrst H, Saba JD (2005) Regulation of sphingosine-1-phosphate lyase gene expression by members of the GATA family of transcription factors. *J Biol Chem*, **280**, 18403-18410.

Ramaswamy S, Ross K, Lander E, Golub T (2003) A molecular signature of metastasis in primary solid tumors. *Nat Genet*, **33**, 49-54.

Reiss U, Oskouian B, Zhou J, Gupta V, Sooriyakumaran P, Kelly S, Wang E, Merrill AH Jr, Saba JD (2004) Sphingosine-phosphate lyase enhances stress-induced ceramide generation and apoptosis. *J Biol Chem*, **279**, 1281-1290.

Saba J (2004) Lysophospholipids in development: miles apart and edging in. *J Cell Biochem*, **92**, 967-992.

Saba JD, Nara F, Bielawska A, Garrett S, Hannun YA (1997) The BST1 gene of *Saccharomyces cerevisiae* is the sphingosine-1-phosphate lyase. *J Biol Chem*, **272**, 26087-26090.

Sims K, Spassieva S, Voit E, Obeid L (2004) Yeast sphingolipid metabolism: clues and connections. *Biochem Cell Biol*, **82**, 45-61.

Skrzypek MS, Nagiec MM, Lester RL, Dickson RC (1999) Analysis of phosphorylated sphingolipid long-chain bases reveals potential roles in heat stress and growth control in Saccharomyces. *J Bacteriol*, **181**, 1134-1140.

Stoffel W, LeKim D, Sticht G (1969) Distribution and properties of dihydrosphingosine-1-phosphate aldolase (sphinganine-1-phosphate alkanal-lyase). *Hoppe Seylers Z Physiol Chem*, **350**, 1233-1241.

Suzuki E, Handa K, Toledo M, Hakomori S (2004) Sphingosine-dependent apoptosis: a unified concept based on multiple mechanisms operating in concert. *Proc Natl Acad Sci U S A*, **101**, 14788-14793.

Tanaka A, Leung P, Kenny T, Au-Young J, Prindville T, Coppel R, Ansari A, Gershwin M (2001) Genomic analysis of differentially expressed genes in liver and biliary epithelial cells of patients with primary biliary cirrhosis. *J Autoimmunity*, **17**, 89-98.

Triola G, Fabrias G, Dragusin M, Niederhausen L, Broere R, Llebaria A, van Echten-Deckert G (2004) Specificity of the dihydroceramide desaturase inhibitor N-[(1R,2S)-2-hydroxy-1-hydroxymethyl-2-(2-tridecyl-1-cyclopropenyl)ethyl]o ctanamide (GT11) in primary cultured cerebellar neurons. *Mol Pharmacol*, **66**, 1671-1678.

van Echten-Deckert G, Zschoche A, Bar T, Schmidt R, Raths A, Heinemann T, Sandhoff K (1997) cis-4-methylsphingosine decreases sphingolipid biosynthesis by specifically interfering with serine palmitoyltransferase activity in primary cultured neurons. *J Biol Chem*, **272**, 15825-15833.

Van Veldhoven PP (2000) Sphingosine-1-phosphate lyase. in *Sphingolipid Metabolism and Cell Signaling, Part A*. (Merrill AH Jr and Hannun YA, eds), pp244-254, Academic Press, New York.

Van Veldhoven PP, Mannaerts GP (1991) Subcellular localization and membrane topology of sphingosine-1-phosphate lyase in rat liver. *J Biol Chem*, **266**, 12502-12507.

Van Veldhoven PP, Gijsbers S, Mannaerts GP, Vermeesch JR, Brys V (2000) Human sphingosine-1-phosphate lyase: cDNA cloning, functional expression studies and mapping to chromosome 10q22. *Biochimica et Biophysica Acta*, **1487**, 128-134.

Zhou J, Saba J (1998) Identification of the first mammalian sphingosine phosphate lyase gene and its functional expression in yeast. *Biochem Biophys Res Commun*, **242**, 502-507.

Part 4

Membrane Domain and Biological Function

4-1 Close Interrelationship of Sphingomyelinase and Caveolin in Triton X-100-Insoluble Membrane Microdomains

Keiko Tamiya-Koizumi[1], Takashi Murate[1], Katsumi Tanaka[2], Yuji Nishizawa[2], Nobuhiro Morone[2], Jiro Usukura[2], and Yoshio Hirabayashi[3]

[1]Nagoya University Graduate School of Medicine, Nagoya 461-8673, Japan, [2]Nagoya University Graduate School of Medicine, Nagoya 466-8550, Japan, [3]RIKEN Brain Science Institute, Saitama 351-0198, Japan

Summary. Much attention has been paid to the roles of sphingomyelin (SM) metabolism in the regulation of various cell functions such as cell growth, differentiation and apoptosis. Sphingomyelinase (SMase) catalyzes the first step of SM-metabolizing pathways that generate the bioactive metabolite, ceramide. SMase present in membrane micro-domains, raft and caveolae may be involved in the agonist-mediated events. However, it is still unclear which molecular species of SMase is stimulated by an agonist to produce ceramide in membrane microdomains, and how agonist-sensitive SMase and SM are topologically localized within the microdomains. Here, we first show the close interaction between neutral SMase and caveolin 1 in 1% Triton X-100-insoluble fractions of plasma membranes isolated from adult rat resting liver and rapidly growing rat ascites hepatoma, AH 7974 cells. Then, we describe the connection between acid SMase and caveolin 2 in cell differentiation or apoptosis induced by all-trans retinoic acid. Finally, we discuss the possible roles for caveolins with respect to the topological distribution of SMase and SM in caveolae.

Keywords. Detergent-insoluble microdomain, plasma membrane, raft, caveolae, neutral sphingomyelinase, acid sphingomyelinase, caveolin 1,2

1. Introduction

Sphingomyelin (SM) is a major constituent of plasma membrane (PM). Up to 90% of the total SM is localized in the outer leaflet of PM of mammalian cells (Koval and Pagano, 1991). SM, in association with Chol, plays an important role in the lateral organization of liquid-ordered and less fluid membrane microdomains, rafts and caveolae (Megha and London, 2004; Ramstedt and Slotte, 2002). These microdomains are also characterized as a detergent-insoluble, floating membrane fractions at the low density in a sucrose gradient (Chamberlain, 2004). More importantly, SM is involved in cell signaling through its metabolites produced in membrane microdomains (Andrieu-Abadie and Levade, 2002; Dobrowsky, 2000; Hannun et al., 2001; Levade and Jaffrezou, 1999).

To generate bioactive metabolites from SM, sphingomyelinase (SMase) is first catalyzed. The degradation of SM to ceramide (Lucero and Robbins) by SMase within the microdomains brings the profound effects on agonist-mediated events. However, it is still unclear which molecular species of SMase is stimulated by an agonist to produce CER in microdomains, or how agonist-sensitive SMase and SM are topologically localized within the microdomains. Here, we describe the close relationships of neutral SMase (nSMase) and caveolin 1 and of acid sphingomyelinase (aSMase) and caveolin 2, respectively, as well as discuss the possible roles of caveolins with respect to the topological distribution of SMase in caveolae and to the regulation of the enzyme activity.

2. Co-existence of SMases and caveolins in microdomains of cell membrane

2.1 Mg^{2+}-dependent, neutral SMase and caveolin 1 in PM microdomains

PM microdomains are conventionally prepared using buoyant density sucrose gradient centrifugation of whole cell homogenate treated with ice-cold 1% Triton X-100. However, the method can introduce contaminants into isolated PM microdomain by microdomains from other organelles or from materials possessing the same density as the PM microdomains. To avoid this, we used isolated and highly purified PMs from the resting liver of adult rat and rat ascites hepatoma, AH7974 cells (Virkamaki et al.) using a method described previously (Tamiya-Koizumi and Kojima, 1986). Liver tissue consists of mainly hepatocyte (85-95% of the hepatic mass). AH is a cancer that originates from a rapidly growing hepatocyte and forms tumor cell islands in ascitic fluid of rat intraperitoneal cavity.

Highly purified PMs were homogenized in ice-cold 1% Triton X-100 by ultra-sonication, adjusted to 40% sucrose. These were subjected to a density equilibrium centrifugation in a continuous linear sucrose gradient (5-30%) at 40,000 rpm for 20 h. Possible contamination by soluble substances was eliminated by spinning down the fractions after they were diluted 3-fold with distilled water. We then performed morphological and biochemical analyses of the obtained pellets as detergent -insoluble membrane fraction (DIM).

Figure 1 shows the representative electron micrographs of DIM. Membrane-like structures were detected in a wide range of fractions, from Fr. 10 to 24 in liver PM and from Fr. 13 to 24 in AH PM. SM-clusters, detected with gold particle-conjugated lysenin (Kiyokawa et al., 2005), were observed from Fr. 10 to 19 in liver and from Fr. 13 to 21 in AH. The adherence junction (AJ) in association with SM-clusters was enriched in Fr. 19 of liver PM. In contrast, neither desmosomes (D) of liver and AH PMs had SM-clusters, as observed in Fr. 20 of liver and Fr. 21 of AH.

Biochemical features of DIM obtained from liver and AH PMs are compared in Fig. 2. Mg^{2+}-dependent, neutral sphingomyelinase (Mg^{2+}-nSMase) activity was detected in a wide range of buoyant fractions (Fr. 10-20) obtained from liver PM. In AH PM, however, the activity distribution covered narrow range at high density (Fr. 21-24). Such a remarkable difference in Mg^{2+}-nSMase distribution in liver and AH PMs correlated well with SM distribution analyzed by thin-layer chromatography (Fig. 2). Caveolin, a specific protein for the microsdomain caveolae, showed interesting distribution. In liver PM, non-phosphorylated caveolin 1β partly overlapped with fractions positive for both SM and Mg^{2+}-nSMase, while in AH PM, phosphorylated caveolin 1β and non-phosphorylated caveolin 1β all associated with such fractions (Fig. 2). These results suggest that Mg^{2+}-nSMase of liver PM is associated with both caveolae and raft, whereas in AH PM, it is exclusively localized in caveolae. In addition, ceramidase activity (CDase) was detected only in liver PM and its distribution in the fractions suggested its possible association with caveolae (Fig. 2). Thus, the higher density of caveolae in AH PM compared with that in liver PM could be due to the phosphorylated caveolin 1 that can bind a number of proteins. The phosphorylation may be catalyzed by src-family tyrosine kinase, Lyn present in AH PM.

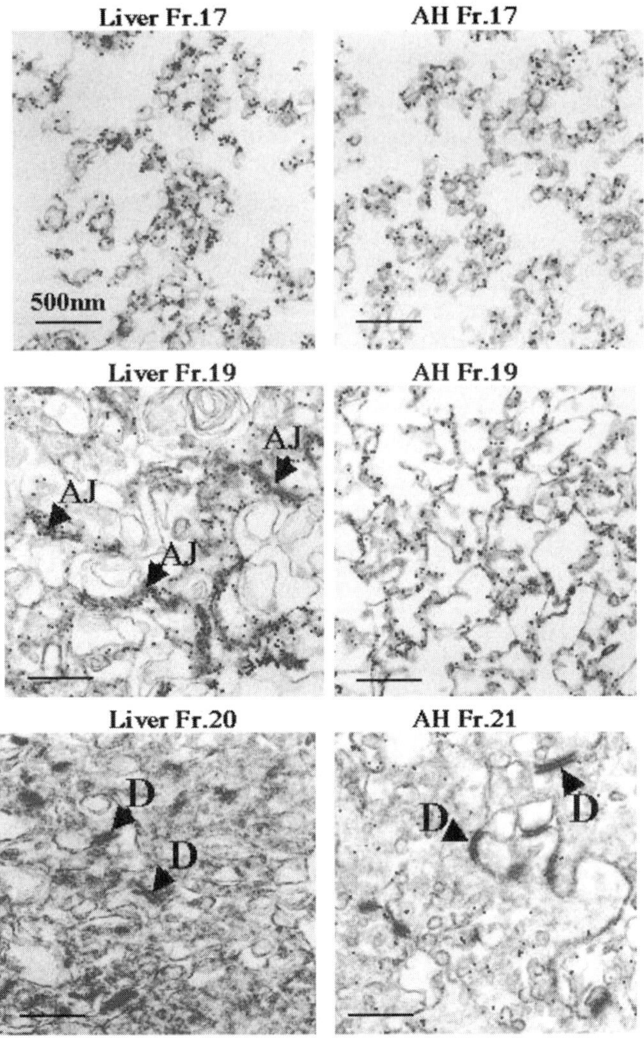

Fig. 1. Electron micrographs of DIM with lysenin obtained from Liver and AH PMs. DIM was obtained from 1% Triton X-100-treated PMs by linear sucrose gradient ultra-centrifugation. DIM incubated with gold particle-conjugated lysenin were fixed, embedded and ultra-thin-sectioned. AJ shows the adherence junction, and D shows desmosome.

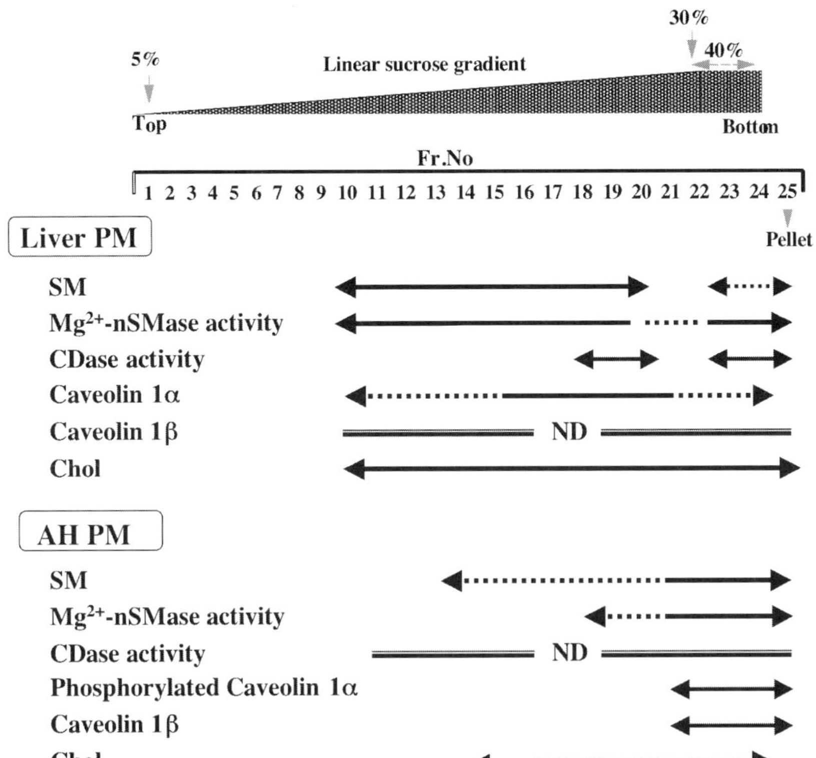

Fig. 2. The distributions of Mg²⁺-nSMase, CDase and caveolins in DIM obtained from liver and AH PMs. SM and Chol were analyzed by thin-layer chromatography. The activities of Mg²⁺-nSMase and CDase were determined by methods previously described (Tsugane et al., 1999). Caveolins were analyzed using various anti-caveolin 1 antibodies. Dotted lines represent fractions with smaller amount or activity than those with solid lines. ND: not detectable.

The molecular species of Mg²⁺-nSMase in PM microdomain likely does not involve nSMase 1 (Mizutani et al., 2001). Yet, since Stoffel et al. reported that all of Mg²⁺-nSMase activity in mouse liver was derived from nSMase 1 and 2 (Stoffel et al., 2005), the PM enzymes of liver and AH may be nSMase 2. However, Marchesini et al. reported that mRNA for nSMase 2 was up-regulated in confluent, contact-inhibited cells (Marchesini et al., 2004). That the specific activities of Mg2+-nSMase in resting liver and rapidly growing AH PMs are approximately the same, 110 verses 90 nmol/h/mg protein of PM, suggests the presence of another novel molecular species of Mg2+-nSMase.

T-Koizumi and Kojima found that Mg^{2+}-nSMase in the PM purified from either liver or AH required membrane acidic phospholipids, especially phosphatidylserine (Sankaram and Thompson) (Tamiya-Koizumi and Kojima, 1986). The delipidized AH PM by 1% Triton X-100 showed only 18% of Mg^{2+}-nSMase activity in the non-treated PM. However, when acidic phospholipids were added, enzyme activity in the delipidized PM was restored to levels that were 77% of that in non-treated PM. Considering the acidic phospholipid composition of liver and AH PMs (Koizumi et al., 1977), we concluded that PS is a natural activator for Mg^{2+}-nSMase. The lipid requirement of Mg^{2+}-nSMase in PM is absolutely limited to the acidic phospholipids, and PE and PC do not affect the enzyme activity. These facts suggest that Mg^{2+}-nSMase in PM of liver and AH differs from the enzyme purified from brain by Bernardo et al., which is activated by PE and PC at the similar efficiency as PS (Bernardo et al., 2000).

Veldman et al. found that neutral sphingomyelinase (nSMase) co-isolated with caveolin 1 was inhibited by a peptide corresponding to the caveolin-scaffolding domain, while nSMase from non-caveolae fraction was not inhibited by the peptide (Veldman et al., 2001). These results suggest a direct molecular interaction between the enzyme and caveolin 1, and a cytosolic orientation of the nSMase in caveolae. However, neither cloned nSMase 1 nor 2 has a caveolin-binding motif, which is commonly seen in caveolae-associated signaling molecules (Veldman et al., 2001). Removal of Chol from micro-domains induces a release of nSMase activity into the detergent-soluble fraction (Veldman et al., 2001). Caveolin is a Chol-binding protein (Fielding and Fielding, 2000; Lucero and Robbins, 2004; Yu et al., 2005), and Chol is preferentially associated with SM (Sankaram and Thompson, 1990; Slotte, 1999; Yu et al., 2005). A hydrogen bond between the amide-linked acyl chain of SM and the free 3β-OH group of Chol stabilizes this interaction (Slotte, 1999; Yu et al., 2005). Head group of SM, phosphocholine, is also important for stablilizing SM-Chol interaction (Yu et al., 2005). Taken together, nSMase, SM, Chol, PS and caveolin 1 may interact each other in the inner leaflet of caveolae as a complex which is resistant to Triton X-100 and is sensitive to stimuli.

2.2 Acid sphingomyelinase and caveolin 2 in microdomains

Murate et al. reported that an all-trans-retinoic acid (ATRA) up-regulated acid sphingomyelinase (aSMase) gene expression during myeloid differentiation of NB4, a human promyelocytic leukemia cell line (Murate et al., 2002). ATRA-induced elevation of aSMase activity was accompanied with an increase of intracellular CER content (Murate et al.,

2002)(Fig. 3, A, B). Interestingly, the increased CER species was limited to C22:0, C24:0 and C24:1 (Fig. 3, B). NB4 possesses caveolin 2, but caveolin 1 is not detectable with immunoblotting. In order to investigate the co-localization in membrane microdomain of aSMase and caveolin 2, 1% Triton X-100-treated NB4 cells were fractionated on a linear density gradient (5-30%) (Fig. 3, C). In the non-treated NB4, caveolin 2 was completely separated from aSMase activity. In ATRA-treated cells, however, caveolin 2 and aSMase overlapped each other at lower density fractions.

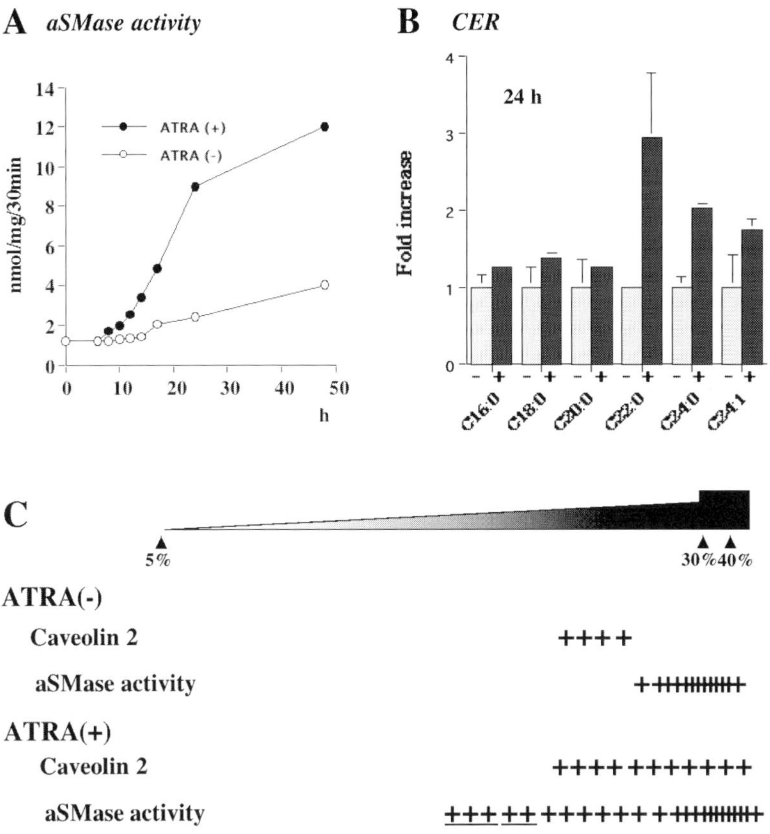

Fig. 3. Up-regulation of aSMase and increase of CER content induced by ATRA, and the different distributions of caveolin 2 and aSMase activity in the membrane fractions with and without ATRA treatment. Figures A (aSMase activity) and B (CER content) were modified from a previous report (Murate et al., 2002). C shows the distributions of caveolin 2 and aSMase activity in fractions obtained from 1% Triton X-100 treated NB4 cells by linear sucrose gradient ultra-centrifugaion. ATRA(-): NB4 cells without ATRA treatment; ATRA(+): NB4 cells with ATRA treatment.

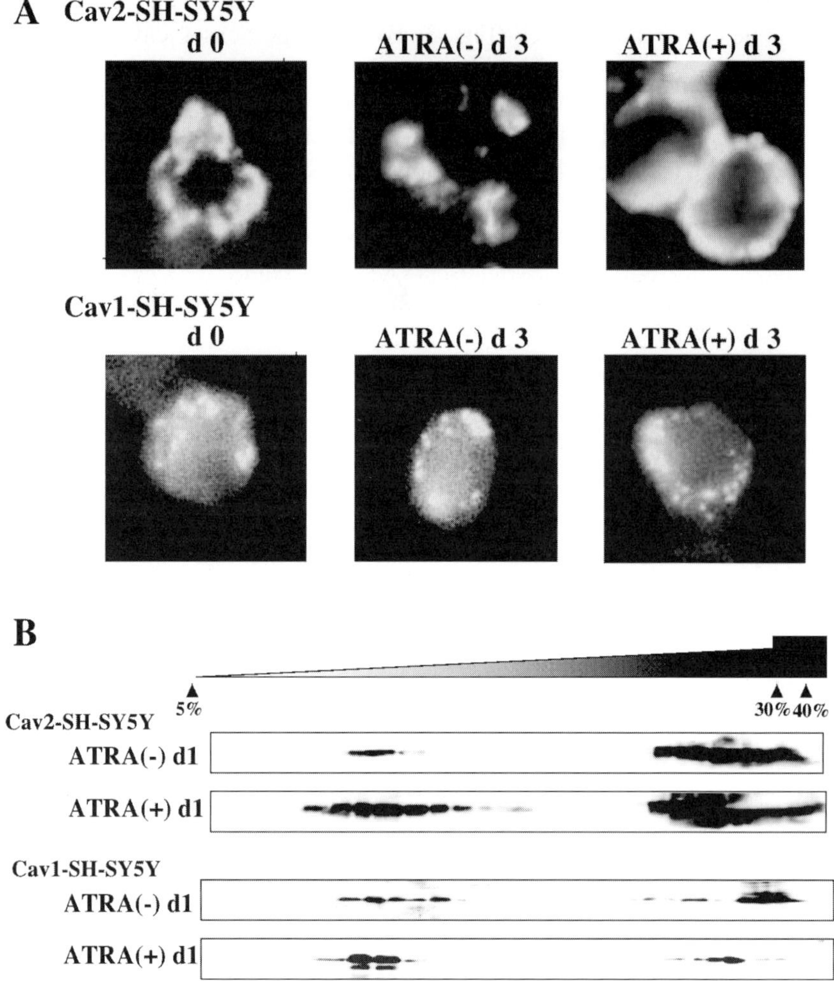

Fig. 4. Immunocytochemical and immunoblot analyses of caveolin 1 and 2 in Cav1-SH-SY5Y and Cav2-SH-SY5Y with and without ATRA. A: Cav2-SH-SY5Y and Cav1-SH-SY5Y were each treated with (ATRA(+)) and without ATRA (ATRA(-)). d refers to the day following ATRA addition. Each cell was immuno-stained with anti-caveolin 2 antibody for Cav2-SH-SY5Y and with anti-caveolin 1 antibody for Cav1-SH-SY5Y. Both cells were then treated with anti-mouse Alexa 488. B: Cav2-SH-SY5Y and Cav1-SH-SY5Y with and without ATRA treatment were each treated with 1% Triton X-100 and subjected to linear sucrose gradient ultra-centrifugation. The distributions of caveolins in the fractions were analyzed by immunoblotting using anti-caveolin 2 antibody for Cav2-SH-SY5Y and anti-caveolin 1 antibody for Cav1-SH-SY5Y. d1 means one day after ATRA addition.

For further investigation, caveolin 1 and 2 were each over-expressed by gene transfection in human neuroblastoama cell line, SH-SY5Y (Cav1-SH-SY5Y and Cav2-SH-SY5Y, respectively), because NB4 cells were not suitable for stable transfection. SH-SY5Y originally expresses s-mall amounts of caveolin 1 and 2. In Cav2-SH-SY5Y, caveolin 2 is localized in Golgi-like organelle. Upon treatment with ATRA, caveolin 2 appeared at the cell surface (Fig. 4, A). Stimulation may have cause Caveolin 2 transfer from Golgi to the PM, while Caveolin 1 in Cav1-SH-SY5Y remained at the cell periphery upon ATRA treatment (Fig. 4, A). Immunoblot analyses of the fractions obtained from the sucrose density gradient clearly supported the notion that caveolin 2 was translocated to caveolae by ATRA treatment (Fig. 4, B).

Our preliminary study shows that the accumulation of CER accompanied with apoptosis in ATRA-stimulated SH-SY5Y under the serum-free condition. Harget et al. reported that the addition of ATRA to teratocarcinoma cell line, PCC7-Mz1 caused an increase in CER level and induced cell differentiation (Herget et al., 2000). Although it is uncertain which occurs first, CER production or translocation of caveolin 2 to cell surface micro-domain, the ATRA-mediated relationship between CER and Caveolin 2 may induce cell apoptosis or differentiation.

Interestingly, scaffolding domains of caveolin 1 and 3 can recognize the same peptide ligand, whereas the corresponding domain within caveolin 2 does not even though these scaffolding domains have extreme homology (Couet et al., 1997). Finding a ligand that is specific to scaffolding domain of caveolin 2 may help elucidate the molecular mechanism of ATRA-induced signaling through sphingomyelin metabolism.

3. Conclusion

Identifying the molecular species of Mg^{2+}-nSMase specifically present in detergent-insoluble microdomain of PM, caveolae may provide insight into CER signaling. The topological distribution and mutual relationships between Mg^{2+}-nSMase, SM, Chol, PS and caveolin 1 in caveolae require more investigation. The scaffolding domain (Okamoto et al., 1998) and Chol-binding site (Fielding and Fielding, 2000) in caveolin 1, the caveolin-binding motif in caveolae-associated proteins (Okamoto et al., 1998) and a novel sphingolipid-binding domain (SBD) recently identified in raft-associated transmembrane protein (Fantini, 2003) all may play roles in the construction of a complex responsible for Mg^{2+}-nSMase-mediated CER signaling. Finally, the translocation of aSMase and caveolin 2 to cell surface must be also elucidated in the light of mechanism and regulation of the CER signaling.

Acknowledgements. We would like to thank Dr. Shonen Yoshida for his valuable corrections and comments.

4. References

Andrieu-Abadie, N., and Levade, T. (2002) Sphingomyelin hydrolysis during apoptosis. *Biochim Biophys Acta*, **1585**, 126-134.

Bernardo, K., Krut, O., Wiegmann, K., Kreder, D., Micheli, M., Schafer, R., Sickman, A., Schmidt, W. E., Schroder, J. M., Meyer, H. E., *et al.* (2000) Purification and characterization of a magnesium-dependent neutral sphingomyelinase from bovine brain. *J Biol Chem*, **275**, 7641-7647.

Chamberlain, L. H. (2004) Detergents as tools for the purification and classification of lipid rafts. *FEBS Lett*, **559**, 1-5.

Couet, J., Li, S., Okamoto, T., Ikezu, T., and Lisanti, M. P. (1997) Identification of peptide and protein ligands for the caveolin-scaffolding domain. Implications for the interaction of caveolin with caveolae-associated proteins. *J Biol Chem*, **272**, 6525-6533.

Dobrowsky, R. T. (2000) Sphingolipid signalling domains floating on rafts or buried in caves? *Cell Signal*, **12**, 81-90.

Fantini, J. (2003) How sphingolipids bind and shape proteins: molecular basis of lipid-protein interactions in lipid shells, rafts and related biomembrane domains. *Cell Mol Life Sci*, **60**, 1027-1032.

Fielding, C. J., and Fielding, P. E. (2000). Cholesterol and caveolae: structural and functional relationships. *Biochim Biophys Acta*, **1529**, 210-222.

Hannun, Y. A., Luberto, C., and Argraves, K. M. (2001) Enzymes of sphingolipid metabolism: from modular to integrative signaling. *Biochemistry*, **40**, 4893-4903.

Herget, T., Esdar, C., Oehrlein, S. A., Heinrich, M., Schutze, S., Maelicke, A., and van Echten-Deckert, G. (2000) Production of ceramides causes apoptosis during early neural differentiation in vitro. *J Biol Chem*, **275**, 30344-30354.

Kiyokawa, E., Baba, T., Otsuka, N., Makino, A., Ohno, S., and Kobayashi, T. (2005) Spatial and functional heterogeneity of sphingolipid-rich membrane domains. *J Biol Chem*, **280**, 24072-24084.

Koizumi, K., Tamiya-Koizumi, K., Fujii, T., and Kiyohide, K. (1977) Phospholipids of plasma membranes isolated from rat ascites hepatomas and from normal rat liver. *Cell Structure and Function*, **2**, 145-153.

Koval, M., and Pagano, R. E. (1991) Intracellular transport and metabolism of sphingomyelin. *Biochim Biophys Acta*, **1082**, 113-125.

Levade, T., and Jaffrezou, J. P. (1999). Signalling sphingomyelinases: which, where, how and why? *Biochim Biophys Acta*, **1438**, 1-17.

Lucero, H. A., and Robbins, P. W. (2004) Lipid rafts-protein association and the regulation of protein activity. *Arch Biochem Biophys*, **426**, 208-224.

Marchesini, N., Osta, W., Bielawski, J., Luberto, C., Obeid, L. M., and Hannun, Y. A. (2004). Role for mammalian neutral sphingomyelinase 2 in confluence-induced growth arrest of MCF7 cells. *J Biol Chem*, **279**, 25101-25111.

Megha, and London, E. (2004) Ceramide selectively displaces cholesterol from ordered lipid domains (rafts): implications for lipid raft structure and function. *J Biol Chem*, **279**, 9997-10004.

Mizutani, Y., Tamiya-Koizumi, K., Nakamura, N., Kobayashi, M., Hirabayashi, Y., and Yoshida, S. (2001) Nuclear localization of neutral sphingomyelinase 1: biochemical and immunocytochemical analyses. *J Cell Sci*, **114**, 3727-3736.

Murate, T., Suzuki, M., Hattori, M., Takagi, A., Kojima, T., Tanizawa, T., Asano, H., Hotta, T., Saito, H., Yoshida, S., and Tamiya-Koizumi, K. (2002) Up-regulation of acid sphingomyelinase during retinoic acid-induced myeloid differentiation of NB4, a human acute promyelocytic leukemia cell line. *J Biol Chem*, **277**, 9936-9943.

Okamoto, T., Schlegel, A., Scherer, P. E., and Lisanti, M. P. (1998) Caveolins, a family of scaffolding proteins for organizing "preassembled signaling complexes" at the plasma membrane. *J Biol Chem*, **273**, 5419-5422.

Ramstedt, B., and Slotte, J. P. (2002) Membrane properties of sphingomyelins. *FEBS Lett*, **531**, 33-37.

Sankaram, M. B., and Thompson, T. E. (1990) Interaction of cholesterol with various glycerophospholipids and sphingomyelin. *Biochemistry*, **29**, 10670-10675.

Slotte, J. P. (1999) Sphingomyelin-cholesterol interactions in biological and model membranes. *Chem Phys Lipids*, **102**, 13-27.

Stoffel, W., Jenke, B., Block, B., Zumbansen, M., and Koebke, J. (2005) Neutral sphingomyelinase 2 (smpd3) in the control of postnatal growth and development. *Proc Natl Acad Sci U S A*, **102**, 4554-4559.

Tamiya-Koizumi, K., and Kojima, K. (1986) Activation of magnesium-dependent, neutral sphingomyelinase by phosphatidylserine. *J Biochem (Tokyo)*, **99**, 1803-1806.

Tsugane, K., Tamiya-Koizumi, K., Nagino, M., Nimura, Y., and Yoshida, S. (1999). A possible role of nuclear ceramide and sphingosine in hepatocyte apoptosis in rat liver. *J Hepatol*, **31**, 8-17.

Veldman, R. J., Maestre, N., Aduib, O. M., Medin, J. A., Salvayre, R., and Levade, T. (2001) A neutral sphingomyelinase resides in sphingolipid-enriched microdomains and is inhibited by the caveolin-scaffolding domain: potential implications in tumour necrosis factor signalling. *Biochem J*, **355**, 859-868.

Virkamaki, A., Ueki, K., and Kahn, C. R. (1999) Protein-protein interaction in insulin signaling and the molecular mechanisms of insulin resistance. *J Clin Invest*, **103**, 931-943.

Yu, C., Alterman, M., and Dobrowsky, R. T. (2005) Ceramide displaces cholesterol from lipid rafts and decreases the association of the cholesterol binding protein caveolin-1. *J Lipid Res*, **46**, 1678-1691.

4-2 Roles of Membrane Domains in the Signaling Pathway for B Cell Survival

Miki Yokoyama[1], Tomoko Kimura[1], Sachio Tsuchida[1], Hiroaki Kaku[2], Kiyoshi Takatsu[2], Yoshio Hirabayashi[3], and Masaki Yanagishita[1]

[1]Department of Hard Tissue Engineering, Biochemistry, Division of Bio-Matrix, Graduate School, Tokyo Medical and Dental University, 1-5-45 Yushima, Bunkyo-ku, Tokyo 113-8549, Japan; [2]Division of Immunology, Department of Microbiology and Immunology, Institute of Medical Science, University of Tokyo, 4-6-1 Shirogane-dai, Minato-ku, Tokyo 108-8639, Japan; [3]Brain Science Institute, The Institute of Physical and Chemical Research (Riken), 2-1 Hirosawa, Wako, Saitama 351-0198, Japan

Summary. B cell response to antigens forms the basis for humoral immune responses. Mature B cells are rescued from spontaneous apoptosis in the presence of survival factors such as anti-IgM antibody, LPS, and anti-CD40 antibody. This in vitro rescue correlates with in vivo B-cell proliferation and differentiation induced by thymus-independent (TI) -2, TI-1, and thymus-dependent (TD) antigens. Crosslinking of B cell receptor (BCR) by anti-IgM antibody promotes a positive feedback loop in the vicinity of BCR to activate those tyrosine kinases, Syk and Btk, crucial for the NF-γB activation. Ligation of CD40 triggers a recruitment of IγB kinase through TNF receptor-activated factors (TRAFs). Sphingolipid- and cholesterol-enriched membrane domains are important for the assembly of the signaling molecules during TI-2 and TD antigens recognition. B cells can be rescued from apoptosis even in the absence of antigen by an agonistic antibody (anti-CD38 antibody), CS/2. An absolute requirement of Syk and Btk in the CS/2 action suggests that an intracellular signaling pathway of CS/2 is similar to that induced by BCR-crosslinking. However, the action of CS/2 does not elicit appreciable tyrosine phosphorylation. Our data implies that CS/2 induces a conformational change of CD38 within the membrane domains, which leads to a generation of sphingolipid-mediated B cell survival signal.

Keywords. B cell survival, B cell receptor, Membrane domains, CD38

1. Proliferation and differentiation of B cells depends on antigen recognition

Upon encountering an antigen, resting B cells exit the G_o phase of the cell cycle, grow in size, and enter S phase (DNA synthesis). Finally, B cells differentiate and secrete antibodies that eliminate the antigen. Both protein and non-protein antigens can induce humoral immune responses.

Membrane-bound IgM or IgD associated with Igγ/Igγ heterodimer act as B cell receptor in mature B cells that have not yet encountered antigens (naïve B cells). Classic thymus-dependent (TD) antigens are monomeric, oligomeric, or haptenated proteins with limited ability to crosslink the B cell receptor (BCR). Presentation of processed peptides from TD antigens to effector T cells stimulates then act on the B cell to drive cell-cycle progression and differentiation. Thymus-independent (TI) non-protein antigens, such as bacterial cell wall components, are polyvalent in structure and have the ability to strongly crosslink the BCR (referred to as TI-2 antigen) or to stimulate B cells through pattern recognition receptors such as Toll-like receptors (referred to as TI-1 antigen).

2. Involvement of membrane domains in the signaling pathway for B cell survival

Mature B cells when cultured in the absence of survival factors that mimic antigen stimulation undergo spontaneous apoptosis. Anti-IgM antibody, LPS, and anti-CD40 antibody correspond to stimulation by TI-2, TI-1, and TD antigens, respectively. Activation of NF-γ B up-regulates anti-apoptotic factors and is therefore crucial for the antigen-dependent B cell survival. However, different antigens will trigger different signaling pathways leading to NF-γ B activation.

The initial consequence of BCR-crosslinking by multivalent antigens is the activation of protein tyrosine kinases (Src-PTKs and Syk) in the vicinity of the cytoplasmic regions of Igγ/Igγ. An immuno-receptor tyrosine-based activation motif (ITAM) within Igγ/Igγ heterodimer is a substrate for the kinases. Syk can be further activated after binding to phosphorylated ITAM tyrosine residues (Rolli et al 2002). A Syk/ITAM positive feedback loop is formed at the receptor and triggers the activation of Btk and phosphoinoside 3-kinase (PI3K), which are crucial for TI antigen-dependent survival (Anderson et al, 1996 and Suzuki et al, 1999). BCAP is an adaptor protein that is phosphorylated by Syk to provide a binding site(s) for PI3K, that activates it (Okada et al. 2000). Syk also phosphorylates another adaptor protein BLNK (also known as SLP-65 or BASH), which recruits Btk and PLCγ2 to BLNK via their SH2 domains (Kurosaki and Tsukada, 2000). Btk is thought to phosphorylate and activate PLCγ2, leading to 1,4,5-trisphosphate production and calcium mobilization.

After BCR-crosslinking, BCR itself, PLCγ2 and PI3K accumulate in detergent-resistant membrane domains (DRMs). Cholesterol removal using filipin inhibits the BCR-mediated tyrosine phosphorylation of PLCγ2 as well as BCR-induced calcium flux (Aman et al. 2000). Co-ligation of BCR to a coreceptor complex (CD19/CD21) enhances BCR signaling and, at the same time, promotes the association of BCR with DRMs via an induced palmitoylation of tetraspanin (Cherukuri et al, 2004). These results suggest that cholesterol- and sphingolipid-enriched membrane domains are necessary for the tyrosine phosphorylation-mediated assembly and activation of the signaling complex upon BCR-crosslinking.

The membrane domains are also important for CD40-mediated B cell survival. They may serve as a scaffold that recruit TRAFs 2 and 3 (Hostager et al. 2000). In fact, constitutive localization of CD40 signalosome within the membrane domains in B cell lymphomas (NHL-B) causes autonomous cell growth (Pham et al, 2002). In the case of LPS-mediated B cell survival, the role of membrane domains has not been evaluated.

3. B cell survival signals induced by an agonistic antibody CS/2 is similar to BCR-crosslinking

CS/2 was raised against bone marrow B cells and isolated as a survival antibody that can rescue B cells from apoptosis in the absence of antigen (Yamashita et al, 1995). CS/2 can induce proliferation, expression of IL-5 receptor γ chain, and IgM and IgG production in mouse splenic B cells (Kikuchi et al 1995 and Mizoguchi et al. 1999). The antigen for CS/2 is lymphocyte cell surface antigen CD38.

CD38 is a 46 kDa, type II transmembrane glycoprotein composed of a short cytoplasmic domain, a transmembrane domain, and a large extracellular domain. The NAD glycohydrolase activity is associated with the extracellular domain of CD38 and is responsible for most tissue-associated NADase activity. This molecule is widely expressed on hematopoietic cells as well as non-hematopoietic cells. CD38-knockout mice show a defect in the humoral immune response (Partida-Sanchez et al. 2001) and the innate inflammatory response (Partida-Sanchez et al. 2004).

CS/2-dependent B cell survival requires Btk and PI3K (what? In reading this sentence a referent noun or action phrase here seems to be missing. What does it require Btk or P13K to be doing to activate Nf-kB?) that activate(s) NF-γ B (Kaku et al, 2002). Thus, downstream pathways of BCR-crosslinking and CS/2-signaling are similar, but not the same. Unlike the BCR-crosslinking, CS/2 ligation does not elicit any increase in the tyrosine phosphorylation of the total lysate. The action of CS/2 implies the presence of an as yet unidentified triggering mechanism within in the B cell survival signal.

4. A possible involvement of lipid-mediated signaling for CS/2-dependent B cell survival

Agonistic action of CS/2 for B cell survival, IL-5 receptor γ chain expression, and IgM and IgG production are appreciably more profound than that of clone 90, another anti-mouse CD38 antibody. Consistently, the activation of NF-γ B by CS/2 was more prominent than that by clone 90 (Fig. 1).

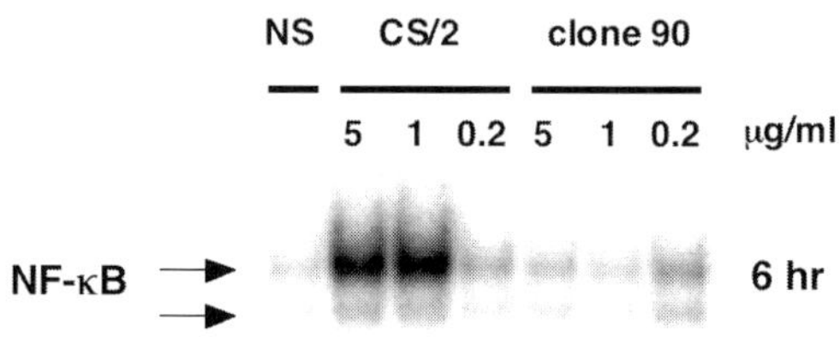

Fig. 1. Activation of NF-γ B by CS/2 was more profound than that by clone 90. Mouse splenic B cells were incubated with CS/2 or clone 90 and then lysed. Nuclear proteins prepared from each treatment were subjected to electrophoretic mobility shift assay using a γ-^{32}P radiolabeled DNA probe containing NF-γ B binding site.

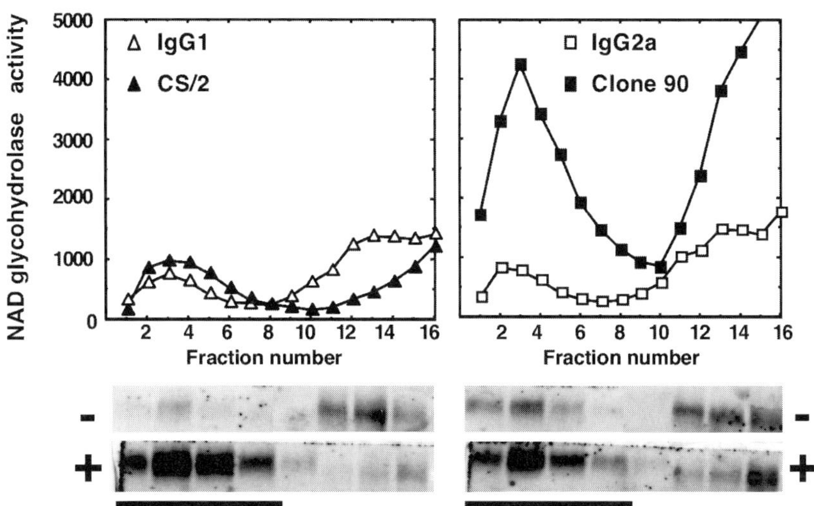

Fig. 2. Specific activity of CD38 within DRMs decreased after CS/2 stimulation. Mouse splenic B cells were incubated with CS/2 or clone 90, or with their isotype control rat IgG1 and IgG2a, respectively, and lysed in 1% Brij-58 lysis buffer. Lysates were subjected to density gradient centrifugation and fractionated into 16 equal volume removed from the top of the column. Bars showed DRMs fractions. CD38 was detected by immunoblotting. NAD$^+$ glycohydrolase activity of each fraction was measured by fluorometric means (arbitrary unit).

The stimulation of B cells by either CS/2 or clone 90 was found to recruit CD38 into DRMs in a subcellular fractionation experiment using ultracentrifugation (Fig. 2). However, the effects of both antibodies on the NADase activity of CD38 were different. The NADase activity in DRMs after stimulation with clone 90 was much higher than that after stimulation with CS/2. Considering the amount of CD38 protein in DRMs, the specific activity of CD38 NADase in DRMs was appreciably reduced in the case of CS/2. On the other hand, in the bottom fractions (#10~16), the amount of CD38 protein was decreased after stimulation with each antibody. However, the NADase activity was increased by clone 90 but decreased by CS/2, indicating that the specific activity of CD38 NADase was elevated in the case of clone 90.

Both CS/2 and clone 90 antibodies enhanced the NADase activity of recombinant CD38 protein. Such intrinsic property was well correlated with the action of clone 90 to increase the NADase activity of CD38 in the bottom fraction (Fig. 2). By contrast, the inhibition of the NADase activity of CD38 in DRMs after CS/2 stimulation was quite different from the effect on recombinant CD38 protein. The discrepancy is probably due to membrane components in DMRs, which directly or indirectly inhibit the NADase activity of CD38 in the presence of CS/2. It is implied that a CS/2-specific interaction of CD38 with DRMs relates to the agonistic action.

We previously reported that gangliosides inhibit the NADase activity of CD38 (Hara-Yokoyama et al, 1996 and 2001). This suggests that CD38 can interact with ganglisoides. However, as the agonistic action of CS/2 was not much impaired in B cells from the mice lacking complex gangliosides (Furukawa and Yokoyama, unpublished results) nor in B cells treated with glycolipid synthesis inhibitors (Inokuchi and Yokoyama, unpublished results), the oligosaccharide moiety of sphingolipids may not be essential to CS/2 action. More recently, we found that the agonistic action of CS/2 was sensitive to dimethylsphingosine, whose pharmacological effects include the inhibition of sphingosine kinase. From these observations, he hypothesize that CS/2 induces an interaction between CD38 and sphingolipids via a conformational change of CD38 within the membrane domains, and this then triggers a generation of sphingolipid-mediated B cell survival signal.

5. References

Aman MJ, Kodimangalam S, Ravichandran S (2000) A requirement for lipid rafts in B cell receptor induced Ca^{2+} flux. *Current Biology,* **10**, 393-396.

Anderson JS, Teutsch M, Dong Z, Wortis HH (1996) An essential role for tyrosine kinase in the regulation of Bruton's B cell apoptosis. *Proc. Natl. Acad. Sci. USA,* **93**, 10966-10971.

Cherukuri A, Carter RH, Brooks S, Bommann W, Finn R, Dowd CS, Pierce SK (2004) B cell signaling is regulated by induced palmitoylation of CD81. *J Biol Chem,* **279**, 31973-31982.

Hara-Yokoyama M, Kukimoto I, Nishina H, Kontani K, Hirabayashi Y, Irie F, Sugiya H, Furuyama S, Katada T (1996) Inhibition of NAD^+ glycohydrolase and ADP-ribosyl cyclase activities of leukocyte cell surface antigen CD38 by gangliosides. *J Biol Chem,* **265**, 12951-12955.

Hara-Yokoyama M, Nagatsuka Y, Katsumata O,Irie F, Kontani K, Hoshino S, Katada T, Ono Y, Fujita-Yoshigaki J, Sugiya H, Furuyama S, and Hirabayashi Y (2001) Complex Gangliosides as Cell Surface Inhibitors for the Ecto-NAD^+ Glycohydrolase of CD38. *Biochemistry,* **40**, 888-895.

Hostager BS, Catlett IM, Bishop GA (2000) Recruitment of CD40 and Tumore Necrosis Factor Receptor-associated Factors 2 and 3 to Membrane Microdomains during CD40 Signaling. *J Biol Chem,* **275**, 15392-15398.

Kaku H, Hirokawa K, Obata Y, Kato I, Okamoto H, Sakaguchi N, Gerondakis S, Takatsu K (2002) NF-γB is required for CD38-mediated induction of $C_\gamma 1$ germline transcripts in murine B lymphocytes. *International Immunology*, **14**, 1-10.

Kikuchi Y, Yasue T, Miyake K, Kimoto M, Takatsu K (1995) CD38 ligation induces tyrosine phosphorylation of Bruton tyrosine kinase and enhanced expression of interleukin 5-receptor alpha chain: synergistic effect with interleukin 5. *Proc. Natl. Acad. Sci. USA*, **92**, 11814-11818.

Kurosaki T, Tsukada S (2000) BLNK:Connecting Syk and Btk to Calcium Signals. *Immunity*, **12**, 1-5.

Okada T, Maeda A, Iwamatsu A, Gotoh K, Kurosaki T (2000) BCAP: the tyrosine kinase substrate that connects B cell receptor to phosphoinositide 3-kinase activation. *Immunity*, **13**, 817-827.

Mizoguchi C, Uehara S, Akira S, and Takatsu K (1999) IL-5 induces IgG1 isotype switch recombination in mouse CD38-activated sIgD-positive B lymphocytes. *J Immunol*, **162**, 2812-2819.

Partida-Sanchez S, Cockayne DA, Monard S, Jakobson EL, Oppenheimer N, Garby B, Kusser K, Goodrich S, Howard M, Harmsen A, Randall T, Lund FE (2001) Cyclic ADP-ribose production by CD38 regulates intracellular calcium release, extracellular calcium influx and chemotaxis in neutrophils and is required for bacterial clearance in vivo. *Nature Medicine*, **7**, 1209-1216.

Partida-Sanchez S, Goodrich S, Kusser K, Oppenheimer N, Randall TD, Lund FE (2004) Regulation of Dendritic Cell Trafficking by the ADP-Ribosyl Cyclase CD38:Impact on the Development of Humoral Immunity. *Immunity*, **20**, 279-291.

Pham LV, Tamayo AT, Yoshimura LC, Lo P, Terry N, Reid PS, Ford RJ (2002) A CD40 Signalosome Anchored in Lipid Rafts Leads to Constitutive Activation of NF-γB and Autonomous Cell Growth in B Cell Lymphomas. *Immunity*, **16**, 37-50.

Rolli V, Gallwitz M, Wossning T, Flemming A, Schamel WW, ZüC, Reth M (2002) Amplification of B Cell Antigen Receptor Signaling by a Syk/ITAM Positive Feedback Loop. *Molecular Cell*, **10**, 1057-1069.

Suzuki H, Terauchi Y, Fujiwara M, Aizawa S, Yazaki Y, Kadowaki T, and Koyasu S (1999) *Xid*-Like Immunodeficiency in Mice with Disruption of the p85γ Subunit of Phosphoinoside 3-Kinase. *Science*, **283**, 390-392.

Yamashita Y, Miyake K, Kikuchi Y, Takatsu K, Noda S, Kosugi A, Kimoto M (1995) A monoclonal antibody against a murine CD38 homologue delivers a signal to B cells for prolongation of survival and protection against apoptosis in vitro :unresponsiveness of X-linked immunodeficient B cells. *Immunology*, **85**, 248-255.

4-3 The Role of Lipid Rafts in Axon Growth and Guidance

Hiroyuki Kamiguchi

Laboratory for Neuronal Growth Mechanisms, RIKEN Brain Science Institute, 2-1 Hirosawa, Wako, Saitama 351-0198, Japan

Summary. Nervous system functions depend on the neuronal networks formed in part by axons. During development, axons migrate long distances to reach particular targets. A variety of environmental cues regulate and guide axon elongation by binding to relevant receptors expressed on the axonal growth cone at its tip. Activated receptors generate specific intracellular signals in a subcellular area of the growth cone, thereby controlling its directional motility. Recent work identified lipid rafts, or cholesterol- and glycosphingolipid- enriched microdomains, in the cell membrane a possible sited for the organization of spatially defined signals. For example, to stimulate axon elongation, cell adhesion molecules require functional rafts in the growth cone periphery. An extracellular gradient of axon guidance cues can induce growth cone turning by recruiting specific receptors to rafts and activating their downstream signals in a polarized manner. This chapter will examine the role of lipid rafts in mediating the directional motility of growth cones.

Keywords. Raft, neuron, axon, adhesion, guidance

1. Introduction

Neuronal network development depends on motile behavior of neuronal processes. Neurites extend from nascent neurons and differentiate into axons or dendrites. The axonal tip, or growth cone, navigates a complicated environment as it guides the axon to its designated target

(Gordon-Weeks, 2000). This growth cone expresses a variety of cell-surface proteins, including cell adhesion molecules (CAMs) and receptors for axon guidance molecules that translate extracellular information into migratory information. Environmental cues provided during this process are integrated by the growth cone in a spatially polarized manner. This polarity is created through the activation and amplification of distinct sets of signaling pathways in specific cell areas. Lipid rafts, which have been characterized as cholesterol- and glycosphingolipid- enriched microdomains in the cell membrane (Simons and Ikonen, 1997), might mitigate these processes to enable movement. Such a role for rafts in cell migration was proposed because: (1) a depletion of membrane cholesterol inhibits cell polarization and migration (Manes et al., 1999; Khanna et al., 2002), (2) cell polarization is accompanied by an asymmetric distribution of rafts and associated receptors in the plasma membrane (Manes et al., 1999; Gomez-Mouton et al., 2001; Seveau et al., 2001; Zhao et al., 2002; van Buul et al., 2003) and (3) a dynamic redistribution of rafts may activate intracellular signals in a polarized manner (Gomez-Mouton et al., 2004). These results suggest that rafts serve as platforms on which spatially defined signals in migrating cells are organized. In this chapter, I will introduce recent advances in studying the functions of membrane rafts during growth cone migration and guidance.

2. Axon growth

Anterograde migration enables a growth cone to elongate its trailing axon (Lamoureux et al., 1989). There are two functionally distinct regions in a growth cone: the peripheral (P-) domain and the central (C-) domain. The P-domain contains actin filaments (F-actin) with retrograde mobility which generate the traction force that pulls the cone forward (Lin and Forscher, 1995). This force can be transmitted to the extracellular environment via CAMs that are mechanically linked to immobile extracellular ligands with a retrograde F-actin flow (Suter and Forscher, 2000; Nishimura et al., 2003). CAMs are translocated into the C-domain by coupling to the F-actin flow, then internalized and recycled at the leading edge of the plasma membrane via vesicular transport along microtubules (Kamiguchi and Lemmon, 2000; Kamiguchi and Yoshihara, 2001). CAMs employ a caterpillar-like movement along the inside and the surface of the growth cone to propel it forward (Figure 1). The movement of CAMs and their dynamic interactions with the cytoskeleton are regulated by signals in the growth cone organized.

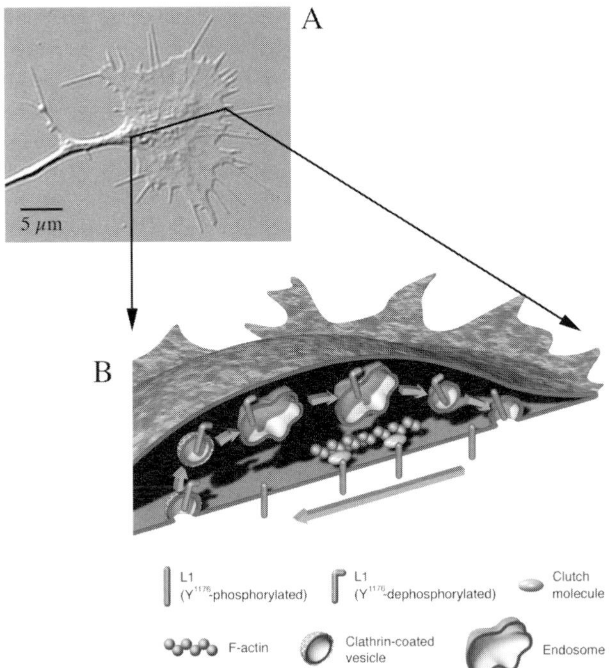

Fig. 1. (A) A differential interference contrast image of a chick dorsal root ganglion neuronal growth cone that migrates in an L1-dependent manner. (B) A schematic view of the growth cone sectioned longitudinally, showing a model of L1 trafficking and axon elongation. Dephosphorylation of Y^{1176} in the L1 cytoplasmic domain triggers L1 internalization at the C-domain via clathrin-mediated pathways. Subsequently, the endocytosed L1 is transported into the P-domain through sorting and recycling endosomes; this process is dependent on the dynamic ends of microtubules (not shown in this figure). Next, the trafficking L1 is reinserted into the plasma membrane at the leading edge of the growth cone, probably after Y^{1176} is re-phosphorylated. Recycled L1 on the cell surface moves toward the C-domain by coupling with the retrograde F-actin flow via clutch molecules. Reprinted, with permission, from Kamiguchi and Lemmon (2000).

Three major classes of CAMs have been identified in the nervous system: (1) integrins, (2) cadherins, and (3) the Ig superfamily of CAMs (IgCAMs). Cadherins and the majority of IgCAM members, including NCAM and L1, mediate cell-cell adhesion and stimulate axon growth via a homophilic binding mechanism (Edelman et al., 1987; Grumet and Edel--man, 1988; Lemmon et al., 1989; Bixby and Zhang, 1990; Doherty et al., 1990; Takeichi, 1991). In contrast, β1 integrin forms a heterodimeric receptor on the growth cone, promoting axon growth upon binding to laminin, an extracellular matrix molecule (Bozyczko and Horwitz, 1986; Tomaselli et al., 1986).

Biochemical and cell biological analyses demonstrated that different CAMs are localized in distinct microdomains of the cell membrane. For example, NCAM, the L1-family members and N-cadherin are expressed in rafts and non-raft membranes, whereas $\beta 1$ integrin is found only in non-raft areas (Olive et al., 1995; Nakai and Kamiguchi, 2002). Localization of NCAM and L1-family members to rafts depends, in part, on palmitoylation of the cytoplasmic domain and the transmembrane region, respectively (Ren and Bennett, 1998; Niethammer et al., 2002). Pharmacological experiments showed that lipid raft components, cholesterol and sphingolipids, are required for axon growth stimulated by L1 and N-cadherin but not by $\beta 1$ integrin (Nakai and Kamiguchi, 2002). Recently, my laboratory developed a method for selectively disrupting membrane rafts within a specific subcellular area (Figure 2). It is based on micro-scale fluorophore-assisted laser inactivation (micro-FALI) (Buchstaller and Jay, 2000). In this way, we found that membrane rafts in growth cones, especially in the P-domain, play a critical role in axon growth (Figure 3). Rafts in the P-domain seem to be involved in the spatial regulation of L1 endocytosis. The L1 cytoplasmic domain contains the tyrosine-based motif YRSL that mediates L1 endocytosis by binding to the clathrin adaptor AP2 (Kamiguchi et al., 1998). The tyrosine (Y^{1176}) in this YRSL motif is subject to phosphorylation by Src-family kinases, which prevents L1 from interacting with AP-2 and being internalized via the clathrin-mediated pathway(Schaefer et al., 2002). Because Src-family kinases are localized to membrane rafts, L1 phosphorylation by Src is likely to depend on whether the L1 is associated with rafts. It is possible that, in the P-domain, the localization of L1 to rafts allows Y^{1176} phosphorylation by Src, which prevents ectopic endocytosis of L1. In contrast, L1 becomes dephosphorylated at Y^{1176} in non-raft regions in the C-domain, leading to its clathrin-mediated endocytosis. In this way, rafts may influence L1-associated signals that are important for axon growth. Such a mechanism may exist in NCAM signaling. Upon homophilic binding, 140-kDa NCAM (NCAM140) triggers distinct signaling cascades through raft and non-raft membranes (Niethammer et al., 2002). NCAM140 most likely phosphorylates the focal adhesion kinase via a non-receptor kinase Fyn in rafts, while non-raft NCAM140 facilitates the fibroblast growth factor receptor signaling. As these pathways converge to activate the mitogen-activated protein kinase pathway. NCAM140 requires activity along the two pathway to be able to stimulate axon growth.

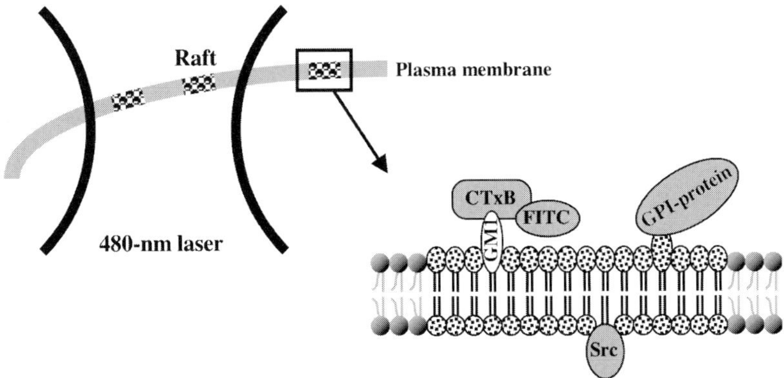

Fig. 2. A schematic diagram showing the micro-FALI-based technique that enables acute and localized inactivation of lipid rafts. A living cell is incubated in the presence of fluorescein isothiocyanate (FITC)-conjugated cholera toxin B-subunit (CTxB) that specifically binds to GM1 gangliosides expressed within rafts. In this situation, membrane rafts can be selectively labeled by FITC. Upon laser irradiation, the FITC generates singlet oxygen that could perturb any molecules located within 4 nm of the FITC (Beck et al., 2002). This spatial specificity of micro-FALI should be sufficient for selective inactivation of rafts. Based on the detergent insolubility of a raft-associated glucosylphosphatidylinositol-anchored molecule (Thy-1), our studies have shown that rafts were inactivated within 30 seconds only at the laser-irradiated area using this method (Nakai and Kamiguchi, 2002).

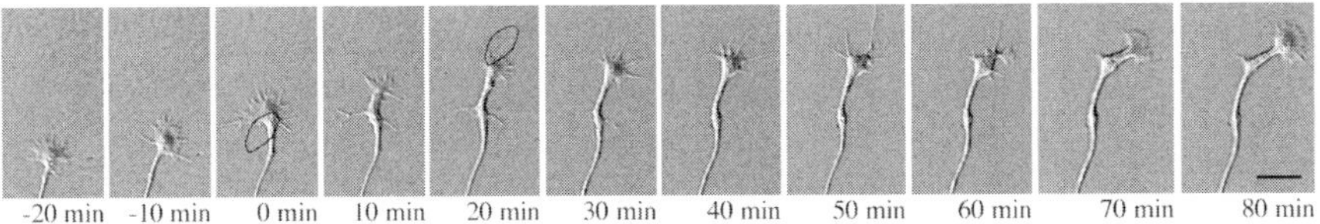

Fig. 3. Micro-FALI-mediated raft disruption in the P-domain, but not in the C-domain, inhibits growth cone migration. Shown are time-lapse image sequences of a dorsal root ganglion neuronal growth cone that migrates in an L1-dependent manner. A part of the growth cone was subject to micro-FALI-mediated raft disruption at the areas outlined in black. Scale bar, 10 µm. Reprinted, with permission of the publisher, from Nakai and Kamiguchi (2002).

3. Axon guidance

During development, attractive or permissive extracellular cues guide growth cones (Tessier-Lavigne and Goodman, 1996). This information is received by axon guidance receptors and translated into localized signals that ultimately regulate directional motility of growth cones (Huber et al., 2003). Netrin-1 is a diffusible laminin-related molecule that attracts or repels axons upon binding to its receptor DCC (Deleted in Colorectal Cancer). DCC is partially associated with lipid rafts, which is mediated by palmitoylation within its transmembrane region (Herincs et al., 2005). DCC localization to rafts is required for netrin-1-induced signaling and axon guidance. Recent work Guirland et al. (2004) further supports a role for membrane rafts in growth cone navigation by mediating actions based on localized signals extracted from extracellular guidance cues. Rafts are required for axon attraction by an extracellular gradient of brain-derived neurotrophic factor as well as for axon repulsion by netrin-1 or semaphorin 3A. Once the gradient is recognized, the growth cone recruits appropriate receptors to the rafts, resulting in asymmetric receptor-raft association. This asymmetry may induce polarized activation of downstream signaling cascades such as the mitogen-activated protein kinase pathway. In this way, a lipid raft might act as a platform that generates specific guidance signals on one side of the growth cone to trigger attractive or repulsive axon turning.

4. Conclusion

The generation and maintenance of cell polarity is vital for the correct development of neurons. The growth cone interacts with its microenvironment and produces specific intracellular signals in a polarized manner that ultimately regulate axon elongation and turning. Recent studies have demonstrated the involvement of lipid rafts in such processes, but much more work will be required to elucidate the mechanisms underlying receptor-raft association and raft-dependent signal transduction. These revelations would help understand the initial steps of cell-polarity generation are key players in directing growing axons to their destinations.

5. References

Beck S, Sakurai T, Eustace BK, Beste G, Schier R, Rudert F, Jay DG (2002) Fluorophore-assisted light inactivation: a high-throughput tool for direct target validation of proteins. *Proteomics*, **2**, 247-255.

Bixby JL, Zhang R (1990) Purified N-cadherin is a potent substrate for the rapid induction of neurite outgrowth. *J Cell Biol*, **110**, 1253-1260.

Bozyczko D, Horwitz AF (1986) The participation of a putative cell surface receptor for laminin and fibronectin in peripheral neurite extension. *J Neurosci*, **6**, 1241-1251.

Buchstaller A, Jay DG (2000) Micro-scale chromophore-assisted laser inactivation of nerve growth cone proteins. *Microsc Res Tech*, **48**, 97-106.

Doherty P, Fruns M, Seaton P, Dickson G, Barton CH, Sears TA, Walsh FS (1990) A threshold effect of the major isoforms of NCAM on neurite outgrowth. *Nature*, **343**, 464-466.

Edelman GM, Murray BA, Mege RM, Cunningham BA, Gallin WJ (1987) Cellular expression of liver and neural cell adhesion molecules after transfection with their cDNAs results in specific cell-cell binding. *Proc Natl Acad Sci U S A*, **84**, 8502-8506.

Gomez-Mouton C, Lacalle RA, Mira E, Jimenez-Baranda S, Barber DF, Carrera AC, Martinez AC, Manes S (2004) Dynamic redistribution of raft domains as an organizing platform for signaling during cell chemotaxis. *J Cell Biol*, **164**, 759-768.

Gomez-Mouton C, Abad JL, Mira E, Lacalle RA, Gallardo E, Jimenez-Baranda S, Illa I, Bernad A, Manes S, Martinez AC (2001) Segregation of leading-edge and uropod components into specific lipid rafts during T cell polarization. *Proc Natl Acad Sci U S A*, **98**, 9642-9647.

Gordon-Weeks PR (2000) *Neuronal growth cones*. Cambridge, UK; New York: Cambridge University Press.

Grumet M, Edelman GM (1988) Neuron-glia cell adhesion molecule interacts with neurons and astroglia via different binding mechanisms. *J Cell Biol*, **106**, 487-503.

Guirland C, Suzuki S, Kojima M, Lu B, Zheng JQ (2004) Lipid rafts mediate chemotropic guidance of nerve growth cones. *Neuron*, **42**, 51-62.

Herincs Z, Corset V, Cahuzac N, Furne C, Castellani V, Hueber AO, Mehlen P (2005) DCC association with lipid rafts is required for netrin-1-mediated axon guidance. *J Cell Sci*, **118**, 1687-1692.

Huber AB, Kolodkin AL, Ginty DD, Cloutier JF (2003) Signaling at the Growth Cone: Ligand-Receptor Complexes and the Control of Axon Growth and Guidance. *Annu Rev Neurosci*, **26**, 509-563.

Kamiguchi H, Lemmon V (2000) Recycling of the cell adhesion molecule L1 in axonal growth cones. *J Neurosci*, **20**, 3676-3686.

Kamiguchi H, Yoshihara F (2001) The role of endocytic L1 trafficking in polarized adhesion and migration of nerve growth cones. *J Neurosci*, **21**, 9194-9203.

Kamiguchi H, Long KE, Pendergast M, Schaefer AW, Rapoport I, Kirchhausen T, Lemmon V (1998) The neural cell adhesion molecule L1 interacts with the AP-2 adaptor and is endocytosed via the clathrin-mediated pathway. *J Neurosci*, **18**, 5311-5321.

Khanna KV, Whaley KJ, Zeitlin L, Moench TR, Mehrazar K, Cone RA, Liao Z, Hildreth JE, Hoen TE, Shultz L, Markham RB (2002) Vaginal transmission of cell-associated HIV-1 in the mouse is blocked by a topical, membrane-modifying agent. *J Clin Invest*, **109**, 205-211.

Lamoureux P, Buxbaum RE, Heidemann SR (1989) Direct evidence that growth cones pull. *Nature*, **340**, 159-162.

Lemmon V, Farr KL, Lagenaur C (1989) L1-mediated axon outgrowth occurs via a homophilic binding mechanism. *Neuron*, **2**, 1597-1603.

Lin CH, Forscher P (1995) Growth cone advance is inversely proportional to retrograde F-actin flow. *Neuron*, **14**, 763-771.

Manes S, Mira E, Gomez-Mouton C, Lacalle RA, Keller P, Labrador JP, Martinez AC (1999) Membrane raft microdomains mediate front-rear polarity in migrating cells. *Embo J*, **18**, 6211-6220.

Nakai Y, Kamiguchi H (2002) Migration of nerve growth cones requires detergent-resistant membranes in a spatially defined and substrate-dependent manner. *J Cell Biol*, **159**, 1097-1108.

Niethammer P, Delling M, Sytnyk V, Dityatev A, Fukami K, Schachner M (2002) Cosignaling of NCAM via lipid rafts and the FGF receptor is required for neuritogenesis. *J Cell Biol*, **157**, 521-532.

Nishimura K, Yoshihara F, Tojima T, Ooashi N, Yoon W, Mikoshiba K, Bennett V, Kamiguchi H (2003) L1-dependent neuritogenesis involves ankyrinB that mediates L1-CAM coupling with retrograde actin flow. *J Cell Biol*, **163**, 1077-1088.

Olive S, Dubois C, Schachner M, Rougon G (1995) The F3 neuronal glycosyl-phosphatidylinositol-linked molecule is localized to glycolipid-enriched membrane subdomains and interacts with L1 and fyn kinase in cerebellum. *J Neurochem*, **65**, 2307-2317.

Ren Q, Bennett V (1998) Palmitoylation of neurofascin at a site in the membrane-spanning domain highly conserved among the L1 family of cell adhesion molecules. *J Neurochem*, **70**, 1839-1849.

Schaefer AW, Kamei Y, Kamiguchi H, Wong EV, Rapoport I, Kirchhausen T, Beach CM, Landreth G, Lemmon SK, Lemmon V (2002) L1 endocytosis is controlled by a phosphorylation-dephosphorylation cycle stimulated by outside-in signaling by L1. *J Cell Biol*, **157**, 1223-1232.

Seveau S, Eddy RJ, Maxfield FR, Pierini LM (2001) Cytoskeleton-dependent membrane domain segregation during neutrophil polarization. *Mol Biol Cell*, **12**, 3550-3562.

Simons K, Ikonen E (1997) Functional rafts in cell membranes. *Nature*, **387**, 569-572.

Suter DM, Forscher P (2000) Substrate-cytoskeletal coupling as a mechanism for the regulation of growth cone motility and guidance. *J Neurobiol*, **44**, 97-113.

Takeichi M (1991) Cadherin cell adhesion receptors as a morphogenetic regulator. *Science*, **251**, 1451-1455.

Tessier-Lavigne M, Goodman CS (1996) The molecular biology of axon guidance. *Science*, **274**, 1123-1133.

Tomaselli KJ, Reichardt LF, Bixby JL (1986) Distinct molecular interactions mediate neuronal process outgrowth on non-neuronal cell surfaces and extracellular matrices. *J Cell Biol*, **103**, 2659-2672.

van Buul JD, Voermans C, van Gelderen J, Anthony EC, van der Schoot CE, Hordijk PL (2003) Leukocyte-endothelium interaction promotes SDF-1-dependent polarization of CXCR4. *J Biol Chem*, **278**, 30302-30310.

Zhao M, Pu J, Forrester JV, McCaig CD (2002) Membrane lipids, EGF receptors, and intracellular signals colocalize and are polarized in epithelial cells moving directionally in a physiological electric field. *Faseb J*, **16**, 857-859.

4-4 Sphingolipids and Multidrug Resistance of Cancer Cells

Gerrit van Meer, Maarten Egmond, and David Halter

Department of Membrane Enzymology, Bijvoet Center and Institute of Biomembranes, Utrecht University, Padualaan 8, 3584 CH Utrecht, The Netherlands

Summary. Multidrug resistance is a dramatic complication that can impede cancer treatment. Some cancer cells can become resistant to a cytostatic agent, survive and develop resistance to most agents available for chemotherapy. As multi-drug resistance is linked to sphingolipid metabolism, manipulating sphingolipid metabolism might be a way to circumvent the sensitization of cancer cells to chemotherapy. Two strategies seem particularly promising. One is to drive sphingolipid metabolism towards the production of proapoptotic lipid ceramide, which leads to cell death, and away from sphingosine-1-phosphate and glucosylceramide, which stimulate proliferation. The other is to alter the expression or activity of multidrug efflux pumps that in many cases supply the molecular basis for multidrug resistance.

Keywords. sphingolipids, glucosylceramide synthase, multidrug resistance, ATP binding cassette transporter

1. Multidrug resistance

Cancer cells that have acquired a resistance to more than one chemotherapeutic agent or anti-cancer drug used to kill cancer cells are multidrug resistant. These cells are able to efficiently remove a drug from inside that develops in a multidrug resistance. This may occur via the expression of transporters that can pump drugs out of the cell or through an increase in drug detoxification or alteration in the cellular drug pharmacokinetics. Multidrug resistance may stem from the cell's ability to compensate for the action of the drugs. The latter may, for example, include an increase in DNA repair or the suppression of drug-induced apoptosis. In general, multidrug resistance involves changes at the level of the DNA and gene transcription that develop over time to provide a cancer cell with an increased ability to survive and proliferate in the presence of chemotherapeutic drugs.

Sphingolipids have been connected to various forms of multidrug resistance. One general finding showed that drug-resistant cancer cells had increased glucosylceramide (GlcCer) levels. GlcCer may stimulate drug transporter expression, or alternatively, promote cell proliferation as its synthesis decreases the level of the proapoptotic ceramide. The same effect may also be caused through increased signalling action of lipid sphingosine-1-phosphate, which acts as an anti-apoptotic signal. Sphingolipid researchers are attempting to see how sphingolipid metabolism might be able to prevent or circumvent multidrug resistance.

2. Sphingolipids and growth control

Sphingolipids are defined by their sphingoid base (Fig. 1). New synthesis produces sphinganine. Sphinganine can be acylated to yield dihydroceramide which is generally desaturated at the C4-C5 bond to yield ceramide carrying sphingosine. Ceramide is a precursor for glycosphingolipid galactosylceramide in the ER lumen of specialized cells, for GlcCer on the cytosolic surface, and for phosphosphingolipid sphingomyelin (SM) in the lumen of the Golgi. GlcCer can be converted to lactosylceramide, the precursor for complex glycosphingolipids.

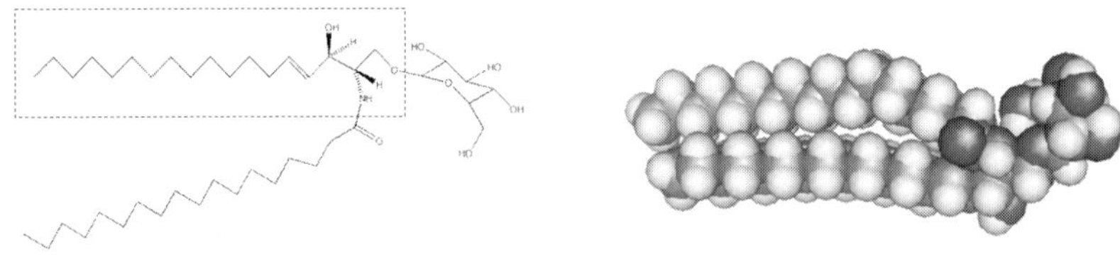

Fig. 1. Glycosphingolipid glucosylceramide. The sphingosine base is boxed.

Sphingolipids are hydrolyzed by a spectrum of lysosomal enzymes, and eventually yield sphingosine, which then leaves the lysosome and can be reutilized at the level of sphinganine. Alternatively, it is phosphorylated to sphingosine-1-phosphate, which can then be dephosphorylated back to sphingosine or degraded by the lyase, the final step in sphingolipid degradation. Apart from lysosomal degradation, SM can be degraded by sphingomyelinases at the plasma membrane during cell signalling. The reverse action at the plasma membrane, SM synthase, produces ceeramide (Huitema et al. 2004). This links ceramide-mediated signalling to diacylglycerol signalling. Finally, a non-lysosomal glucocerebrosidase (van Weely et al. 1993), also produces ceramide, but the enzyme's nature and location remain to be determined. Sphingosine-1-phosphate, a potent signalling lipid, can activate G-protein coupled receptors on the cell surface that induce mitogenesis and cell survival. But even in the absence of receptors, e.g. in yeast, sphingosine blocks cell-cycle progression, whereas this block is relieved by sphingosine-1-phosphate, a regulatory system called the sphingolipid rheostat (Spiegel and Milstien 2003). In contrast to sphingosine-1-phosphate, ceramide is proapoptotic (Futerman and Hannun 2004), and because ceramide-1-phosphate has been reported to promote survival (Gómez-Muñoz et al. 2005), a similar sphingolipid rheostat exists for this lipid pair.

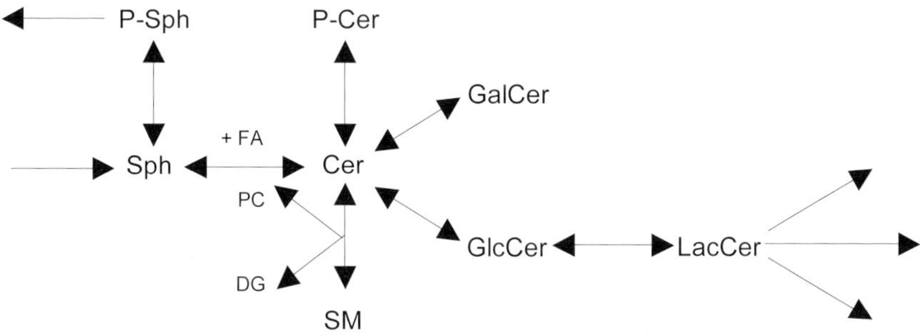

Fig. 2. Metabolic pathways connecting the signalling sphingolipids. Cer, ceramide; DG, diacylglycerol; FA, fatty acid; Gal, galactosyl; Glc, glucosyl; Lac, lactosyl; P, phosphate; P-Sph, sphingosine-1-phosphate; PC, phosphatidylcholine; SM, sphingomyelin; Sph, sphingosine.

The activity of a signalling lipid is determined by its local concentration. Therefore, one must understand how enzyme activities that generate and remove the signalling sphingolipids are regulated (Fig. 2). The synthesis of GlcCer, for example, most likely abrogates apoptosis by reducing the ceramide concentration (Kohyama-Koganeya et al. 2004). Remarkably,

GlcCer synthase was transcriptionally upregulated when drug-resistant cells were treated with doxorubicin, but not in drug sensitive cells (Uchida et al. 2004) This implies that drug-resistant cells use this anti-apoptotic mechanism next to expressing high levels of the doxorubicin transporter MRP1.

3. Sphingolipids and multidrug transporters

30 years ago Ling et al realised that multidrug resistance in cells can be caused by transporters in the plasma membrane (Ling 1975). Later, these transporters were identified as belonging to the family of ATP-binding cassette transporters, 49 in the human genome. Functional transporters span the membrane at least twelve times (and some family members are half transporters that dimerize for activity) and possess two ATP binding loops on the cytosolic side of the membrane (Borst and Oude Elferink 2002). Theories about a mechanism of action of a prototype multidrug transporter, the MDR1 P-glycoprotein (ABCB1), are derived from available data on structure obtained by electron microscopy (Rosenberg et al. 2005) and by comparing the structures observed when the ABCB1 sequence is projected onto available structures of the bacterial transporter MsbA (Fig. 3). ABCB1 transports a wide variety of amphiphilic compounds from the cytosol out of the cell (Seelig et al. 2000), and may employ a flippase mechanism to move the drug from the cytosolic leaflet of the bilayer to the outer leaflet from where it would be released (Higgins and Gottesman 1992).

Cabot et al. found that the capacity to synthesize GlcCer is correlated with ABCB1 gene expression. For example, inhibition of GlcCer synthesis by antisense or small interfering RNAs for GlcCer synthase, or by the chemical inhibitors PPMP and PDMP, all led to a reduction in ABCB1 expression to 15-40% of the control values (see Gouazé et al. 2005).Two other inhibitors, N-alkylated iminosugars, did not have this effect (Norris-Cervetto et al. 2004). Possibly, the latter inhibitors functionally mimick GlcCer. However, GlcCer synthesis can only be considered a regulator of ABCB1 expression if its concentration is subject to regulation. Evidence for this regulation at the transcriptional (e.g. Watanabe et al. 1998; Uchida et al. 2004) or post-translational level (Boldin and Futerman 2000) have been reported. How it is regulated: through its hydrolysis, directly via glucocerebrosidases, or by sorting towards the lysosomes, is not known.

ABC transporters, including multidrug transporters ABCB1 and MRP-1 (ABCC1), have been found in DRMs, the membranes remaining after detergent extraction (Bacso et al. 2004; Ghetie et al. 2004; Klappe et al. 2004; Bucher et al. 2005). Therefore, they reside in sphingolipid and cholesterol enriched regions of the plasma membrane, called lipid rafts. It is likely that transporter activity depends on unique physical properties of such domains that are modulated by their proteins (caveolin; Lavie et al. 2001) and their lipid composition just as gangliosides might activate ABCB1 (Plo et al. 2002).

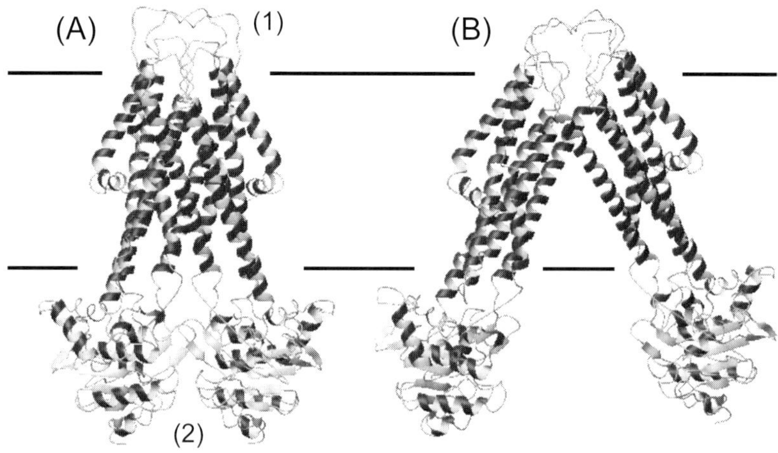

Fig. 3. Projection of the sequence of multidrug transporter ABCB1 (MDR1 P-glycoprotein) onto that of the bacterial ABC transporter MsbA (Chang and Roth 2001; Reyes and Chang 2005). The open conformation (B) would allow the substrate to enter the transport site, whereas the nucleotide bound form (A) would actually move the molecule across. Notably, the interface between the two halves of ABCB1 is much more extensive than in the MsbA dimer (Rosenberg et al. 2005). Differences between ABCB1 and ABCB4 are largest at the very top (1) and bottom (2) of the molecule.

Changes in the metabolism of sphingolipids have been reported for drug-resistant cells in the presence (Norris-Cervetto et al. 2004) or absence (Prinetti et al. 2003) of multidrug transporter overexpression. ABCB1 overexpression might increase higher glycolipid synthesis by translocating newly synthesized GlcCer from the cytosolic surface to the lumenal surface of the Golgi (de Rosa et al. 2004). Roles for ABCB1 and ABCC1 in transmembrane translocation of GlcCer was originally proposed from studies on fluorescent GlcCer (van Helvoort et al. 1996; Raggers et al. 1999). Also sphingoid bases have been proposed to be a substrate for extrusion by

ABCB1 (Sugawara et al. 2004). In contrast, inward transport of sphingosine-1-phosphate specifically correlated with expression of ABCC7 (the cystic fibrosis transmembrane conductance regulator; Boujaoude et al. 2001). In support of a lipid translocator activity, ABCB1 displays 75% identity and 86% homology to ABCB4, presumed to be an outward translocator for the phospholipid phosphatidylcholine (Borst and Oude Elferink 2004).

4. Sphingolipids and multidrug resistance

A complex relationship between sphingolipid metabolism and multidrug resistance exists. In order to exploit these connections for improving the efficiency of chemotherapy, both processes require detailed molecular analysis. While near complete identification of the proteins involved has had a tremendous impact on the field, elucidating the structural details of the drug transporters still needs to be tackled before educated guesses can be made to counter multidrug resistance. That effort will show how to stimulate drug pumping via the sphingolipids, and, second, how to apply proteomics and metabolics approaches to identify all components of the system and to establish their structural and functional relationships. The agents that we have presently available for influencing sphingolipid metabolism and transport may then be explored in our quest for developing drugs that can circumvent the deleterious effects in multidrug resistance in cancer therapy.

5. References

Bacso Z, Nagy H, Goda K, Bene L, Fenyvesi F, et al. (2004) Raft and cytoskeleton associations of an ABC transporter: P-glycoprotein. *Cytometry A*, **61**, 105-116.

Boldin SA, Futerman AH (2000) Up-regulation of glucosylceramide synthesis upon stimulation of axonal growth by basic fibroblast growth factor. Evidence for post-translational modification of glucosylceramide synthase. *J Biol Chem*, **275**, 9905-9909.

Borst P, Oude Elferink R (2002) Mammalian ABC transporters in health and disease. *Annu Rev Biochem*, **71**, 537-592.

Boujaoude LC, Bradshaw-Wilder C, Mao C, Cohn J, Ogretmen B, et al. (2001) Cystic fibrosis transmembrane regulator regulates uptake of sphingoid base phosphates and lysophosphatidic acid: modulation of cellular activity of sphingosine 1-phosphate. *J Biol Chem*, **276**, 35258-35264.

Bucher K, Besse CA, Kamau SW, Wunderli-Allenspach H, Kramer SD (2005) Isolated rafts from adriamycin-resistant P388 cells contain functional ATPases and provide an easy test system for P-glycoprotein-related activities. *Pharm Res*, **22**, 449-457.

De Rosa MF, Sillence D, Ackerley C, Lingwood C (2004) Role of multiple drug resistance protein 1 in neutral but not acidic glycosphingolipid biosynthesis. *J Biol Chem*, **279**, 7867-7876.

Futerman AH, Hannun YA (2004) The complex life of simple sphingolipids. *EMBO Rep*, **5**, 777-782.

Ghetie MA, Marches R, Kufert S, Vitetta ES (2004) An anti-CD19 antibody inhibits the interaction between P-glycoprotein (P-gp) and CD19, causes P-gp to translocate out of lipid rafts, and chemosensitizes a multidrug-resistant (MDR) lymphoma cell line. *Blood*, **104**, 178-183.

Gouazé V, Liu YY, Prickett CS, Yu JY, Giuliano AE, Cabot MC (2005) Glucosylceramide synthase blockade down-regulates P-glycoprotein and resensitizes multidrug-resistant breast cancer cells to anticancer drugs. *Cancer Res*, **65**, 3861-3867.

Gómez-Muñoz A, Kong JY, Parhar K, Wang SW, Gangoiti P, et al. (2005) Ceramide-1-phosphate promotes cell survival through activation of the phosphatidylinositol 3-kinase/protein kinase B pathway. *FEBS Lett*, **579**, 3744-3750.

Higgins CF, Gottesman MM (1992) Is the multidrug transporter a flippase? *Trends Biochem Sci*, **17**, 18-21.

Huitema K, Van Den Dikkenberg J, Brouwers JF, Holthuis JC (2004) Identification of a family of animal sphingomyelin synthases. *Embo J*, **23**, 33-44.

Klappe K, Hinrichs JW, Kroesen BJ, Sietsma H, Kok JW (2004) MRP1 and glucosylceramide are coordinately over expressed and enriched in rafts during multidrug resistance acquisition in colon cancer cells. *Int J Cancer*, **110**, 511-522.

Kohyama-Koganeya A, Sasamura T, Oshima E, Suzuki E, Nishihara S, et al. (2004) Drosophila glucosylceramide synthase: a negative regulator of cell death mediated by proapoptotic factors. *J Biol Chem*, **279**, 35995-36002.

Lavie Y, Fiucci G, Liscovitch M (2001) Upregulation of caveolin in multidrug resistant cancer cells: functional implications. *Adv Drug Deliv Rev*, **49**, 317-323.

Ling V (1975) Drug resistance and membrane alteration in mutants of mammalian cells. *Can J Genet Cytol*, **17**, 503-515.

Norris-Cervetto E, Callaghan R, Platt FM, Dwek RA, Butters TD (2004) Inhibition of glucosylceramide synthase does not reverse drug resistance in cancer cells. *J Biol Chem*, **279**, 40412-40418.

Plo I, Lehne G, Beckstrom KJ, Maestre N, Bettaieb A, et al. (2002) Influence of ceramide metabolism on P-glycoprotein function in immature acute myeloid leukemia KG1a cells. *Mol Pharmacol*, **62**, 304-312.

Prinetti A, Basso L, Appierto V, Villani MG, Valsecchi M, Loberto N, Prioni S, Chigorno V, Cavadini E, Formelli F, Sonnino S (2003) Altered sphingolipid metabolism in N-(4-hydroxyphenyl)-retinamide-resistant A2780 human ovarian carcinoma cells. *J Biol Chem*, **278**, 5574-5583.

Raggers RJ, van Helvoort A, Evers R, van Meer G (1999) The human multidrug resistance protein MRP1 translocates sphingolipid analogs across the plasma membrane. *J Cell Sci*, **112**, 415-422.

Reyes CL, Chang G (2005) Structure of the ABC transporter MsbA in complex with ADP.vanadate and lipopolysaccharide. *Science*, **308**, 1028-1031.

Rosenberg MF, Callaghan R, Modok S, Higgins CF, Ford RC (2005) Three-dimensional structure of P-glycoprotein: the transmembrane regions adopt an asymmetric configuration in the nucleotide-bound state. *J Biol Chem*, **280**, 2857-2862.

Seelig A, Blatter XL, Wohnsland F (2000) Substrate recognition by P-glycoprotein and the multidrug resistance- associated protein MRP1: a comparison. *Int J Clin Pharmacol Ther*, **38**, 111-121.

Spiegel S, Milstien S (2003) Sphingosine-1-phosphate: an enigmatic signalling lipid. *Nat Rev Mol Cell Biol*, **4**, 397-407.

Sugawara T, Kinoshita M, Ohnishi M, Tsuzuki T, Miyazawa T, et al. (2004) Efflux of sphingoid bases by P-glycoprotein in human intestinal Caco-2 cells. *Biosci Biotechnol Biochem*, **68**, 2541-2546.

Uchida Y, Itoh M, Taguchi Y, Yamaoka S, Umehara H, Ichikawa S, Hirabayashi Y, Holleran WM, Okazaki T (2004) Ceramide reduction and transcriptional up-regulation of glucosylceramide synthase through doxorubicin-activated Sp1 in drug-resistant HL-60/ADR cells. *Cancer Res*, **64**, 6271-6279.

van Helvoort A, Smith AJ, Sprong H, Fritzsche I, Schinkel AH, et al. (1996) MDR1 P-glycoprotein is a lipid translocase of broad specificity, while MDR3 P-glycoprotein specifically translocates phosphatidylcholine. *Cell*, **87**, 507-517.

van Weely S, Brandsma M, Strijland A, Tager JM, Aerts JM (1993) Demonstration of the existence of a second, non-lysosomal glucocerebrosidase that is not deficient in Gaucher disease. *Biochim. Biophys Acta*, **1181**, 55-62.

Watanabe R, Wu K, Paul P, Marks DL, Kobayashi T, et al. (1998) Up-regulation of glucosylceramide synthase expression and activity during human keratinocyte differentiation. *J Biol Chem*, **273**, 9651-9655.

Part 5

Membrane Lipid Domain and Human Pathobiology

5-1 A New Pathological Feature of Insulin Resistance and Type 2 Diabetes: Involvement of Ganglioside GM3 and Membrane Microdomains

Jin-ichi Inokuchi[1,2], Kazuya Kabayama[2], Takashige Sato[1] and Yasuyuki Igarashi[1]

[1]Department of Biomembrane and Biofunctional Chemistry and [2]CREST, Japan Science and Technology Agency, Graduate School of Pharmaceutical Sciences, Frontier Research Center for Post-Genomic Science and Technology, Hokkaido University, Kita 21, Nishi 11, Kita-ku, Sapporo 001-0021, Japan

Summary. Membrane microdomains (lipid rafts), which are critical for proper compartmentalization of insulin signaling, also play a role in the pathogenesis of insulin resistance, and yet this role has not been investigated. Detergent-resistant membrane microdomains (DRMs), isolated in low density fractions, are rich in cholesterol, glycosphingolipids and various signaling molecules. TNFϕ induces insulin resistance in type 2 diabetes, but its action mechanism is not fully understood. We found a selective increase in the acidic glycosphingolipid ganglioside GM3 in 3T3-L1 adipocytes treated with TNFϕ, suggesting a specific function for GM3. We extended these *in vitro* observations to living animals using obese Zucker *fa/fa* rats and *ob/ob* mice, in which the GM3 synthase mRNA levels in the white adipose tissues are significantly higher than in their lean controls. In DRMs from TNFϕ-treated 3T3-L1 adipocytes, GM3 levels were doubled those of normal adipocytes. Additionally, insulin receptor (IR) accumulations in DRMs were diminished, while caveolin and flotillin levels were unchanged. GM3 depletion counteracted the TNFϕ-induced inhibition of IR accumulation into DRMs. Together, these findings provide compelling evidence an insulin metabolic signaling defect can be attributed to a loss of IRs in the microdomains due to an accumulation of GM3. Therefore, it is likely that life-style related diseases, such as type 2 diabetes, are membrane microdomain disorders caused by aberrant expression of glycosphingolipids.

Keywords. insulin receptor, insulin resistance, ganglioside GM3, lipid rafts, detergent-resistant microdomain (DRM)

1. Introduction

Insulin resistance, defined as the decreased ability of cells or tissues to respond to physiological levels of insulin, is considered the primary pathophysiological defect of type 2 diabetes (Virkamaki et al., 1999). Numerous studies implicate TNFφ in insulin resistance in cultured adipocyte and whole-animal models (Hotamisligil et al., 1995; Hotamisligil et al., 1993; Uysal et al., 1997). In adipocytes cultured in relatively low concentrations of TNFφ (levels that do not cause a generalized suppression of gene expression) interference with insulin action occurs. This effect requires prolonged treatment (at least 72 h), unlike many acute responses to this cytokine. This protracted effect suggests that TNFφ triggers the synthesis of an inhibitor that is the actual effector.

One clue into the mechanism of this hormone's unique actions lies in the compartmentalization of its signaling molecules. Cellular membranes contain sub-domains called detergent resistant microdomains (DRMs), that are detergent-insoluble and rich in cholesterol and glycosphingolipids (GSLs) but lacking phospholipids (Hakomori 2000; Simons and Toomre 2000). Over the past decade, many laboratories have shown that these lipid microdomains are critical for proper compartmentalization of insulin signaling in adipocytes (see reviews by (Bickel 2002) and (Cohen et al., 2003a)).

Gangliosides, a family of sialic acid-containing GSLs, are an important component of DRMs. In adipose tissues from various species, including human and mouse, GM3 is the most abundant ganglioside (Ohashi 1979). Recently, we reported that in mouse 3T3-L1 adipocytes, insulin resistance induced by TNFφ was accompanied by increased GM3 expression. We extended these in vitro observations to living animals using obese Zucker *fa/fa* rats and *ob/ob* mice, in which the GM3 synthase mRNA levels in the white adipose tissues are significantly higher than in their lean controls (Tagami et al., 2002). Moreover, we examined the effect of TNFφ on the composition and function of DRMs in adipocytes and demonstrated that increased GM3 levels result in the elimination of IRs from the DRMs while caveolin and flotillin remain in the DRMs (Kabayama et al., 2005).

Thereby we uncovered a new pathological feature of insulin resistance in adipocytes induced by TNFφ.

2. Effects of TNFφ and D-PDMP on Insulin Signaling through IR to IRS-1

We examined the effects of low TNFφ concentrations on levels of IR and IRS-1 proteins, since high concentrations of TNFφ induce de-differentiation of adipocytes and reduce gene expression along the IR signal transduction pathway, including on IRS-1 and GLUT4 genes (Hotamisligil et al., 1993; Hotamisligil et al., 1994a; Stephens et al., 1997). Fully differentiated 3T3-L1 adipocytes were treated with 0.1 nM TNFφ for 96 h. TNFφ induced a moderate decrease of insulin stimulated phosphorylation of IR and a more pronounced inhibition of insulin-promoted

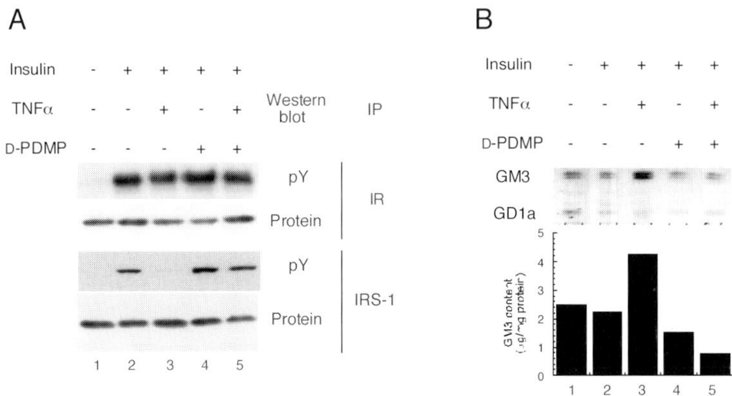

Fig. 1. TNFφ increases the expression of GM3. The prevention of GM3 synthesis reverses TNFφ induced suppression insulin signaling in adipocytes. (A) 3T3-L1 adipocytes were cultured in maintenance medium without (lanes 1,2 and 4) or with (lanes 3 and 5) 0.1 nM TNFφ for 96 h and, in order to deplete GM3, 20 μM D-PDMP was also included (lanes 4 and 5). Before insulin stimulation (100 nM for 3 min), cells were starved in serum-free media containing 0.5% bovine serum albumin in the absence or presence of TNFφ and D-PDMP as above for 8 h. Proteins in cell lyzates were immunoprecipitated with antiserum to IR and IRS-1, fractionated SDS-PAGE, and transferred to Immobilon-P. Western blot testing with anti-phosphotyrosine monoclonal antibody, stripped and reproved with antiserum to IR and IRS-1. (B) 3T3-L1 adipocytes were incubated in the absence or presence of TNFφ and D-PDMP as in (A) and ganglioside fraction was visualized by resorcinol staining.

phosphorylation of IRS-1 without affecting the expression of either IR or IRS-1 (Fig. 1A). Marked accumulation of GM3 occurred with TNFɸ treatment (Fig. 1B). One approach to depleting cellular gangliosides is to use a glucosylceramide synthase inhibitor, D-PDMP which is a well-known tool to study the functional roles of endogenous glycosphin-golipids including gangliosides (Inokuchi et al., 1987; Radin et al., 1993). D-PDMP lowered GM3 content in 3T3-L1 adipocytes with or without TNFɸ treatment (Fig. 1B), with a concomitant increase of tyrosine phos-phorylation of IRS-1 in response to insulin stimulation (Fig. 1A). The binding affinity of insulin to adipocytes treated with 20 μM D-PDMP for 96 h was similar to that of the non-treated control cells, indicating that GM3 depletion by D-PDMP treatment does not affect insulin binding to its receptor and that the D-PDMP treatment affects the insulin signaling pathway through IR to IRS-1.

3. Enhanced Expression of GM3 Synthase mRNA in Adipose Tissues from Typical Rodent Models with Insulin Resistance

Adipose tissues of the obese-diabetic *db/db*, *ob/ob*, KK-A^y mice, and the Zucker *fa/fa* rat produced significant levels of TNFɸ (Hotamisligil et al., 1993). Much less expression was seen in fat obtained from the lean control animals. Interestingly, those obese-diabetic animals showed no evidence of altered expression of other cytokines, such as TNFɸ, IL-1 or IFNɸ (Hotamisligil et al., 1993; Hotamisligil et al., 1994b). So we meas-ured the expression of GM3 synthase mRNA in the epididymal fat of

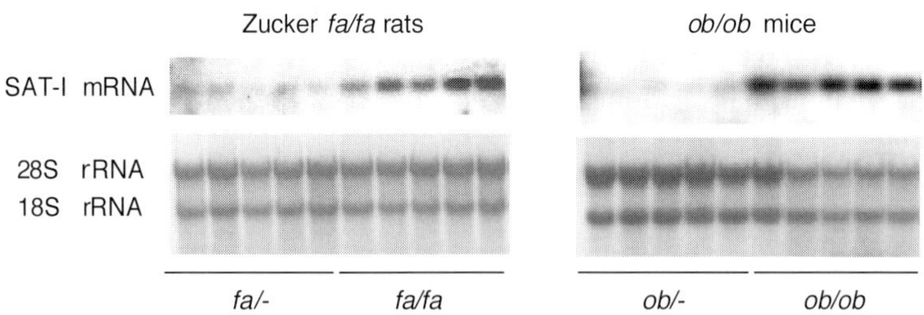

Fig. 2. Increased GM3 synthase mRNA in adipose tissue of typical rodent models of insulin resistance. Northern blot analysis of GM3 synthase mRNA was performed using total mRNA form adipose tissues of Zucker *fa/fa* rats and *ob/ob* mice, and their lean counterparts.

Zucker *fa/fa* rats and *ob/ob* mice. Northern blot analysis of GM3 synthase mRNA contents in adipose tissues from these two models of insulin resistance exhibited significantly higher levels of the protein when compared to their leaner counterparts (Fig. 2).

4. IR is a component of DRMs and is selectively eliminated during insulin resistance induced by TNFφ

Studies of the presence of IRs in DRMs/caveolae have provided conflicting data (Gustavsson et al., 1999; Kimura et al., 2002; Mastick and Saltiel 1997; Muller et al., 2001). Localization of IRs in a flotation assay followed extraction with increasing concentrations of Triton X-100 or under hypertonic alkaline conditions (500 mM sodium carbonate) (Fig. 3). In the carbonate buffer, IRs from normal 3T3-L1 cells were found exclusively in low density, insoluble fractions 4 and 5, which carry DRMs, whereas a small portion of IRs in TNFφ-treated cells shifted to fractions 6-8.On the other hand, there was no accumulation in cellular DRMs, untreated or treated with TNFφ, when examined using an extraction buffer containing 1% (data not shown) or 0.1% Triton X-100, confirming a previous report using higher levels of detergent (Gustavsson et al., 1999). However, when the flotation assay was performed using buffer with lower concentrations of Triton X-100 (0.08% or 0.05%), DRMs held on to IRs in untreated samples while IRs in TNFφ-treated cells tended to shift to higher density fractions. Next, using membranes extracted with the 0.08% Triton X-100 buffer, we analyzed the distribution of GM3 and cholesterol, and of proteins normally associated with DRMs (e.g. caveolin, flotillin, fyn), in each fraction of the sucrose density gradient (Fig. 3). GM3 was preferentially localized at the DRMs in both normal and TNFφ-treated 3T3-L1 adipocytes. Remarkably, though, the accumulation of GM3 observed in the DRMs was 2-fold higher in the TNFφ-treated cells (Fig. 3A). There were no distinct differences in expression levels or the distribution of cholesterol (Fig. 3B), caveolin, flotillin, or fyn (Fig. 3C). Taken together, these results demonstrate the selective elimination of IRs from the DRMs and the accumulation of GM3 in adipocytes under a chronic state of TNFφ-induced insulin resistance.

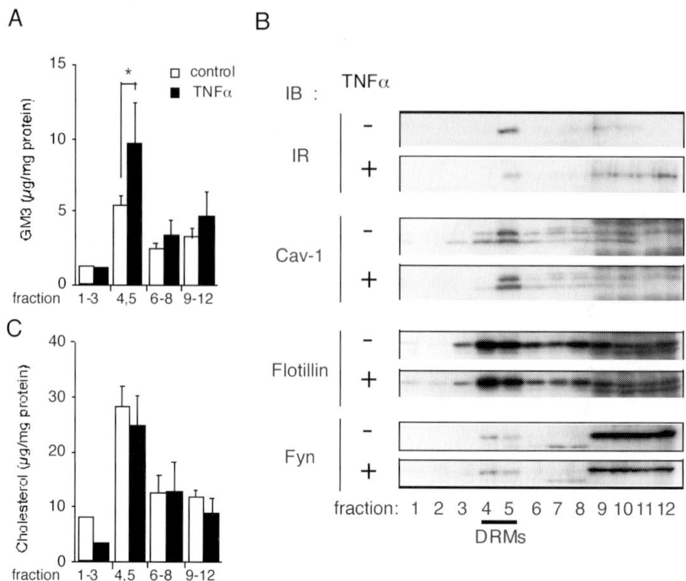

Fig. 3. Accumulation of GM3 in the DRMs and selective dissociation of IRs from DRMs in a state of insulin resistance. Adipocytes, untreated (open bars) or treated with 100 pM TNFϕ for 96 h (solid bars), were lysed with buffer containing 0.08% Triton X-100 and subjected to a sucrose density gradient floatation assay. (A) GM3 levels in the sucrose gradient fractions were examined by HPTLC analysis. (B) Cholesterol levels in the sucrose gradient fractions. (C) The samples were subjected to SDS-PAGE and immunoblotted with antibodies against IRϕ, caveolin-1, flotillin, or fyn.

5. GSL depletion attenuates TNFϕ-induced inhibition of insulin-stimulated IR internalization and dissociation of IR from DRMs

To determine the role of GM3 increases in the inhibition of insulin-dependent IR internalization and IR dissociation from the microdomains in TNFϕ-treated cells, we used a glucosylceramide synthase inhibitor, D-PDMP. After GM3 depletion by D-PDMP, the suppression of IR internalization was indeed partially recovered Additionally, the dissociation of IRs from DRMs was blocked (Kabayama et al. 2005). There was no obvious change in the accumulation of IR in DRMs after insulin stimulation. These results indicate direct involvement of GM3 in the chronic state of insulin resistance in adipocytes.

6. Discussion

Caveolae are a subset of membrane microdomains particularly abundant in adipocytes. Insulin metabolic signal transduction has been shown to be critically dependent on caveolae/microdomains in adipocytes. Disruption of microdomains by cholesterol extraction with ϕ-cyclodextrin resulted in progressive inhibition of tyrosine phosphorylation of IRS-1 and the activation of glucose transport in response to insulin, although the autophosphorylation of IR and activation of MAP kinase were not impaired (Parpal et al., 2001). Similarities exist between these cell culture results and the findings in many cases of clinical insulin resistance (Virkamaki et al., 1999). This suggests a potential role for microdomains in the pathogenesis of this disorder. Gangliosides are also known as structurally and functionally important components in microdomains, however, their role(s) in the microdomains of adipocytes remain unknown.

This transformation to an insulin resistant state induced in adipocytes by TNFϕ may depend on increased ganglioside GM3 biosynthesis following upregulated GM3 synthase gene expression. Additionally, GM3 may function as an inhibitor of insulin signaling during chronic exposure to TNFϕ (Tagami et al., 2002). These findings are further supported by the report that mice lacking GM3 synthase exhibit enhanced insulin signaling (Yamashita et al., 2003). Since GSLs, including GM3, are important components of DRMs/caveolae, we pursued the possibility that increased GM3 levels in DRMs confer insulin resistance upon TNFϕ-treated 3T3-L1 adipocytes.

Evidence suggesting that caveolae and caveolins play a major role in insulin signaling initially came from experiments using rat adipocytes, in which gold-labeled insulin was endocytosed by a mechanism involving clathrin-independent, uncoated invaginations (Goldberg et al., 1987). Immunogold electron (Gustavsson et al., 1999) and immunofluorescence microscopy (Kimura et al., 2002) further demonstrated that IRs are highly concentrated in caveolae. Additionally, Couet et al., revealed the presence of a caveolin binding motif (ϕXXXXϕXXϕ) in the ϕ subunit of IRs that could bind to the scaffold domain of caveolin (Couet et al., 1997). Moreover, mutation of this motif resulted in the inhibition of insulin signaling (Nystrom et al., 1999). Recently, Lisanti's laboratory reported that caveolin-1-null mice developed insulin resistance when placed on a high fat diet (Cohen et al., 2003a). Interestingly, insulin signaling, as measured by IR phosphorylation and its downstream targets, was selectively decreased in the adipocytes of these animals while signaling in both muscle and liver cells was normal (Cohen et al., 2003a). This signaling defect was attributed to a 90% decrease in IR protein content in the adipocytes, with no changes in mRNA levels, indicating that caveolin-1 serves to stabilize the IR protein (Cohen et al., 2003a,b). These studies clearly indicate the critical importance of the interaction between caveolin and IR in executing successful insulin signaling in adipocytes.

Saltiel and colleagues found that insulin stimulation of 3T3-L1 adipocytes was associated with tyrosine phosphorylation of caveolin-1 (Mastick and Saltiel 1997). However, since only trace levels of IR were recovered in the caveolae microdomains in assays with a buffer of 1% Triton X-100, they speculated on the presence of intermediate molecule(s) bridging IR and caveolin. Gustavsson et al., also observed the dissociation of IRs from caveolin-containing DRMs after treatments of 0.3 and 0.1% Triton X-100 (Gustavsson et al. 1999). It has been reported that comparison of protein and lipid contents of DRMs prepared with a variety of detergents exhibited the considerable differences in their ability to selectively solubilize membrane proteins and to enrich sphingolipids and cholesterol over glycerophospholipids, and Triton was the most reliable detergent (Schuck et al., 2003). Therefore, we performed a flotation assay with a wide range of Triton X-100 concentrations to identify the protein of interest which might weakly associate with DRMs (Figs. 3 and 4) In an assay system containing less than 0.08% Triton X-100, we showed that in normal adipocytes IRs can localize to DRMs., However, when TNFϕ was present, IR was selectively eliminated from DRMs even while caveolin-1 remained (Fig. 3C). Thus, by employing low detergent concentrations we successfully demonstrated the presence of IR in DRMs. We believe that elimination of IR from the DRMs by TNFϕ treatment is due to an excessive accumulation of GM3 in these microdomains, since preventing GM3 biosynthesis using D-PDMP attenuated the elimination of IR from DRMs (Kabayama et al. 2005). Reportedly, localization in DRMs of several proteins (including receptor protein tyrosine kinases) is affected by changes in the expression levels of GSLs. For example, overexpression of the ganglioside GM1 in Swiss 3T3 cells removed b type platelet-derived growth factor receptors from DRMs (Mitsuda et al., 2002). Similarly, genetically enhanced accumulation of endogenous GM3 in keratinocytes caused the dissociation of caveolin-1 from DRMs, thereby changing the signaling of epidermal growth factor receptor (Wang et al., 2002). In HuH7 hepatoma cells, which lack caveolin, IRs associate with DRMs in response to insulin stimulation, but crosslinking of GM2 by its antibody eliminated this association (Vainio et al., 2002). Such results support the likelihood that localization of IRs to the DRMs is affected by the presence of not only caveolin but also GSLs, especially gangliosides. Additionally, tyrosine

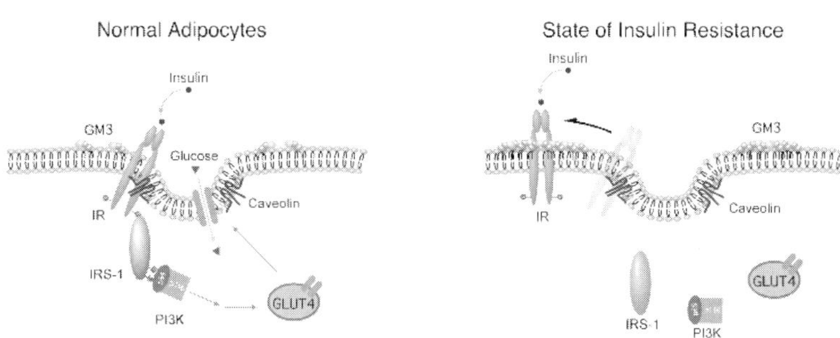

Fig. 4. Proposed model of caveolae microdomains in the state of insulin resistance.

phosphorylation of IRS-1 in response to insulin was selectively impaired without affecting the activation of IR and MAPK (Tagami et al., 2002). The observed impairment of IR-IRS-1 signaling by TNFφ may be attributed to the elimination of IR from microdomains due to the excess accumulation of GM3 (Fig. 3A). Although the localization of IRs to DRMs may be maintained through association with caveolin-1 as mentioned above, excess accumulation of GM3 in DRMs may weaken IR-caveolin interaction. Indeed, IR, not caveolin-1, was co-immunoprecipitated with anti-GM3 antibody (unpublished observation). A current view of microdomains in the state of insulin resistance is depicted in Fig 4. Further work will elucidate the mechanisms for interactions between the ganglioside GM3, IR and caveolin in the microdomains.

Acknowledgements. This study was supported in part by Grants-in Aid 15390019 (to J.I.) and 15040202 (Scientific Research on Priority Area to J.I.) from the Ministry of Education, Culture, Sports, Science and Technology. This work was also supported by the Mizutani Foundation for Glycoscience (to J.I.).

7. References

Bickel, P. E. (2002). Lipid rafts and insulin signaling. *Am J Physiol Endocrinol Metab*, **282**, E1-E10.

Cohen, A. W., Razani, B., Wang, X. B., Combs, T. P., Williams, T. M., Scherer, P. E., and Lisanti, M. P. (2003a). Caveolin-1-deficient mice show insulin resistance and defective insulin receptor protein expression in adipose tissue. *Am J Physiol Cell Physiol*, **285**, C222-235.

Cohen, A. W., Combs, T. P., Scherer, P. E., and Lisanti, M. P. (2003b). Role of caveolin and caveolae in insulin signaling and diabetes. *Am J Physiol Endocrinol Metab*, **285**, E1151-1160

Couet, J., Li, S., Okamoto, T., Ikezu, T., and Lisanti, M. P. (1997). Identification of peptide and protein ligands for the caveolin-scaffolding domain. Implications for the interaction of caveolin with caveolae-associated proteins. *J Biol Chem*, **272**, 6525-6533.

Goldberg, R. I., Smith, R. M., and Jarett, L. (1987). Insulin and alpha 2-macroglobulin-methylamine undergo endocytosis by different mechanisms in rat adipocytes: I. Comparison of cell surface events. *J Cell Physiol*, **133**, 203-212.

Gustavsson, J., Parpal, S., Karlsson, M., Ramsing, C., Thorn, H., Borg, M., Lindroth, M., Peterson, K. H., Magnusson, K. E., and Stralfors, P. (1999). Localization of the insulin receptor in caveolae of adipocyte plasma membrane. *Faseb J*, **13**, 1961-1971.

Hakomori, S. I. (2000). Cell adhesion/recognition and signal transduction through glycosphingolipid microdomain. *Glycoconj J*, **17**, 143-151.

Hotamisligil, G. S., Shargill, N. S., and Spiegelman, B. M. (1993). Adipose expression of tumor necrosis factor-alpha: direct role in obesity-linked insulin resistance. *Science*, **259**, 87-91.

Hotamisligil, G.S., Murray, D.L., Choy, L.N. and Spiegelman, B.M. (1994a) Tumor necrosis factor alpha inhibits signaling from the insulin receptor. *Proc Natl Acad Sci U S A*, **91**, 4854-4858.

Hotamisligil, G.S. and Spiegelman, B.M. (1994b) Tumor necrosis factor alpha: a key component of the obesity-diabetes link. *Diabetes*, **43**, 1271-1278.

Hotamisligil, G.S., Arner, P., Caro, J.F., Atkinson, R.L. and Spiegelman, B.M. (1995) Increased adipose tissue expression of tumor necrosis factor-alpha in human obesity and insulin resistance. *J Clin Invest*, **95**, 2409-2415.

Inokuchi, J., Momosaki, K., Shimeno, H., Nagamatsu, A. and Radin, N.S. (1989) Effects of D-threo-PDMP, an inhibitor of glucosylceramide synthetase, on expression of cell surface glycolipid antigen and binding to adhesive proteins by B16 melanoma cells. *J Cell Physiol*, **141**, 573-583.

Inokuchi, J. and Radin, N. (1987) Preparation of the active isomer of 1-phenyl-2-decanoylamino-3-morpholino-1-propanol, inhibitor of murine glucocerebroside synthetase. *J Lipid Res*, **28**, 565-571.

Iwanishi, M., Haruta, T., Takata, Y., Ishibashi, O., Sasaoka, T., Egawa, K., Imamura, T., Naitou, K., Itazu, T. and Kobayashi, M. (1993) A mutation (Trp1193-->Leu1193) in the tyrosine kinase domain of the insulin receptor associated with type A syndrome of insulin resistance. *Diabetologia*, **36**, 414-422.

Kabayama, K., Sato, T., Kitamura, F., Uemura, S., Kang, B.W., Igarashi, Y. and Inokuchi, J. (2005) TNFalpha-induced insulin resistance in adipocytes as a membrane microdomain disorder: involvement of ganglioside GM3. *Glycobiology*, **15**, 21-29.

Kimura, A., Mora, S., Shigematsu, S., Pessin, J. E., and Saltiel, A. R. (2002). The insulin receptor catalyzes the tyrosine phosphorylation of caveolin-1. *J Biol Chem*, **277**, 30153-30158.

Mastick, C. C., and Saltiel, A. R. (1997). Insulin-stimulated tyrosine phosphorylation of caveolin is specific for the differentiated adipocyte phenotype in 3T3-L1 cells. *J Biol Chem*, **272**, 20706-20714.

Mitsuda, T., Furukawa, K., Fukumoto, S., Miyazaki, H., and Urano, T. (2002). Overexpression of ganglioside GM1 results in the dispersion of platelet-derived growth factor receptor from glycolipid-enriched microdomains and in the suppression of cell growth signals. *J Biol Chem*, **277**, 11239-11246.

Muller, G., Jung, C., Wied, S., Welte, S., Jordan, H., and Frick, W. (2001). Redistribution of glycolipid raft domain components induces insulin-mimetic signaling in rat adipocytes. *Mol Cell Biol*, **21**, 4553-4567.

Nystrom, F. H., Chen, H., Cong, L. N., Li, Y., and Quon, M. J. (1999). Caveolin-1 interacts with the insulin receptor and can differentially modulate insulin signaling in transfected Cos-7 cells and rat adipose cells. *Mol Endocrinol*, **13**, 2013-2024.

Ohashi, M. (1979). A comparison of the ganglioside distributions of fat tissues in various animals by two-dimensional thin layer chromatography. *Lipids*, **14**, 52-57.

Parpal, S., Karlsson, M., Thorn, H., and Stralfors, P. (2001). Cholesterol depletion disrupts caveolae and insulin receptor signaling for metabolic control via insulin receptor substrate-1, but not for mitogen-activated protein kinase control. *J Biol Chem*, **276**, 9670-9678.

Radin, N.S., Shayman, J.A. and Inokuchi, J. (1993) Metabolic effects of inhibiting glucosylceramide synthesis with PDMP and other substances. *Adv Lipid Res*, **26**, 183-213.

Schuck, S., Honsho, M., Ekroos, K., Shevchenko, A., and Simons, K. (2003). Resistance of cell membranes to different detergents. *Proc Natl Acad Sci U S A*, **100**, 5795-5800.

Simons, K., and Toomre, D. (2000). Lipid rafts and signal transduction. *Nat Rev Mol Cell Biol*, **1**, 31-39.

Stephens, J.M., Lee, J. and Pilch, P.F. (1997) Tumor necrosis factor-alpha-induced insulin resistance in 3T3-L1 adipocytes is accompanied by a loss of insulin receptor substrate-1 and GLUT4 expression without a loss of insulin receptor-mediated signal transduction. *J Biol Chem*, **272**, 971-976.

Tagami, S., Inokuchi Ji, J., Kabayama, K., Yoshimura, H., Kitamura, F., Uemura, S., Ogawa, C., Ishii, A., Saito, M., Ohtsuka, Y., *et al.* (2002). Ganglioside GM3 participates in the pathological conditions of insulin resistance. *J Biol Chem*, **277**, 3085-3092.

Uysal, K. T., Wiesbrock, S. M., Marino, M. W., and Hotamisligil, G. S. (1997). Protection from obesity-induced insulin resistance in mice lacking TNF-alpha function. *Nature*, **389**, 610-614.

Vainio, S., Heino, S., Mansson, J. E., Fredman, P., Kuismanen, E., Vaarala, O., and Ikonen, E. (2002). Dynamic association of human insulin receptor with lipid rafts in cells lacking caveolae. *EMBO Rep*, **3**, 95-100.

Virkamaki, A., Ueki, K., and Kahn, C. R. (1999). Protein-protein interaction in insulin signaling and the molecular mechanisms of insulin resistance. *J Clin Invest*, **103**, 931-943.

Wang, X. Q., Sun, P., and Paller, A. S. (2002). Ganglioside induces caveolin-1 redistribution and interaction with the epidermal growth factor receptor. *J Biol Chem*, **277**, 47028-47034.

Yamashita, T., Hashiramoto, A., Haluzik, M., Mizukami, H., Beck, S., Norton, A., Kono, M., Tsuji, S., Daniotti, J. L., Werth, N., *et al.* (2003). Enhanced insulin sensitivity in mice lacking ganglioside GM3. *Proc Natl Acad Sci USA*, **100**, 3445-3449.

5-2 Neuronal Cell Death in Glycosphingolipidoses

Yaacov Kacher and Anthony H. Futerman

Department of Biological Chemistry, Weizmann Institute of Science, Rehovot 76100, Israel

Summary. Glycosphingolipid (GSL) storage diseases (GSDs) are a group of monogenic disorders among a large family of lysosomal storage diseases, which include more than 40 different human disorders. GSDs are caused by the impaired catalytic activity of the lysosomal hydrolases responsible for GSL degradation, which leads to accumulation of undegraded GSLs in lysosomes. The clinical presentation of GSDs varies between disorders and within each disorder. The reason for this heterogeneity is not known, but almost certainly depends on the level and nature of the accumulating GSL and the down-stream biochemical and cellular pathways activated upon GSL accumulation. We now review what little is known about the correlation of the clinical progression and pathology with the underlying biochemistry of the GSDs.

Keywords. Glycosphingolipidoses, calcium, phospholipids, apoptosis, inflammation

1. Clinical description of the GSDs

The occurrence of GSDs is about 1 in 18,000 live births (Meikle et al., 1999), but, due to a lack of good diagnostic tools and non-specific and variable clinical presentations, incidents of GSDs may have been diagnosed as either symptoms from concurrent but different diseases or as another disease sharing similar symptoms. This would result in an underestimation of the disease prevalence. All GSDs except type 1 Gaucher disease, in which only visceral systems are affected, display neurological involvement. This neurological involvement is not surprising as GSLs are known to play a variety of essential roles in the central nervous system where they are highly enriched (Ledeen, 1992). The clinical symptoms by themselves do not provide a mechanistic connection between GSL accumulation and disease progression, but the similarity in many of the clinical symptoms between individual diseases may suggest common mechanisms of pathology. The major clinical symptoms of GSDs are summarized in Raas-Rothschild et al. (2004).

2. Neuronal cell dysfunction and death in GSDs

2.1 Altered calcium homeostasis

Over the past few years, our laboratory has outlined a mechanistic description for a major cellular pathway, intracellular Ca^{2+}-homeostasis, that is altered in at least three GSDs and in Niemann-Pick A disease (NPD-A), an LSD (lipid storage disease) caused by abnormal activity of acid sphingomyelinase. We initially observed that in those neurons with glucosylceramide (GlcCer) accumulation in neurons, a symptom of neuronopathic forms of Gaucher disease, more Ca^{2+} was released from intracellular stores in response to caffeine-treatment than untreated cells (Korkotian et al., 1999). This Ca^{2+}-release enhanced sensitivity to neurotoxic agents (Korkotian et al., 1999; Pelled et al., 2000). Since caffeine is an agonist of the ryanodine receptor (RyaR), which is a Ca^{2+}-release channel located in the endoplasmic reticulum (ER), we concluded that GlcCer accumulation in neurons enhances agonist-induced Ca^{2+}-release from the ER via the RyaR (Lloyd-Evans et al., 2003a), which is subsequently responsible for neuronal death. In addition, analysis of Ca^{2+}-release *in vitro* from microsomes demonstrated that only GlcCer among all the sphingolipids tested was able to stimulate agonist-induced Ca^{2+}-release (Lloyd-Evans et al., 2003a). Lyso-GSLs (Lloyd-Evans et al., 2003b) also stimulated Ca^{2+}-release, but via a different mechanism to GlcCer. Finally, a study on human brain tissue obtained post mortem from neuronopathic Gaucher disease patients demonstrated a correlation between levels of GlcCer accumulation, levels of Ca^{2+}-release, and disease phenotype (Pelled et al., 2005). Together, these findings suggest that defective Ca^{2+}-homeostasis might be one of the mechanisms responsible for at least some of the neuropathophysiology in neuronopathic forms of Gaucher disease.

We also observed altered brain Ca^{2+}-homeostasis in a Sandhoff disease mouse model, the Hexb mouse, although the mechanism is different to that observed in Gaucher disease. In brain microsomes obtained from Hexb-/- mice, the rate of Ca^{2+}-uptake into the ER, via the sarco/endoplasmic reticulum ATPase (SERCA), was dramatically reduced, but no difference in the rate of Ca^{2+}-release via the RyaR was observed (Pelled et al., 2003a). The reduced activity of SERCA was reversed when Hexb-/- mice were fed *N*-butyldeoxynojirimycin, an inhibitor of glycolipid synthesis that reduces levels of GM2 storage (Lachmann, 2003), which correlated with reduced levels of GM2 accumulation, delayed symptom onset and the increased life expectancy of these mice (Jeyakumar et al., 1999). This study suggests a mechanistic link between GM2 accumulation, reduced SERCA activity, neuronal cell death, and the survival of these mice. Since GM2 accumulates in a number of other LSDs is a secondary storage product (see below), this pathway may be involved in the pathology of other unrelated disorders.

Ca^{2+}-homeostasis is also altered in the ASM mouse, a model of Niemann-Pick A disease (Ginzburg and Futerman, 2005). The mechanism responsible for defective Ca^{2+}-homeostasis here is completely different from those observed in the other two disease models. Levels of SERCA protein expression are significantly reduced in the ASM-/- cerebellum, as are levels of the inositol 1,4,5-triphosphate receptor (IP_3R), the major Ca^{2+}-release channel in the cerebellum. These results suggest that dysregulated Ca^{2+}-homeostasis plays a role in neurodegeneration observed in a specific Purkinje cell population in NPD-A, but does not identify the lipid that initiates the death of this cell population.

Ca^{2+} helps regulate a variety of neuronal processes, including cell death, and axonal and dendritic growth, which require an optimal narrow range of Ca^{2+} levels. Our results indicate an important role for Ca^{2+} in the pathology of GSDs (Fig. 1). This hypothesis is supported by a study on a mouse model of GM1 gangliosidosis (Tessitore et al., 2004), in which neuronal cell death, initiated by GM1 accumulation in the ER, resulted in depletion of Ca^{2+} stores and subsequent activation of the unfolded protein response (UPR). This kind of apoptotic pathway might also be activated in other GSDs. In the UPR, global protein synthesis is suppressed, stress gene expression is activated, and apoptosis is induced (Kaneko and Nomura, 2003; Rutkowski and Kaufman, 2004). Although there is no evidence supporting UPR activation in other GSDs, this is an attractive mechanism to explain neuronal dysfunction and death.

2.2 Altered phospholipid metabolism

We also demonstrated that another major cellular pathway, phospholipid metabolism, is altered in some GSDs (Fig. 1). A significant elevation of phosphatidylcholine (PC) synthesis was observed in a mouse model of Gaucher disease, in a chemically-induced model of neuronopathic Gaucher disease (Bodennec et al., 2002), and in a chemically-induced model of Gaucher macrophages (Trajkovic-Bodennec et al., 2004). In the case of neurons, we suggest that GlcCer accumulation leads to enhanced PC synthesis which subsequently results in increased axonal growth rates. In contrast, reduced levels of phospholipid synthesis were observed in the Sandhoff mouse model (Buccoliero et al., 2004), which correlated with decreased rates of axonal growth (Pelled et al., 2003b). Our data therefore suggest a direct correlation between the type of accumulating GSL, its effect on phospholipid synthesis, and rates of axonal growth, and imply that altered phospholipid metabolism might also play a role in disease pathogenesis.

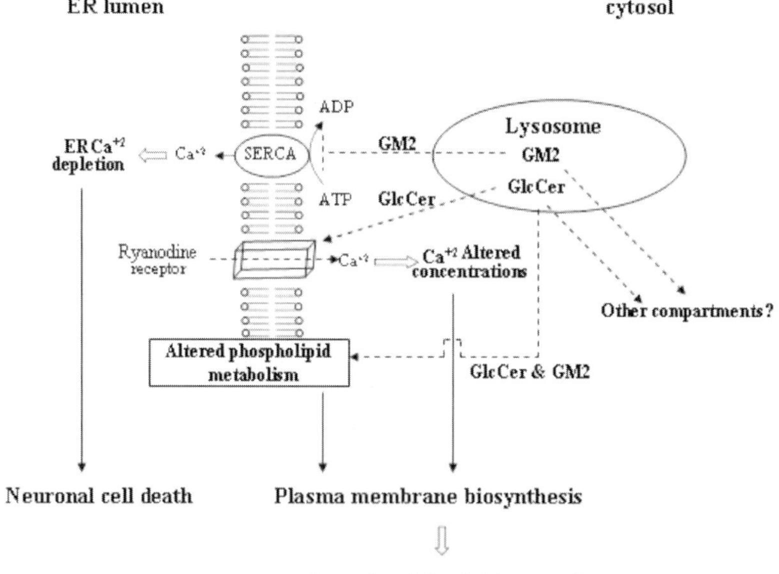

Fig. 1. The relationship between GSL storage and down-stream biochemical pathways.

2.3 Secondary GSL storage in GSDs

GSLs also accumulate as a secondary storage process in some LSDs (reviewed in Walkley, 2004). For instance, brain storage of gangliosides GM2 and GM3 was documented in NPD-A, in which the primary storage material is sphingomyelin, and in Niemann-Pick type C disease where the primary storage material is cholesterol. In addition, GM2 and GM3 accumulate in mucopolysaccharidoses types I and III, where the primary storage materials are dermatan sulphate and heparan sulphate. Interestingly, neurons in many with secondary ganglioside accumulation display ectopic dendrites and axonal spheroids resembling those observed in the gangliosidoses (Walkley, 1998; 2004). This raises the possibility that disrupting normal ganglioside processing can affect neurons, even when those molecules are not the primary accumulated storage material. However, the reason that gangliosides GM2 and GM3 may accumulate is largely unknown.

3. Other potential pathological mechanisms

Ceramide regulates apoptosis and a variety of other pathways involved in cell survival. Thus, altering levels of metabolites in the pathways of ceramide formation could, in principle, have dramatic effects on neuronal function (Buccoliero and Futerman, 2003). However, there is little evidence that ceramide levels are altered in the GSDs. Nevertheless, apoptosis has been observed in neuronal and other tissues in many GSDs (Tardy et al., 2004). For example, in Gaucher disease, apoptotic neurons were detected in the anterior horn and brainstem (Finn et al., 2000). In a mouse model of Sandhoff disease, DNA fragmentation was observed in the cerebral cortex, brainstem, cerebellum and spinal cord, and apoptosis was also observed in the brain of a human Krabbe disease patient (Jatana et al., 2002). Although some of the GSDs display activation of apoptotic pathways, it is unclear whether a direct relationship linking apoptotic cell death to the clinical symptoms exists. Moreover, it does not appear likely, based on the current state of research in this area, that ceramide is involved in the initiation of these apoptotic pathways.

Inflammation, a local response to cellular insult, may also play a role in pathology (Wada et al., 2000), and markers indicating inflammatory responses have been detected in a number of GSDs. Overexpression of genes associated with activated macrophages/microglia and astrocytes have been detected in mouse models of Sandhoff and Tay-Sachs diseases (Myerowitz et al., 2002; Wada et al., 2000), and progressive CNS inflammation is observed in Sandhoff, Tay-Sachs and GM1-gangliosidosis mouse models (Jeyakumar et al., 2003). Whether inflammation is a direct response to intracellular GSL accumulation or a response to apoptotic or injured cells (Jeyakumar et al., 2003) is still unknown. The latter could be caused by the changes described above in intracellular Ca^{2+}-levels, which would lead to neuronal cell death.

Another possible pathological mechanism could stem from altered lysosomal stability, integrity or permeability (Futerman and van Meer, 2004). Some lysosomal components, such as storage materials and hydrolases, might leak into the cytoplasm or extracellular space in GSDs (reviewed in Futerman and van Meer, 2004). This could then cause cell damage. Interestingly, Ca^{2+} plays an important role (Jaiswal et al., 2002) in regulating lysosomal fusion with the plasma membrane, which then secretes several lysosomal proteins, such as cathepsin B (Linke et al., 2002).

Yet another explanation for GSD pathology implicates altered intracellular transport to, or from, lysosomes. Recent data supports this. Chen et al showed that short-acyl chain derivatives of lactosylceramide specifically targeted endosomes and lysosomes in cells from sphingolipidoses patients, but targeted the Golgi in normal cells (Chen et al., 1999). This indicates that there may be a common defect in lipid sorting and transport among the different sphingolipidoses (Marks and Pagano, 2002; Sillence and Platt, 2003). Little is known about possible defective protein sorting in GSDs. However, intracellular trafficking in GSDs might play an important role in GSD pathogenesis.

In summary, there are many unanswered questions regarding the biochemical triggers and down-stream pathways that cause GSD pathology. However, novel biochemical pathways are identified and clarify the neuronal role of GSLs that lead to altered rates of cell growth or even neuronal death are beginning to emerge. Additional research should provide an answer relating these pathways to the known neurological symptoms in the GSDs.

4. References

Bodennec, J., Pelled, D., Riebeling, C., Trajkovic, S., and Futerman, A. H. (2002). Phosphatidylcholine synthesis is elevated in neuronal models of Gaucher disease due to direct activation of CTP:phosphocholine cytidylyltransferase by glucosylceramide. *Faseb J*, **16**, 1814-1816.

Buccoliero, R., Bodennec, J., Van Echten-Deckert, G., Sandhoff, K., and Futerman, A. H. (2004). Phospholipid synthesis is decreased in neuronal tissue in a mouse model of Sandhoff disease. *J Neurochem*, **90**, 80-88.

Buccoliero, R., and Futerman, A. H. (2003). The roles of ceramide and complex sphingolipids in neuronal cell function. *Pharmacol Res*, 47, 409-419.

Chen, C. S., Patterson, M. C., Wheatley, C. L., O'Brien, J. F., and Pagano, R. E. (1999). Broad screening test for sphingolipid-storage diseases. *Lancet*, **354**, 901-905.

Finn, L. S., Zhang, M., Chen, S. H., and Scott, C. R. (2000). Severe type II Gaucher disease with ichthyosis, arthrogryposis and neuronal apoptosis: molecular and pathological analyses. *Am J Med Genet*, **91**, 222-226.

Futerman, A. H., and van Meer, G. (2004). The cell biology of lysosomal storage disorders. *Nat Rev Mol Cell Biol*, **5**, 554-565.

Ginzburg, L. and Futerman, A.H. Defective calcium homeostasis in the cerebellum in a mouse model of Niemann-Pick A disease. Submitted for publication.

Jaiswal, J. K., Andrews, N. W., and Simon, S. M. (2002). Membrane proximal lysosomes are the major vesicles responsible for calcium-dependent exocytosis in nonsecretory cells. *J Cell Biol*, **159**, 625-635.

Jatana, M., Giri, S., and Singh, A. K. (2002). Apoptotic positive cells in Krabbe brain and induction of apoptosis in rat C6 glial cells by psychosine. *Neurosci Let*, **330**, 183-187.

Jeyakumar, M., Butters, T. D., Cortina-Borja, M., Hunnam, V., Proia, R. L., Perry, V. H., Dwek, R. A., and Platt, F. M. (1999). Delayed symptom onset and increased life expectancy in Sandhoff disease mice treated with N-butyldeoxynojirimycin. *Proc Natl Acad Sci U S A*, **96**, 6388-6393.

Jeyakumar, M., Thomas, R., Elliot-Smith, E., Smith, D. A., van der Spoel, A. C., d'Azzo, A., Perry, V. H., Butters, T. D., Dwek, R. A., and Platt, F. M. (2003). Central nervous system inflammation is a hallmark of pathogenesis in mouse models of GM1 and GM2 gangliosidosis. *Brain*, **126**, 974-987.

Kaneko, M., and Nomura, Y. (2003). ER signaling in unfolded protein response. *Life Sci*, **74**, 199-205.

Korkotian, E., Schwarz, A., Pelled, D., Schwarzmann, G., Segal, M., and Futerman, A. H. (1999). Elevation of intracellular glucosylceramide levels results in an increase in endoplasmic reticulum density and in functional calcium stores in cultured neurons. *J Biol Chem*, **274**, 21673-21678.

Lachmann, R. H. (2003). Miglustat. Oxford GlycoSciences/Actelion. *Curr Opin Investig Drugs*, **4**, 472-479.

Ledeen RW, W. G. (1992). Ganglioside function in the neuron. *Trends Glycosci Glycotechnol*, **4**, 174-187.

Linke, M., Herzog, V., and Brix, K. (2002). Trafficking of lysosomal cathepsin B-green fluorescent protein to the surface of thyroid epithelial cells involves the endosomal/lysosomal compartment. *J Cell Sci*, **115**, 4877-4889.

Lloyd-Evans, E., Pelled, D., Riebeling, C., Bodennec, J., de-Morgan, A., Waller, H., Schiffmann, R., and Futerman, A. H. (2003a). Glucosylceramide and glucosylsphingosine modulate calcium mobilization from brain microsomes via different mechanisms. *J Biol Chem*, **278**, 23594-23599.

Lloyd-Evans, E., Pelled, D., Riebeling, C., and Futerman, A. H. (2003b). Lyso-glycosphingolipids mobilize calcium from brain microsomes via multiple mechanisms. *Biochem J*, **375**, 561-565.

Marks, D. L., and Pagano, R. E. (2002). Endocytosis and sorting of glycosphingolipids in sphingolipid storage disease. *Trends Cell Biol*, **12**, 605-613.

Meikle, P. J., Hopwood, J. J., Clague, A. E., and Carey, W. F. (1999). Prevalence of lysosomal storage disorders. *JAMA*, **281**, 249-254.

Myerowitz, R., Lawson, D., Mizukami, H., Mi, Y., Tifft, C. J., and Proia, R. L. (2002). Molecular pathophysiology in Tay-Sachs and Sandhoff diseases as revealed by gene expression profiling. *Hum Mol Genet*, **11**, 1343-1350.

Pelled, D., Lloyd-Evans, E., Riebeling, C., Jeyakumar, M., Platt, F. M., and Futerman, A. H. (2003a). Inhibition of calcium uptake via the sarco/endoplasmic reticulum Ca2+-ATPase in a mouse model of Sandhoff disease and prevention by treatment with N-butyldeoxynojirimycin. *J Biol Chem*, **278**, 29496-29501.

Pelled, D., Riebeling, C., van Echten-Deckert, G., Sandhoff, K., and Futerman, A. H. (2003b). Reduced rates of axonal and dendritic growth in embryonic hippocampal neurones cultured from a mouse model of Sandhoff disease. *Neuropathol Appl Neurobiol*, **29**, 341-349.

Pelled, D., Shogomori, H., and Futerman, A. H. (2000). The increased sensitivity of neurons with elevated glucocerebroside to neurotoxic agents can be reversed by imiglucerase. *J Inherit Metab Dis*, **23**, 175-184.

Pelled, D., Trajkovic-Bodennec, S., Lloyd-Evans, E., Sidransky, E., Schiffmann, R., and Futerman, A. H. (2005). Enhanced calcium release in the acute neuronopathic form of Gaucher disease. *Neurobiol Dis*, **18**, 83-88.

Raas-Rothschild, A., Pankova-Kholmyansky, I., Kacher, Y., and Futerman, A. H. (2004). Glycosphingolipidoses: beyond the enzymatic defect. *Glycoconj J*, **21**, 295-304.

Rutkowski, D. T., and Kaufman, R. J. (2004). A trip to the ER: coping with stress. *Trends Cell Biol*, **14**, 20-28.

Sillence, D. J., and Platt, F. M. (2003). Storage diseases: new insights into sphingolipid functions. *Trends Cell Biol*, **13**, 195-203.

Tardy, C., Andrieu-Abadie, N., Salvayre, R., and Levade, T. (2004). Lysosomal storage diseases: is impaired apoptosis a pathogenic mechanism? *Neurochem Res*, **29**, 871-880.

Tessitore, A., del, P. M. M., Sano, R., Ma, Y., Mann, L., Ingrassia, A., Laywell, E. D., Steindler, D. A., Hendershot, L. M., and d'Azzo, A. (2004). GM1-ganglioside-mediated activation of the unfolded protein response causes neuronal death in a neurodegenerative gangliosidosis. *Mol Cell*, **15**, 753-766.

Trajkovic-Bodennec, S., Bodennec, J., and Futerman, A. H. (2004). Phosphatidylcholine metabolism is altered in a monocyte-derived macrophage model of Gaucher disease but not in lymphocytes. *Blood Cells Mol Dis*, **33**, 77-82.

Wada, R., Tifft, C. J., and Proia, R. L. (2000). Microglial activation precedes acute neurodegeneration in Sandhoff disease and is suppressed by bone marrow transplantation. *Proc Natl Acad Sci U S A*, **97**, 10954-10959.

Walkley, S. U. (1998). Cellular pathology of lysosomal storage disorders. *Brain Pathol*, **8**, 175-193.

Walkley, S. U. (2004). Secondary accumulation of gangliosides in lysosomal storage disorders. *Semin Cell Dev Biol*, **15**, 433-444.

5-3 Endocytic Trafficking of Glycosphingolipids in Sphingolipidoses

Amit Choudhury, David L. Marks, and Richard E. Pagano

Department of Biochemistry and Molecular Biology, Thoracic Diseases Research Unit, Mayo Clinic and Foundation, Rochester, Minnesota 55905, USA

Summary. Previous studies from our laboratory demonstrated that membrane transport along the endocytic pathway is perturbed in a broad collection of sphingolipid storage disease (SLSD) fibroblasts. Namely, we studied a fluorescent analog of lactosylceramide (LacCer), which is endocytosed from the plasma membrane (PM) and subsequently transported to the Golgi complex of normal human skin fibroblasts (HSFs); however, in SLSD cells, transport to the Golgi apparatus is blocked and the lipid accumulates in endosomes. These findings led us to study the endocytic itinerary of LacCer (and other molecules) in normal *vs* SLSD fibroblasts. In this Chapter we describe the endocytic itinerary of SLs in normal fibroblasts and the perturbation of this lipid trafficking in SLSD fibroblasts. We also summarize our results demonstrating that the observed disruption of membrane trafficking in SLSDs is a consequence of elevated cholesterol which inhibits the function of selected rab proteins. Finally we highlight our attempts to correct the SLSD phenotype by stimulation of membrane traffic in the disease fibroblasts.

Keywords. glycosphingolipids, trafficking, rab GTPases, sphingolipid storage diseases

1. Introduction

Sphingolipids (SLs) are essential components of all animal cell membranes—they are synthesized at the endoplasmic reticulum and Golgi apparatus and are subsequently transported to the PM where they are highly enriched (Chatterjee 1998; Simons and Vaz 2004). SLs and cholesterol interact with each other and appear to be co-regulated. For example, if SLs at the PM are degraded by an enzymatic treatment, cholesterol redistributes from the cell surface to intracellular membranes until SLs are repleted at the PM (Slotte and Bierman 1988). Once delivered to the PM, these molecules undergo endocytosis and recycling, but over time a portion of these lipids is also degraded in endosomes/lysosomes. SL storage diseases (SLSDs) are a subset of lysosomal storage disorders in which SLs accumulate in various tissues in the body, often resulting in neurological (e.g., dementia; developmental delay) as well as somatic (e.g., hepatomegaly) pathology (Scriver et al. 2001). In most of these disorders, this abnormal lipid accumulation is due to a defect in a hydrolytic enzyme or activator protein required for the catabolism of a particular lipid. For example, in Niemann Pick disease, type A (NP-A), there is a deficiency in acid sphingomyelinase, resulting in an accumulation of sphingomyelin. In Niemann Pick, type C (NP-C) and mucolipidosis, type IV (ML-IV) diseases, lipids also accumulate; however, lysosomal hydrolase activities are normal in these individuals and the primary defects appear to be in proteins involved in the transport of lipids and proteins through the cell (Bach 2001; Chen et al. 1998; Patterson et al. 2001). In many SLSD cell types, cholesterol often accumulates along with SLs, presumably due to interactions of SLs with cholesterol (Marks and Pagano 2002).

Recent studies in our laboratory have focused on the endocytosis and intracellular transport of a fluorescent analog of lactosylceramide (BODIPY-LacCer). In the course of these studies we identified several membrane trafficking defects that were associated with multiple SLSDs. In the present Chapter, we present an overview of our studies on the endocytic itinerary of this GSL analog in these cell types. We also focus on the mechanisms underlying the defective intracellular sorting and transport in SLSDs. Finally, we highlight our attempts to stimulate membrane trafficking in cells by rab protein over-expression as a potential method for correcting the SLSD phenotype *in vitro* and discuss the prospects for extending this approach *in vivo*.

2. Endocytic itinerary of GSLs (Fig. 1)

Endocytosis occurs when a portion of the PM pinches off from the cell surface, forming an intracellular vesicle. This process serves many important cellular functions including the uptake of nutrients, the regulation of cell surface receptor levels, and the regulation of cell polarity. Multiple mechanisms of endocytosis exist, the most extensively characterized mechanism being clathrin-mediated endocytosis, in which coat adaptors recruit receptors into coated areas of the PM and the clathrin lattice forces the deformation of the membrane and vesicle budding (Traub 2003). Recently, alternative and distinct mechanisms of endocytosis that are *clathrin-independent* have come under study (Johannes and Lamaze 2002; Kirkham and Parton 2005; Marks et al. 2005). One non-clathrin mechanism is uptake *via* caveolae—flask shaped invaginations (40-80 nm in diameter) at the PM that are enriched in SLs and cholesterol, and are associated with the protein caveolin-1 (Cav1) (Cohen et al. 2004; Mineo and Anderson 2001). Caveolar endocytosis appears to be important for cell entry and intracellular delivery of some toxins, viruses, and bacteria, as well as some growth factors and other circulating proteins (Duncan et al. 2002; Lencer et al. 1999; Norkin 2001; Pelkmans and Helenius 2002; Smith and Helenius 2004). Other clathrin-independent mechanisms include uptake of the interleukin 2 receptor (IL-2R) which is internalized by a clathrin-independent, dynamin-dependent, RhoA-regulated mechanism (Lamaze et al. 2001), and fluid phase endocytosis which is clathrin-independent, dynamin-independent, and Cdc42-dependent (Sabharanjak et al. 2002).

An important tool for studying the endocytic itinerary of SLs has been the use of wild type (WT) and dominant negative (DN) constructs of several rab proteins. Rab proteins are a family of small GTPases involved in tethering an incoming vesicle to a target organelle. Human cells contain more than 60 different Rabs, each involved in specific steps in vesicle trafficking (Zerial and McBride 2001). These proteins and their "effectors," are thought to be primary determinants for maintaining compartmental specificity in the docking and fusion reactions which occur when an intracellular vesicle interacts with other organelles [reviewed in (Seabra et al. 2002; Zerial and McBride 2001)]. In our studies, we expressed DN rab proteins known to block specific steps of vesicular transport in order to delineate which steps are involved in the transport of a SL analog as it moves through the cell (Choudhury et al. 2002; Choudhury et al. 2004).

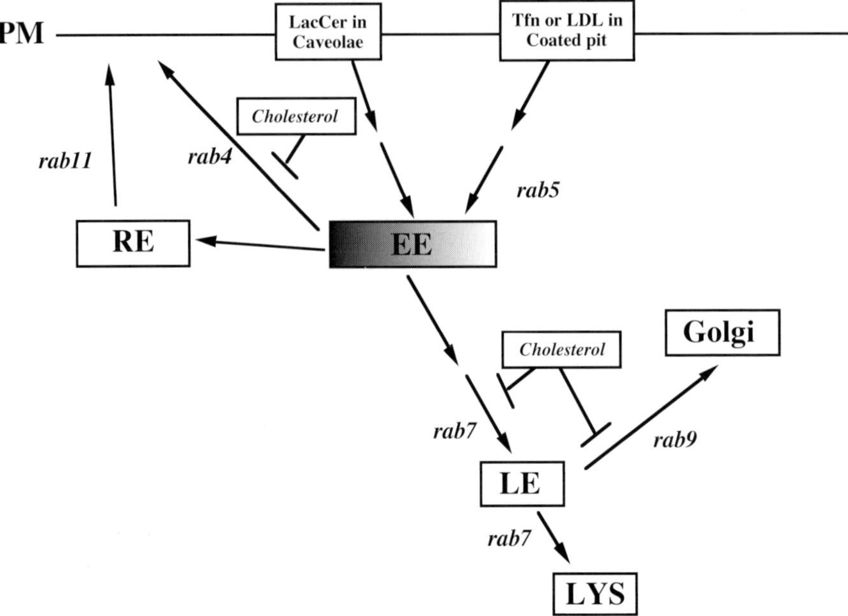

Fig. 1. Endocytic itinerary of fluorescent lactosylceramide (LacCer) in cultured fibroblasts. EE, early endosome; RE, recycling endosome; LE, late endosome; LYS, lysosome. Following internalization *via* caveolae (Step I), LacCer merges with Tfn internalized through the coated pit pathway (Step II) at the EEs where it is fractionated into two pools. One pool recycles (Step III) back to the plasma membrane (PM), while the second pool is transported to the Golgi apparatus (Step IV). The Rab GTPases, which participate in the intracellular transport of LacCer are also shown. Rab 4, 7, and 9 are specifically inhibited by elevated cholesterol in several lipid storage disease fibroblasts (see Text).

2.1 Initial internalization and transport to early endosomes

To study the endocytic mechanism of fluorescent SLs in cultured fibroblasts, we typically incubate cells with the lipid analog at low temperature to label the PM, and then raise the temperature to 37°C to allow endocytosis to occur [see (Marks et al. 2005) for methodologies]. The samples are then chilled to 10°C and any fluorescent lipid that was not endocytosed from the PM is removed by incubating the cells with defatted BSA in a process termed "back-exchange" (Martin and Pagano 1994). Internalization of fluorescent SLs can then be quantified by fluorescence microscopy or by biochemical analysis of cell-associated fluorescence (Chen et al. 1998; Chen et al. 1997; Martin et al. 1993). Using pathway-specific inhibitors, dominant negative (DN) proteins, and co-localization with markers known to be internalized by a particular mechanism, we found that BODIPY-LacCer is internalized almost exclusively *via* caveolae in multiple cell types including HSFs (Fig. 1, Step I) (Puri et al. 2001; Sharma et al. 2003; Singh et al. 2003). Other fluorescently tagged glycosphingolipid analogs (e.g., trihexosylceramide, GM_1 ganglioside) were similarly found to be internalized via caveolae (Singh et al. 2003). Internalization of the endogenous glycosphingolipid, GM_1 ganglioside, from the PM can be studied using fluorescently tagged cholera toxin B subunit (CtxB), which binds to GM_1 at the PM. Following endocytosis, non-internalized CtxB is removed from the cell surface by acid stripping (Singh et al. 2003). CtxB has been shown to be internalized via caveolae in some cell types (Orlandi and Fishman 1998; Puri et al. 2001), but may be endocytosed by other mechanisms in other cell types, especially those deficient in Cav1 (Kirkham and Parton 2005; Singh et al. 2003; Torgersen et al. 2001).

Following internalization of BODIPY-LacCer *via* caveolae, the lipid is rapidly delivered (within ~ 5 min) to EEA1 positive early endosomes (EEs) where it merges with markers internalized *via* clathrin-mediated endocytosis (e.g., Tfn; see Fig. 1, Step II) (Sharma et al. 2003). Within EEs, endocytosed lipids and proteins can segregate into distinct domains—in particular, the fluorescent LacCer fractionates into two distinct pools that can be readily distinguished from one another on the basis of their concentration-dependent fluorescence emission (Sharma et al. 2003). In normal HSFs, the more concentrated pool of BODIPY-LacCer (fluoresces red due to excimer formation) recycles back to the PM along with the endocytosed Tfn, while the less concentrated pool (fluoresces green due to monomers) is transported to the late endosomes and Golgi complex. These two pathways are discussed in the subsequent sections.

2.2 Transport along the endocytic recycling pathway (Fig. 1, Step III)

The pathways taken by endocytosed lipids and proteins for recycling back to the PM can involve several sorting steps, distinct organelles, and specific rab GTPases (Fig. 1). After delivery to EEs, ~50% of endocytosed LacCer analog is recycled back to the cell surface, while the rest is delivered via the late endosome to the Golgi complex (Choudhury et al. 2002; Choudhury et al. 2004; Sharma et al. 2003). Of the recycling pool of BODIPY-LacCer, in normal HSFs the majority is rapidly returned directly to the PM from EEs by a rab4-dependent process, whereas the remaining fraction is delivered to the recycling endosome (RE), and is recycled to the PM by rab11-dependent transport (Choudhury et al. 2004). In normal HSFs the $t_{1/2}$ for LacCer recycling is ~ 8 min, while in NP-A and NP-C fibroblasts (hereafter, NPFs) recycling is delayed ($t_{1/2}$ is ~30-40 min). We expressed DN rab proteins to inhibit endogenous rab function and showed that in NPFs, the rab4-dependent pathway is blocked and instead recycling occurs only via the rab11-dependent pathway (Choudhury et al. 2004).

The molecular mechanism responsible for determining whether LacCer utilizes the rapid (rab4) or slow (rab11) recycling pathway is not known. One intriguing clue comes from studying the organization of rab proteins on EEs. In a seminal study, Zerial and co-workers (Sonnichsen et al. 2000) demonstrated that multiple rab proteins localized to the same endosome can segregate into distinct subdomains resulting in a "rab mosaic." For example, while both rab4 and rab5 are found on EEs, they are segregated into distinct subdomains of the endosome. In our studies using NPFs, cells were co-stained for EEA1 (to identify EEs) and endogenous rab4, and then examined by confocal microscopy. The distribution of rab4 on EEs of NPFs frequently had a globular appearance, whereas rab4 was generally present in tubular extensions in normal HSFs (Choudhury et al. 2004). No perturbation of rab11 distribution was seen in NPFs *vs* HSFs. These data strongly suggest that the organization of rab4 (but not rab11) was perturbed on the endosomes of NPFs. Additional studies demonstrated that cholesterol depletion of NPFs restored the organization of rab4 on EEs to that seen in normal HSFs. Cholesterol depletion also enhanced the recycling of LacCer in NPFs. In contrast, elevation of cholesterol in *normal HSFs* decreased LacCer recycling. Elevated intracellular cholesterol has emerged as an important factor in the disruption of membrane trafficking in SLSDs as will be further described below.

2.3 Transport of SL analogs from EEs to the Golgi apparatus (Fig. 1, Step IV)

The "second pool" of LacCer, which is not recycled back to the PM, is transported to the Golgi apparatus. In HSFs, the targeting to the Golgi complex is dependent on microtubules and PI3-kinases, and is inhibited by expression of DN-rab7 and -rab9 (rab7 is involved in transport from EEs to LEs and LEs to lysosomes, while rab9 is involved in LE to Golgi transport). Furthermore, Golgi targeting of GSLs is independent of rab11. These results indicate that LacCer passes through LEs *en route* to the Golgi apparatus. In SLSD cells, Golgi targeting is blocked, leading to accumulation of fluorescent LacCer in endosomes.

Since most SLSDs have a unique biochemical defect with respect to SL degradation (except for ML- IV and NP-C cells; see above), it is not obvious what common feature is responsible for the altered intracellular transport of LacCer in multiple SLSD cell types. We speculated that stored cholesterol might be responsible for the defects in intracellular sorting of BODIPY-LacCer since cholesterol is known to co-accumulate with SLs in some SLSDs (Patterson et al. 2001; Puri et al. 1999; Schuchman and Desnick 2001) and since different SLs interact with cholesterol to various extents (Brown and London 1998; Brown and London 2000). Indeed there is increased filipin staining (indicative of high levels of intracellular cholesterol) in all of the SLSD cell types that accumulate BODIPY-LacCer in endosomes, except for GM_2 gangliosidosis cells in which filipin staining is similar to that in control fibroblasts (Puri et al. 1999). To examine the possible effects which cholesterol might have on the intracellular targeting of LacCer, SLSD cells were grown under conditions that deplete cellular cholesterol and then pulse-labeled with the fluorescent lipid. Cholesterol depletion restores Golgi targeting of both fluorescent LacCer and endogenous GM_1 (as monitored by fluorescent CtxB) (Puri et al. 1999; Puri et al. 2001). Depletion of cellular cholesterol also reduces the punctate endosomal pattern of LacCer fluorescence seen in SLSD cells. Interestingly, incubation of normal HSFs with high levels of LDL resulted in an elevation of cellular cholesterol, a loss of BODIPY-LacCer targeting to the Golgi, and a concomitant appearance of punctate endosomal structures similar to that seen in SLSD cells (Puri et al. 1999). These results suggest that cholesterol plays a major role in modulating the intracellular sorting and transport of the LacCer and endogenous GM_1 in normal *vs* SLSD fibroblasts.

3. Perturbation of the function of selected Rab GTPases by cholesterol

As discussed above, stored cholesterol in SLSD cells perturbs rab4-dependent lipid recycling of LacCer as well as the rab7- and rab9-dependent transport of LacCer and CtxB to the Golgi apparatus. Furthermore, high levels of cholesterol alter the rab mosaic on EEs. Interestingly, no effect was seen on rab5 or rab11 function in the same cells (Choudhury et al. 2004), suggesting that only selected rab proteins are affected by stored cholesterol. One possible explanation for these findings is that the levels of rab4, 7, and 9 were reduced in SLSD cells. To examine this possibility, we compared the levels of rabs 4, 5, 7, 9, and 11 in normal HSFs with those found in SLSD (NP-A, NP-C, and GM$_1$ gangliosidosis) fibroblasts.   No differences were found in the *cytosolic levels* of these rabs, however, rabs 4, 7 and 9 were significantly elevated in membrane fractions from GM$_1$ gangliosidosis cells (Fig. 2) and from NP-A and NP-C fibroblasts (data not shown).   This result suggests that lipid storage increases the amount of rabs 4, 7, and 9 on the endosomes even though their function is inhibited.

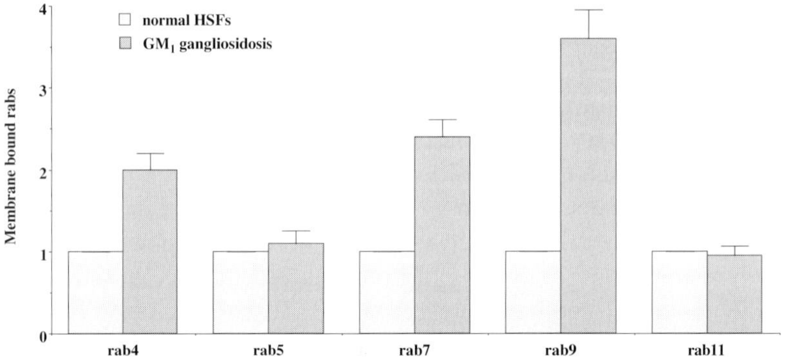

Fig. 2. Levels of membrane bound rabs in normal human skin (HSFs) *vs* **GM$_1$ gangliosidosis fibroblasts.**   A membrane fraction was prepared from cell homogenates and the amount of each rab was then quantified by Western blotting and expressed relative to that found in normal HSFs.   The levels of rabs4, 7 and 9 were elevated 2-4 fold in the membrane fractions of the GM$_1$ gangliosidosis cells relative to control cells, whereas no changes were seen for rab5 and rab11. Values represent mean ± SD (n=3, independent experiments).

Rab proteins in the active (GTP-bound) state are primarily membrane-associated and regulate vesicular transport from donor to target membranes where the GTP is then hydrolyzed to GDP.   Subsequently, the GDP-bound (inactive) rab is extracted from the target membrane into the cytosol by a chaperone known as rab GDP dissociation inhibitor (rab-GDI) (Seabra et al. 2002).   Rab GDI maintains the retrieved rabs in the GDP-bound state in the cytosol and facilitates their delivery back to the donor membrane where they are recharged with GTP.   Gruenberg and colleagues (Lebrand et al. 2002) previously showed that accumulation of cholesterol in Les, induced by treatment with the drug U18666A, increased the association of rab7 (but not rab5) on endosomes and interfered with its extraction from the target membrane by the GDI, perhaps as a result of altered membrane fluidity.

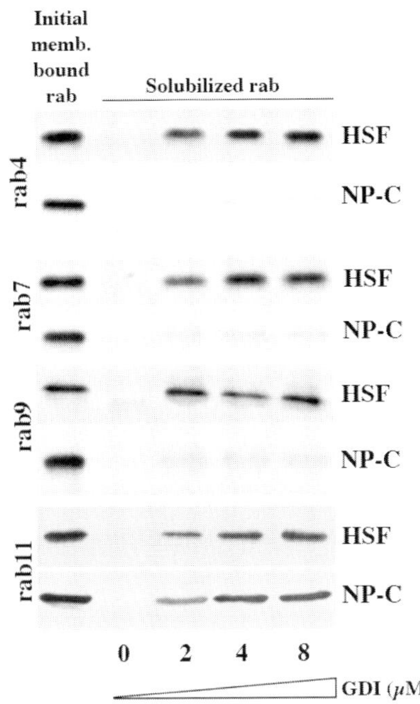

Fig. 3. Extraction of membrane bound rabs from endosomes by GDI. An enriched endosome fraction was prepared from HSFs and NP-C fibroblasts and the preparations normalized for equal amounts of rabs4, 7, 9 and 11. Samples were subsequently incubated with 0, 2, 4, or 8 µM GST-GDI. The GST-GDI bound rab protein was recovered on glutathione beads and analyzed by Western blotting using antibodies against the indicated rab. Note the resistance of rabs 4, 7, 9 to extraction by GST-GDI in NP-C compared to control HSFs.

These observations led us to examine the role of stored cholesterol in NPFs on the GDI-mediated extraction of various rab proteins (Choudhury et al. 2004). We focused on rabs4, -7, and -9 since LacCer transport mediated by these rabs was disrupted in several different SLSD cell types including NPFs (Fig. 1). Enriched endosome fractions were prepared from NPC and normal HSF cell lysates. We then measured the ability of rabs4, -7, -9 and -11 to be extracted from these endosome membranes by increasing concentrations of rab-GDI (Fig. 3). Extraction of rabs4, 7, and 9 was severely inhibited in NP-C endosomes relative to endosomes isolated from control HSFs. In contrast, the GDI-mediated extraction of rab5 (data not shown) and rab11 (Fig. 3) was not affected. We also pretreated *NP-C endosomes* with methyl β-cyclodextrin (mβ-CD) to extract cholesterol prior to incubating the endosomes with GDI. This treatment restored the extraction of rab4 by GDI (Choudhury et al. 2004). Finally, we increased the cholesterol content of endosomes from *normal HSFs* by treating them with a complex of mβ-CD and cholesterol. This treatment dramatically reduced the extraction of rab4 (without affecting rab11 extraction) (Choudhury et al. 2004). Together, these results suggest that elevated membrane cholesterol selectively inhibits GDI-mediated extraction of rab4, -7, and -9 from endosomes. Further studies will be required to determine the molecular basis for this inhibition. In principle, this could result from a direct effect of endosomal cholesterol on the affinity of the affected rabs for the endosome. Alternatively, cholesterol may exert its effect on selected rabs indirectly by modulating the activity of a rab effector protein, which regulates GDI extraction. Experiments are currently in progress to test these and other possibilities.

4. Rab over-expression and correction of the SLSD phenotype

Since the function of selected rab proteins was inhibited by elevated cholesterol in SLSD cells (Fig. 1), we examined the effect of over expressing wild type (WT) rabs4, 7, or 9 as a potential method to overcome the block in membrane transport in SLSD cells [(Choudhury et al. 2004) and unpublished studies). When NP-C cells were transfected with WT rab7 or rab9 (but not rab11) and then pulse-labeled with fluorescent LacCer, there was a restoration of fluorescent LacCer and endogenous GM_1 targeting to the Golgi apparatus and a reduction in endosomal accumulation of the fluorescent lipid in 60–70% of the transfected cells (Choudhury et al. 2004). Importantly, NP-C cells transfected with WT rab7 or rab9 constructs showed a dramatic reduction in cellular cholesterol, while no such effects were seen in cells over-expressing DN constructs of rab7 or rab9. In recent unpublished studies, we have also seen a dramatic reduction in cellular cholesterol in NP-C cells following over-expression of WT rab4. These results suggest that elevated levels of WT rab4, 7 or 9 may alleviate the block in membrane transport, at least in NPC cells. Stimulation of membrane transport by rab over-expression could facilitate transport of excess endosomal cholesterol to the ER for esterification, or promote cholesterol (and cholesterol ester) secretion into the medium of transfected cells.

In recent experiments we used protein transduction to elevate the levels of rab9 in NP-C cells (Narita et al. 2005). This treatment was much more efficient than DNA-based transfection of rab9. Nearly 100% of the NPC fibroblasts in cell culture were transduced with rab9 and showed a reduction in cholesterol. This method allowed us to examine the fate of cholesterol in rab9 over-expressing cells by biochemical analysis. These studies indicated that some of the reduction in cellular cholesterol could be accounted for by cholesterol esterification; however, most of the reduction was due to secretion of cholesterol (and cholesterol ester) into the bathing medium of the cells. The transporters involved in cholesterol secretion from the rab9 over-expressing cells are not currently known.

Our studies of rab over-expression demonstrate that stimulation of membrane transport by rab over-expression is effective in clearing stored cholesterol from several different SLSD cell types. We are currently evaluating the principle that modulation of membrane traffic is a useful strategy for treatment of lipid storage disorders by mating transgenic mice over-expressing rab9 to an SLSD mouse model.

Acknowledgement. Supported by USPHS Grant GM60934 to REP and a grant from the National Niemann Pick Disease Foundation. AC was supported by a fellowship from the American Heart Association.

5. References

Bach, G. (2001) Mucolipidosis type IV. *Mol Genet Metab*, **73**, 197-203.

Brown, D. A. and London, E. (1998) Functions of lipid rafts in biological membranes. *Annu Rev Cell Dev Biol*, **14**, 111-136.

Brown, D. A. and London, E. (2000) Structure and function of sphingolipid- and cholesterol-rich membrane rafts. *J Biol Chem*, **275**, 17221-17224.

Chatterjee, S. (1998) Sphingolipids in atherosclerosis and vascular biology. *Arterioscler Thromb Vasc Biol*, **18**, 1523-1533.

Chen, C. S., Bach, G. and Pagano, R. E. (1998) Abnormal transport along the lysosomal pathway in mucolipidosis, type IV disease. *Proc Natl Acad Sci USA*, **95**, 6373-6378.

Chen, C. S., Martin, O. C. and Pagano, R. E. (1997) Changes in the spectral properties of a plasma membrane lipid analog during the first seconds of endocytosis in living cells. *Biophys J*, **72**, 37-50.

Chen, C. S., Patterson, M. C., Wheatley, C. L., O'Brien, J. F. and Pagano, R. E. (1999) Broad screening test for sphingolipid-storage diseases. *Lancet*, **354**, 901-905.

Choudhury, A., Dominguez, M., Puri, V., Sharma, D. K., Narita, K., Wheatley, C. W., Marks, D. L. and Pagano, R. E. (2002) Rab proteins mediate Golgi transport of caveola-internalized glycosphingolipids and correct lipid trafficking in Niemann-Pick C cells. *J Clin Invest*, **109**, 1541-1550.

Choudhury, A., Sharma, D. K., Marks, D. L. and Pagano, R. E. (2004) Elevated Endosomal Cholesterol Levels in Niemann-Pick Cells Inhibit Rab4 and Perturb Membrane Recycling. *Mol Biol Cell*, **15**, 4500-4511.

Cohen, A. W., Hnasko, R., Schubert, W. and Lisanti, M. P. (2004) Role of caveolae and caveolins in health and disease. *Physiol Rev*, **84**, 1341-1379.

Duncan, M. J., Shin, J. S. and Abraham, S. N. (2002) Microbial entry through caveolae: variations on a theme. *Cell Microbiol*, **4**, 783-791.

Johannes, L. and Lamaze, C. (2002) Clahrin-dependent or not: Is it still the question? *Traffic, 3*, 443-451.

Kirkham, M. and Parton, R. G. (2005) Clathrin-independent endocytosis: New insights into caveolae and non-caveolar lipid raft carriers. *Biochim Biophys Acta*

Lamaze, C., Dujeancourt, A., Baba, T., Lo, C. G., Benmerah, A. and Dautry-Varsat, A. (2001) Interleukin 2 receptors and detergent-resistant membrane domains define a clathrin-independent endocytic pathway. *Mol Cell, 7*, 661-671.

Lebrand, C., Corti, M., Goodson, H., Cosson, P., Cavalli, V., Mayran, N., Faure, J. and Gruenberg, J. (2002) Late endosome motility depends on lipids via the small GTPase Rab7. *Embo J, 21*, 1289-1300.

Lencer, W. I., Hirst, T. R. and Holmes, R. K. (1999) Membrane traffic and the cellular uptake of cholera toxin. *Biochim Biophys Acta, 1450*, 177-190.

Marks, D. L. and Pagano, R. E. (2002) Endocytosis and sorting of glycosphingolipids in sphingolipid storage disease. *Trends Cell Biol, 12*, 605-613.

Marks, D. L., Singh, R. D., Choudhury, A., Wheatley, C. L. and Pagano, R. E. (2005) Use of fluorescent sphingolipid analogs to study lipid transport along the endocytic pathway. *Methods, 36*, 186-195.

Martin, O. C., Comly, M. E., Blanchette-Mackie, E. J., Pentchev, P. G. and Pagano, R. E. (1993) Cholesterol deprivation affects the fluorescence properties of a ceramide analog at the Golgi apparatus of living cells. *Proc Natl Acad Sci USA, 90*, 2661-2665.

Martin, O. C. and Pagano, R. E. (1994) Internalization and sorting of a fluorescent analog of glucosylceramide to the Golgi apparatus of human skin fibroblasts: utilization of endocytic and nonendocytic transport mechanisms. *J Cell Biol, 125*, 769-781.

Mineo, C. and Anderson, R. G. (2001) Potocytosis. Robert Feulgen Lecture. *Histochem Cell Biol, 116*, 109-118.

Narita, K., Choudhury, A., Dobrenis, K., Sharma, D., Holicky, E., Marks, D., Walkley, S. and Pagano, R. (2005) Protein transduction of Rab9 in Niemann-Pick C cells reduces cholesterol storage. *FASEB J*, in press.

Norkin, L. C. (2001) Caveolae in the uptake and targeting of infectious agents and secreted toxins. *Adv Drug Deliv Rev, 49*, 301-315.

Orlandi, P. A. and Fishman, P. H. (1998) Filipin-dependent inhibition of cholera toxin: evidence for toxin internalization and activation through caveolae-like domains. *J Cell Biol, 141*, 905-915.

Patterson, M., Vanier, M., Suzuki, K., Morris, J., Carstea, E., Neufeld, E., Blanchette-Machie, E. and Pentchev, P. (2001) Niemann-Pick disease type C: a lipid trafficking disorder. in *The metabolic and molecular bases of inheritied disease (Vol. III)*, (C. R. Scriver, A. L. Beaudet, W. S. Sly and D. Valle, Eds), pp.3611-3634, McGraw Hill, New York.

Pelkmans, L. and Helenius, A. (2002) Endocytosis via caveolae. *Traffic, 3*, 311-320.

Puri, V., Watanabe, R., Dominguez, M., Sun, X., Wheatley, C. L., Marks, D. L. and Pagano, R. E. (1999) Cholesterol modulates membrane traffic along the endocytic pathway in sphingolipid storage diseases. *Nature Cell Biol,* **1**, 386-388.

Puri, V., Watanabe, R., Singh, R. D., Dominguez, M., Brown, J. C., Wheatley, C. L., Marks, D. L. and Pagano, R. E. (2001) Clathrin-dependent and -independent internalization of plasma membrane sphingolipids initiates two Golgi targeting pathways. *J Cell Biol,* **154**, 535-547.

Sabharanjak, S., Sharma, P., Parton, R. G. and Mayor, S. (2002) GPI-anchored proteins are delivered to recycling endosomes via a distinct cdc42-regulated, clathrin-independent pinocytic pathway. *Develop Cell,* **2**, 411-423.

Schuchman, E. H. and Desnick, R. J. (2001) Niemann-Pick Disease Types A and B: Acid Sphingomyelinase Deficiencies. in *The Metabolic & Molecular Bases of Inherited Disease (vol. III),* (C. R. Scriver, A. I. Beaudet, W. S. Sly and D. Valle, Eds), pp.3589-3610, McGraw-Hill, New York.

Scriver, C. R., Beaudet, A. L., Sly, W. S. and Valle, D. D. (2001) Part 16: Lysosomal enzymes. in *The Metabolic and Molecular Bases of Inherited Disease (vol. III),* (C. R. Scriver, A. I. Beaudet, W. S. Sly and D. Valle, Eds), pp.3371-3894, McGraw-Hill, New York.

Seabra, M. C., Mules, E. H. and Hume, A. N. (2002) Rab GTPases, intracellular traffic and disease. *Trends Molec Med,* **8**, 23-30.

Sharma, D. K., Choudhury, A., Singh, R. D., Wheatley, C. L., Marks, D. L. and Pagano, R. E. (2003) Glycosphingolipids internalized via caveolar-related endocytosis rapidly merge with the clathrin pathway in early endosomes and form microdomains for recycling. *J Biol Chem,* **278**, 7564-7572.

Simons, K. and Vaz, W. L. (2004) Model systems, lipid rafts, and cell membranes. *Annu Rev Biophys Biomol Struct,* **33**, 269-295.

Singh, R. D., Puri, V., Valiyaveettil, J. T., Marks, D. L., Bittman, R. and Pagano, R. E. (2003) Selective caveolin-1-dependent endocytosis of glycosphingolipids. *Mol Biol Cell,* **14**, 3254-3265.

Slotte, J. P. and Bierman, E. L. (1988) Depletion of plasma-membrane sphingomyelin rapidly alters the distribution of cholesterol between plasma membranes and intracellular cholesterol pools in cultured fibroblasts. *Biochem J,* **250**, 653-658.

Smith, A. E. and Helenius, A. (2004) How viruses enter animal cells. *Science,* **304**, 237-242.

Sonnichsen, B., De Renzis, S., Nielsen, E., Rietdorf, J. and Zerial, M. (2000) Distinct membrane domains on endosomes in the recycling pathway visualized by multicolor imaging of Rab4, Rab5, and Rab11. *J Cell Biol,* **149**, 901-914.

Torgersen, M. L., Skretting, G., van Deurs, B. and Sandvig, K. (2001) Internalization of cholera toxin by different endocytic mechanisms. *J Cell Sci,* **114**, 3737-3742.

Traub, L. M. (2003) Sorting it out: AP-2 and alternate clathrin adaptors in endocytic cargo selection. *J Cell Biol,* **163**, 203-208.

Zerial, M. and McBride, H. (2001) Rab proteins as membrane organizers. *Nat Rev Mol Cell Biol,* **2**, 107-117.

5-4 Ganglioside and Alzheimer's Disease

Katsuhiko Yanagisawa

Department of Alzheimer's Disease Research, National Institute for Longevity Sciences, National Center for Geriatrics and Gerontology, 36-3 Gengo, Morioka, Obu 474-8522, Japan

Summary. The assembly and deposition of amyloid ß-protein (Aß) in the brain are fundamental pathological events in Alzheimer's disease (AD) and cerebral amyloid angiopathy (CAA). However, it remains to be determined how a nontoxic, monomeric Aß converts to its assembled toxic form.. We identified a unique Aß species that tightly binds to GM1 ganglioside and appears during the early pathological changes of AD. Based on the molecular characteristics of GM1 ganglioside-bound Aß (GAß), we hypothesized that Aß adopts an altered conformation through binding to GM1 ganglioside and acts as a seed for Aß aggregation. We confirmed GAß generation in the brain using a novel monoclonal antibody against purified GAß. While, the specific mechanism of GAß generation in the brain remains to be clarified, results from animal models suggest that GM1 expression and distribution are altered in AD brains. Furthermore, our recent finding shows that specific, selectively expressed gangliosides preferentially facilitate assembly of some hereditary variations of Aß that can lead to accumulation.   The pivotal roles of gangliosides in the induction of Aß assembly and deposition of Aß in the brain are discussed.

Keywords. amyloid ß protein, Alzheimer's disease, ganglioside, seed, microdomain.

1. Introduction

One of the fundamental questions about the pathogenesis of Alzheimer's disease (AD) is how nontoxic, a monomeric amyloid ß-protein (Aß) converts to a toxic aggregate. The expression of specific genes in familial AD likely facilitates Aβ assembly and deposition by accelerating Aß generation. However, for sporadic AD, the major form of the disease, enhanced Aß generation has not been confirmed. Expression of *amyloid precursor protein (APP)* genes with mutations in the Aß sequence that are implicated in familial AD or hereditary cerebral amyloid angiopathy (CAA) does not enhance Aß generation. It suppresses Aβ production. Moreover, the hereditary variant Aß generated from these *APP* mutations are preferentially deposited in the brain in a region-specific manner. This line of evidence suggests that accelerated Aß production alone is insufficient for Ab accumulation and deposition. Environmental factors also probably play critical roles in Aß pathology.

2. Identification of ganglioside-bound form of Aß in human and monkey brains

To discover the molecular mechanism underlying Aß pathology in AD brain, we attempted to identify which Aß species is initially deposited in the brain. (Yanagisawa et al.1995). We fractionated autopsied AD and non-AD brains as well as autopsied Down syndrome brains. In this way, we identified a unique Aß species, characterized by tight binding to GM1 ganglisoide, in brains manifesting early pathological changes of AD, namely abundant, diffuse plaques without neurofibrillary tangles. Interestingly, several antibodies raised against synthetic Aß peptides failed to recognize GM1 ganglioside-bound Aß (GAß). Furthermore, GAß could be isolated only by electrophoretic fractionation, typical protein-isolation techniques failed because large aggregates of soluble Aßs tend to form in a sample. GAß's altered immunoreactivity and extreme potency to induce Aß assembly, we assumed that Aß adopts an altered conformation through binding to GM1 which then triggers the aggregation of soluble Aß in the brain. Following our initial report on GAß, several investigators performed in vitro studies of GAß and their results support our hypothesis. (McLaurin and Chakrabartty 1996, Choo-Smith and Surewicz 1997, Choo-Smith et al. 1997, Matsuzaki and Horikiri 1999, Kakio et al. 2001).

3. GAß scaffolding for Aß aggregation

Pathological assembly of constituent proteins in the brain, including Aß and prion proteins, may occur via a type of "seeded polymerization". To confirm GAß role in the aggregation of soluble Aß, we performed kinetic and morphological studies. As Aß contains many hydrophobic amino acids, it is considered an aggregation-prone peptide. However, Aß will remain soluble longer than expected in a solution prepared by removing undissolved peptides that draw other proteins out of solution. "Seed-free" Aß will assemble immediately after GM1-containing liposomes are added (Fig. 1). Kinetic analysis of Aß assembly in this environment yields a perfectly linear plot using a semilogarithmic calculation (Fig. 1). This suggests that Aß assembles in the presence of GM1-containing liposomes following a first-order kinetic model (Naiki and Nakakuki 1996). Fibril formation from soluble Aß in the presence of GM1-containing liposomes has also been shown by electron microscopy (Fig. 2). Thus, we conclude that Aß initially binds to GM1, leading to the formation of GAß which then attracts soluble Aβ via consecutive binding that extends into growing fibrils (Hayashi et al. 2004).

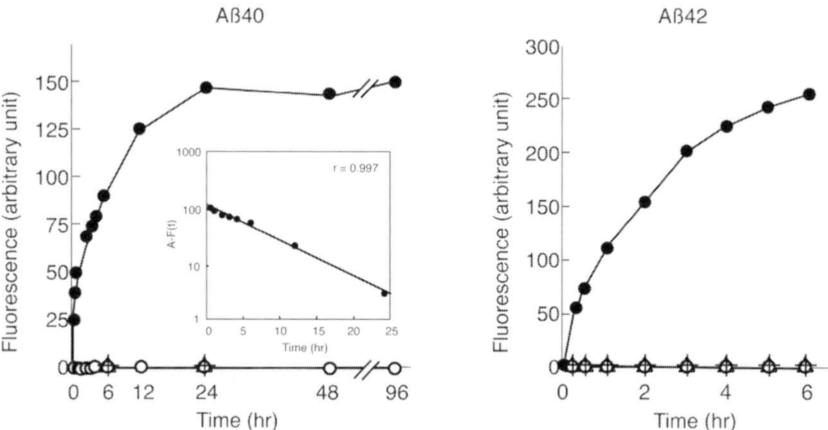

Fig. 1. Kinetics of Aß fibrillogenesis in the absence or presence of GM1-containing liposomes. Aß(Aß40 and Aß42) solutions, after the removal of undissolved peptides, were incubated at 50 μM and 37°C in the presence of GM1-containing liposomes (filled circle) or GM1-lacking liposomes (plus), or incubated in the absence of liposomes (open circle). GM1-containing liposomes alone were also incubated in the absence of Aß (triangle). The fluorescence intensity of thioflavin T was obtained by excluding background activity at 0 hr. Inset: Semilogarithmic plot of the difference, A-F(t), versus incubation time (0-24 hr). F(t) represents the increase in fluorescence intensity as a function of time in the case of Aß incubated with GM1-containing liposomes and A is tentatively determined as F (infinity). Linear regression and correlation coefficient values were calculated (r=0.997). F(t) is described by a differential equation: F'(t)=B-CF(t). (Copyright 2004, the Society for Neuroscience)

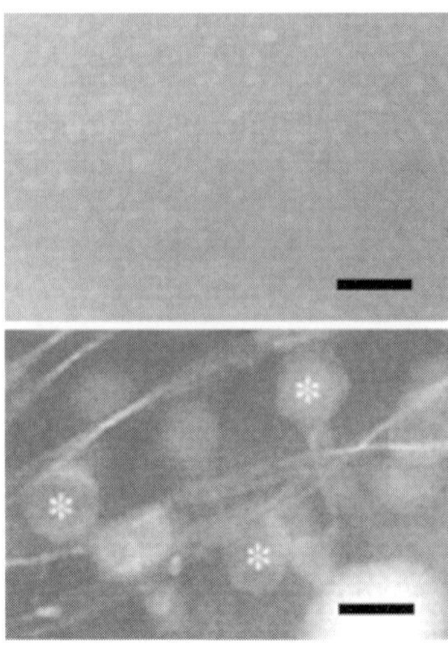

Fig. 2. Electron micrographs of the Aß40 solutions incubated for 24 hr at 50 μM and 37°C with GM1-containing liposomes (lower panel) or without liposomes (upper panel). The liposomes are indicated by asterisks. Bar, 50 nm. (Copyright 2004 by the Society for Neuroscience)

Another important aspect in Aß formation is whether it is the oligomers or polymers that are neurotoxic. We examined the viability of neurons following treatment with incubation mixtures containing Aß and GM1-containing liposomes. We found that neuronal viability markedly decreased only in the culture with Aß and GM1-containing liposomes together (Hayashi et al. 2004). This suggests that GM1 plays a critical role in the production of neurotoxic Aß assemblies.

4. Validation of GAß generation in human and monkey brains

To validate our hypothesis, we attempted to directly detect GAß in the brain using a monoclonal antibody specific to Aß with an altered conformation induced by binding to GM1 (Hayashi et al. 2004). GAß was purified from the human brain by electrophoretic fractionation and subjected to in vitro immunization to generate monoclonal antibodies. The IgG monoclonal antibody 4396C was produced by a genetic class-switch technique from the original IgM hybridomas that were obtained by in vitro immunization. Immunohistochemistry of the sections of Alzheimer brain, fixed with Kryofix, showed numerous neurons stained by 4396C. Neuronal immunostaining with 4396C was also demonstrated in the brains of aged nonhuman primates, which naturally show Aß deposition with a granular pattern after age 20 (Fig. 3a). Intraneuronal staining with 4396C distinctly colocalized with that with an antibody specific to amino terminus of Aß and that with cholera toxin, a natural ligand for GM1. Furthermore, GAß was readily immunoprecipitated from the brain of old nonhuman primate (Fig. 3b). These results indicate that GAß is generated in the brain.

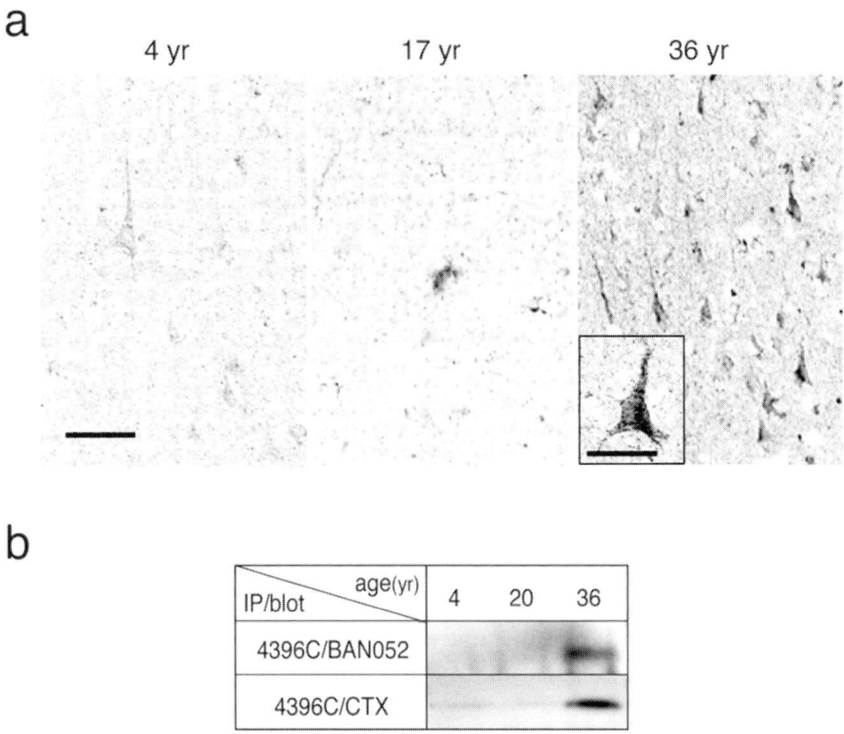

Fig. 3. Immunohistochemistry and immunoprecipitation of GAß in sections of nonhuman primate brains. (a) Immunostaining by 4396C of sections of the cerebral cortices of primate brains, which were fixed in paraformaldehyde, from animals of different ages. Bar, 50 μm. Inset: higher magnification. Bar, 20 μm. (b) Immunoprecipitation of GAß by 4396C from cerebral cortices of primates at different ages. Immunoprecipitates were blotted and reacted with BAN052 or HRP-conjugated cholera toxin subunit B (CTX). (Copyright 2004 by the Society for Neuroscience)

5. Putative molecular mechanisms underling GAß generation

The mechanism of GAß generation in the brain remains to be clarified. Although Aß is physiologically secreted from neurons and GM1 is also synthesized and expressed on the surface of neurons, GAß cannot be detected in brains without pathological Aß deposition. In this regard, Matsuzaki and his colleagues previously reported that an increase in cholesterol content of the host membrane facilitates the binding of soluble Aß to GM1 (Kakio et al. 2001). Moreover, it was also suggested that the cholesterol-dependent acceleration of GAß generation is likely through the formation of GM1 clusters in a cholesterol-rich environment. The next question was whether an increase in the local cholesterol content can occur under biological conditions. Wood and his colleagues previously reported that the cholesterol content of the exofacial leaflet of synaptic plasma membranes increases approximately twofold in aged mice and apolipoprotein E (apoE)-knock-out mice (Igbavboa et al. 1996). We extended their study using human-apoE3- and human-apoE4- knock-in mice. Notably, the cholesterol content of the exofacial leaflet of synaptic plasma membranes significantly increases in apoE4- but not apoE3- knock-in mice. (Hayashi et al. 2002). Because aging and apoE4 expression are strong risk factors for the development of AD, an increase in the local cholesterol content may be closely associated with the Aß pathology through GAß generation. Our previous study also suggested the putative role of cholesterol in the generation of seed Aß. Seed Aß is secreted only from the apical surface of epithelial cells transfected with the *APP* gene. (Mizuno et al. 1999). With evidence that the cholesterol content of the apical plasma membrane is higher than that of the basolateral plasma membrane (Van Helvoort and Van Meer 1995), the generation of seed Aß in cells is likely to be dependent on the presence of not only GM1 but also cholesterol.

6. Pathological implications of risk factors for AD to GAß generation

To directly ascertain the possibility that GM1 content is increased in neuronal membranes by biological factors, including aging and apoE4 expression, we performed lipid chemical analysis of neuronal membranes from three different groups of apoE3- and apoE4-knock-in mice based on age. (Yamamoto et al. 2004). In that study, we focused on detergent-resistant membrane microdomains (DRMs), which are rich in cholesterol and GM1.

Interestingly, the GM1 content of the DRMs increased with age in both the apoE3- and apoE4-knock-in mice; however, this increase was significantly more pronounced in the apoE4-knock-in mice. Interestingly, the age-dependent increase in the GM1 content observe only in the DRM fraction, not in other membrane fractions. At this point, it remains to be determined how GM1 accumulates in specific membrane microdomains, such as DRMs of synaptic plasma membranes, in age-dependent manner and expression of apoE-isoform-dependent manner. One possible explanation is that aging and apoE4 expression induce an increase in the cholesterol content of local membrane microdomains of synaptic plasma membranes, leading to the accumulation and stabilization of GM1 in these domains. Our studies may provide new insights into the implications of risk factors for AD development.

The study of the accumulation of GM1 in neuronal membranes by Fredman and her colleagues (Molander-Melin M et al. 2005) showed that DRMs isolated from the frontal cortex of AD brains had significantly higher GM1 and GM2 levels. This is significant because the frontal cortex is believed to be the area first affected in AD. Although further studies are required before the pathological significance of increased local GM1 content can be ascertained, accumulating data strongly suggest that an alteration in the lipid composition of neuronal membranes initiates Aß assembly and deposition through the generation of seed Aß, such as GAß.

7. Pivotal role of ganglioside in brain-region-specific assembly and Aß deposition

Several different studies of the pathological significance of Aß binding to gangliosides have recently identified that several mutations within the Aß sequence are responsible for familial AD or hereditary cerebral amyloid angiopathy. Among the mutations, Arctic-type mutation (E22G) and Dutch-type mutation (E22Q) show distinct pathological and clinical phenotypes. E22G induces a predominant Aß deposition in the brain parenchyma and E22Q induces a predominant Aß deposition in the cerebral vessel wall. It seems that local environmental factors, such as the presence of particular gangliosides in the brain, play a pivotal role in the deposition of hereditary Aß variants in a brain-region-specific manner. To clarify this, we incubated hereditary Aß variants, including Arctic- Dutch-, Italian-, and Flemish-type Aßs, in the presence of various gangliosides. We found that the assembly of Aßs was significantly accelerated in the presence of specific gangliosides. (Yamamoto et al. 2005). Notably, the assemblies of

Arctic-, Dutch-, Italian-, and Flemish-type Aßs were markedly accelerated in the presence of GM1, GM3, GM3, and GD3, respectively. The specific requirement for GM3 by Dutch- and Italian-type Aßs suggests that human cerebrovascular smooth muscle cells (HCSMs), which form the vessel wall where these variants are selectively deposited, predominantly express GM3. Alternatively, the specific requirement for GD3 by Flemish-type Aß suggests that a perivascular environment, which is likely the site for the assembly and deposition of the variant, selectively provides GD3. To verify this, we determined the ganglioside that is expressed by cultured HCSMs and by cocultured human astrocytes and endothelial cells. The latter forms a perivascular environment. Notably, the cultured HCSMs predominantly expressed GM3, whereas the astrocytes cocutured with epithelial cells expressed GD3. We therefore concluded that gangliosides play a critical role in the brain-region-specific assembly and deposition of Aß.

8. Conclusions

Our studies suggest that gangliosides contribute to the assembly and deposition of Aß as an environmental factor and that Aß binding to GM1 initiates the amyloid cascade of AD. However, GM1 is expressed rather broadly in the brain beyond the preferential sites of Aß deposition. Thus, a future challenge will be to determine how the expression and distribution of GM1 is altered in AD brains, leading to GAß generation.

9. References

Choo-Smith LP, Surewicz WK (1997) The interaction between Alzheimer amyloid ß(1-40) peptide and ganglioside GM1-containing membranes. *FEBS Let*, **402**, 95-98.

Choo-Smith LP, Garzon-Rodriguez W, Glabe CG, Surewicz WK (1997) Acceleration of amyloid fibril formation by specific binding of Aß -(1-40) peptideto ganglioside-containing membrane vesicles. *J Biol Chem*, **272**, 22987-22990.

Hayashi H, Igbavboa U, Hamanaka H, Kobayashi M, Fujita SC, Wood WG, Yanagisawa K (2002) Cholesterol is increased in the exofacial leaflet of synapticplasma membranes of human apolipoprotein E4 knock-in mice. *Neuroreport*, **13**, 383-386.

Hayashi H, Kimura N, Yamaguchi H, Hasegawa K, Yokoseki T, Shibata M, Yamamoto N, Michikawa M, Yoshikawa Y, Terao K, Matsuzaki K, Lemere CA,

Selkoe DJ, Naiki H, Yanagisawa K (2004) A seed for Alzheimer amyloid in the brain. *J Neurosci*, **24**, 4894-4902.

Igbavboa U, Avdulov NA, Schroeder F, Wood WG (1996) Increasing age alterstransbilayer fluidity and cholesterol asymmetry in synaptic plasma membranes ofmice. *J Neurochem*, **66**, 1717-1725.

Kakio A, Nishimoto SI, Yanagisawa K, Kozutsumi Y, Matsuzaki K (2001) Cholesterol-dependent formation of GM1 ganglioside-bound amyloid ß-protein, an endogenous seed for Alzheimer amyloid. *J Biol Chem*, **276**, 24985-24990.

Matsuzaki K, Horikiri C (1999) Interactions of amyloid β-peptide (1-40) with ganglioside-containing membranes. **38**, 4137-4142.

McLaurin J, Chakrabartty A (1996) Membrane disruption by Alzheimer ß-amyloid peptides mediated through specific binding to either phospholipids or gangliosides. Implications for neurotoxicity. *J Biol Chem*, **271**, 26482-26489.

Mizuno T, Nakata M, Naiki H, Michikawa M, Wang R, Haass C, Yanagisawa K (1999) Cholesterol-dependent generation of a seeding amyloid ß-protein in cell culture. *J Biol Chem*, **274**, 15110-15114.

Molander-Melin M, Blennow K, Bogdanovic N, Dellheden B, Mansson JE, Fredman P (2005) Structural membrane alterations in Alzheimer brains found to be associated with regional disease development; increased density of gangliosides GM1 and GM2 and loss of cholesterol in detergent-resistant membrane domains. *J Neurochem*, **92**, 171-182.

Naiki H, Nakakuki K (1996) First-order kinetic model of Alzheimer's ß-amyloid fibril extension in vitro. *Lab Invest*, 74, 374-383.

van Helvoort A, van Meer G (1995) Intracellular lipid heterogeneity caused by topology of synthesis and specificity in transport. Example: sphingolipids. *FEBS Lett*, **369**, 18-21.

Yamamoto N, Igbabvoa U, Shimada Y, Ohno-Iwashita Y, Kobayashi M, Wood WG, Fujita SC, Yanagisawa K (2004) Accelerated Aß aggregation in the presence of GM1-ganglioside-accumulated synaptosomes of aged apoE4-knock-in mouse brain. *FEBS Lett*, **569**, 135-139.

Yamamoto N, Hirabayashi Y, Amari M, Yamaguchi H, Romanov G, Van Nostrand WE, Yanagisawa K (2005) Assembly of hereditary amyloid ß-protein variant in the presence of favorable gangliosides. *FEBS Lett*, **579**, 2185-2190.

Yanagisawa K, Odaka A, Suzuki N, Ihara Y (1995) GM1 ganglioside-bound amyloid ß-protein (Aß): a possible form of preamyloid in Alzheimer's disease. *Nat Med*, **1**, 1062-1066.

5-5 Modulation of Proteolytic Processing by Glycosphingolipids Generates Amyloid ß-Peptide

Irfan Y. Tamboli[1], Kai Prager[1], Esther Barth[1], Micheal Heneka[1], Konrad Sandhoff[2], and Jochen Walter[1]

[1]Department of Neurology, University of Bonn, Sigmund-Freud-Str. 25, 53127 Bonn, Germany; [2]Kekulé-Institute for Organic Chemistry and Biochemistry, University of Bonn, Gerhard-Domagk-Strasse 1, 53121 Bonn, Germany

Summary. Extracellular amyloid ß-peptide (Aß) deposits in the brain are characteristic of Alzhemier's disease. Proteolytic cleaving of amyloid precursor protein (APP) by ß- and α-secretases generate these deposits. The cleavage by those secretases occurs predominantly in post-Golgi secretory and endocytic compartments and is influenced by cholesterol, indicating a role of the membrane lipid composition in APP processing. To analyze the function of glycosphingolipids (GSLs) in the proteolytic processing of APP and the generation of Aß, we inhibited glycosylceramide synthase, the first enzyme in GSL biosynthesis pathway. The depletion of GSLs markedly reduced the secretion of endogenous APP in different cell types, including human neuroblastoma SH-SY5Y cells. Conversely, the addition of exogenous brain gangliosides to cultured cells increased the levels of both cellular and secreted APP. Importantly, depletion of GSLs strongly decreased the secretion of Aß. Thus, enzymes involved in GSL metabolism might represent targets to inhibit Aß production.

Keywords. Alzheimer's disease, ß-amyloid precursor protein, glycosphingolipids, protein transport, secretase

1. Introduction

The deposition of amyloid ß-peptides (Aß) as extracellular plaques is an invariant neuropathological feature of Alzheimer's disease (AD) (Aguzzi and Haass, 2003; Selkoe, 2001). Aß derives from the ß-amyloid precursor protein (APP) through proteolytic processing involving sequential cleavages by proteases called ß- and α-secretases (Selkoe, 2001; Walter et al., 2001). APP is a type I membrane protein that is transported from the endoplasmic reticulum (ER) via the Golgi compartment to the cell surface. During transport it undergoes maturation by N'- and O'-glycosylation (Annaert and de Strooper, 2002). Within the secretory pathway and at the cell surface, APP is predominantly cleaved by α-secretase resulting in the secretion of soluble APP (APP$_S$) (Sisodia et al., 1990). Since α-secretase cleaves APP within the Aß domain, this cleavage precludes the generation of Aß. Alternatively, APP can be cleaved by ß-secretase. The cleavage of APP by ß-secretase occurs predominantly in endosomal and lysosomal compartments after internalization from the cell surface (Koo et al., 1996). The C-terminal fragments (CTFs) of APP, resulting from α- or ß-secretase cleavage can then be cleaved within the transmembrane domain by α-secretase to release p3 and Aß, respectively (Selkoe, 2001; Steiner and Haass, 2000).

The proteolytic processing of APP is influenced by the lipid composition of cellular membranes, as demonstrated by pharmacological modulation of cellular cholesterol and cholesterol ester levels (Hartmann, 2001; Hutter-Paier et al., 2004; Puglielli et al., 2003; Wolozin, 2004). It has also been shown that the levels of several gangliosides are altered in AD brains (Cutler et al., 2004). In addition, ganglioside GM1 binds to Aß and might contribute to early deposition of the peptide in amyloid plaques (Hayashi et al., 2004; Yanagisawa et al., 1995).

The biosynthesis of GSLs starts with the generation of glucosylceramide from UDP-glucose and ceramide by glycosylceramide synthase (GCS; Fig. 1A). Glucosylceramide represents the precursor of a large variety of GSLs that are transported in the secretory pathway from the Golgi to the cell surface (Kolter et al., 2002). The physiological functions of GSLs include the regulation of cell adhesion, cell differentiation, and signal transduction (Simons and Ehehalt, 2002; Sprong et al., 2001a). Here, we sought to analyze the role of GSLs in the proteolytic processing of APP and the generation of Aß. Our data indicate that GSLs mediate transport of APP in the secretory pathway and expression at the cell surface, thereby affecting proteolytic processing by secretases.

2. Results

To analyze the function of GSLs in the proteolytic processing of APP and Aß generation, GSL biosynthesis was inhibited with PDMP, a competitive inhibitor of GCS that has been shown previously to efficiently decrease GSL biosynthesis in cultured cells (Kok et al., 1998; Mutoh et al., 1998; Naslavsky et al., 1999; Rosenwald and Pagano, 1994). Treatment of HEK293 cells with PDMP for 48 hrs led to a significant decrease in GSL biosynthesis as demonstrated by a strong reduction of GM1 levels (Fig. 1B). The treatment of cells with PDMP at these concentrations did not affect cell viability (data not shown). As demonstrated in Fig. 1C, D, the secretion of soluble APP (APP$_S$) into conditioned media was markedly decreased after treatment. In contrast, addition of exogenous GSLs significantly increased the cellular levels of endogenous APP as well as the secretion of APP$_S$ (Fig. 1E).

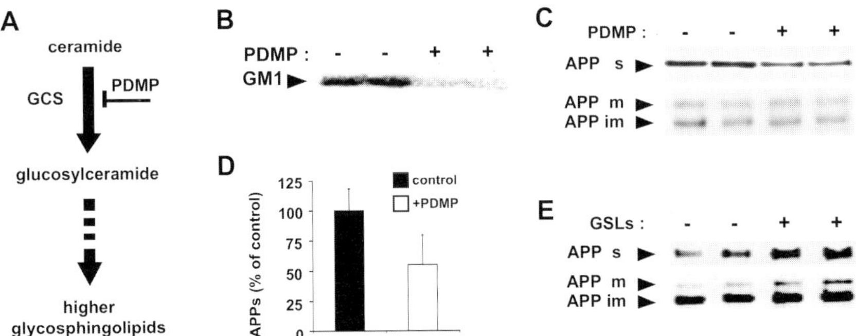

Fig. 1. GSLs regulate APP processing. *A*, Schematic showing of the biosynthesis pathway of glycosphingolipids and targeted inhibition of glucosylceramide synthase (GCS) by PDMP. *B*, HEK293 cells cultured in the absence (-) or presence (+) of 10 μM PDMP for 48 hrs, cellular membranes were separated by SDS-PAGE and GM1 was detected by western immunoblotting with cholera toxin. *C - E*, APP was immunoprecipitated from conditioned media (C, E, upper panels) and cell lysates (C, E, lower panels) after 48 hrs treatment with 10μM PDMP (C) or 50 μg/ml purified ganglioside mixture from bovine brain (E) , and separated by SDS-PAGE. Secreted (APP$_S$) and cellular APP detected by western immunoblotting. Migration of APP$_S$, mature (m) and immature (im) APP indicated by arrowheads. Secretion of APP$_S$ was quantified by ECL imaging and normalization to cellular APP expression (*D*). Values represent means of three independent experiments ± s.d. (solid bar, no PDMP; open bar, 10 μM PDMP).

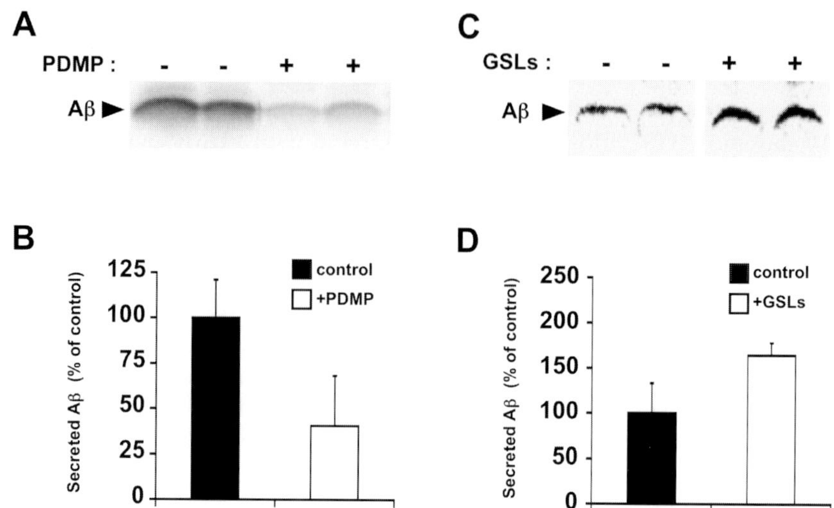

Fig. 2. Human SH-SY5Y cells were cultured in the absence (-) or presence (+) of 10 μM PDMP (A) or   50 μg/ml purified ganglioside mixture from bovine brain (C) for 48 hrs. Endogenously generated Aß was immunoprecipitated from conditioned media and detected by western immunoblotting. Quantification of relative amounts of secreted Aß performed by ECL imaging (B, D). Values represent means of three independent experiments ± s.d.

Because the levels of endogenous Aß in the conditioned media of HEK293 cells were below detection limits, (data not shown), we used human neuroblastoma SH-SY5Y cells that secrete higher amounts of Aß. The decrease in APP$_S$ secretion after PDMP was observed in this cell type (data not shown). Moreover, inhibition of GSL biosynthesis significantly reduced the secretion of Aß (Fig. 2A and B). On the other hand, addition of exogenous GSLs increased the secretion of Aß (Fig. 2C and D), indicating that GSLs facilitated Aß generation. Proteolytic processing of APP has been shown to occur predominantly in post Golgi secretory and endocytic compartments, and at the cell surface (Haass et al., 1992; Koo and Squazzo, 1994; Sisodia et al., 1990). We, therefore, assessed the cell surface expression of APP in the presence, or absense, of PDMP by biotinylation.

In GSL depleted cells, the levels of biotin-labeled APP were markedly reduced (Fig. 3A), demonstrating that suppression of GSL biosynthesis reduces the expression of APP at the cell surface. We then assessed the effect of PDMP on the general expression of glycoproteins at the cell surface. Cell staining with TRITC-labeled wheat germ agglutinin (WGA), a lectin that binds to glycoproteins did not reveal any significant effects after GSL depletion (Fig. 3B), indicating a selective inhibition of APP transport to the cell surface. To address this, we performed pulse-chase experiments and analyzed APP maturation in the Golgi compartment. GSL depletion reduced the transport of APP to or within the Golgi compartment by decreasing APP maturation in PDMP treated cells (Fig. 3C and D). We also observed decreased levels in total APP in GSL depleted cells after 60 and 90 min (Fig. 3C and E).

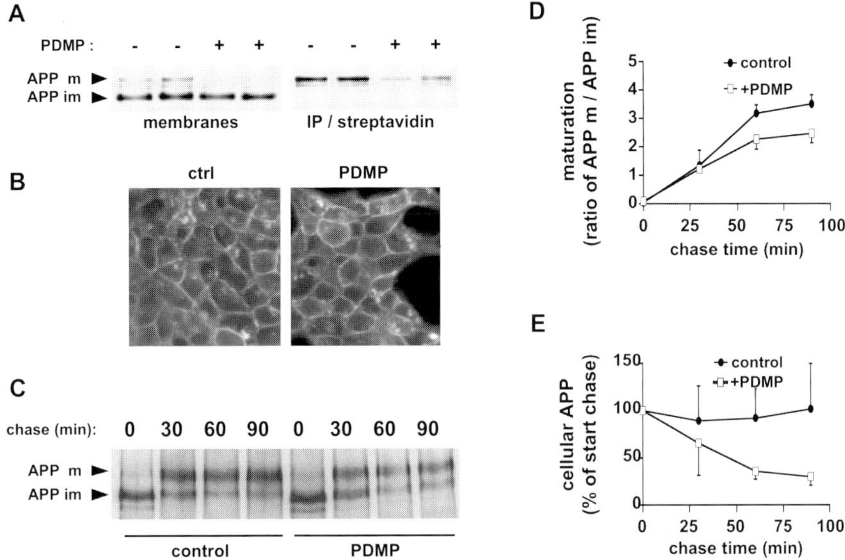

Fig. 3. **GSLs facilitate forward transport of APP.** *A*, Cell surface proteins of control (-) and PDMP-treated (+) HEK293 cells were labelled with sulfo-NHS-biotin and isolated with streptavidin-conjugated agarose beads. Precipitates were separated by SDS-PAGE and APP was detected by western blotting (right panel). As a control, cellular levels of APP were also analyzed by western immunoblotting of isolated cell membranes with the respective antibodies (left panel). *B*, Control or PDMP-treated cells were stained with TRITC-labeled WGA to detect cell surface-located glycoproteins and analyzed by fluorescence microscopy. *C*, After culturing in the absence or presence of 10 μM PDMP for 48 hrs, cells were labeled with [^{35}S]methionine for 10 min and chased for the indicated time periods . APP was immunoprecipitated from cell lysates, separated by SDS-PAGE and detected by phosphoimaging. The migration of mature (m) and immature (im) APP is indicated by arrow heads. *D, E*, Quantification of APP maturation and stability. In PDMP treated cells (open squares) the maturation of APP is significantly decreased as compared to untreated cells (closed circles) (D). In addition, the stability of cellular APP is reduced in PDMP treated cells (E). Values represent means of three independent experiments ± s.d.

To confirm these findings in an independent cellular model, we used mouse melanoma cell lines B16 and GM95. While B16 cells produce GSLs, GM95 cells have defective GSL biosynthesis due to decreased activity of GCS. The latter are commonly used as a model of GSL deficient cells (Ichikawa et al., 1994; Komori et al., 1999; Sprong et al., 2001b).

As expected, very little GM1 could be detected in GM95 cells, while B16 cells express robust amounts of GM1 (Fig. 4A). To investigate the maturation of APP in both cell types, we performed pulse-chase experiments. In B16 cells, endogenous APP undergoes maturation as indicated by the appearance of a slower migrating band during the chase period (Fig. 4B, left panel). In contrast, the GSL deficient GM95 cell line revealed significantly reduced maturation of APP (Fig. 4B, right panel), which is consistent with the data obtained with pharmacological inhibition of GSL biosynthesis (see Fig. 3). In addition, steady state levels of cellular APP were strongly decreased in GM95 cells as compared to B16 cells (Fig. 4C), while expression of APP mRNA was similar in both cell lines (Fig. 4D).

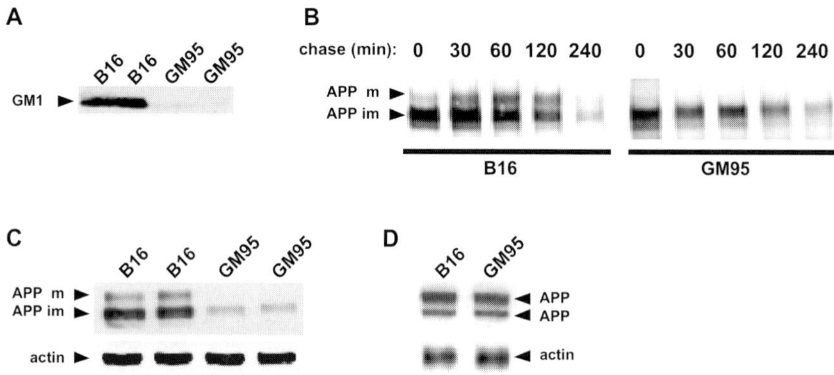

Fig. 4. Decreased maturation and cellular levels of APP in GSL deficient cells. *A*, Membranes of B16 and GM95 cells separated by SDS-PAGE, and GM1 was detected by western immunoblotting with cholera toxin. *B*, Pulse-chase experiment for APP in B16 (left panel) and GM95 (right panel) cells. The maturation of APP was significantly decreased in GSL-deficient GM95 cells as compared to B16 cells. *C*, Steady state levels of APP in B16 and GM95 cells were compared by western immunoblotting. Actin was used as a loading control. *D*, RT-PCR analysis of APP expression (upper panel) and actin (lower panel) in B16 and GM95 cells.

3. Discussion

By targeting GSL biosynthesis, which occurs in the early secretory pathway, we demonstrated that GSLs are implicated in the maturation of APP in the Golgi and its transport to the cell surface. The inhibition of GSL biosynthesis reduced the secretion of APP$_S$ and Aß, while the addition of exogenous brain gangliosides reversed these effects. However, our data do not exclude the possibility that GSLs might have additional roles in proteolytic processing of APP at the cell surface, like direct or indirect modulation of secretase activities (Sawamura et al., 2004; Zha et al., 2004). Of note, the effects are observed selectively for endogenous APP, while overexpressed APP revealed unaltered maturation and secretion (data not shown). This might be due to tightly regulated interactions of APP with specific lipids within cellular membranes. The overexpression of membrane proteins might therefore lead to altered interactions with lipids probably by saturation effects (Opekarova and Tanner, 2003).

Recently, it has been shown that GSLs are involved in the forward transport of membrane proteins in yeast and mammalian cells (Sprong et al., 2001b; Sutterlin et al., 1997). In agreement with our data, these studies also demonstrated that inhibition of GSL biosynthesis does not generally impair protein transport or secretion (Sprong et al., 2001b; Sutterlin et al., 1997; Watanabe et al., 2002). Thus, GSLs appear to mediate transport of individual proteins, probably at distinct steps in the secretory pathway. The attenuated transport of APP to the cell surface in GSL deficient cells is consistent with decreased secretion of APP by α-secretase, which is known to occur during transport to or directly at the cell surface (Chyung and Selkoe, 2003; Koo et al., 1990). In contrast, ß-secretase cleavage likely occurs predominantly in endocytic compartments after reinternalization of APP from the cell surface (Koo and Squazzo, 1994). Thus, the decreased generation of Aß upon depletion of cells from GSLs might be due to decreased access of ß-secretase to APP in endocytic compartments.

Taken together, our data indicate that GSLs, and the respective enzymes involved in their biosynthesis, might represent targets to decrease formation of Aß in therapeutic strategies for AD.

Acknowledgements. We thank Dr. C. Haass for generous gift of antibodies. We also thank the Riken Cell Bank for B16 and GM95 cell lines and Dr. G. Thinakaran for advice with GM1 detection. This work was supported by the Deutsche Forschungsgemeinschaft (DFG) and the BONFOR program to J.W.

Abbreviations. Aß, amyloid ß-peptide; AD, Alzheimer's disease; APP, ß-amyloid precursor protein; CTF, C-terminal fragment; GCS, glycosylceramide synthase; GSL, glycosphingolipid; HEK, human embryonic kidney; PDMP, d-threo-1-phenyl-2-decanoylamino-3-morpholino-1-propanol.

4. References

Aguzzi A, Haass C (2003) Games played by rogue proteins in prion disorders and Alzheimer's disease. *Science*, **302**, 814-818.

Annaert W, de Strooper B (2002) A cell biological perspective on Alzheimer's disease. *Annu Rev Cell Dev Biol*, **18**, 25-51.

Chyung JH, Selkoe DJ (2003) Inhibition of receptor mediated endocytosis demonstrates generation of amyloid beta -protein at the cell surface. *J.Biol.Chem*, **278**, 51035-51043.

Cutler RG, Kelly J, Storie K, Pedersen WA, Tammara A, Hatanpaa K, Troncoso JC, Mattson MP (2004) Involvement of oxidative stress-induced abnormalities in ceramide and cholesterol metabolism in brain aging and Alzheimer's disease. *Proc Natl Acad Sci U S A*, **101**, 2070-2075.

Haass C, Koo EH, Mellon A, Hung AY, Selkoe DJ (1992) Targeting of cell-surface beta-amyloid precursor protein to lysosomes: alternative processing into amyloid-bearing fragments. *Nature*, **357**, 500-503.

Hartmann T (2001) Cholesterol, A beta and Alzheimer's disease. *Trends Neurosci*, **24**, S45-S48.

Hayashi H, Kimura N, Yamaguchi H, Hasegawa K, Yokoseki T, Shibata M, Yamamoto N, Michikawa M, Yoshikawa Y, Terao K, Matsuzaki K, Lemere CA, Selkoe DJ, Naiki H, Yanagisawa K (2004) A seed for Alzheimer amyloid in the brain. *J Neurosci*, **24**, 4894-4902.

Hutter-Paier B, Huttunen HJ, Puglielli L, Eckman CB, Kim DY, Hofmeister A, Moir RD, Domnitz SB, Frosch MP, Windisch M, Kovacs DM (2004) The ACAT inhibitor CP-113,818 markedly reduces amyloid pathology in a mouse model of Alzheimer's disease. *Neuron*, **44**, 227-238.

Ichikawa S, Nakajo N, Sakiyama H, Hirabayashi Y (1994) A mouse B16 melanoma mutant deficient in glycolipids. *Proc Natl Acad Sci U S A*, **91**, 2703-2707.

Kok JW, Babia T, Filipeanu CM, Nelemans A, Egea G, Hoekstra D (1998) PDMP blocks brefeldin A-induced retrograde membrane transport from golgi to ER:

evidence for involvement of calcium homeostasis and dissociation from sphingolipid metabolism. *J Cell Biol,* **142**, 25-38.

Kolter T, Proia RL, Sandhoff K (2002) Combinatorial ganglioside biosynthesis. *J Biol Chem,* **277**, 25859-25862.

Komori H, Ichikawa S, Hirabayashi Y, Ito M (1999) Regulation of intracellular ceramide content in B16 melanoma cells. Biological implications of ceramide glycosylation. *J Biol Chem,* **274**, 8981-8987.

Koo EH, Sisodia SS, Archer DR, Martin LJ, Weidemann A, Beyreuther K, Fischer P, Masters CL, Price DL (1990) Precursor of amyloid protein in Alzheimer disease undergoes fast anterograde axonal transport. *Proc Natl Acad Sci U S A,* **87**, 1561-1565.

Koo EH, Squazzo SL (1994) Evidence that production and release of amyloid beta-protein involves the endocytic pathway. *J Biol Chem,* **269**, 17386-17389.

Koo EH, Squazzo SL, Selkoe DJ, Koo CH (1996) Trafficking of cell-surface amyloid beta-protein precursor. I. Secretion, endocytosis and recycling as detected by labeled monoclonal antibody. *J Cell Sci,* **109**, 991-998.

Mutoh T, Tokuda A, Inokuchi J, Kuriyama M (1998) Glucosylceramide synthase inhibitor inhibits the action of nerve growth factor in PC12 cells. *J Biol Chem,* **273**, 26001-26007.

Naslavsky N, Shmeeda H, Friedlander G, Yanai A, Futerman AH, Barenholz Y, Taraboulos A (1999) Sphingolipid depletion increases formation of the scrapie prion protein in neuroblastoma cells infected with prions. *J Biol Chem,* **274**, 20763-20771.

Opekarova M, Tanner W (2003) Specific lipid requirements of membrane proteins--a putative bottleneck in heterologous expression. *Biochim Biophys Acta,* **1610**, 11-22.

Puglielli L, Tanzi RE, Kovacs DM (2003) Alzheimer's disease: the cholesterol connection. *Nat Neurosci,* **6**, 345-351.

Rosenwald AG, Pagano RE (1994) Effects of the glucosphingolipid synthesis inhibitor, PDMP, on lysosomes in cultured cells. *J Lipid Res,* **35**, 1232-1240.

Sawamura N, Ko M, Yu W, Zou K, Hanada K, Suzuki T, Gong JS, Yanagisawa K, Michikawa M (2004) Modulation of amyloid precursor protein cleavage by cellular sphingolipids. *J Biol Chem,* **279**, 11984-11991.

Selkoe DJ (2001) Alzheimer's disease: genes, proteins, and therapy. *Physiol Rev,* **81**, 741-766.

Simons K, Ehehalt R (2002) Cholesterol, lipid rafts, and disease. *J Clin Invest,* **110**, 597-603.

Sisodia SS, Koo EH, Beyreuther K, Unterbeck A, Price DL (1990) Evidence that beta-amyloid protein in Alzheimer's disease is not derived by normal processing. *Science,* **248**, 492-495.

Sprong H, Degroote S, Claessens T, van Drunen J, Oorschot V, Westerink BH, Hirabayashi Y, Klumperman J, van der SP, van Meer G (2001b) Glycosphingolipids are required for sorting melanosomal proteins in the Golgi complex. *J Cell Biol,* **155**, 369-380.

Sprong H, van der Sluijs P., van Meer G (2001a) How proteins move lipids and lipids move proteins. *Nat Rev Mol Cell Biol,* **2**, 504-513.

Steiner H, Haass C (2000) Intramembrane proteolysis by presenilins. *Nat Rev Mol Cell Biol*, **1**, 217-224.

Sutterlin C, Doering TL, Schimmoller F, Schroder S, Riezman H (1997) Specific requirements for the ER to Golgi transport of GPI-anchored proteins in yeast. *J Cell Sci*, **110** (**Pt 21**), 2703-2714.

Walter J, Kaether C, Steiner H, Haass C (2001) The cell biology of Alzheimer's disease: uncovering the secrets of secretases. *Curr Opin Neurobiol*, **11**, 585-590.

Watanabe R, Funato K, Venkataraman K, Futerman AH, Riezman H (2002) Sphingolipids are required for the stable membrane association of glycosyl-phosphatidylinositol-anchored proteins in yeast. *J Biol Chem*, **277**, 49538-49544.

Wolozin B (2004) Cholesterol and the biology of Alzheimer's disease. *Neuron*, **41**, 7-10.

Yanagisawa K, Odaka A, Suzuki N, Ihara Y (1995) GM1 ganglioside-bound amyloid beta-protein (A beta): a possible form of preamyloid in Alzheimer's disease. *Nat Med*, **1**, 1062-1066.

Zha Q, Ruan Y, Hartmann T, Beyreuther K, Zhang D (2004) GM1 ganglioside regulates the proteolysis of amyloid precursor protein. *Mol Psychiatry*, **9**, 946-952.

5-6 Hereditary Sensory Neuropathy

Simon J. Myers[1] and Garth A. Nicholson[1,2]

[1]Northcott Neuroscience Laboratory, ANZAC Research Institute, Hospital Rd, Concord, Sydney, NSW 2139, Australia
[2]Department of Molecular Medicine, Concord Repatriation General Hospital, Hospital Rd, Concord, Sydney, NSW 2139, Australia

Summary. Hereditary neuropathies are the most prevalant group of hereditary disorders appearing in genetic counseling clinics, affecting approximately 1 in 2500 people. Sensory neuropathies account for some 10% of these disorders. Although rarely fatal, they do cause lifelong disabilities with significant social and economic impacts. No specific treatment currently available; however, we have developed a comprehensive management approach that has been widely adopted to compensate for distal weakness and sensory loss.

Although the causes of hereditary neuropathy are diverse, a common mechanism is responsible for the manifested disabilities: axonal degeneration. Axonal degeneration is significant in several disorders of the central nervous system, including multiple sclerosis and other common neurodegenerative diseases such as Alzheimer's and Parkinson's diseases. Axonal degeneration may stem from defects in axonal transport (Kamholz et al., 2000), failure of neurotrophin uptake (Bartlett et al., 1999, Ginty and Segal, 2002), or changes in sodium transport.

Dominantly inherited neurodegenerative disorders are generally caused by protein mutations that lead to cell toxicity, a gain of toxic function. We identified mutations in the protein, serine palmitoyltransferase (SPT) long chain subunit 1 (SPTLC1), that cause hereditary sensory neuropathy (HSN). In HSN, the mutant SPT enzyme is toxic, progressively impairing long nerve function first distally but eventually spreading to the whole nerve causing cell death. This is also described as a distal sensory and motor axonopathy.

Keywords. Hereditary Sensory Neuropathy (HSN), Axonal Degeneration, Aggregation, Protein Transport, SPTLC1

1. Many gene defects (mutations) cause hereditary neuropathies

Although initially classified as a single pathology, we now know that there are more than 50 different molecular genetic forms of hereditary neuropathies. Mutations from at least 30 genes are known to cause these disorders. These genes have diverse functions, mostly related to intracellular membranes and organelles, but no function is common among them.

Hereditary sensory neuropathy type 1 (HSN1) was originally classified as simply a sensory neuropathy but, as motor neuron degeneration is also involved, it is actually a dominantly inherited sensory and motor neuropathy. We have shown that it is caused by mutations in SPTLC1. This is subunit one of the heterodimeric enzyme SPT, which is one of the 30 genes that cause motor and sensory neuropathies. HSN1 is clinically an axonal neuropathy affecting distal long nerves that spreads proximally over a lifetime. It is a "dying-back" neuropathy that degenerates the motor and sensory axons distally to proximally.

2. Hereditary Sensory Neuropathy type 1

Hereditary axonal neuropathies include a disparately large group of disorders that primarily affect axons of peripheral nerves with distal axonal degeneration. Defects in axonal transport are most commonly implicated in this distal distribution, however, the assortment of protein mutations contributing to axonal degeneration indicates that this degeneration is a final common path, or the Achilles' heel of peripheral neurons.

HSN1 is an autosomal-dominant disorder of peripheral sensory neurons with progressive degeneration of dorsal root ganglia. The onset of clinical symptoms usually occurs during the second or third decades of life. The initial symptoms are sensory loss in the feet, followed by distal muscle wasting and weakness, as the longest nerves are involved first. The loss of pain sensation leads to painless injuries and chronic skin ulcers. Eventually distal amputations may be needed for osteomyelitis. HSN1 is the most common and best characterized of the degenerations of sensory neurons. Degeneration of dorsal root ganglia (DRG) and ventral horn neurons is reported as well as one report of amyloid-like material accumulating in the DRG (Denny-Brown, 1951). These complications cause long-term disabilities with economic and social repercussions. Affected individuals are unable to work through most of their adult life and require lifelong medical and economic support. Treatment is entirely symptomatic (Dyck, 1993; Thomas, 1993).

3. Discovery of the HSN1 gene mutations

In 1996 our group first tracked the suspect mutation to chromosome 9, and subsequently narrowed the candidate region (Blair et al., 1997; Blair et al., 1998; Hulme et al., 2000; Nicholson et al. 2001; Nicholson et al. 1996) to locate mutations in the transcription region of SPTLC1 (Dawkins et al. 2001). These were confirmed by Bejaoui et al., (2001). No mutations were found in SPT subunit 2 (SPTLC2) (Dawkins et al., 2002). The most common mutation within HSN1 families of Australian/English origin (eight families) was a single base mutation 399T→G in exon 5 of SPT1 coding region, which was predicted to result in a single amino acid substitution of cysteine by tryptophan at the position 133 of SPTLC1. Two families had mutations 431T→A and another had mutation 398G→A, resulting in substitution of valine by aspartate at position 144 and cysteine by tyrosine at position 133, respectively.   Recently, Verhoeven et al., (2004) described a novel missense *SPTLC1* mutation in twin sisters with HSN1. This base mutation occurs at 1160G→C in exon 13, changing   glycine for alanin at the position 387 of SPTLC1. How this mutation effects SPT activity is still unknown. It will be interesting to know whether this mutation also inhibits SPT activity.

4. Rationale for exploring mechanisms of distal axonopathy

The distal nature of the disorder suggests that a length dependent process is affected. Specifically axonal transport of essential substances could be impaired or distal growth factors needed for maintenance of the ends of nerves may fail (Kruttgen et al., 2003). A lack of nerve growth factor (NGF) uptake or transport causes dying-back of nerves and cell death of pain and temperature sensitive neurons. Alterations in NGF levels in the target tissues for sensory neurons profoundly alter nociceptive transmission in mice. Over-expression leads to hyperalgesia (Molliver et al., 2005) and reduced levels producing profound hypoalgesia (Davis et al., 1993). Selective decrease in axonal NGF was found in axonopathies with an unknown cause (Fressinaud et al., 2003). There are also congenital developmental disorders of pain and temperature fibers due to mutations in NGF and one of its receptors. HSN's primary defect is the loss of pain and temperature fibers and is a distal dying-back neuropathy, so SPT mutations may affect axonal transport of NGF and other neurotrophins.

5. Do the mutations affect SPT activity?

In more recent studies (Bejaoui et al. 2002; Gable et al. 2002), the effect of HSN1 mutations on cellular physiology was examined in several non-human cell culture models. Earlier results on human HSN1 lymphocytes suggested substantial increased levels of glucosyl ceramide levels in HSN1 patients compared to that of controls (Dawkins et al., 2001). This however, was not supported by further evidence (Dedov et al., 2004). Attempts to over-express mutant *LCB1* in yeast or CHO cells resulted in loss of SPT activity and cell death (Bejaoui et al. 2002; Gable et al. 2002). Such significant toxic effects may be due to uncontrolled over-expression of mutant protein. In these CHO cells, expression of the enzyme was ten-fold over normal (Hanada et al. 2000). Bejaoui et al. (2002) and Gable et al. (2002) also reported that in transformed lymphocytes from HSN1 patients SPT enzyme activity was inhibited by 50%, while sphingolipid synthesis was reduced. HSN1-associated mutations likely confer dominant negative effects on SPT activity and thereby cause the disease.

However, fundamental arguments against the 'loss of activity' model of the disease exist. First, a 50% loss of enzyme activity does not cause any known autosomal dominant disease. In fact, mutations in both copies of a gene with a nearly complete loss of enzyme function are prerequisites for autosomal dominant diseases. This is not the case for HSN1. Moreover, the disease occurs after the second decade of life and individuals carrying the mutations develop normally. If a defect in enzyme activity causes cell death, this should manifest early in life. In transformed lymphocytes from HSN1 patients we found normal SPT activity and sphingolipid profiles. This data correlated with normal cell proliferation and cell death (apoptotic and necrotic) rates in HSN1 patient cells. Moreover, inhibition of SPT activity by ISP-1, a highly specific SPT inhibitor, did not affect cell proliferation and cell death rates. We also checked for any previously undiscovered changes in haematopoiesis that might be caused by HSN1 mutations. There were no significant differences in whole blood counts of control donors and HSN1 patients (Dedov et al., 2004). It is possible that it is not the loss of SPT enzyme activity but rather the accumulation of mutant protein that leads to cell death, which supports the gain of toxic function model.

6. Protein instability, aggregation and interference with protein trafficking

The mechanism of neuronal cell damage in HSN1 may be similar to protein aggregation or accumulation in other neurodegenerative disorders that has been implicated in several neuronopathies, such as Parkinson's and Alzheimer's diseases. However, how such a mechanism might cause axonal transport defects prior to neuronal cell death in HSN has yet to be determined. Figure 1 shows schematically the differences between axonopathies and neuronopathies in the neuronal cell.

Neural cell death could be precipitated by either (a) aggregation of a mutant protein in the neuronal cell or (b) by secondary protein interactions with a mutant SPTLC1 protein causing toxic effects either in the cell body or the distal axon. As, SPTLC1 is predicted to be an inherently unstable protein (database ref: au.expasy.org/tools/protparam.html), it may require specific conditions, for example, interactions with a chaperone-like protein to maintain its normal conformation or its normal function. Mutations are therefore likely to destabilize the protein and cause abnormal protein interactions.

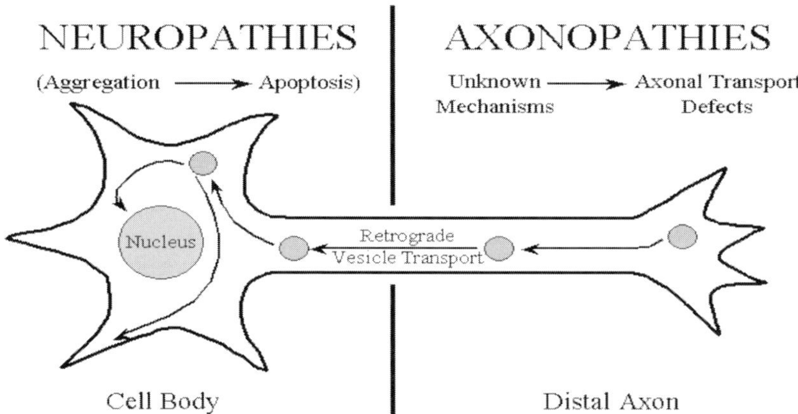

Fig. 1. A schematic of neuronal cell, showing retrograde vesicle transport from distal axon to the cell body.

7. Final Remarks

How mutations in a ubiquitously expressed housekeeping enzyme damage HSN1 sensory neurons and motor nerves is completely unexplored, but the process may bear similarities with SOD1 mutations that cause familial motor neuron disease (ALS). HSN1 models will provide a precise mechanism to determine the pathway that turns a mutant protein into a cell toxin. The link between SPTLC1 mutations and distal axonal degeneration is yet unexplored and may be the common mechanism in many other neurological disorders. If so, this would indicated targets on which to develop effective therapies.

8. References

Bartlett SE, Reynolds AJ, Hendry IA (1999) The regulation of the retrograde axonal transport of ^{125}I-beta nerve growth factor is independent of calcium. *Brain Res*, **837**, 8-14.

Bejaoui K, Wu C, Scheffler MD, Haan G, Ashby P, Wu L, de Jong P, Brown RH Jr. (2001) SPTLC1 is mutated in hereditary sensory neuropathy, type 1. *Nat Genet*, **27**, 261-262.

Bejaoui K, Uchida Y, Yasuda S, Ho M, Nishijima M, Brown RH, Jr., Holleran WM (2002) Hereditary sensory neuropathy type 1 mutations confer dominant negative effects on serine palmitoyltransferase, critical for sphingolipid synthesis. *J Clin Invest*, **110**, 1301-1308.

Blair IP, Dawkins JL, Nicholson GA (1997) Fine mapping of the hereditary sensory neuropathy type I locus on chromosome 9q22.1-]q22.3 - exclusion of GAS1 and XPA. *Cytogenet Cell Genet*, **78**, 140-144.

Blair IP, Hulme D, Dawkins JL, Nicholson GA (1998) A yac-based transcript map of human chromosome 9q22.1-q22.3 encompassing the loci for hereditary sensory neuropathy type I and multiple self-healing squamous epithelioma. *Genomics*, **51**, 277-281.

Davis BM, Lewin GR, Mendell LM, Jones ME, Albers KM (1993) Altered expression of nerve growth factor in the skin of transgenic mice leads to changes in response to mechanical stimuli. *Neurosci*, **56**, 789-792.

Dawkins JL, Brahmbhatt S, Auer-Grumbach M, Wagner K, Hartung HP, Verhoeven K, Timmerman V, De Jonghe P, Kennerson M, LeGuern E, Nicholson GA (2002) Exclusion of serine palmitoyltransferase long chain base subunit 2 (SPTLC2) as a common cause for hereditary sensory neuropathy. *Neuromuscul Disord*, **12**, 656-658.

Dawkins JL, Hulme DJ, Brahmbhatt SB, Auer-Grumbach M, Nicholson GA (2001) Mutations in SPTLC1, encoding serine palmitoyltransferase, long chain base subunit-1, cause hereditary sensory neuropathy type I. *Nat Genet*, 27, 309-312.

Dedov VN, Dedova IV, Merrill AH Jnr, Nicholson GA (2004) Activity of partially inhibited serine palmitoyltransferase is sufficient for normal sphingolipid metabolism and viability of HSN1 patient cells. *Biochim Biophys Acta*, **1688**, 168-175.

Denny-Brown D (1951) Hereditary sensory radicular neuropathy. *J Neurochem*, **14**, 237-252.

Dyck PJ (1993) Neuronal atrophy and degeneration predominantly affecting peripheral sensory and autonomic neurones. in *Peripheral Neuropathy* (Dyck PJ, Thomas PK, Griffin JW, Low PA & Poduslo JF, Eds.), W B Saunders, Philadelphia

Fressinaud C, Jean I, Dubas F (2003) Selective decrease in axonal nerve growth factor and insulin-like growth factor 1 immunoreactivity in axonopathies of unknown etiology. *Acta Neurpathol*, **105**, 477-483.

Gable K, Han G, Monaghan E, Bacikova D, Natarajan M, Williams R, Dunn TM (2002) Mutations in the yeast LCB1 and LCB2 genes, including those corresponding to the hereditary sensory neuropathy type I mutations, dominantly inactivate serine palmitoyltransferase. *J Biol Chem*, **277**, 10194-10200.

Ginty DD, Segal RA (2002) Retrograde neurotrophin signaling: Trk-ing along the axon. *Curr Opin Neurobiol*, **12**, 268-274.

Hanada K, Hara T, Nishijima M (2000) Purification of the serine palmitoyltransferase complex responsible for sphingoid base synthesis by using affinity peptide chromatography techniques. *J Biol Chem*, **275**, 8409-8415.

Hulme DJ, Blair IP, Dawkins JL, Nicholson GA (2000) Exclusion of NFIL3 as the gene causing hereditary sensory neuropathy type I by mutation analysis. *Hum Genet*, **106**, 594-596.

Kamholz J, Menichella D, Garbern JA, Lewis RA, Krajewski KM, Lilien J, Scherer SS, Shy ME (2000) Charcot-Marie-Tooth disease type 1: molecular pathogenesis to gene therapy. *Brain*, **123**, 222-233.

Kruttgen A, Saxena S, Evangelopoulos ME, Weis J (2003) Neurotrophins and neurodegenerative diseases: receptors stuck in traffic? *J Neuropathol Exp Neurol*, **62**, 340-350.

Molliver DC, Lindsay J, Albers KM, Davis BM (2005) Overexpression of NGF or GDNF alters transcriptional plasticity evoked by inflammation. *Pain*, **113**, 277-284.

Nicholson GA, Dawkins JL, Blair IP, Auer-Grumbach M, Brahmbratt SB, Hulme DJ (2001) Hereditary sensory neuropathy type I: Haplotype analysis shows founders in southern England and Europe. *Am J Hum Genet*, **69**, 655-659.

Nicholson GA, Dawkins JL, Blair IP, Kennerson ML, Gordon MJ, Cherryson AK, Nash J, Bananis T (1996) The gene for hereditary sensory neuropathy type I (HSN1) maps to chromosome 9q22.1-q22.3. *Nat Genet*, **13**, 101-104.

Thomas PK (1993) Hereditary sensory neuropathies. Brain Pathol, 157-163.

Verhoeen K, Coen K, De Vriendt E, Jacobs A, Van Gerwen V, Smouts I, Pou-Serradell A, Martin JJ, Timmerman V, De Jonghe P (2004) SPTLC1 mutation in twin sisters with hereditary sensory neuropathy type I. *Neurology*, **62**, 1001-1002.

5-7 Ceramide, Ceramide Kinase and Vision Defects: A BLIND Spot for LIPIDS

Roser Gonzàlez-Duarte, Miquel Tuson, and Gemma Marfany

Departament de Genètica, Facultat de Biologia, Universitat de Barcelona, Spain

Keywords. Retinitis Pigmentosa (RP), photoreceptors, ceramide, CERKL, ceramide kinase

Retinitis Pigmentosa (RP), the most prevalent inherited retinal disorder in humans (1 in 3000), is characterized by progressive, bilateral and symmetrical degeneration of photoreceptors. Prominent clinical features include night blindness and constriction of visual fields, generally leading to complete blindness (Hims et al., 2003). RP is a monogenic disorder with extremely high clinical and genetic heterogeneity. Genetic studies to date have identified around 40 genes and loci (RetNet, Retinal Information Network, http://www.sph.uth.tmc.edu/Retnet/) responsible for the phenotype, yet the causative genes for more than half of the diagnosed cases remain unknown. Allelic heterogeneity, when different mutations of the same gene cause the same phenotype such as *ABCA4* (Martínez-Mir et al., 1998) which lead to retinal disease, further complicates the genetic scenario. From photon reception to amplified synaptic transmission and final image integration in the brain, vision is a complex biological process. It is hardly surprising, therefore, that there are so many genes involved in RP. Candidate genes for RP normally participate in a variety of activities, including the phototransduction cascade and the visual cycle, and encode transcription and splicing factors. Hence, identifying RP genes requires a multifaceted, collaborative set of strategies drawing on direct mutational analysis to laborious positional cloning that involves genome-wide linkage analyses to map the target gene chromosomal regions Subsequent gene sequencing eventually exposes pathogenic mutations.

There are six types of neurons layering the human retina: rod- and cone-photoreceptors, bipolar cells, horizontal, amacrine and ganglion cells, which form the optic nerve with their axons. The outer segments of photoreceptors have stacks of membranous disks that are generated by plasma membrane invaginations. These disk contain most of the structural and enzymatic proteins involved in light reception, phototransduction, and the visual cycle, and those disk continuously regenerate. The retinal pigment epithelium (RPE) cells, that protect photoreceptors, ingest the last disk of a rod or cone.

Several mouse model studies revealed that apoptosis triggers retinal degeneration. (Chang et al., 1993; Portera-Cailliau et al., 1994). While a normal event during development, apoptosis in the adult (Young, 1984) is tightly regulated to maintain the exquisite architecture of the retina (Chow and Lang, 2001). Regulation failure leads to disease: escape from apoptosis sets off tumorigenesis (retinoblastoma) whereas uncontrolled cell death causes neurodegeneration (reviewed in (Remé et al., 1998)).

Photoreceptor apoptosis has various catalysts, including genetic mutations from physical or chemical injury or prolonged light exposure (reviewed in (Wenzel et al., 2005)). Once activated, the apoptotic program follows distinct signal transduction pathways that converge on the same effector cascade. Two separate pathways can be activated. One depends on the excess of bright light, which induces the transcription factor AP-1 downstream, and the other depends on the permanent activation of the phototransduction cascade (Hao et al., 2002). Although the events that induce apoptosis have been identified, the primary cellular players, particularly those that monitor a cell's stress level and commits to apoptosis, are yet to be elucidated (Jacobson and McInnes, 2002). Reports on the balance between death and survival in photoreceptor cells implicate the phosphatidylinositol 3-kinase (PI3-K)/Akt kinase pathway in retinal neuron survival (Johnson et al., 2005).

Overall, several exciting issues emerge. First, modest light exposures, well within visible intensity, may damage rod cells (it is worth noting that phototoxic wavelength coincides with the absorbance spectrum of rhodosin, rod visual pigment, see minireview in (Jacobson and McInnes, 2002)). Therefore, several protective mechanisms needed to evolve to avoid massive photoreceptor cell death in daylight. Second, RP genetic mutations make photoreceptors more susceptible to damage, diminishing their resilience, thereby tilting the balance towards apoptosis over survival (Wenzel et al., 2005). This explains the progressive depletion of photoreceptors. Third, susceptibility to light damage and retinal degeneration is influenced by many genetic and environmental factors as evidenced by i) phenotypic diversity observed in patients sharing the same mutation (even within the same family) and, ii) high prevalence of age-related Macular Degeneration (AMD), a multifactorial retinal dystrophy. Fourth, as already stated, despite uncovering some of the molecular events initiating photoreceptor apoptosis, most transduction signal mediators are still unknown (Hao et al., 2002). Finally, if all the genetic mutations of RP (and other retinal dystrophies) lead to photoreceptor apoptosis, one potential treatment strategy would be to promote photoreceptor survival, either by addition of neurotrophic factors (Chaum, 2003) or by prevention of apoptosis (Wenzel et al., 2005) in these diseases using in a mutation-independent manner. To this end, understanding the mechanisms and pathways by which photoreceptor cells commit to survival or death is crucial.

When we identified the role of CERKL, a presumptive ceramide kinase, in several familial cases of RP another link was added to the chain. This finding showed that sphingolipid molecules function as mediator signals for photoreceptor apoptosis and survival. It also showed that mutations in genes encoding sphingolipid-metabolism enzymes could lead to susceptibility for retinal degeneration disorders. Studies of *Drosophila* retinal degeneration mutants showed fly photoreceptor cells could be rescued from apoptosis by modulating ceramide levels via ceramidase overexpression in rhabdomeres (Acharya et al., 2003; Rohrbough et al., 2004). Moreover, oxidative-stress induced cell death in cultured human RPE cells showed that ceramide functions as a second messenger in that apoptosis (Barak et al., 2001). Further connections between ceramide levels and photoreceptor apoptosis were made from a study of two genes associated with Batten disease, a juvenile neurodegenerative disorder characterized by cerebral atrophy and retinitis pigmentosa (Luberto et al., 2002).

In the sphingolipid metabolism network, ceramide is the hub connecting all other sphingolipids (Hannun and Obeid, 2002). Ceramide is a regulator of eukaryotic stress response and high ceramide levels can arrest cell growth and cause apoptosis (Hannun and Luberto, 2000; Pettus et al., 2002). Yet, the picture is much more complex. A comprehensive view should include not only interconvertible lipids, ceramide, sphingosine, but also their phosphate forms as the latter serve antagonistic functions of the former (Futerman and Hannun, 2004). Considering the role of these closely related lipids as cellular biosensors (Spiegel et al., 2003), those enzymes involved generating, modifying or removing these lipids (and in our case, a ceramide kinase-like enzyme) must be finely attuned to the cell state. In addition to regulating apoptosis and cell growth (Gómez-Muñoz, 2004), ceramide and ceramide-1-phosphate are also involved in phagocytosis (Hinkovska-Galcheva et al., 2005), endocytosis (Acharya et al., 2004), membrane fusion of synaptic vesicles (Rohrbough et al., 2004), and angiogenesis and inflammation (reviewed in (Pettus et al., 2004) (Baumruker et al., 2005)). All these processes are relevant to either the aetiology, progression and pathological manifestations of these retinal dystrophies.

Being one of the most hydrophobic cellular lipids, ceramide cannot move freely around the cell. It is confined to the particular membrane organelles where it is generated, unless suitable transfer chaperon proteins are provided (Hanada et al., 2003), thus hinting at new means for regulation (Hannun and Luberto, 2004). Other groups have proposed that cells respond differently to ceramide generated in distinct cellular compartments (Bionda et al., 2004; Futerman and Hannun, 2004). Thus, response threshold and cell fate may depend on when and how (by *de novo* or breakdown synthesis) ceramide is produced. It is worth noting that   all the proteins involved in light reception and transduction in photoreceptors are imbedded in, or associated with, the membranes. Thus, it is conceivable that the first stress sensors in photoreceptors would be the lipids in membranes housing phototransduction machinery. Ceramide production may be the retinal cell response to photic and oxidative stress (as well as to misfolded or dysfunctional proteins due to RP genetic mutations). Depending on the overall health of the cell, ceramide production at various sites may be determined by the availability of enzymes, such as ceramide kinase or ceramidase (Fig.1) ultimately effecting the photoreceptor's fate. Consistent with this, ceramide levels are increased in different *Drosophila* retinal degeneration mutants and targeted ceramidase over-expression rescues photoreceptors from apoptosis (Acharya et al., 2003). In this context, phosphorylation of ceramide would reduce levels of ceramide and increase pools of ceramide-1-phosphate, a signalling molecule that promotes cell survival. This protective role seems to be exerted in macrophages through direct induction of the PI3-K/Akt pathway (Gómez-Muñoz et al., 2005) and indirectly, by inhibition of acid sphingomyelinase, thereby reducing ceramide production (Gómez-Muñoz et al., 2004). It is then feasible that ceramide-1-phosphate also mediates cell survival in the retina, as the PI3-K/Akt pathway has already been shown to be active in retinal neurons (Johnson et al., 2005).

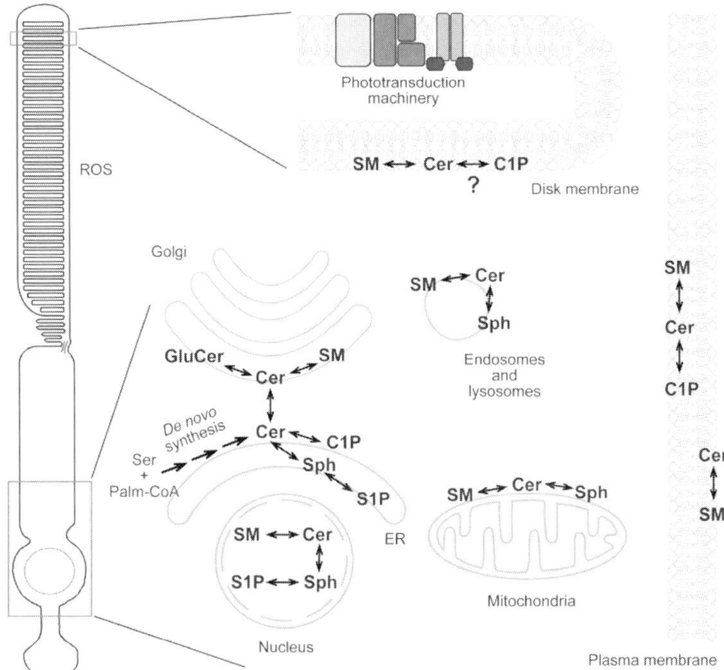

Fig. 1. Subcellular compartimentalization of ceramide metabolism. Scheme showing photoreceptor rod cell with the membranous disks stacked in the rod outer segment (ROS) and the hypothesized sites of ceramide (Cer) and ceramide-1-phosphate (C1P) production according to references cited in the text. Note the location of the proteins involved in photon reception and phototransduction in the disk membrane. Other abbreviations: ER: endoplasmic reticulum; GluCer: glucosylceramide; SM: sphingomyelin; Sph: sphingosine; S1P: sphingosine-1-phosphate; Ser: serine; Palm-CoA: palmitoyl-CoA.

Although confirmation that CERKL is a *bona fide* ceramide kinase is still lacking and its substrate/s are proving elusive (Bornancin et al., 2005); our unpublished results), it is clearly a lipid kinase, phylogenetically clustering with ceramide kinases and shares most of the distinctive signature domains of the ceramide kinase family (Tuson et al., 2004; Baumruker et al., 2005; Bornancin et al., 2005). CERKL appears to have been an evolutionary invention of vertebrates, as no homologues have been identified in invertebrates. Orthologues of CERK are found in pluricellular eukaryotes, however. At least in human and mouse, CERK and CERKL are coexpressed in some tissues. Determining whether their function is complementary or partially redundant is not a trivial question.

Standard protocols for assaying ceramide kinase activity appear ineffective with CERKL. This requires that several issues concerning

CERKL be next addressed, including a) the correct subcellular location of CERKL is relevant to the enzymatic activity; b) whether unknown partners are needed for CERKL to be active or recognise its substrate, c) whether, as happens with many other enzymes acting in signalling pathways, CERKL activity is regulated by specific modifications (phosphorylation, ubiquitylation or sumoylation), which may occur only as a response to particular stimuli and, d) if the substrate is a sphingolipid similar to ceramide. All of these are exciting avenues worth exploring in the near future.

What is then the function of CERKL in the retina? What can we learn from the role of ceramide in retinal disorders? Our preliminary analyses of CERKL expression reveal a complex pattern of alternative splicing, generating several in-frame isoforms, some of which are devoid of the lipid-kinase domain. It is tempting to speculate about different functions for each isoform, some of them with kinase functions, others more related to ceramide traffic or storage. The generation of an adequate animal model may hold most of the answers. If CERKL is indeed a ceramide kinase, its function could be directly related to neuronal survival that phosphorylates ceramide generated in response to photoreceptor stress thereby halting the apoptosis signals and allowing the cell to survive. If ceramide is one of the first mediators of photoreceptor apoptosis in RP and other retinal genetic dystrophies, the modulation of the sphingolipid metabolism in photoreceptors, irrespective of the underlying genetic defect, could constitute a universal therapeutic strategy.

Acknowledgements. We apologise for not citing all of the relevant contributions due to space constraints. We thank all the patients and their families, as well as all past and present members of our group, whose work set the foundations for and encouraged us to continue future work despite all difficulties. We thank Robin Rycroft for revising the English. This work has received financial support by grants from FUNDALUCE, BIDONS EGARA, BMC2003-05211 (MCyT) to R. G-D.

References

Acharya, U., Mowen, M. B., Nagashima, K., and Acharya, J. K. (2004). Ceramidase expression facilitates membrane turnover and endocytosis of rhodopsin in photoreceptors. *Proc Natl Acad Sci U S A*, **101**, 1922-1926.

Acharya, U., Patel, S., Koundakjian, E., Nagashima, K., Han, X., and Acharya, J. K. (2003). Modulating sphingolipid biosynthetic pathway rescues photoreceptor degeneration. *Science*, **299**, 1740-1743.

Barak, A., Morse, L. S., and Goldkorn, T. (2001). Ceramide: a potential mediator of apoptosis in human retinal pigment epithelial cells. *Invest Ophthalmol Vis Sci*, **42**, 247-254.

Baumruker, T., Bornancin, F., and Billich, A. (2005). The role of sphingosine and ceramide kinases in inflammatory responses. *Immunol Lett*, **96**, 175-185.

Bionda, C., Portoukalian, J., Schmitt, D., Rodriguez-Lafrasse, C., and Ardail, D. (2004). Subcellular compartmentalization of ceramide metabolism: MAM (mitochondria-associated membrane) and/or mitochondria? *Biochem J*, **382**, 527-533.

Bornancin, F., Mechtcheriakova, D., Stora, S., Graf, C., Wlachos, A., Devay, P., Urtz, N., Baumruker, T., and Billich, A. (2005). Characterization of a ceramide kinase-like protein. *Biochim Biophys Acta*, **1687**, 31-43.

Chang, B., Heckenlively, J. R., Hawes, N. L., and Roderick, T. H. (1993). New mouse primary retinal degeneration (rd-3). *Genomics*, **16**, 45-49.

Chaum, E. (2003). Retinal neuroprotection by growth factors: a mechanistic perspective. *J Cell Biochem*, **88**, 57-75.

Chow, R. L., and Lang, R. A. (2001). Early eye development in vertebrates. *Annu Rev Cell Dev Biol*, **17**, 255-296.

Futerman, A. H., and Hannun, Y. A. (2004). The complex life of simple sphingolipids. *EMBO Rep*, **5**, 777-782.

Gómez-Muñoz, A. (2004). Ceramide-1-phosphate: a novel regulator of cell activation. *FEBS Lett*, **562**, 5-10.

Gómez-Muñoz, A., Kong, J. Y., Parhar, K., Wang, S. W., Gangoiti, P., Gonzalez, M., Eivemark, S., Salh, B., Duronio, V., and Steinbrecher, U. P. (2005). Ceramide-1-phosphate promotes cell survival through activation of the phosphatidylinositol 3-kinase/protein kinase B pathway. *FEBS Lett*, **579**, 3744-3750.

Gómez-Muñoz, A., Kong, J. Y., Salh, B., and Steinbrecher, U. P. (2004). Ceramide-1-phosphate blocks apoptosis through inhibition of acid sphingomyelinase in macrophages. *J Lipid Res*, **45**, 99-105.

Hanada, K., Kumagai, K., Yasuda, S., Miura, Y., Kawano, M., Fukasawa, M., and Nishijima, M. (2003). Molecular machinery for non-vesicular trafficking of ceramide. *Nature*, **426**, 803-809.

Hannun, Y. A., and Luberto, C. (2000). Ceramide in the eukaryotic stress response. *Trends Cell Biol*, **10**, 73-80.

Hannun, Y. A., and Luberto, C. (2004). Lipid metabolism: ceramide transfer protein adds a new dimension. *Curr Biol*, **14**, R163-165.

Hannun, Y. A., and Obeid, L. M. (2002). The Ceramide-centric universe of lipid-mediated cell regulation: stress encounters of the lipid kind. *J Biol Chem*, **277**, 25847-25850.

Hao, W., Wenzel, A., Obin, M. S., Chen, C. K., Brill, E., Krasnoperova, N. V., Eversole-Cire, P., Kleyner, Y., Taylor, A., Simon, M. I., *et al.* (2002).

Evidence for two apoptotic pathways in light-induced retinal degeneration. *Nat Genet*, **32**, 254-260.

Hims, M. M., Diager, S. P., and Inglehearn, C. F. (2003). Retinitis pigmentosa: genes, proteins and prospects. *Dev Ophthalmol*, **37**, 109-125.

Hinkovska-Galcheva, V., Boxer, L. A., Kindzelskii, A., Hiraoka, M., Abe, A., Goparju, S., Spiegel, S., Petty, H. R., and Shayman, J. A. (2005). Ceramide 1-Phosphate, a Mediator of Phagocytosis. *J Biol Chem*, **280**, 26612-26621.

Jacobson, S. G., and McInnes, R. R. (2002). Blinded by the light. *Nat Genet*, **32**, 215-216.

Johnson, L. E., van Veen, T., and Ekstrom, P. A. (2005). Differential Akt activation in the photoreceptors of normal and rd1 mice. *Cell Tissue Res*, **320**, 213-222.

Luberto, C., Kraveka, J. M., and Hannun, Y. A. (2002). Ceramide regulation of apoptosis versus differentiation: a walk on a fine line. Lessons from neurobiology. *Neurochem Res*, **27**, 609-617.

Martínez-Mir, A., Paloma, E., Allikmets, R., Ayuso, C., del Rio, T., Dean, M., Vilageliu, L., Gonzàlez-Duarte, R., and Balcells, S. (1998). Retinitis pigmentosa caused by a homozygous mutation in the Stargardt disease gene ABCR. *Nat Genet*, **18**, 11-12.

Pettus, B. J., Chalfant, C. E., and Hannun, Y. A. (2002). Ceramide in apoptosis: an overview and current perspectives. *Biochimica et Biophysica Acta (BBA) - Molecular and Cell Biology of Lipids*, **1585**, 114-125.

Pettus, B. J., Chalfant, C. E., and Hannun, Y. A. (2004). Sphingolipids in inflammation: roles and implications. *Curr Mol Med*, **4**, 405-418.

Portera-Cailliau, C., Sung, C. H., Nathans, J., and Adler, R. (1994). Apoptotic photoreceptor cell death in mouse models of retinitis pigmentosa. *Proc Natl Acad Sci U S A*, **91**, 974-978.

Remé, C. E., Grimm, C., Hafezi, F., Marti, A., and Wenzel, A. (1998). Apoptotic cell death in retinal degenerations. *Prog Retin Eye Res*, **17**, 443-464.

Rohrbough, J., Rushton, E., Palanker, L., Woodruff, E., Matthies, H. J., Acharya, U., Acharya, J. K., and Broadie, K. (2004). Ceramidase regulates synaptic vesicle exocytosis and trafficking. *J Neurosci*, **24**, 7789-7803.

Spiegel, S., Milstien, S., Shabbits, J. A., and Mayer, L. D. (2003). Sphingosine-1-phosphate: an enigmatic signalling lipid. *Nat Rev Mol Cell Biol*, **4**, 397-407.

Tuson, M., Marfany, G., and Gonzalez-Duarte, R. (2004). Mutation of CERKL, a novel human ceramide kinase gene, causes autosomal recessive retinitis pigmentosa (RP26). *Am J Hum Genet*, **74**, 128-138.

Wenzel, A., Grimm, C., Samardzija, M., and Reme, C. E. (2005). Molecular mechanisms of light-induced photoreceptor apoptosis and neuroprotection for retinal degeneration. *Prog Retin Eye Res*, **24**, 275-306.

Young, R. W. (1984). Cell death during differentiation of the retina in the mouse. *J Comp Neurol*, **229**, 362-373.

5-8 Fumonisin Inhibition of Ceramide Synthase: A Possible Risk Factor for Human Neural Tube Defects

Ronald T. Riley[1], Kenneth A. Voss[1], Marcy Speer[2], Victoria L. Stevens[3], and Janee Gelineau-van Waes[4]

[1]Toxicology and Mycotoxin Research Unit, United States Department of Agriculture-Agricultural Research Service, PO Box 5677, Athens, Georgia 30604, USA; [2]Duke Center for Human Genetics, School of Medicine, Duke University, Durham, North Carolina 27710, USA; [3]Epidemiology and Surveillance Research, American Cancer Society, Atlanta, Georgia, USA;[4]Department of Genetics, Cell Biology & Anatomy, University of Nebraska Medical Center, Omaha, Nebraska 68198-5455, USA

Summary. Fumonisins are carcinogenic mycotoxins that cause farm animal diseases. They commonly contaminate maize and are suspected, but not proven, to cause human disease. Their mode of action involves inhibition of the enzyme (ceramide synthase) that controls the formation of sphingolipids, important regulators of pathways involved in cell death and survival. Sphingolipids are needed for the proper function of receptors associated with lipid rafts (for example, the high affinity folate-binding protein). Fumonisin disruption of folate transport interferes with neural tube closure in animal models in vitro, and this effect is reduced by folate supplementation. In vivo studies in the LMBc mouse strain have shown that maternal fumonisin administration during pregnancy causes a dose-related increase in the frequency of neural tube defects (NTDs) in the embryos. Co-exposing the dams to folate or ganglioside GM_1 is protective, suggesting that fumonisin alters sphingolipid-dependent lipid raft function. In addition, altered expression of cytokines, inducible nitric oxide synthase, elevated levels of sphinganine and sphinganine 1-phosphate and several genes involved in redox homeostasis are observed in affected embryos. NTDs are the second most common birth defect in humans. The etiology of human NTDs is complex. Increased risk has been associated with genetic predisposition, dietary exposure to environmental contaminants, and reduced intake of folate and other vitamins/nutrients. Human clinical and epidemiological studies show folate supplementation reduces the risk for NTDs. While there is no direct evidence for fumonisin as a cause of NTD in humans, the incidence of NTD is higher where maize consumption is high, and both fumonisin exposure and folate deficient diets are likely. Thus, it has been hypothesized that fumonisin inhibition of ceramide synthase is a risk factor for NTDs in humans with folate deficient diets who consume large quantities of low quality maize.

Keywords. Fumonisin, Ceramide Synthase, Folate, Neural Tube Defects

1. Introduction

Fumonisins are water soluble fungal toxins produced by the fungus *Fusarium verticillioides* (formerly *F. moniliforme*) which is frequently found on maize wherever it is grown. The most prevalent of the fumonisins is fumonisin B_1 which has been shown to cause liver and kidney cancer in rats and mice (NTP, 2001; Gelderblom et al., 1991). The International Agency for Research on Cancer (IARC) evaluated fumonisin B_1 as a Group 2B carcinogen, i.e. possibly carcinogenic to humans (IARC, 2002). In 2002, the Joint FAO/WHO Expert Committee on Food Additives allocated a group provisional maximum tolerable daily intake (PMTDI) of 2 µg/kg body weight for fumonisins B_1, B_2 and B_3, alone or in combination (WHO, 2001). In countries where maize provides a substantial percentage of the daily caloric intake, the PMTDI is often exceeded. Exposure varies because of differences in levels of contamination and in dietary habits in various parts of the world. For example, consumption of maize in certain parts of Africa is on the order of 400 g/person/day (Shephard, 2004). If the maize contains 2 ppm of total fumonisins (the USFDA Industry Guidance for degermed dry milled maize products), then dietary exposure for a 60 kg adult would be about 650 percent of the PMTDI. In areas were maize intake is low, the PMTDI is seldom exceeded.

Recently, it has been suggested that fumonisin exposure could be a risk factor for neural tube defects (NTDs) (Marasas et al., 2004). NTDs are common congenital malformations that occur when the embryonic neural tube, which ultimately forms the brain and spinal cord, fails to close properly during the first few weeks of development (Campbell et al., 1986). There are three lines of evidence that support the contention that the fungal toxin, fumonisin, could be a risk factor for the high incidence of NTDs in certain parts of the world (Marasas et al., 2004). First, fumonisins disrupt folate transport (Stevens and Tang, 1997) through inhibition of ceramide synthase (Wang et al., 1991), and folate deficient diets increase the risk of NTDs in humans (Czeizel and Dudas, 1992). Second, fumonisin disruption of sphingolipid metabolism is a known cause of disease in farm animals that consume large amounts of maize-based feeds (Bolger et al., 2001). Further, fumonisins have been shown to cause NTDs in mouse models *in vitro* and *in vivo*. NTD incidence in these mouse models is reduced by folate supplementation and supplementation with ganglioside GM_1 (Sadler et al., 2002; Gelineau van Waes et al., 2005). Third, fumonisins occur worldwide in maize, and high incidences of NTDs have been reported in areas of the world where maize consumption is high (reviewed in Marasas et al., 2004). The remainder of this review will briefly summarize the data supporting the three lines of evidence.

2. Fumonisin inhibition of ceramide synthase and folate transport

The structural similarity between sphinganine and fumonisin B_1 led to the hypothesis that its mechanism of action was via disruption of sphingolipid metabolism or a function of sphingolipids (Wang et al., 1991). Fumonisin potently inhibits the acylation of sphinganine and sphingosine in every cell line, animal, plant and fungus in which it has been tested (Riley et al., 2001) (Fig. 1). Inhibition of ceramide synthase is competitive with respect to both the long-chain (sphingoid) base and fatty acyl-CoA (Merrill et al., 2001). The inhibition of ceramide synthase by fumonisins causes decreased biosynthesis of complex sphingolipids and marked increases intracellular sphinganine concentration in all species tested (Riley et al., 2001). Recently, it has been shown that sphinganine 1-phosphate accumulates in blood of fumonisin exposed pigs and horses (Piva et al., 2005; Constable et al., 2005), where it is hypothesized to act directly on S1P receptors (formerly EDG receptors). Sphingosine- and sphinganine 1-phosphate also increase in their tissues (Riley et al., 2005a). The amount of free sphinganine and sphinganine 1-phosphate that accumulates is probably dependent on the relative activities of enzymes in the sphingolipid biosynthetic and turnover pathways. Once accumulated, free sphingoid bases can persist in tissues (especially kidney), and sub-threshold doses can prolong the elevation of free sphinganine caused by a higher dose (Wang et al., 1999; Enongene et al., 2002). Disrupted sphingolipid metabolism also leads to imbalances in phosphoglycerolipid metabolism, and inhibition of ceramide biosynthesis prevents the formation of more complex sphingolipids (Merrill et al., 2001). The adverse effects of fumonisin-induced depletion of more complex sphingolipids have been demonstrated in numerous studies (for review see IPCS, 2000; Bolger et al., 2001). The loss of complex sphingolipids also plays a role in the abnormal behavior, altered morphology, and altered proliferation of fumonisin-treated cells (for review see IPCS, 2000; Bolger et al., 2001). Thus, fumonisin inhibition of ceramide synthase can cause a wide spec trum of changes in lipid metabolism and associated lipid-dependent signaling pathways, and consequently there are many ways that disruption of sphingolipid metabolism can account for the cell damage and animal diseases caused by fumonisins.

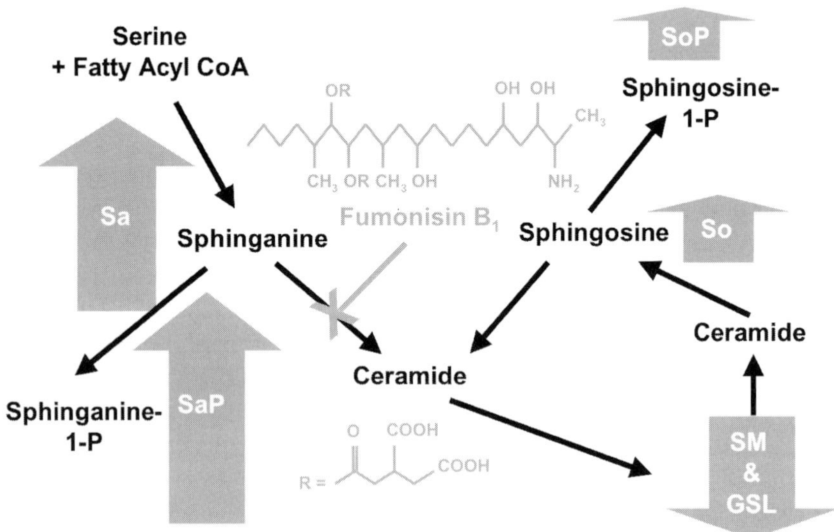

Fig. 1. *De novo* sphingolipid biosynthetic pathway and turnover pathway showing point of disruption by fumonisins B$_1$ (structure inset) and consequences. Arrows indicate the known effect in animal models on various sphingolipid pools; the size and direction of the arrows indicate an increase or decrease in the pool size and the relative magnitude of the change. CSL=complex sphingolipids (gangliosides, sphingomyelin, and others).

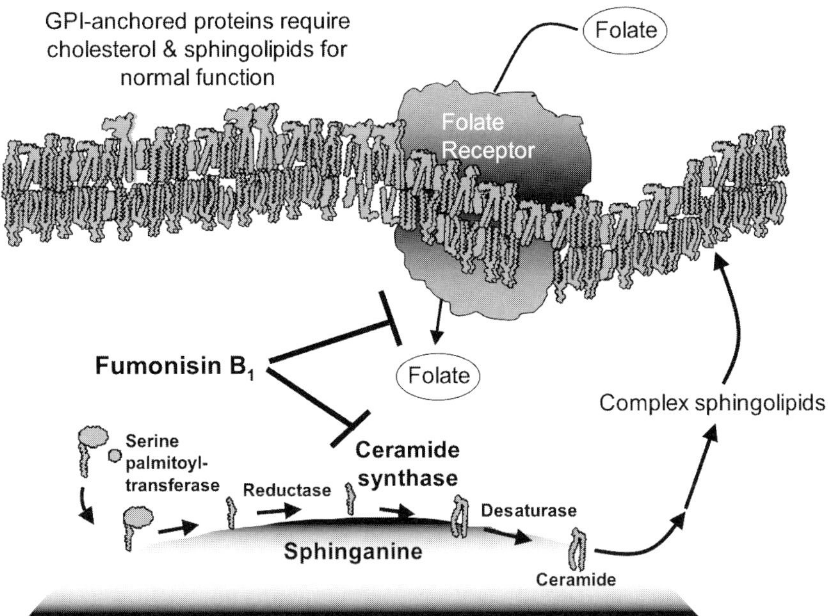

Fig. 2. Disruption of sphingolipid metabolism and folate transport by fumonisins. The scheme shows the step where fumonisins inhibit sphingolipid biosynthesis (the acylation of sphinganine by ceramide synthase in the endoplasmic reticulum), thereby reducing the formation of complex sphingolipids (sphingomyelin, glycosphingolipids), which are important components of the plasma membrane and required for the proper function of glycosylphosphatidylinositol (GPI)-anchored proteins, such as the folate transporter (after Merrill, 2001).

Several early *in vivo* investigations failed to find evidence that fumonisins were teratogenic when orally administered to the dams, or that they crossed the placenta (for review see Voss et al., 2005a). However, fetal toxicity and NTDs were induced in pre-somite rat embryos in culture by hydrolyzed fumonisin B₁ (Flynn et al., 1997). Hydrolyzed fumonisin B₁ is the alkaline hydrolysis product of fumonisin B₁ and, along with other hydrolyzed fumonisins, occurs in tortillas and other nixtamalized food products (Humpf et al., 2004; Palencia et al., 2004). Fumonisin B₁ was significantly more fetotoxic than hydrolyzed fumonisin B₁ but, unlike hydrolyzed fumonisin B₁, did not cause NTDs (Flynn et al., 1996). The relevance of these *in vitro* results to the *in vivo* situation was considered questionable as a dose-response between hydrolyzed fumonisin B₁ concentration in the culture medium and number of NTDs induced in the embryos was not established. Furthermore, multiple *in vivo* experiments had failed to demonstrate increased tissue sphingoid base concentrations, a biomarker of fumonisin exposure, in fetuses of fumonisin-exposed dams (for review see Voss et al., 2005a), suggesting that the placenta effectively prevented *in utero* exposure. Recently, oral administration of hydrolyzed fumonisin B_1 at up to 120 mg/kg bw/day was found to be less toxic than fumonisin B_1, had no effect on the sphinganine to sphingosine ratio, and was not teratogeic in rats (Collins et al., 2005). The additional finding that ^{14}C-fumonisin B_1 did not accumulate in placentae or fetuses following intravenous administration to pregnant rats on gestation day 15, provided further evidence that fumonisins do not cross the rat placenta (Voss et al., 2005a).

The first indication that fumonisins might affect fetuses through an indirect mechanism was the report by Stevens and Tang (1997) that fumonisin B_1 inhibited receptor-mediated transport of folate in CaCo-2 cells. This was an important finding because the folate binding protein 1, *folbp1* (mouse), and folate receptor α (FRα, human), high-affinity placental folate transporters, are glycosylphosphatidylinositol (GPI)-anchored proteins (Luhrs and Slomiany, 1989) that are associated with sphingolipid-rich lipid rafts in cell membranes (Brown and London, 1998). Thus, it was hypothesized that fumonisins could contribute to NTD development by a mechanism (Fig. 2) involving in sequence ceramide synthase inhibition, decreased complex sphingolipids in lipid rafts, disrupted placental folate receptor function, and fetal folate deficiency at the critical time for neural tube closure.

Folate is an essential nutrient critical for a variety of cellular processes involving one-carbon metabolism. Folate deficiency has been linked to an increased risk of NTD in the developing human fetus, and heart disease and cancer in mature individuals. Adequate folate nutrition, as well as competent mechanisms for uptake of this vitamin into cells, is required to maintain the necessary levels of intracellular folate. FRα plays a critical role in neural tube closure; NTDs have been demonstrated in mouse models in which *Folbp1* has been inactivated either through genetic manipulation (Piedrahita et al., 1999), or through modulation with antisense oligonucleotides (Hansen et al., 2003). In addition, maternal administration of antibodies against FRα has been shown to cause NTDs in a rat model (da Costa et al., 2003), and autoantibodies against FRα have recently been described in a subset of women with NTD-complicated pregnancies (Rothenberg et al., 2004). Formerly referred to as the folate binding protein, the folate receptor mediates vitamin uptake into the cell via fluid phase endocytosis. Like other GPI-anchored proteins, the folate receptor resides in membrane microdomains commonly referred to as lipid rafts.

These rafts are enriched in cholesterol and sphingolipids and can affect the function of proteins found in them. In cultured cells, fumonisin B_1-induced inhibition of ceramide synthase significantly altered cellular sphingolipid levels and decreased folate receptor-mediated vitamin uptake into the cells (Fig. 2). Alterations in both the endocytic trafficking and the amount of the receptor that is available for transport contributed to the inhibition of the folate receptor. These findings suggest a possible model which accounts for exposure to fumonisin B_1 increasing the risk of NTDs by disrupting sphingolipid-dependent lipid rafts and, consequently, folate receptor function.

3. NTDs in mouse models, folate and sphingolipids

In vitro concentration-response (25-100 µM fumonisin B_1) studies with cultured neurulating mouse embryos exposed to fumonisin B_1 for 26 h showed that folinic acid supplementation reduced the incidence of NTDs (Sadler et al., 2002). Supplementation also improved embryo growth but did not did not reverse the fumonisin B_1-induced disruption of sphingolipid metabolism in the embryos. In a subsequent experiment it was found that exposing the cultured embryos to 50 µM fumonisin B_1 for two hours resulted in an even higher rate of NTDs (67% of embryos affected) and in the appearance of facial defects (83% of embryos affected); folinic acid supplementation was again protective. These results lent additional support to the mechanistic hypothesis of Stevens and Tang (1997) by showing that fumonisins could affect neuroepithelial and neural crest cells to cause NTDs and craniofacial malformations in a manner that was folate or folate receptor dependent.

The findings of Stevens and Tang (1997) and Sadler et al. (2002) were extended by Gelineau van-Waes when they showed that early gestational exposure of the inbred LMBc mouse strain to fumonisin B_1 (20 mg/kg bw/day by intraperitoneal injection) just prior to neural tube closure (GD 7.5-8.5) resulted in NTDs in 80% of the fetuses (Gelineau van Waes et al., 2005). The lowest intraperitoneal dose to cause NTD was 5 mg/kg bw/day which was equivalent to an oral dose of approximately 50 mg/kg bw. NTDs were also induced by gavage on days GD 7.5 and 8.5 (20 mg/kg bw/day). While this is a high oral dose of fumonisin, it should be noted that the 'no observable effect' level has not been determined and that the number of mice used in the study was small. In humans, an increased incidence of NTD from 10/10,000 to 20/10,000 would be a cause for concern. In addition, it should be noted that the window of susceptibility (period of neural tube closure) could be as short as one or two days and in humans occurs before pregnancy is even suspected. Immunohistochemical studies in the LMBc mice, (gestation day 10.5) revealed that GM_1 and the *Folbp1* folate receptor were co-localized in the yolk sac membrane and the embryonic neuroepithelium, and that staining for *Folbp1* in these structures was noticeably reduced by fumonisin B_1. Daily maternal folate supplementation reduced the incidence of NTDs, while replacement therapy with GM_1 partially restored folate uptake and rescued the phenotype (Fig. 3). Although levels of folate in the embryos were not restored to control levels, folate uptake was apparently adequate to provide protection from fumonisin B_1 toxicity and promote normal neural tube closure. These findings provide unequivocal evidence that compromised lipid raft function is a risk factor for NTD.

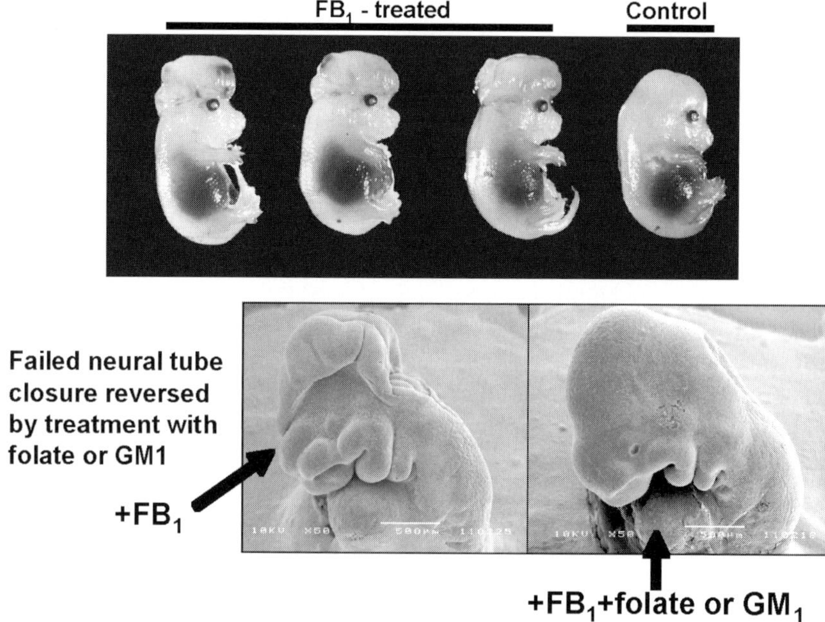

Fig. 3. Effects of fumonisin B_1 (FB_1) in mouse embryonic development. Upper panel - control fetus or fumonisin B_1 (FB_1)-treated fetuses from pregnant mice of the LMBc strain injected i.p. with 20 mg/kg fumonisin B_1 on gestational days 7.5 and 8.5 and sacrificed on day 17.5; the control fetus is at the same gestational age. Lower panel - scanning electron micrograph showing the inhibition of normal neural tube closure in a fetus (GD 10.5) from a fumonisin-treated LMBc dam and a fetus from one treated with fumonisin B_1 and also either folate or ganglioside GM_1 (see text for additional details).

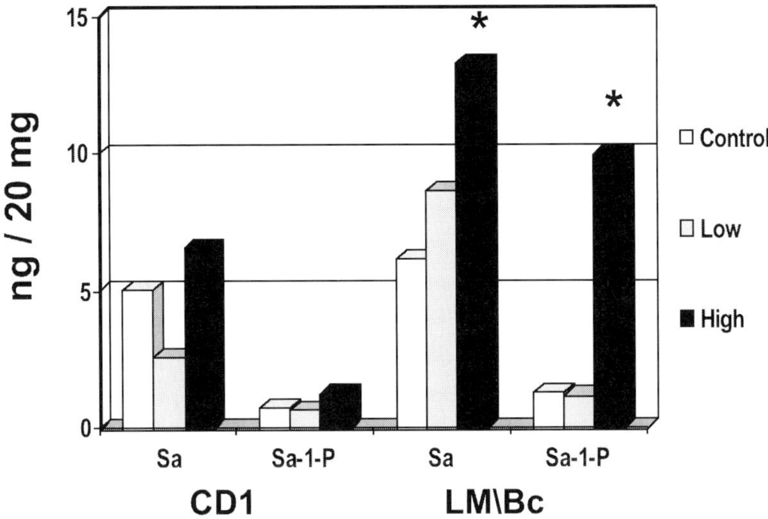

Fig. 4. Concentrations of sphinganine (Sa) and its metabolite Sa-1-phosphate in fetal livers of CD1 and LMBc litters. Values are group means; asterix denotes a significant difference ($p<0.05$) from the control group of the same strain; n=10 CD1, n=4-5 LM/Bc. Low = low dose of 50 ppm fumonisin B_1 in the diet; High = high dose of 150 ppm fumonisin B_1 in the diet.

More recently (Gelineau van Waes et al., 2006), embryos with fumonisin B_1-induced NTDs demonstrated an increase in TNFα expression in the ventricular zone of the open neural tube that corresponds to positive staining for inducible nitric oxide synthase (iNOS) and TUNEL (apoptosis). Microarray data indicate altered expression of cytokines, as well as numerous genes involved in sphingolipid metabolism and redox homeostasis.

Earlier studies showed that oral administration of fumonisin B_1 to CD1 mice during gestation did not cause any teratogenic effects, so investigations to compare the susceptibility of the LMBc and CD1 mouse strains to fumonisin-induced NTDs using a modified dietary exposure regimen have been initiated (Voss et al., 2005b). At the high dose level of 150 ppm FB_1, estimated daily fumonisin B_1 intakes during the five-weeks before mating of 25 mg/kg body weight and 38 mg/kg body weight were calculated for the LMBc and CD1 strains, respectively. Daily intakes of 8 mg/kg and 13 mg /kg were calculated for LMBc and CD1 dams, respectively, fed the low-dose diets. Microscopic examination of the livers established that the high-dose diet was maternally toxic to both strains. One of five (20%) LMBc high-dose litters was positive for NTDs although only one of the 10 fetuses in this litter was affected. NTDs were not found in any CD1 or in low-dose LMBc fetuses. Preliminary analysis of fetal livers for sphinganine and sphinganine 1-phosphate concentrations indicate that sphingolipid metabolism in the LMBc fetuses, but not the CD1 fetuses, was significantly affected as a consequence of maternal dietary fumonisin exposure (Fig. 4). This finding is important because it strongly supports the conclusion that dietary fumonisin can cross the placenta and be taken up by the developing LMBc fetus.

These findings are preliminary and require confirmatory studies, but nonetheless suggest that: 1) the dietary 'no observed adverse effect' level for fumonisin B_1-induced NTDs is ≥ 50 ppm; 2) fetotoxicity and NTDs develop at maternally toxic doses; 3) strain-dependent differences in sensitivity to fetotoxicity and NTDs in mice are likely; 4) compared to the placentae of LMBc mice, those of CD1 mice provide a more effective barrier to prevent *in utero* fumonisin exposure.

4. NTDs in humans, maize consumption and fumonisin exposure

About 3-4% of all babies are born with some type of birth defect (March of Dimes Perinatal Data Center, 2001). The second most common category of birth defects is NTDs, which include spina bifida, anencephaly, and other neural crest cell-related anomalies such as craniofacial malformations (i.e. cleft palate) (Detrait et al., 2005). Spina bifida refers to incomplete development and fusion of one or more vertebral arches, with associated involvement of the posterior neural tube. Anencephaly results from failure of closure of the anterior neural tube, with absence of the bones of the cranial vault, and absent or rudimentary cerebral and cerebellar hemispheres and brainstem. Incidence of these birth defects varies throughout the world. The etiology of these birth defects is complex, with both genetic and environmental influences involved (Detrait et al., 2005; Mitchell, 2005). The environmental risks for NTDs are largely nutritional: adequate periconceptional folate supplementation is known to reduce NTD risk by 50-70% (Czeizel and Dudas, 1992; Green, 2002; MRC, 1991; Shaw et al., 1995; Werler et al., 1993). Other known risks include maternal diseases and/or exposures including obesity, valproate use, and diabetes (Detrait et al., 2005; Mitchell, 2005).

NTD risk varies across ethnic groups suggesting that underlying genetic variation may predispose groups to NTD (Feuchtbaum et al., 1999; CDC, 2000; Mitchell, 2005). American Caucasians have an approximate

1/1000 incidence; whereas, the incidence appears to be higher in the Hispanic population (Detrait et al., 2005). The rate among Hispanics living along the American-Mexican border in Texas is about 15/10,000 live births, thus representing an increasing public health concern for the United States. NTD incidence "spiked" in the Hispanic population living in the border region near Brownsville, Texas in 1990-91, reaching a rate of 27/10,000 live births (Hendricks, 1999) (Fig. 5).

US "Outbreak"

Southeastern Texas
- 1989 Fumonisins high in feeds (to 70 ppm)
- 1990-91 NTD
- Incidence = 27 per 10,000
  - Average = 10 per 10,000

"High Fumonisin" Areas

Southern Africa
- Transkei = 61 per 10,000
- Limpopo = 35 per 10,000

Northern China
- Hebei = 57 per 10,000

Guatamala
- Quetzaltenango = 106 per 10,000

Fig. 5. Hypothesized involvement of fumonisins in a spike ("Outbreak") in NTD incidence in Hispanics living along the Texas/Mexico border during a period when fumonisins were also reported to be high in animal feeds made from maize (Hendricks, 1999). NTD incidence in areas of where maize is consumed in large amounts, and where fumonisin exposure is potentially high (Marasas et al., 2004).

This NTD cluster in 1990-1991 was important because it focused renewed attention on a possible role of fumonisins as reproductive toxins (Marasas et al., 2004; Hendrick, 1999). Hendricks (1999) led an investigation into this NTD cluster and noted that the fumonisin concentrations in locally grown maize in 1989-1991 were higher than average, and hypothesized that fumonisins might have been involved. The epidemiological findings of Missmer et al. (unpublished data summarized in Marasas et al., 2004) were consistent with this hypothesis. Specifically, they reported an association between an increased likelihood of NTD-affected pregnancy and eating an "intermediate" (defined as 301-400 tortillas) amount of homemade tortillas during the first trimester of pregnancy. Their results also suggested that women consuming commercially manufactured tortillas were less likely to suffer an NTD-affected pregnancy than women who had eaten only home-made tortillas. It is not known why an association was also not found between NTDs and even greater tortilla consumption (>400 during the first three months of pregnancy), but one possibility, as speculated by Hendricks (1999), is that there might have been a higher rate of *in utero* fetal death in these women than in those consuming fewer tortillas. It must be noted that the traditional processing of maize, as is practiced in Mexico and Central America, is known to reduce the level of total fumonisins and to convert a significant proportion of fumonisins to less toxic hydrolyzed fumonisins in the nixtamalized products (Palencia et al., 2003); however, the natural levels of important nutrients such as folate, riboflavin, and vitamins are also reduced by over 50% (Figueroa Cardenas et al., 2001).

Other populations exhibiting high NTD rates are found in southern Africa (35-61/10,000), including parts of the Transkei, rural northern China (57-73/10,000), and Guatamala (up to 106/10,000 in some areas) (Cifuentes, 2000; Kromberg and Jenkins, 1982; Lian et al., 1987; Melnick and Marazita, 1998; Moore et al., 1997; Ncayiyana, 1986; Venter et al., 1995; Xiao et al., 1990). Large amounts of maize or maize products are consumed in many of these areas and some of the maize is likely at times to be contaminated with high levels of fumonisins (Rheeder et al., 1992; Riley et al., 2005b; Yoshizawa et al., 1994).

The etiology of NTDs worldwide is not well understood and a variety of factors related to heredity, local environment, and nutrition are involved. Folate is a particularly significant nutrient and its levels in maize are very low. In addition, maize is frequently infected with *F. verticillioides* and fumonisin in areas were NTD incidence is high, and thus exposure to a proven inhibitor of folate metabolism, may also at times be high. It is therefore not unreasonable to suspect that fumonisin exposure could be an environmental risk factor for NTD.

Acknowledgements. The sources of financial and research support necessary to accumulate the data contained within this brief review were numerous. The authors thank all those who have worked so diligently to accumulate all the information that has contributed to the understanding of fumonisins and we apologize to all those who deserved to be cited for their original contributions but were subsumed by a review article.

5. References

Bolger M, Coker RD, Dinovi M, Gaylor D, Gelderblom MO, Paster N, Riley RT, Shephard G, and Speijers JA (2001) Fumonisins. In: *Safety evaluation of cer-*

tain mycotoxins in food. Food and Agriculture Organization of the United Nations, paper 74. World Health Organization Food Additives, Series 47, 103-279.

Brown DA and London E (1998) Functions of lipid rafts in biological membranes. *Annu Rev Cell Dev Biol,* **14,** 111-136.

Campbell LR, Dayton DH and Sohal GS (1986) Neural tube defects: A review of human and animal studies on the etiology and neural tube defects. *Teratology,* **34,** 171-187.

CDC (2000) Neural tube defect surveillance and folic acid intervention – Texas-Mexico border, 1993-1998. *Morbidity and Mortality Weekly Report,* **49(1),** 1-4.

Cifuentes G (2002) Perfil epidemiologico de las anomalias del tubo neural en Guatemala, durante el año 2000. in *Graduation Thesis of the School of Medicine of Universidad San Carlos de Guatemala.*

Collins TFX, Sprando RL, Black TN, Olejnik N, Eppley RM, Shackelford ME, Howard PC, Rorie JI, Bryant M, Ruggles DI (2005) Effects of aminopentol on in utero development in rats. *Food Chem Toxicol,* in press.

Constable PD, Riley RT, Waggoner AL, Hsiao S-H, Foreman J H, Tumbleson ME and. Haschek WM (2005) Serum sphingosine-1-phosphate and sphinganine-1-phosphate are elevated in horses exposed to Fumonisin B_1. in *AOAC International Midwest Section Final Program,* pp. 63-64.

Czeizel AE and Dudas I (1992) Prevention of the first occurrence of neural-tube defects by periconceptual vitamin supplementation. *N Engl J Med,* **327,** 1832-1835.

da Costa M, Sequeira JM, Rothenberg SP, Weedon J. (2003) Antibodies to folate receptors impair embryogenesis and fetal development in the rat. *Birth Defects Res Part A Clin Mol Teratol,* **67,** 837-847.

Detrait ER, George TM, Etchevers HC, Gilbert JR, Vekemans M, Speer MC (2005) Human neural tube defects: Developmental biology, epidemiology, and genetics. *Neurotoxicology and Teratology,* **27,** 515-524.

Enongene EN, Sharma RP, Bhandari N, Meredith FI, Voss KA and Riley RT (2002) Persistence and reversibility of the elevation in free sphingoid bases induced by fumonisin inhibition of ceramide synthase. *Tox Sci,* **67,** 173-181.

Feuchtbaum LB, Currier RJ, Riggle S, Robertson M, Lorey FW, Cunningham GC (1999) Neural tube defect prevalence in California (1990-1994): Eliciting patterns by type of defect and maternal race/ethnicity. Genetic Testing, 3(3), 265-272.

Figueroa Cardenas JD, Acero Godinez MG, Vasco Mendez NL, Lozano Guzman A, Flores Acosta LM, Gonzalez-Hernandez J (2001) Fortification and evaluation of the nixtamal tortillas. *Arch Latinoam Nutr,* **51,** 293-302.

Flynn TJ, Pritchard D, Bradlaw J, Eppley R and Page S (1996) In vivo embryotoxicity of fumonisin B_1 evaluated with cultured postimplantation staged rat embryos. *In vitro Toxicology,* **9,** 271-279.

Flynn TJ, Stack ME, Troy AL and Chirtel SJ (1997) Assessment of the embryotoxic potential of the total hydrolysis product of fumonisin B1 using cultured organogenesis-staged rat embryos. *Food Chem Toxicol,* **35,** 1135-1141.

Gelderblom WCA, Kriek NPJ, Marasas WFO, and Thiel PG (1991) Toxicity and carcinogenicity of the *Fusarium moniliforme* metabolite, fumonisin B_1, in rats. *Carcinogenesis*, **12**, 1247-1251.

Gelineau-van Waes J, Starr L, Maddox JR, Aleman F, Voss KA, Wilberding J and Riley RT (2005) Maternal fumonisin exposure and risk for neural tube defects: Disruption of sphingolipid metabolism and folate transport in an *in vivo* mouse model. *Birth Defects Research (A)*, **73**, 487-497.

Gelineau-van Waes J, Maddox J, Starr L, Wilberding J, Bauer L, Voss K, and Riley R (2006) Maternal fumonisin exposure and neural tube defects: Mechanisms in a mouse model. *Tox Sci (Suppl. The Toxicologist)*, **85 (S1)**, accepted.

Green NS (2002) Folic acid supplementation and prevention of birth defects. *J Nutr*, **132**, 2356S-2360S.

Hendricks K (1999) Fumonisins and neural tube defects in South Texas. *Epidemiol*, **10**, 198-200.

Hansen, DK, Streck, RD and Antony, AC (2003) Antisense modulation of the coding or regulatory sequence of the folate receptor (folate binding protein-1) in mouse embryos leads to neural tube defects. *Birth Defects Res Part A Clin. Mol Teratol*, **67**, 475-487.

Humpf H-U, and Voss KA (2004) Effects of food processing on the chemical structure and toxicity of fumonisin mycotoxins. *Mol Nutr Food Res*, **48**, 255-269.

IARC (2002) (International Agency for Research on Cancer). Fumonisin B_1. IARC Monographs on the Evaluation of Carcinogenic Risks to Humans: Some Traditional Medicines, Some Mycotoxins, Naphthalene and Styrene. *IARC*, **82**, 301-366.

IPCS (International Programme on Chemical Safety). (2000) Fumonisin B_1, Environmental Health Criteria 219. World Health Organization, Geneva, 150 pp

Kromberg JG and Jenkins T. (1982) Common birth defects in South African Blacks. *S Afr Med J*, **62(17)**, 599-602.

Lian Z-H, Yang H-Y and Li Z (1987) Neural tube defects in Beijing-Tianjin area of China. Urban-rural distribution and some other epidemiological characteristics. *J Epidemiol Community Health*, **41**, 259-262.

Luhrs CA and Slomiany BL (1989) A human membrane-associated folate binding protein is anchored by a glycosyl-phosphatidylinositol tail. *J Biol Chem*, **264**, 21446-21449.

Marasas WFO, Riley RT, Hendricks KA, Stevens VL, Sadler TW, Gelineau-van Waes J, Missmer, SA, Cabrera Valverde J, Torres OL, Gelderblom W, Allegood J, Martínez de Figueroa AC, Maddox J, Miller JD, Starr L, Sullards MC, Roman Trigo AV, Voss KA, Wang E, Merrill AH, Jr (2004) Fumonisins disrupt sphingolipid metabolism, folate transport and development of neural crest cells in embryo culture and *in vivo*: A risk factor for human neural tube defects among populations consuming fumonisin-contaminated maize? *J Nutr*, **134**, 711-716.

March of Dimes Perinatal Data Center (2001) Maternal, Infant, and Child Health in the United States.

Melnick M and Marazita ML (1998). Neural tube defects, methylenetetrahydrofolate reductase mutation, and north/south dietary differences in China. *J Craniofac Genet Dev Biol*, **18**, 233-235.

Meredith F I, Torres O R, Saenz de Tejada S, Riley R T and Merrill A H Jr (1999) Fumonisin B$_1$ and hydrolyzed fumonisin B$_1$ levels in nixtamalized maize (*Zea mays* L.) and tortillas from two different geographical locations in Guatemala. *J Food Prot*, **62**, 1218-1222.

Merrill AH Jr, Sullards MC, Wang E, Voss KA and Riley RT (2001) Sphingolipid metabolism:Roles in signal transduction and disruption by fumonisins. *Environ Health Perspect*, **109**, 283-289.

Merrill AH Jr. (2002) De novo sphingolipid biosynthesis: A necessary, but dangerous, pathway. *J Biol Chem*, **277**, 25843-25846.

Mitchell LE (2005) Epidemiology of neural tube defects. *American Journal of Medical Genetics Part C*, **135C**, 88-94.

Moore CA, Li S, Gu H, Berry RJ, Mulinare J and Erickson JD (1997) Elevated rates of severe neural tube defects in a high-prevalence area in Northern China. *Amer J Med Genetics*, **73**, 113-118.

MRC (1991) MRC Vitamin Study Research Group Prevention of neural tube defects: results of the Medical Research Council vitamin study. Lancet II, 131-137.

NTP (2001) (National Toxicology Program Technical Report). (2002) Toxicology and Carcinogenesis Studies of Fumonisin B1 in F344/N rats and B6C3F1 mice. Publication no. 99-3955, NIH, Washington, DC

Ncayiyana DJ (1986) Neural tube defects among rural blacks in a Transkei district. *S A Med J*, **69**, 618-620.

Palencia E, Torres O, Hagler W, Meredith FI, Williams L D, and Riley RT (2003) Total fumonisins are reduced in tortillas using the traditional nixtamalization method of Mayan communities. *J Nutr*, **133**, 3200-3203.

Piva A, Casadei G, Pagliuca G, Cabassi E, Galvano F, Solfrizzo M, Riley RT and Diaz DE (2005) Inability of activated carbon to prevent the toxicity of culture material containing fumonisin B1 when fed to weaned piglets. *J Animal Sci*, **83**, 1939-1947

Piedrahita JA, Oetama B, Bennett GD, van Waes J, Kamen BA, Richardson J, Lacey SW, Anderson RG and Finnell RH (1999) Mice lacking the folic acid-binding protein *Folbp1* are defective in early embryonic development. *Nature Genetics*, **23**, 228-232.

Rheeder J P, Marasas W F O, Thiel P G, Sydenham E W, Shephard G S and Van Schalkwyk D J (1992) *Fusarium moniliforme* and fumonisins in corn in relation to human esophageal cancer in Transkei. *Phytopathology*, **82**, 353-357.

Riley RT, Enongene E, Voss KA, Norred WP, Meredith FI, Sharma RP, Spitsbergen J, Williams DE, Carlson DB and Merrill AH Jr (2001) Sphingolipid perturbations as mechanisms for fumonisin carcinogenesis. *Environ Health Perspect*, **109**, 301-308.

Riley RT, Showker JL, Voss KA (2005a) Time- and dose-dependent changes in sphingoid base 1 phosphates in tissues from rats fed diets containing fumonisins. *Tox Sci (Suppl. The Toxicologist)*, **84 (S1)**, 284-285.

Riley RT, Torres OA, Palencia E, Lopez de Pratdesaba L, Glenn AE, O'Donnell K and Fuentes M. (2005b) Fumonisins in highland and lowland maize in Guatemala and a preliminary exposure estimate. *AOAC International Midwest Section Final Program*, pp. 34-35.

Rothenberg SP, Da Costa MP, Sequeira JM, Cracco J, Roberts JL, Weedon J, and Quadros EV (2004) Autoantibodies against folate receptors in women with a pregnancy complicated by a neural tube defect. *Obstet Gynecol Surv*, **59**, 410-411.

Sadler TW, Merrill AH, Jr, Stevens VL, Sullards MC, Wang E and Wang P (2002) Prevention of fumonisin B_1-induced neural tube defects by folic acid. *Teratology*, **66**, 169-176.

Shaw GM, Schaffer D, Velie E, Morland KM and Harris JA (1995) Periconceptional vitamin use, dietary folate, and the occurrence of neural tube defects. *Epidemiol*, **6**, 219-226.

Shephard GS (2004). Mycotoxins worldwide: Current issues in Africa. *In* Barug D, Van Egmond, H.P., López Garciá, R., Van Osenbruggen, W.A. & Visconti, A. *Meeting the mycotoxin menace*. Wageningen Academic Publishers, the Netherlands, pp. 81-88.

Stevens VL and Tang J (1997) Fumonisin B_1-induced sphingolipid depletion inhibits vitamin uptake via the glycosylphosphatidylinositol-anchored folate receptor. *J Biol Chem*, **272**, 18020-18025.

Venter PA, Christianson AL, Hutamo CM, Makhura MP and Gericke GS (1995) Congenital anomalies in rural black South African neonates-a silent epidemic? *S A Med J*, **83**, 15-20.

Voss KA, Riley RT, and Gelineau-van Waes J (2005a) Trends in fumonisin research: Recent studies on the developmental effects of fumonisins and *Fusarium verticillioides*. *Mycotoxins*, **55**, 91-100.

Voss KA, Gelineau van-Waes JB, Riley RT, Burns TD, Bacon CW. (2005b) Developmental toxicity of diets containing fumonisin B_1 to LM/Bc and CD1 mice: a comparative study. *Tox Sci (Suppl. The Toxicologist)*, **84 (S1)**, 1396.

Wang E, Norred WP, Bacon CW, Riley RT, and Merrill AH Jr (1991) Inhibition of sphingolipid biosynthesis by fumonisins: Implications for diseases associated with *Fusarium moniliforme*. *J Biol Chem*, **266**, 14486-14490.

Wang E, Riley RT, Meredith FI and Merrill AH Jr (1999) Time course, dose-dependence, and reversibility of increases in urinary sphinganine and sphingosine in animals fed defined diets containing fumonisin B_1: characteristics of urinary biomarkers for exposure to fumonisins. *J Nutr*, **129**, 214-220.

Werler MM, Sharpio S and Mitchell AA (1993) Periconceptual folic acid exposure and risk of occurrent neural tube defects. *JAMA*, **269**, 1257-1261.

WHO (2001) World Health Organization. Evaluation of Certain Mycotoxins in Food. Fifty-sixth Report of the Joint FAO/WHO Expert Committee on Food Additives (JECFA). *WHO Technical Report*, **906**, 16-27.

Xiao KZ, Zhang ZY, Su YM, Liu FQ, Yan ZZ, Jiang ZQ, Zhou SF, He WG, Wang BY, Jiang HP, (1990). Central nervous system congenital malformations, especially neural tube defects in 29 provinces, metropolitan cities and autonomous regions of China: Chinese Birth Defects Monitoring Program. *Int J Epidemiol*, **19**, 978-982.

Yoshizawa T, Yamashita A and Luo Y (1994) Fumonisin occurrence in corn from high- and low-risk areas for human esophageal cancer in China. *Appl Environ Microbiol*, **60**, 1626-1629.

5-9 Sphingolipids and Cancer

Eva M. Schmelz and Holly Symolon

Karmanos Cancer Institute, Wayne State University School of Medicine, HWCRC 608, 110 E. Warren Avenue, Detroit, MI 48201, USA, and School of Biology and the Petit Institute for Bioengineering and Bioscience, Georgia Institute of Technology, Atlanta, GA 30332, USA

Summary. Sphingolipids regulate processes that are dysregulated in cancer such as cell growth and death, adhesion and motility. Changes in sphingolipid metabolism in cancer cells that result in an altered sphingolipid composition may contribute to unlimited proliferation in cancer cells and aide progression, invasion and metastasis. Thus, exogenous sphingolipid metabolites such as ceramides and sphingoid bases have been used both *in vivo* and *in vitro* to reverse aberrant cell behavior and suppress tumorigenesis. Here we review the role of sphingolipids and sphingolipid metabolites in cancer development, prevention and treatment.

Keywords. cancer, carcinogenesis, sphingolipid metabolites

1. Introduction

Successful preventions and treatment regimens for cancer are now available, but they are not universally applicable to all types of cancers. Conventional chemotherapeutic agents induce apoptosis through mechanisms that are not specific for cancer cells. Its toxic effects also kill healthy cells. Sphingolipids which can induce death, senescence, and differentiation in cells as well as regulate cell attachment, migration, angiogenesis and multi-drug resistance in a dose-dependent manner. All these activities strongly suggest that sphingolipids may serve as chemopreventive and chemotherapeutic agents.

2. Cancer development: multiple stages that can be targeted by sphingolipids

Carcinogenesis is a complex multi-stage process that can emerge from the aberrant expression in a single gene, or from the cumulative build up of such genes, due to point mutations, deletions, translocations, amplifications, chromosome duplications or other such losses. These aberrations can transform epithelial cells into, initially benign, tumors via clonal expansion in the promotion phase. In this phase, potentially malignant tumors form when growth is accelerated and apoptosis inhibited, and these tumors may eventually migrate and colonize other areas. Genetic changes interfere with normal processes in the cell cycle, including dys-regulated apoptosis as well as altered cell adhesion and migration properties. Although these cell-type specific genetic changes are diverse, these changes also point to multiple points where intervention with sphingolipids, as shown in Figure 1, may prove effective. The range of known processes (with unknown, but potentially beneficial, functions in gray) as well as the specific targets regulated by sphingolipids indicate that sphingolipids can be potential anti-cancer drugs at all stages of the disease.

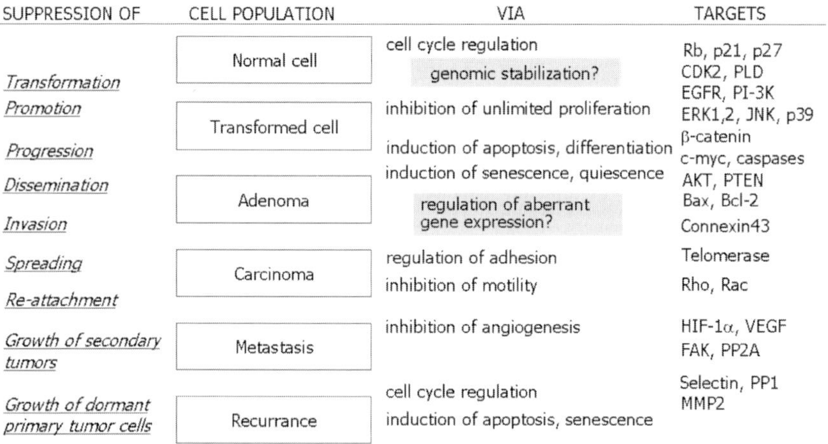

Fig. 1. Targets of sphingolipids in carcinogenesis. Multiple targets to prevent or suppress malignancy and metastasis exist within the multi-step process of carcinogenesis by regulating proliferation, cell death, motility and invasion. Some targets, such as cell cycle regulation, may have potential benefits at more than one stage of the disease.

2.1 Enzymatic expression and activity in sphingolipid metabolism and modulation of bioactive metabolites in cancer cells

Changes in the expression and activity of enzymes that participate in sphingolipid metabolism can alter the composition of sphingolipid metabolites in cancer cells. This activity could shift cells to a pro-proliferative and anti-apoptotic state, critically altering their phenotype and their response to regulatory stimuli and treatment. Reduced expression and activity of sphingomyelinases in the colonocytes of carcinogen-treated rats (Dudeja et al. 1986; Notterman et al. 2001) and in human colon tumors (Hertervig et al. 1997; Di Marzio et al. 2005) appear to protect cells from cell death by not generate ceramide in response to extracellular stimuli. A lack of sphingomyelinase activation underlies the mechanism of resistance in some tumor cells to β-irradiation (Michael et al. 1997) and chemotherapeutic agents (Cai et al. 1997). Defects in the organization, number or function of membrane microdomains, such as rafts (Gulbins and Kolesnick 2003), may also account for the lack of ceramide generated by sphingomyelinases.

Ceramidases are highly expressed in blood and spleen, moderately expressed in prostate and ovaries, and weakly expressed in thymus and colon. Levels of acid ceramidase mRNA were reduced in benign and malignant thyroid tumors (Maeda et al. 1999), and therefore may be a marker for the early stages in thyroid cancer. In contrast, increased ceramidase mRNA levels were reported in prostate cancer (Seelan et al. 2000) and, while not essential for colon, breast or ovarian cancer, the expression of ceramidase does appear to be crucial for melanomas (Musumarra et al. 2003). Overexpression of acid ceramidase in tumor cells can protect against TNF-β-induced apoptosis (Strelow et al. 2000) by removing cytotoxic ceramide and releasing sphingosine as substrate for sphingosine kinase. This is a cellular mechanism that prevents ceramide-induced apoptosis.

Overexpression of sphingosine kinase was observed in lung tumors compared to matched normal lung tissue (Johnson et al. 2005). Increased mRNA levels (up to five-fold) were also detected in breast, ovarian, uterine, stomach, colon, small intestinal, kidney and rectal tumors (French et al. 2003; Johnson et al. 2005). An increase in sphingosine kinase expression or activity has been correlated to hyperproliferation, transformation and may discriminate between benign and malignant fate of transformed epidermal cells (Wang et al. 2002). Transfection of NIH3T3 fibroblasts with human sphingosine kinase enhanced cell growth enabled anchorage-independent cell growth but reduced serum dependence and contact inhibition in transfected cells. These cells as well as MCF7 cells transfected with sphingosine kinase can readily form tumors in nude mice (Xia et al. 2000, Nava et al. 2002). Transforming growth factor-beta (TGF-ß) treatment (Yamanaka et al. 2004) is a potent member of a polypeptide family that induces the expression of extracellular matrix genes. Sphingosine kinase is activated and sphingosine-1-phosphate is elevated in dermal fibroblasts by TGF-β to promote invasion and metastasis. This implicates sphingosine kinase in late stages of cancer.

Enzymatic changes can modify the intracellular composition of bioactive sphingolipids. Many of these changes can deplete growth inhibitory/cytotoxic metabolites. These can inhibit the generation of cytotoxic sphingolipid metabolites, trigger removal via degradation, or synthesize complex sphingolipids all of which increase the resistance of cancer cells to treatment. Accordingly, levels of ceramide are decreased in head and neck squamous cell carcinomas (Koybasi et al. 2004), colon cancer (Selzner et al. 2001), larynx carcinoma (Chi et al. 2004), astrocytomas (Riboni et al. 2002) and leukemia cells (Itoh et al. 2003). These low levels were associated with unlimited proliferation, chemoresistance, and a poor outcome for the patients. In contrast, levels of glucosylceramide, galactosylceramide, and gangliosides are elevated in some tumor cells (Riboni et al. 2002; Prinetti et al. 2003). Sphingosine-1-phosphate is significantly elevated in several human glioma cell lines (Sullards and Merrill 2003) and ovarian tumors (Sutphen et al. 2004), but it is also secreted into ascites fluid, affecting ovarian cancer cell dissemination and attachment of cells at distant sites (Hong et al. 1999). Still, others found increases in the ceramide and dihydroceramide content in sarcomas, melanomas and Lewis lung carcinomas (Koyanagy et al. 2003), and suggest that these increases may be required for tumor survival.

Identifying sphingolipid compositional changes in cancer cells may be useful approach for making diagnoses and treatment decisions. For example, elevated sphingosine-1-phosphate levels in tumors indicate that the patient may be treated with sphingosine kinase inhibitors. Since the reversal of these changes can make cancer cells less resistant to treatment, thereby enhancing radiation treatment, tumor formation and metastasis may be suppressed. Hence, sphingolipids composition, expression or activity of specific metabolic enzymes may also indicate the efficacy of specific drugs to tumors.

3. Targeting enzymes of the sphingolipid metabolism for cancer prevention and treatment

Using activated sphingolipid metabolites to direct cancer cells towards growth inhibition and apoptosis is not a new concept. It has been extensively documented in cancer cells treated with novel and established anti-cancer drugs. Several natural compounds such as beta-sitosterol or tea catechins also generate ceramide and facilitate the apoptotic response of cancer cells. Most studies focus on ceramide generation by the activation of a sphingomyelinase, *de novo* synthesis, or a combination of the two that subsequently induces apoptosis in cancer cells. Table 1 lists examples of anti-cancer agents or treatments that generate ceramide *in vitro*.

However, by limiting the scope of the investigations to the generation ceramide, other metabolites that might contribute beneficially to treatment may have been missed. Furthermore, other endpoints, such as cell cycle arrest, induction of differentiation or senescence, that may be important events in cancer inhibition should be evaluated. For example, activating acid sphingomyelinase by blocking matrix ligation of intergrins is proposed mechanism to produce an anti-angiogenic effect of intergrin-blocking peptides and induce apoptosis in endothelial cells (Erdreich-Epstein et al. 2002). Enzyme regulated removal of bioactive metabolites via synthesis of complex sphingolipids such as glucosylceramide synthase or lactosylceramide synthase may prevent the inactivation of cytotoxic ceramide (see chapter 4-4). Enzymes that generate mitotic metabolites are also the focus of increased attention. Phenoxodiol, a synthetic isoflavone analog, is the first sphingosine kinase inhibitor to be developed. In preliminary studies, phenoxodiol induced apoptosis and reversed the resistance of refractory ovarian cancer to cisplatin and taxol (Kamsteeg et al. 2003). It also inhibited the growth of breast cancer xenografts (Constantinou et al. 2003). Although not yet approved by the FDA, several pre-clinical trials showed a disease-stabilizing effect of phenoxodiol (see www.phenoxodiol.com for past and current clinical trials). The enzymes of the sphingolipid metabolism that are actual or potential targets of intervention attempts are shown in Figure 2.

3.1 Use of exogenous sphingolipids in vitro

Instead of activating cellular enzymes to generate bioactive sphingolipid metabolites, elevating intracellular levels by delivering the sphingolipids directly to the cells is possible (see Figure 2). Many in vitro studies show that this elicits the same effects as endogenous metabolites in cell systems.

Reports on the effects of exogenous sphingolipids on cancer cells in vitro are too numerous to be mentioned here. The endpoint of many of these studies was the induction of apoptosis in cancer cells. These studies revealed the signaling pathways and specific targets for many cell lines. However, numerous studies have also used non-toxic doses to affect cell cycle regulation, as a potential measurement for long-term induction of cellular senescence, to reduce cancer cell proliferation and viability, and to remove of growth advantages and drug resistance.

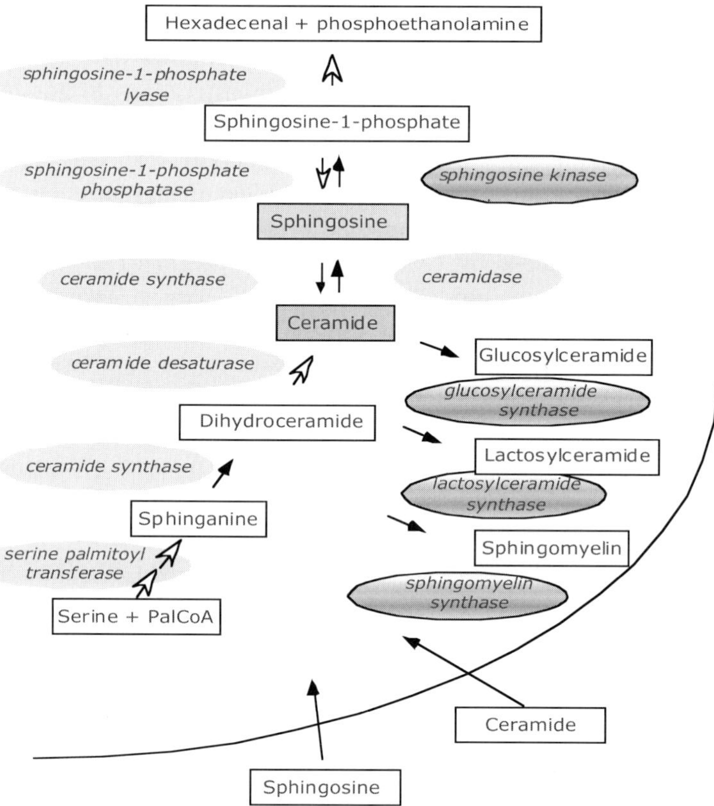

Fig. 2. Sphingolipid metabolites are attractive targets for chemotherapeutic and chemopreventive drugs. Enzymes of the sphingolipid metabolism pathway would elevate intracellular levels of growth inhibitory/cytotoxic metabolites after activation (open arrows) or inhibition (shaded). Alternatively, bioactive metabolites can be delivered directly to the cancer cells.

Table 1. Generation of bioactive sphingolipid metabolites by selected agents used for cancer prevention and treatment.

Route of Sphingolipid Generation	Cells	Reference
De Novo synthesis		
Etoposide	prostate, leukemia	Sumimoto 2002; Perry 2000; Wang 2002
Captothecins	colon	Chauvier 2002
Anthracyclines	prostate, leukemia, breast, lung	Wang 2002; Bose 1995; Cullivier 2001; Chalfant 2002
Nucleoside analogs	leukemia, lung	Chalfant 2002; Biswal 2000
Fenretinamide	prostate, breast, ovarian, neuroblastoma	Wang 2003; Andrews 2005; Appierto 2004; Wang 2001
Resveratrol	breast	Scarlatti 2003
β-Tocopherol	prostate	Jiang 2004
Catabolism		
Etoposide	prostate, glioma	Sumimoto 2002, Sawada 2000
Anthracyclines	leukemia	Allouche 1997
Nucleoside analogs	pancreas, leukemia	Modrak 2004, Grazide 2002
Vinca alkaloids	leukemia	Zhang 1996
Platinum compounds	kidney, glioma	Asakuma 2003; Noda 2001
Inhibition of Sphingomyelin Synthase		
Tricylin xanthate	leukemia	Meng 2004
Inhibition of Glucosylceramide Synthase		
Tamoxifen	breast, melanoma	Cabot 1996; Lavie 1997
Verapamil	breast	Lavie 1997

Combinatorial cancer treatments can be used to lower the effective doses of anti-cancer drugs, thereby reducing toxic side effects. Co-treatment of resistant breast cancer cells with safingol (L-threo sphinganine) enhanced the toxicity of vinca alkaloids and anthracyclines independent of p-glycoprotein expression (Sachs et al. 1995). Similar results were found with a combination of taxol and ceramide (Myrick et al, 1999) and sphingomyelin and doxorubicin (Veldmann et al. 2004). Ore-treatment with ceramide or sphingosine sensitized prostate cancer cells to β-irradiation (Nava et al. 2000; Kimura et al. 2003). Specific targets of sphingolipids may cause that increased sensitivity to an apoptotic insult; however, the use of non-toxic doses of safingol or sphingomyelin significantly increased drug influx in cancer cells to increase their toxicity (Sachs et al. 1995; Veldman et al. 2004). It is likely that membrane properties change or that drug transporters are activated through some unspecific effect. Although intracellular sphingolipid levels can be altered by exogenous sphingolipids, exogenous complex sphingolipids are endocytosed and degraded before reaching signaling pools (Chigorno et al. 2004). This sphingolipid treatment may produce a membrane-associated action that causes this degradation. Therefore altering membrane properties, such as membrane fluidity, assembly of lipid rafts or caveolae, to regulate active or passive drug uptake may be an attractive target for cancer treatment.

Elevating the number of bioactive sphingolipids metabolites by regulating the enzymes involved in ceramide accumulation (i.e., inhibitors of ceramidase, sphingosine kinase, glucosylceramide synthase and lactosylceramide synthase) is another approached used to sensitize cells to the drug treatment. A combination of these inhibitors reversed the resistance of squamous carcinoma cells to β-irradiation (Alphonse et al. 2004). Apoptosis has also been successfully induced in cancer cells using doxorubicin, taxol and vincristine, drugs that generate ceramide by activating sphingomyelinase or de novo synthesis, in combination with inhibitors of glucosylceramide synthase (Lucci et al. 1999; Sietsma et al. 2000; Olshefski and Ladisch 2001).

Manipulating sphingosine-1-phosphate levels by over-expressing sphingosine phosphate lyase (to degrade to hexadecenal and phosphoethanolamine) (Min et al. 2005) or activating sphingosine phosphate phosphatase (to generate sphingosine that is rapidly acylated to ceramide) (Le Stunff, et al. 2002) are other novel approaches to inhibit growth, induce apoptosis or enhance drug sensitivity in cancer cells.

3.2 Use of exogenous sphingolipids in vivo

3.2.1 Topical Administration

Direct application of sphingosine, methylsphingosine and N-acetylsphingosine preparations onto the skin of Sencar mice treated with Dimethylbenz-[a]-anthracene did not inhibit the development of papillomas. At high doses, this application enhanced papillomas formation.

However, both N-methylsphingosine and N-acetylsphingosine increased cancer-free survival and suppressed tumor progression (Envet-chakul et al. 1992; Birt et al. 1998). The only human study used a 1% mixture of short-chain ceramides to treat patients with cutaneous breast cancer. A partial response in 1/26 patients was seen. This was insufficient to conduct further studies (Jatoi et al. 2003).

3.2.2 Intraperitoneal, intravenous, and subcutaneous injections

Novel, less polar, ceramide analogs were injected i.p. into nude mice carrying human colon cancer xenografts. Apoptosis resulted in the xenografts, significantly reducing their sizes (Macchia et al. 2001). A pronounced and sustained inhibition of tumor growth was also seen after i.p. injections of sphingosine, dimethylsphingosine or trimethylsphingosine in mice that had been inoculated with MKN74 human gastric cancer cells (Endo et al. 1991). Injection of sphingomyelin (i.v.) did not affect the growth of human colon cancer xenografts; however, in combination with 5-fluorouracil, a widely used drug used to treat colon cancer, there was a significant reduction of tumor size. This reduction may be associated with the increase of apoptosis in the xenografts (Modrak et al. 2000).

The systemic delivery of sphingolipids is difficult because of their limited stability in vivo. Incorporating ceramide into liposomes to protect ceramide from degradation or precipitation, and generate a maximum ceramide increase in cancer cells was very effective in cell culture and significantly elevated ceramide stability in vivo by 20% (Shabbits et al. 2003; Shabbits et al. 2003a; Stover et al. 2003). Treatment with these liposomes, injected i.v., suppressed ovarian cancer cell dissemination (Shabbits et al. 2003a) and reduced the size of breast cancer xenografts without causing severe toxic side effects (Stover et al. 2005). Ceramides dissolved in soybean oil and injected i.v. into nude mice reduced pulmonary metastases derived from Meth A-T tumor cells (Takenaga et al. 1999). Mice with established prostate cancer xenografts were s.c. injected with a ceramide analog. This reduced the growth of the tumors when compared to untreated controls but the combination with β-irradiation completely blocked tumor growth but did not induce tumor shrinkage (Samsel et al. 2004).

3.2.3 Oral administration

Intestinal cells are exposed to bioactive sphingolipid metabolites when dietary complex sphingolipids are hydrolyzed to ceramide and sphingosine by intestinal enzymes. Ceramides and sphingoid bases are efficiently absorbed by small intestinal cells, however, about 10% of the complex sphingolipids reach the colon intact. These are then hydrolyzed by the colonic microflora to ceramides and sphingoid bases (Schmelz et al. 1994). Oral administration of complex sphingolipids, therefore, may be a convenient way to deliver bioactive sphingolipids to the colon to prevent colon cancer. When natural and synthetic complex sphingolipids were fed to carcinogen-treated CF1 mice, aberrant crypt foci (ACF), precursors of colonic adenomas and adenocarcinomas, diminished by 50 to 70% (Dillehay et al. 1994; Schmelz et al. 1996, 1997, 2000, 2001). Oral administration also suppressed these foci in carcinogen-treated rats (Exon and South 2003). Synthetic dihydrosphingomyelin and glucosylceramide derived from soy (which contains a sphingoid base with an 8,9-trans double bond) also significantly inhibited ACF formation (Schmelz et al. 1997; Symolon et al. 2004), indicating that it may be the sphingoid bases, not ceramides, in bioactive metabolites that protect against colon cancer. The number and localization of double bonds may also not be critical.

When sphingolipids are orally administered, colon tumors in CFI mice were suppressed by more than 70% (Lemonnier et al. 2003). Administering sphingolipids after tumor initiation, after colonic cells are damage, also suppressed tumor formation in the study, suggesting that the mechanisms of tumor prevention by sphingolipids may not be to prevent cell damage but to regulate and eliminate predisposed and premalignant cells. The number of intestinal tumors was reduced significantly in treated Min mice (multiple intestinal neoplasia mice) with Adenomatous Polyposis Coli (APC) mutations. These mutations are typically found in patients with familial adenomatous polyposis and in 40-80% of patients with sporadic colon cancer. Feeding Min mice on a diet that included 0.1% complex milk or soy sphingolipids for 65 days reduced tumor formation throughout the intestinal tract by 40-50%. (Schmelz et al. 2001, Symolon et al. 2004). with no adverse side effects noted in any of these studies. Notably, orally delivered sphingolipids down regulated cytosolic and nuclear ß-catenin in Min mice and in cell lines with APC mutations (Schmelz et al. 2001). This is critical because the removal of cytosolic and nuclear ß-catenin reverses transformed properties of cells (Kim et al. 2002). Hence, reversing early changes in colon carcinogenesis may be a key event in cancer prevention by sphingolipids.

A constant elevation in bioactive sphingolipid metabolites is not seen in whole blood after several weeks of sphingomyelin administration in mice (Schmelz et al. 2001). Yet, sufficient amounts sphingolipid metabolites derived from orally administered complex sphingolipids are absorbed and transported to the body to significantly reduce the number and size of diethylnitrosamine-induced pre-neoplastic lesions in the livers of rats (Silins et al. 2003). Whether this could also be prevent cancer in other organs remains to be seen.

4. The future of sphingolipids in cancer prevention and treatment

Sphingolipids regulate many intracellular processes in cells through various signaling processes that alter cell type, concentration, and structure. They can also function in a time-dependent manner. Cancer cells can modify sphingolipid signaling to promote growth and survival. This modification may involve a single enzyme, as shown above, or several as shown in epidermal growth factor receptor (EGFR) signaling. As many cancer cells overexpress members of the EGFR family, these proteins are thought to be causal factors in many human malignancies. Cytokine-upregulated sphingomyelinase activity can be blocked by activated EGFR (Garcia-Lloret et al. 1996), which then attenuates or completely blocks ceramide production and cell response. In parallel, activation of EGFR activates ceramidase (Payne et al. 1999), further reducing endogenous ceramide and generating sphingosine as substrate for sphingosine kinase. Activated EGFR can also activate sphingosine kinase (Meyer zu Heringdorf et al. 1999). This suggests that EGFR-induced cell growth and protection against cell death in cancer cells entails, at least in part, the activation of enzymes of sphingolipid metabolism and accumulation of sphingosine-1-phosphate as well as a depletion in growth inhibitory and/or cytotoxic sphingolipid metabolites (Carpio et al. 2000). On the other hand, key proteins that regulate EGFR-activated proliferation and survival pathways, such as the EGFR itself, ERK1,2, JNK, PI-3K, AKT, and PKC isozymes, are directly or indirectly regulated by sphingolipid metabolites. Elevating the growth inhibitory sphingolipid pool can therefore counteract proliferation signals by EGFR at multiple points. This would be more effective than anti-cancer drugs that target specific proteins, like EGFR, that have specific roles at one development stage and that can be replaced by redundant signaling pathways (i.e., constitutively activated PI-3K), mutated, silenced, overexpressed or otherwise altered and thereby rendered them ineffective.

Regulating aberrant signaling pathways could remove growth advantages, eliminate cancer cells and inhibit progression of cancer cells in ways other than induced cell death. This is a challenging approach to cancer prevention. Sphingolipids treatment may also reduce adverse side effects in vivo to enhance patient compliance and improve their quality of life during treatment. Regulating dormant primary cells, which may not cycle, to prevent tumor formation is a particular challenge. Targets such as β-catenin need to be identified for non-toxic doses of sphingolipids to prevent and treat cancer. The efficacy of these targets as cell cycle regulators, adhesion molecules, growth factor receptors will also need to be evaluated. Designer sphingolipids, modified to adapt physio-chemical properties within cells of specific organs are promising approaches. For example, modifications in ceramide glucuronide (contains ß-glucuronic acid as headgroup on position 1 of the ceramide moiety) can eliminate hydrolysis and uptake in the small intestine to increase the amount of lipids entering the colon and thereby reduce ACF in carcinogen-treated mice (Schmelz et al. 1999). More research is necessary to determine the effective in vivo doses, delivery methods to distant sites, limitations of usefulness, and possible interactions with other drugs before a safe and effective prevention and treatment strategy with sphingolipids can be designed. However, as serious side effects are not observed in most applications along with apparent specificity for cancer cells within an identifiable effective treatment window (not restricted to prevention of transformation), the potential of sphingolipids as anti-cancer drugs is significant.

Acknowledgements. Our research has been supported by grants CA101125 and CA109019 from the National Cancer Institute, and grants 04B071from the AICR and BCT0503453 from the Susan B. Komen Foundation (to EMS).

5. References

Alphonse G, Bionda C, Aloy MT, Ardail D, Rousson R, Rodriguez-Lafrasse C (2004) Overcoming resistance to c-rays in squamous carcinoma cells by poly-drug elevation of ceramide levels. *Oncogene*, **23**, 2703–2715.

Allouche M, Bettaieb A, Vindis C, Rousse A, Grignon C, Laurent G (1997) Influence of Bcl-2 overexpression on the ceramide pathway in daunorubicin-induced apoptosis of leukemic cells. *Oncogene*, **14**, 1837-1845.

Andrews WJ, Winnett G, Rehman F, Shanmugasundaram P, Hagen D, Schrey MP (2005) Aromatase inhibition by 15-deoxy-prostaglandin J(2) (15-dPGJ(2)) and N-(4-hydroxyphenyl)-retinamide (4HPR) is associated with enhanced ceramide production. *J Steroid Biochem Mol Biol*, **94**, 159-165.

Appierto V, Villani MG, Cavadini E, Lotan R, Vinson C, Formelli F (2004) Involvement of c-Fos in fenretinide-induced apoptosis in human ovarian carcinoma cells. *Cell Death Diff*, **11**, 270-279.

Asakuma J, Sumitomo M, Asano T, Asano T, Hayakawa M (2003) Selective Akt inactivation and tumor necrosis actor-related apoptosis-inducing ligand sensitization of renal cancer cells by low concentrations of paclitaxel. *Cancer Res*, **63**, 1365-1370.

Birt DF, Merrill AH, Jr, Barnett T, Enkvetchakul B, Pour PM, Liotta DC, Geisler V, Menaldino DS, Schwartzbauer J (1998) Inhibition of skin papillomas by sphingosine, N-methyl sphingosine, and N-acetyl sphingosine. *Nutr Cancer*, **31**, 119-126.

Biswal SS, Datta K, Acquaah-Mensah GK, Kehrer JP (2000) Changes in ceramide and sphingomyelin following fludarabine treatment of human chronic B-cell leukemia cells. *Toxicol*, **154**, 45-53.

Bose R, Verheij M, Haimovitz-Friedman A, Scotto K, Fuks Z, Kolesnick R (1995) Ceramide synthase mediates daunorubicin-induced apoptosis: an alternative mechanism for generating death signals. *Cell*, **82**, 405-414.

Cabot MC, Giuliano AE, Volner A, Han TY (1996) Tamoxifen retards glycosphingolipid metabolism in human cancer cells. *FEBS Letters*, **394**, 129-131.

Cai Z, Bettaieb A, Mahdani NE, Legres LG, Stancou R, Masliah J, Chouaib S (1997) Alteration of the sphingomyelin/ceramide pathway is associated with resistance of human breast carcinoma MCF7 cells to tumor necrosis factor-alpha-mediated cytotoxicity. *J Biol Chem*, **272**, 6918-6926.

Carpio LC, Shiau H, Dziak R (2000) Changes in sphingolipid levels induced by epidermal growth factor in osteoblastic cells. Effects of these metabolites on cytosolic calcium levels. *Prostaglandins Leukot Essent Fatty Acids*, **62**, 225-232.

Chalfant CE, Rathman K, Pinkerman RL, Wood RE, Obeid LM, Ogretmen B, Hannun YA (2002) De novo ceramide regulates the alternative splicing of caspase 9 and Bcl-x in A549 lung adenocarcinoma cells. Dependence on protein phosphatase-1. *J Biol Chem*, **277**, 12587-12595.

Chauvier D, Morjani H, Manfait M (2002) Ceramide involvement in homocamptothecin- and camptothecin-induced cytotoxicity and apoptosis in colon HT29 cells. *Int J Oncol*, **20**, 855-863.

Chi FL, Yuan YS, Wang SY, Wang ZM (2004) Study on ceramide expression and DNA content in patients with healthy mucosa, leukoplakia, and carcinoma of the larynx. *Arch Otolaryngol Head Neck Surg*, **130**, 307-310.

Chigorno V, Giannotta C, Ottico E, Sciannamblo M, Mikulak J, Prinetti A, Sonnino S (2004) Sphingolipid uptake by cultured cells: complex aggregates of cell sphingolipids with serum proteins and lipoproteins are rapidly catabolized. *J Biol Chem*, **280**, 2668-2675.

Constantinou AI, Mehta R, Husband A (2003) Phenoxodiol, a novel isoflavone derivative, inhibits dimethylbenz[a]anthracene (DMBA)-induced mammary carcinogenesis in female Sprague-Dawley rats. *Eur J Cancer*, **39**, 1012-108.

Cuvillier O, Nava VE, Murthy SK, Edsall LC, Levade T, Milstien S, Spiegel S (2001) Sphingosine generation, cytochrome c release, and activation of caspase-7 in doxorubicin-induced apoptosis of MCF7 breast adenocarcinoma cells. *Cell Death Different*, **8**, 162-171.

Dillehay DL, Webb SK, Schmelz EM, Merrill AH Jr. (1994) Dietary sphingomyelin inhibits 1,2-dimethylhydrazine-induced colon cancer in CF1 mice. *J Nutr*, **124**, 615-620.

Di Marzio L, Di Leo A, Cinque B, Fanini D, Agnifili A, Berloco P, Linsalata M, Lorusso D, Barone M, De Simone C, Cifone MG (2005) Detection of alkaline sphingomyelinase activity in human stool: proposed role as a new diagnostic and prognostic marker of colorectal cancer. *Cancer Epidemiol Biomarkers Prev*, **14**, 856-862.

Dudeja PK, Dahiya R, Brasitus TA (1986) The role of sphingomyelin synthase and sphingomyelinase in 1,2-dimethylhydrazine-induced lipid alterations of rat colonic plasma membranes. *Biochem Biophys Acta*, **863**, 309-312.

Endo K, Igarashi Y, Nisar M, Zhou Q, Hakomori S-I (1991) Cell membrane signaling as target in cancer therapy: inhibitory effect of N,N-dimethyl and N,N,N-trimethyl sphingosine derivatives on in vitro and in vivo growth of human tumor cells in nude mice. *Cancer Res*, **51**, 1613-1618.

Enkvetchakul B, Barnett T, Liotta DC, Geisler V, Menaldino DS, Merrill AH Jr, Birt DF (1992) Influences of sphingosine on two-stage skin tumorigenesis in SENCAR mice. *Cancer Lett*, **62**, 35-42.

Erdreich-Epstein A, Tran LB, Bowman NN, Wang H, Cabot MC, Durden DL, Vlckova J, Reynolds CP, Stins MF, Groshen S, Millard M (2002) Ceramide signaling in fenretinide-induced endothelial cell apoptosis. *J Biol Chem*, **277**, 49531-49537.

Exon JH, South EH (2003) Effects of sphingomyelin on aberrant colonic crypt foci development, colon crypt cell proliferation and immune function in an aging rat tumor model. *Food Chem Toxicol*, **41**, 471-476.

French KJ, Schrecengost RS, Lee BD, Zhuang Y, Smith SN, Eberly JL, Yun JK, Smith CD (2003) Discovery and evaluation of inhibitors of human sphingosine kinase. *Cancer Res*, **63**, 5962-5969.

Garcia-Lloret M, Yui J, Winkler-Lowen B, O'Brien L, Brindley DN (1996) Epidermal growth factor inhibits cytokine-induced apoptosis of primary human trophoblasts. *J Cell Physiol*, **167**, 324-333.

Grazide S, Maestre N, Veldman RJ, Bezombes C, Maddens S, Levade T, Laurent G, Jaffrezou JP (2002) Ara-C- and daunorubicin-induced recruitment of Lyn in sphingomyelinase-enriched membrane rafts. *FASEB J*, **16**, 1685-1687.

Gulbins E, Kolesnick R (2003) Raft ceramide in molecular medicine. *Oncogene*, **22**, 7070-7077.

Hertervig E, Nilsson A, Nyberg L, Duan RD (1997) Alkaline sphingomyelinase activity is decreased in human colorectal carcinoma. Cancer, 79, 448-453.

Hong G, Baudhuin LM, Xu Y (1999) Sphinogsine-1-phosphate modulates cell growth and adhesion of ovarian cancer cells. *FEBS Lett*, **460**, 513-518.

Itoh M, Kitano T, Watanabe M, Kondo T, Yabu T, Taguchi Y, Iwai K, Tashima M, Uchiyama T, Okazaki T (2003) Possible role of ceramide as an indicator of chemoresistance: decrease of ceramide content via activation of glucosylceramide synthase and sphingomyelin synthase in chemoresistant leukemia. *Clin Cancer Res*, **8**, 415-423.

Jatoi A, Suman VJ, Schaefer P, Block M, Loprinzi C, Roche P, Garneau S, Morton R, Stella PJ, Alberts SR, Pittelkow M, Sloan J, Pagano R (2003) A phase II study of topical ceramides for cutaneous breast cancer. *Breast Cancer Res Treat*, **80**, 99-104.

Jiang Q, Wong J, Fyrst H, Saba JD, Ames BN (2004) gamma-Tocopherol or combinations of vitamin E forms induce cell death in human prostate cancer cells by interrupting sphingolipid synthesis. *Proc Natl Acad Sci USA*, **101**, 17825-17830.

Johnson KR, Johnson KY, Crellin HG, Ogretmen B, Boylan AM, Harley RA, Obeid LM (2005) Immunohistochemical Distribution of Sphingosine Kinase 1 in Normal and Tumor Lung Tissue. *J Histochem Cytochem*, **59**, 1159-1166.

Kamsteeg M, Rutherford T, Sapi E, Hanczaruk B, Shahabi S, Flick M, Brown D, Mor G. (2003) Phenoxodiol--an isoflavone analog--induces apoptosis in chemoresistant ovarian cancer cells. *Oncogene*, **22**, 2611-2620.

Kim JS, Crooks H, Foxworth A, Waldman T (2002) Proof-of-principle: oncogenic ß-catenin is a valid molecular target for the development of pharmacological inhibitors. *Mol Cancer Ther*, **1**, 1355-1359.

Kimura K, Markowski M, Edsall LC, Spiegel S, Gelmann EP (2003) Role of ceramide in mediating apoptosis of irradiated LNCaP prostate cancer cells. *Cell Death Different*, **10**, 240–248.

Koyanagi S, Kuga M, Soeda S, Hosoda Y, Yokomatsu T, Takechi H, Akiyama T, Shibuya S, Shimeno H (2003) Elevation of de novo ceramide synthesis in tumor masses and the role of microsomal dihydroceramide synthase. *Int J Cancer*, **105**, 1-6.

Koybasi S, Senkal CE, Sundararaj K, Spassieva S, Bialewski J, Osta W, Day TA, Jiango JC, Jazwinskio SM, Hannun YA, Obeid LM, Ogretmen B (2004) Defects in cell growth regulation by C18:0-ceramide and longevity assurance gene 1 (LAG1) in human head and neck squamous cell carcinomas (HNSCC). *J Biol Chem*, **279**, 44311-44319.

Lavie Y, Cao H, Volner A, Lucci A, Han TY, Geffen V, Giuliano AE, Cabot MC (1997) Agents that reverse multidrug resistance, tamoxifen, verapamil, and cyclosporin A, block glycosphingolipid metabolism by inhibiting ceramide glycosylation in human cancer cells. *J Biol Chem*, **272**, 1682-1687.

Lemonnier LA, Dillehay DL, Vespremi MJ, Abrams J, Brody E, Schmelz EM (2003) Sphingomyelin in the suppression of colon tumors: prevention versus intervention. *Arch Biochem Biophys*, **419**, 129-138.

Le Stunff H, Galve-Roperh I, Peterson C, Milstien S, Spiegel S (2002) Sphingosine-1-phosphate phosphohydrolase in regulation of sphingolipid metabolism and apoptosis. *J Cell Biol*, **158**, 1039-1049.

Lucci A, Han TY, Liu YY, Giuliano AE, Cabot MC (1999) Multidrug resistance modulators and doxorubicin synergize to elevate ceramide levels and elicit apoptosis in drug-resistant cancer cells. *Cancer*, **86**, 300-311.

Macchia M, Barontini S, Bertini S, Di Bussolo V, Fogli S, Giovannetti E, Grossi E, Minutolo F, Danesi R (2001) Design, synthesis, and characterization of the antitumor activity of novel ceramide analogues. *J Med Chem*, **44**, 3994-4000.

Maeda I, Takano T, Matsuzuka F, Maruyama T, Higashiyama T, Liu G, Kuma K, Amino N (1999) Rapid screening of specific changes in mRNA in thyroid carcinomas by sequence specific-differential display: decreased expression of acid ceramidase mRNA in malignant and benign thyroid tumors. *Int J Cancer*, **81**, 700-704.

Meyer zu Heringdorf D, Lass H, Kuchar I, Alemany R, Guo Y, Schmidt M, Jakobs KH (1999) Role of sphingosine kinase in Ca2+ signaling by epidermal growth factor receptor. *FEBS Lett*, **461**, 217-222.

Meng A, Luberto C, Meier P, Bai A, Yang X, Hannun YA, Zhou D (2004) Sphingomyelin synthase as a potential target for D609-induced apoptosis in U937 human monocytic leukemia cells. *Exp Cell Res*, **292**, 385-392.

Michael JM, Lavin MF, Watters DJ (1997) Resistance to radiation-induced apoptosis in Burkitt's lymphoma cells is associated with defective ceramide signaling. *Cancer Res*, **57**, 3600-3605.

Min J, Van Veldhoven PP, Zhang L, Hanigan MH, Alexander H, Alexander S (2005) Sphingosine-1-phosphate lyase regulates sensitivity of human cells to select chemotherapy drugs in a p38-dependent manner. *Mol Cancer Res*, **3**, 287-296.

Modrak DE, Lew W, Goldenberg DM, Blumenthal R (2000) Sphingomyelin potentiates chemotherapy of human cancer xenografts. *Biochem Biophys Res Commun*, **268**, 603-606.

Modrak DE, Cardillo TM, Newsome GA, Goldenberg DM, Gold DV (2004) Synergistic interaction between sphingomyelin and gemcitabine potentiates ceramide-mediated apoptosis in pancreatic cancer. *Cancer Res*, **64**, 8405-8410.

Musumarra G, Barresi V, Condorelli DF, Scire S (2003) A bioinformatic approach to the identification of candidate genes for the development of new cancer diagnostics. *Biol Chem*, **384**, 321-327.

Myrick D, Blackinton D, Klostergaard J, Kouttab N, Maizel A, Wanebo H, Mehta S (1999) Paclitaxel-induced apoptosis in Jurkat cells, a leukemic T cell line, is enhanced by ceramide. *Leuk Res*, **23**, 569-578.

Nava VE, Cuvillier O, Edsall LC, Kimura K, Milstien S, Gelmann EP, Spiegel S (2000) Sphingosine Enhances Apoptosis of Radiation-resistant Prostate Cancer Cells. *Cancer Res,* **60**, 4468–4474.

Nava VE, Hobson JP, Murthy S, Milstien S, Spiegel S (2002) Sphingosine kinase type 1 promotes estrogen-dependent tumorigenesis of breast cancer MCF-7 cells. *Exp Cell Res*, **281**, 115-127.

Noda S, Yoshimura S, Sawada M, Naganawa T, Iwama T, Nakashima S, Sakai N (2001) Role of ceramide during cisplatin-induced apoptosis in C6 glioma cells. *J Neurooncol*, **52**, 11-21.

Notterman DA, Alon U, Sierk AJ, Levine AJ (2001) Transcriptional gene expression profiles of colorectal adenoma, adenocarcinoma, and normal tissue examined by oligonucleotide arrays. *Cancer Res*, **61**, 3124-3130.

Olshefski RS, Ladisch S (2001) Glucosylceramide synthase inhibition enhances vincristine-induced cytotoxicity. *Int J Cancer*, **93**, 131-138.

Payne SG, Brindley DN, Guilbert LJ (1999) Epidermal growth factor inhibits ceramide-induced apoptosis and lowers ceramide levels in primary placental trophoblasts. *J Cell Physiol*, **180**, 263-270.

Perry DK, Carton J, Shah AK, Meredith F, Uhlinger DJ, Hannun YA (2000) Serine palmitoyltransferase regulates de novo ceramide generation during etoposide-induced apoptosis. *J Biol Chem*, **275**, 9078-9084.

Prinetti A, Basso L, Appierto V, Villani MG, Valsecchi M, Loberto N, Prioni S, Chigorno V, Cavadini E, Formelli F, Sonnino S (2003) Altered sphingolipid metabolism in N-(4-hydroxyphenyl) retinamide-resistant A2780 human ovarian carcinoma cells. *J Biol Chem*, **278**, 5574-5583.

Riboni I, Campanella R, Bassi R, Villani R, Gaini SM, Martinelli-Boneschi F, Viani P, Tettamanti G (2002) Ceramide levels are inversely associated with malignant progression of glial tumors. *Glia*, **39**, 105-113.

Sachs CW, Safa AR, Harrison SD, Fine RL (1995) Partial inhibition of multidrug resistance by safingol is independent of modulation of P-glycoprotein substrate activities and correlated with inhibition of Protein Kinase C. *J Biol Chem*, **270**, 26639–26648.

Samsel L, Zaidel G, Drumgoole HM, Jelovac D, Drachenberg C, Rhee JG, Brodie AM, Bielawska A, Smyth MJ (2004) The ceramide analog, B13, induces apoptosis in prostate cancer cell lines and inhibits tumor growth in prostate cancer xenografts. *Prostate*, **58**, 382-393.

Sawada M, Nakashima S, Banno Y, Yamakawa H, Hayashi K, Takenaka K, Nishimura Y, Sakai N, Nozawa Y (2000) Ordering of ceramide formation, caspase activation, and Bax/Bcl-2 expression during etoposide-induced apoptosis in C6 glioma cells. *Cell Death Differ*, **7**, 761-772.

Scarlatti F, Sala G, Somenzi G, Signorelli P, Sacchi N, Ghidoni R (2003) Resveratrol induces growth inhibition and apoptosis in metastatic breast cancer cells via de novo ceramide signaling. *FASEB J*, **17**, 2339-2341.

Schmelz EM, Crall KL, LaRocque R, Dillehay DL, Merrill AH Jr. (1994) Uptake and metabolism of sphingolipids in isolated intestinal loops of mice. *J Nutr*, **124**, 702-712.

Schmelz EM, Dillehay DL, Webb SK, Reiter A, Adams J, Merrill AH Jr. (1996) Sphingomyelin consumption suppresses aberrant colonic crypt foci and increases the proportion of adenomas versus adenocarcinomas in CF1 mice treated with 1,2-dimethylhydrazine: implications for dietary sphingolipids and colon carcinogenesis. *Cancer Res*, **56**, 4936-4941.

Schmelz EM, Bushnev AB, Dillehay DL, Liotta DC, Merrill AH Jr. (1997) Suppression of aberrant colonic crypt foci by synthetic sphingomyelins with saturated or unsaturated sphingoid base backbones. *Nutr Cancer*, **28**, 81-85.

Schmelz EM, Bushnev AS, Dillehay DL, Sullards MC, Liotta DC, Merrill AH Jr. (1999) Ceramide-ß-glucuronide: synthesis, digestion, and suppression of early markers of colon carcinogenesis. *Cancer Res*, **59**, 5768-5772.

Schmelz EM, Sullards MC, Dillehay DL, Merrill AH Jr. (2000) Inhibition of colonic cell proliferation and aberrant crypt foci formation by dairy glycosphingolipids in 1,2 dimethylhydrazine-treated CF1 mice. *J Nutr*, **130**, 522-527.

Schmelz EM, Roberts PC, Kustin EM, Lemonnier LA, Sullards MC, Dillehay DL, Merrill AH Jr. (2001) Modulation of intracellular ß-catenin localization and intestinal tumorigenesis in vivo and in vitro by sphingolipids. *Cancer Res*, **61**, 6723-6729.

Seelan RS, Qian C, Yokomizo A, Bostwick DG, Smith DI, Liu W (2000) Human acid ceramidase is overexpressed but not mutated in prostate cancer. *Genes Chromosomes Cancer*, **29**, 137-146.

Selzner M, Bielawska A, Morse MA, Rudiger HA, Sindram D, Hannun YA, Clavien PA (2001) Induction of apoptotic cell death and prevention of tumor growth by ceramide analogues in metastatic human colon cancer. *Cancer Res*, **61**, 1233-1240.

Shabbits JA, Mayer LD (2003) Intracellular delivery of ceramide lipids via liposomes enhances apoptosis in vitro. Biochim. Biophys. Acta. 1612: 98-106.

Shabbits JA, Mayer LD (2003a) High ceramide content liposomes with in vivo antitumor activity. *Anticancer Res*, **23**, 3663-3669.

Sietsma H, Veldman RJ, Kolk D, Ausema B, Nijhof W, Kamps W, Vellenga E, Kok JW (2000) 1-phenyl-2-decanoylamino-3-morpholino-1-propanol chemosensitizes neuroblastoma cells for taxol and vincristine. *Clin Cancer Res*, **6(3)**, 942-948.

Silins I, Nordstrand, M, Hogberg J, Stenius U (2003) Sphingolipids suppress preneoplastic rat hepatocytes in vitro and in vivo. *Carcinogenesis*, **24**, 1077-1083.

Stover TC, Kester M (2003) Liposomal delivery enhances short-chain ceramide-induced apoptosis of breast cancer cells. *J Pharmacol Exp Ther*, **307**, 468-475.

Stover TC, Sharma A, Robertson GP, Kester M (2005) Systemic delivery of liposomal short-chain ceramide limits solid tumor growth in murine models of breast adenocarcinoma. *Clin Cancer Res*, **11**, 3465-3474.

Strelow A, Bernardo K, Adam-Klages S, Linke T, Sandhoff K, Kronke M, Adam D (2000) Overexpression of acid ceramidase protects from tumor necrosis factor-induced cell death. *J Exp Med*, **192**, 601-612.

Sullards MC, Merrill AH Jr. (2003) Analysis of sphingosine 1-phosphate, ceramides, and other bioactive sphingolipids by high-performance liquid chromatography-tandem mass spectrometry. *Sci STKE*, **67**, PL1.

Sumitomo M, Ohba M, Asakuma J, Asano T, Kuroki T, Asano T, Hayakawa M. (2002) Protein kinase Cdelta amplifies ceramide formation via mitochondrial signaling in prostate cancer cells. *J Clin Invest*, **109**, 827-836.

Sutphen R, Xu Y, Wilbanks GD, Fiorica J, Grendys EC, LaPolla JP, Arango H, Hoffman MS, Martino M, Wakely K, Griffin D, Blanco RW, Cantor AB, Xiao YJ, Krischer JP (2004) Lysophospholipids are potential biomarkers of ovarian cancer. *Cancer Epidemiol Biomarkers Prev*, **13**, 1185-1191.

Symolon H, Schmelz EM, Dillehay DL, Merrill AH Jr. (2004) Dietary soy sphingolipids suppress tumorigenesis and gene expression in 1,2-

dimethylhydrazine-treated CF1 mice and ApcMin/+ mice. *J Nutr*, **134**, 1157-1161.

Takenaga M, Igarashi R, Matsumoto K, Takeuchi J, Mizushima N, Nakayama T, Morizawa Y, Mizushima, Y (1999) Lipid microsphere preparation of a lipophilic ceramide derivative suppresses colony formation in a murine experimental metastasis model. *J Drug Target*, **7**, 187-195.

Veldman RJ, Zerp S, van Blitterswijk WJ, Verheij M (2004) N-hexanoyl-sphingomyelin potentiates in vitro doxorubicin cytotoxicity by enhancing its cellular influx. *Br J Cancer*, **90**, 917-925.

Wang H, Maurer BJ, Reynolds CP, Cabot MC (2001) N-(4-hydroxyphenyl)retinamide elevates ceramide in neuroblastoma cell lines by coordinate activation of serine palmitoyltransferase and ceramide synthase. *Cancer Res*, **61**, 5102-5105.

Wang H Giuliano AE, Cabot MC (2002) Enhanced de novo ceramide generation through activation of serin palmitoyltransferase by the P-glycoprotein antagonist SDZ PCS 833 in breast cancer cells. *Mol Cancer Therapeutics*, **1**, 719-726.

Wang H, Charles AG, Frankel AJ, Cabot MC, (2003) Increasing intracellular ceramide: an approach that enhances the cytotoxic response in prostate cancer cells. *Urology*, **61**, 1047-1052.

Wang Z, Liu Y, Mori M, Kulesz-Martin M (2002) Gene expression profiling of initiated epidermal cells with benign or malignant tumor fates. *Carcinogenesis*, **23**, 635-643.

Xia P, Gamble JR, Wang L, Pitson SM, Moretti PA, Wattenberg BW, D'Andrea RJ, Vadas MA.(2000) An oncogenic role of sphingosine kinase. *Curr Biol*, **10**, 1527-1530.

Yamanaka M, Shegogue D, Pei H, Bu S, Bielawska A, Bielawski J, Pettus B, Hannun YA, Obeid L, Trojanowska M (2004) Sphingosine kinase 1 (SPHK1) is induced by transforming growth factor-beta and mediates TIMP-1 up-regulation. *J Biol Chem*, **279**, 53994-534001.

Zhang J, Alter N, Reed JC, Borner C, Obeid LM, Hannun YA (1996) Bcl-2 interrupts the ceramide-mediated pathway of cell death. *Proc Natl Acad Sci USA*, **93**, 5325-5328.

Part 6

S1P Signaling and SIPR

6-1 Sphingosine-1-Phosphate and the Regulation of Immune Cell Trafficking

Maria Laura Allende and Richard L. Proia

Genetics of Development and Disease Branch, National Institutes of Diabetes and Digestive and Kidney Diseases, National Institutes of Health, Bethesda, MD 20892, USA

Summary. Sphingosine-1-phosphate (S1P) is an extracellular sphingolipid signaling molecule that acts through a family of G-protein coupled receptors. The signaling pathways stimulated by S1P receptors profoundly affect the activities of lymphocytes and endothelial cells. Pharmacological and genetic experiments have clearly established the S1P-S1P receptor system as a dominant regulatory axis for the trafficking of lymphocytes. Manipulating this regulatory axis might lead to the development of therapies that target immune system dysfunctions.

Keywords. Sphingolipids, Sphingosine-1-phosphate, immunity

Sphingolipids are a structurally diverse class of cellular lipids that function as organizers of membrane structure and as signaling molecules (Hla et al. 2001, Allende and Proia 2002, Hakomori 2003, Degroote et al. 2004, Futerman and Hannun 2004). Sphingosine-1-phosphate (S1P), a sphingolipid metabolite, has emerged as a potent lipid mediator with wide-ranging effects. S1P stimulated receptor-mediator cell signaling pathways regulate key processes including proliferation, apoptosis, and cell mobility (Spiegel and Milstien 2003, Hla 2004). This review will focus on the role of S1P and its G-protein coupled receptors (GPCRs) in immunity.

1. Generation of S1P

S1P is produced from phosphorylated sphingosine that is produced when ceramide is degraded by ceramidase (El Bawab et al. 2000). Sphingosine kinases 1 and 2 phosphorylated sphingosine to yield S1P (Kohama et al. 1998, Liu et al. 2000a, Olivera and Spiegel 2001). Levels of S1P can be controlled by recycling S1P back to sphingosine by S1P phosphatase (De Ceuster et al. 1995, Mao et al. 1997, Mandala et al. 1998) or SIP lyase degradation to form phosphoethanolamine and hexadecenal (van Veldhoven and Mannaerts 1993, Saba et al. 1997, Zhou and Saba 1998). Alternatively, S1P may be generated from the hydrolysis of sphingosylphosphorylcholine by autotaxin, an ecto-nucleotide pyrophosphatase/ phosphodiesterase that functions primarily as a lysophospholipase D (Clair et al. 2003). The physiological role of autotaxin in S1P production is uncertain since the levels of sphingosylphosphorylcholine in plasma and serum are three orders of magnitude lower than the reported Km for this activity (van Meeteren et al. 2005).

2. S1P family of receptors

S1P interacts with the S1P receptor family of GPCRs (also known as endothelial differentiation gene - Edg - receptors) and triggers multiple signaling pathways (Spiegel and Milstien 2003, Hla 2004). The first receptor identified was S1P1, which was isolated from activated endothelial cells (Hla and Maciag 1990). Five structurally related receptors in the family for S1P: S1P1 (Edg-1), S1P2 (Edg-5), S1P3 (Edg-3), S1P4 (Edg-6) and S1P5 (Edg-8). Although sequentially similar, these receptors have different cell and tissue distribution, signal through different G proteins, and mediate different biological activities (Kluk and Hla 2002, Sanchez and Hla 2004, Taha et al. 2004). S1P1, S1P2, and S1P3 receptors are widely expressed whereas the distribution of S1P4 is restricted to cells in the immune systems and S1P5 to cells of the nervous system (MacLennan et al. 1994, Liu and Hla 1997, Graler et al. 1998, Yamaguchi et al. 1999, Zhang et al. 1999, Im et al. 2000, Chae et al. 2004). S1P1 and S1P4 mainly activate Gi but the other receptors couple to multiple G proteins. Both S1P2 and S1P3 are linked to Gi, Gq and G12/13. S1P5 couples to Gi and G12/13 (Kluk and Hla 2002, Spiegel and Milstien 2003). Targets linked to S1P receptor signaling pathways include phospholipase C, adenylyl cyclase-cyclic AMP, phosphatidylinositol 3-kinase, the small GTPases of the Rho family, particularly Rac and Rho, and mitogen-activated protein kinase (Zhang et al. 1999, Spiegel and Milstien 2003, Sanchez and Hla 2004).

Disruption of the S1P1 receptor gene in mice causes embryonic lethality, demonstrating that the S1P1 signaling pathway is essential for development (Liu et al. 2000b, Allende et al. 2003). In contrast, genetic ablation of the genes encoding S1P2, S1P3 or S1P5 receptors in mice results in viable offspring and are not needed for development (Ishii et al. 2001, MacLennan et al. 2001, Kono et al. 2004, Jaillard et al. 2005). The vascular phenotypes of S1P1/S1P2, S1P2/S1P3 double-knockout embryos and S1P1/S1P2/S1P3 triple-knockout embryos are substantially more severe than S1P1 deficient embryos, indicating overlapping or redundant functions between these receptors (Kono et al. 2004).

3. S1P is a normal component of circulating blood and acutely released during inflammation

Considering its how concentrations in the blood, S1P is an important extracellular mediator. In plasma and serum, S1P is present at approximately micromolar concentrations (Table 1). The majority of S1P in blood is bound to high and low density lipoproteins as well as to albumin (Okajima 2002). Levels in tissue are substantially lower than those found in blood (Table 1). The "free" S1P concentration is unknown, in part because it is rapidly degradation by phosphatases, but bulk levels indicate that distinct S1P concentrations are compartmentalized within the body, which, in the steady state, can produce concentration gradients of the signaling molecule with highest concentrations in blood and lowest concentrations in tissues.

The production of hematopoietic cells appears to be a major source of S1P in blood (Yang et al. 1999). Lymphocytes rapidly convert sphingosine to S1P, which is then secreted (Yang et al. 1999). S1P is also produced by platelets, macrophages, mast cells, neutrophils and dendritic cells (Yatomi et al. 2000, Goetzl and Graler 2004, Olivera and Rivera 2005).

S1P concentrations may increase acutely. Platelets, which store abundant amounts of S1P as the result of a highly active sphingosine kinase and the lack of S1P lyase, can release S1P after stimulation with thrombin during platelet aggregation (Yatomi et al. 2001). Mast cells synthesize and release S1P after stimulation with allergen, an event that regulates degranulation, leukotriene release and migration towards antigen gradients (Prieschl et al. 1999, Jolly et al. 2004). Sphingosine kinase activity in several cell types is rapidly stimulated by specific inflammatory signals, such as tumor necrosis factor γ and IL 1γ (Xia et al. 1998, Xia et al. 2002, Pettus et al. 2003).

4. Immune cells express S1P receptors and are regulated by S1P

Cells in the immune system express S1P family of receptors according to individual cell type and state of differentiation. The S1P1 receptor is the most widely and abundantly expressed of the five members of the family (Table 2).

S1P potently regulates immune cell migration (Muller et al. 2003, Goetzl and Rosen 2004, Payne et al. 2004, Cyster 2005, Olivera and Rivera 2005, Rosen and Goetzl 2005). Mature CD4+ and CD8+ single positive T cells, which primarily express S1P1 and S1P4 receptors, migrate toward S1P in chemotactic assays in vitro, and are recruited into S1P-injected subcutaneous pouches in vivo (Graeler and Goetzl 2002, Graeler et al. 2002). The expression of S1P1 is up-regulated during the differentiation of T cells and down-regulated after their activation (Allende et al. 2004a, Matloubian et al. 2004). S1P, in a concentration dependent manner, also regulates the chemotactic responses of T cells to chemokines such as CCL-21 (Exodus-2) and CCL-5 (RANTES) (Graeler et al. 2002). Interestingly, the S1P4 receptors on T cells may control a set of responses to S1P, independent of migration, that regulate the suppression of T cell proliferation as well as the generation of trophic- and positive- effector cytokines such as IL-2, IL-4 and IFN-γ and the inhibitory cytokine, IL-10 (Wang et al. 2005). S1P, when administrated locally to rats, can cause edema that is accompanied by an eosinophil infiltrate (Roviezzo et al. 2004). S1P stimulation strongly upregulates the CCR3 receptor on eosinophils, which may trigger S1P directed recruitment (Roviezzo et al. 2004). S1P regulates mast cell migration via S1P1 and S1P2 receptors (Jolly et al. 2004). S1P induces the chemotaxis of natural killer cells, which express S1P1, S1P4 and S1P5 receptors (Kveberg et al. 2002). In immature dendritic cells, S1P also has chemotactic activity, but this response is lost in mature dendritic cells (Idzko et al. 2002). In mouse peritoneal macrophages, S1P induces the expression of IL-1γ and TNFγ. These cytokines are known to regulate the expression of adhesion molecules on endothelial cells that are necessary for platelets and leukocyte recruitment during inflammation, atherosclerosis and immune responses (Lee et al. 2002).

Table 1. S1P concentration in tissues

Tissue	S1P concentration*				
	Mouse [a]	Rat [b]	Rat [c]	Rat [d]	Human
Brain	2.98	35	5.8	nd	nd
Heart	0.39	10	0.5	76.4	nd
Kidney	0.36	15	0.7	24.4	nd
Liver	0.42	5	0.7	64.4	nd
Muscle	0.24	5	nd	24.2	nd
Spleen	0.31	40	2.5	545.4	nd
Testis	0.65	100	0.5	27.4	nd
Serum	4.12	nd	nd	nd	0.5-0.8 [e], 0.98 [f], 0.48 [g]
Plasma	1.9	nd	nd	nd	0.19 [g], 0.2-0.4 [e]
Ammiotic fluid	nd	nd	nd	nd	0.019 [h]
Ovarian cancer ascites	nd	nd	nd	nd	0.051 [i]

* S1P concentrations are given as nmol per ml for serum, plasma, ammiotic fluid and ovarian cancer ascites and as nmol per gram of wet tissue for the rest of the listed tissues.

[a] (Allende et al. 2004b), [b] (Yatomi et al. 1997a), [c] (Edsall and Spiegel 1999), [d] (Min et al. 2002), [e] (Okajima 2002), [f] (Ruwisch et al. 2001), [g] (Yatomi et al. 2001), [h] (Kim et al. 2003), [i] (Xu et al. 2003).

Table 2. Expression of S1P family of receptors in cells from the immune system

Immune cell	S1P receptor expression	Chemotaxis toward S1P	Functional effects of S1P	Genetic manipulation of S1P receptors
T cells	S1P1, S1P4, (low levels of S1P2, S1P5 and S1P3) [a]	Increases at low [S1P] than plasma level, decreases at plasma [S1P] [b]	· Decreases proliferation [c] · Decreases generation (IFNα, IL-4) [c] · S1P-S1P1 inhibits chemotaxis to chemokine CCL21 and CCL5 [c]	Deletion of S1P1 [d, e] · blocks the emigration of T cells from thymus and the exit from peripheral lymphoid organs Overexpression of S1P1 [f, g] · mature T cells display increased chemotaxis towards S1P · mature T cells are found in higher numbers in blood · decreases proliferation · causes defective contact hypersensitivity reaction and local Ag-induced response as a consequence of the reduced number of activated T cells in lymph nodes · causes increase in IgE Ab and less IgG2 Ab production in plasma
B cells	S1P1, S1P4 [h] S1P1, S1P3 (on MZB cells) [i]	Increases at low [S1P], decreases at high [S1P] [i]	· S1P-S1P1 inhibits chemotaxis to chemokine CXCL13 [i]	Deletion of S1P1 [e, i] · blocks the exit from periphal lymphoid organs affects MZB cell localization in spleen Deletion of S1P3 [j] · causes the reduction of follicular and MZ B cells and mislocalization of the MOMA1(+) macrophages and MAdCAM-1(+) endothelial cells along the marginal sinus · defective immune responses to thymus-independent antigen type

Immune cell	S1P receptor expression	Chemotaxis toward S1P	Functional effects of S1P	Genetic manipulation of S1P receptors
Macrophages /Monocytes	S1P1, S1P2, S1P4 (human) [k] S1P1, S1P2 (mouse) [l]		· Enhances expression of IL-1α, TNF-α [l] · Induces CD32 expression, which is related to phagocitic activity [k]	nd
Dendritic cells	S1P1, S1P2, S1P3, S1P4 [m]	Promotes chemotaxis [m,n,o]	· Inhibits secretion of IL-12 and TNF-α and enhances secretion of IL-10 (human) [m]	nd
Mast cells	S1P1, S1P2 [p]	Promotes chemotaxis [p]	· Regulates degranulation and leukotriene release [p,q]	nd
Natural killer cells	S1P1, S1P4, S1P5 [r]	Promotes chemotaxis [r]		nd
Eosinophils	S1P1 [s]	Promotes chemotaxis [s]		nd

[a] (Graeler and Goetzl 2002), [b] (Graeler et al. 2002), [c] (Dorsam et al. 2003), [d] (Allende et al. 2004a), [e] (Matloubian et al. 2004), [f] (Chi and Flavell 2005), [g] (Graler et al. 2005), [h] (Goetzl and Graeler 2004), [i] (Cinamon et al. 2004), [j] (Girkontaite et al. 2004), [k] (Duong et al. 2004), [l] (Lee et al. 2002), [m] (Idzko et al. 2002), [n] (Renkl et al. 2004), [o] (Czeloth et al. 2005), [p] (Prieschl et al. 1999), [q] (Jolly et al. 2004), [r] (Kveberg et al. 2002), [s] (Roviezzo et al. 2004).

5. Insights from Pharmacology

The studies using FTY720, an immunosuppressant, highlighted the importance of S1P-S1P receptor control of lymphocyte trafficking during homeostasis and diseases. FTY720 (2-amino-2-[2-(4-octylphenyl)ethyl]-1, 3-propanediol) was derived from ISP-1, also known as myriocin, a compound isolated from the fungus *Isaria sinclairii* (Kiuchi et al. 2000). ISP-1 suppresses lymphocyte proliferation and inhibits serine palmitoyl transferase activity, the first enzyme in the sphingolipid biosynthetic pathway (Miyake et al. 1995). ISP-1 was modified to reduce its toxicity and identify the structure essential for its immunosuppressive activity. This derivative, named FTY720, is a more potent immunosuppressant compound that, unlike ISP-1, neither inhibits serine palmitoyl transferase nor functions as a anti-proliferative (Kiuchi et al. 2000). FTY720 functions by producing lymphopenia caused by the sequestration of lymphocytes within lymph nodes and Peyer's patches (Chiba et al. 1998, Brinkmann et al. 2000). FTY720 has been used in clinical trials to suppress rejection after transplantation and is effective in numerous autoimmune disease models in rodents including type I diabetes and experimental autoimmune encephalomyelitis (Brinkmann et al. 2002, Brinkmann and Lynch 2002, Yang et al. 2003).

FTY720 produces its immunosuppressive activity through S1P receptors (Brinkmann et al. 2002, Mandala et al. 2002). As a structural sphingosine analogue, FTY720 is phosphorylated primarily by sphingosine kinase 2 to form FTY720-P, which then functions as an agonist for four of the S1P receptors (S1P1, S1P3, S1P4, S1P5, not S1P2) (Brinkmann et al. 2002, Mandala et al. 2002, Billich et al. 2003, Paugh et al. 2003, Sanchez et al. 2003, Kharel et al. 2005). After binding to S1P1 receptor, FTY720-P induces receptor internalization and partial degradation inside the cell to block S1P signaling (Graler and Goetzl 2004).

In lymph nodes from FTY720-treated mice, lymphocytes are retained on the adluminal side of the lymphatic endothelium and are absent from the lymphatic sinus (Mandala et al. 2002). FTY720 induced-blood lymphopenia is temporally associated with a decrease in thoracic duct lymphocytes numbers (Mandala et al. 2002). Altogether, these findings suggest that FTY720 causes the sequestration of lymphocytes in lymph nodes by inhibiting their egress and, consequently, blocking their presence in the periphery. This sequestration by FTY720 causes the reduction of naïve T cell recirculation and blocks the release of Ag-activated T cells from draining lymph nodes (Xie et al. 2003).

FTY720 has profound effects on thymocyte development. When administered in low doses, and within a short period of time, FTY720 triggers the maturation of single positive thymocytes to a late medullary phenotype, ultimately causing the inhibition of T cell egress from thymus (Rosen et al. 2003). With prolonged FTY720 treatment, the amount of mature, single positive T cells increases in the thymus because T cells are unable to emigrate to the periphery (Yagi et al. 2000).

B cell migration is also affected by FTY720 treatment. FTY720 alters the location of marginal zone B cells. These B cells, located at the border between the red and white pulp, are exposed to much of the blood that enters the spleen. This location is the marginal zone and ensures rapid exposure to an antigen. FTY720 and FTY720-P induces the migration of marginal zone B cells into follicles, a process thought to occur after cell activation (Cinamon et al. 2004, Vora et al. 2005).

FTY720 has potent effects on endothelial cells through a Gi linked pathway that includes junction assembly and blocking vascular permeability (Sanchez et al. 2003, Paik et al. 2004). S1P has similar activity in producing endothelial barrier enhancement and in counteracting vascular permeability (McVerry and Garcia 2005).

6. Insights from Genetics

Genetic ablations of S1P1 in mice reveals the receptor's role in the vascular system: mice with no S1P1 receptor die as embryos from vascular system collapse (Liu et al. 2000b). The critical function of the S1P1 receptor during vascular development is located within endothelial cells where it participates in the maturation of nascent vessels (Allende et al. 2003).

To determine the precise role of the S1P1 receptor in immunity two approaches were taken. For one, chimeras were produced by transplanting fetal liver cells from S1P1 receptor mutant mice into immunodeficient recipients to yield adult mice with S1P1 deleted from cells of the immune system (Matloubian et al. 2004). In the other, the S1P1 gene was deleted in early thymocytes using the Cre-LoxP paradigm (Allende et al. 2004a). The thymocytes in both models lacking the S1P1 receptor could not leave the thymus, demonstrating that the S1P1 receptor is essential for thymic egress. Furthermore, the S1P1 receptor was up-regulated on thymocytes as they matured as they acquired chemotactic responsiveness to S1P. The results suggest that high blood levels of S1P would provide a chemotactic signal via the S1P receptor on mature thymocytes to migrate from the thymus into the blood.   Another possible explanation is that S1P1 receptor signaling on thymocytes modifies those other chemotactic signals required for to leave the thymus. Proper egress of T cells from lymph nodes also requires lymphocyte expression of the S1P1 receptor. In fact, S1P1 receptor expression levels appear to regulate the amount of time it take a lymphocyte to proceed through the lymph node (Matloubian et al. 2004, Lo et al. 2005).   These observations suggest that cyclical S1P expression that regulates lymphocyte trafficking through lymph nodes.   In this model, high levels of S1P in blood cause the S1P1 receptor on lymphocytes to be down-regulated.   After entering lymph nodes, which may have lower levels of S1P, the cell surface S1P1 receptor expression recovers.   Directed chemotaxis out of the node through the endothelium can then proceed into lymph, which may contain a relatively high S1P concentration.   Lymphocyte activation, producing a down-relation of S1P1 receptor expression, would be one mechanism contributing to extended retention in lymph nodes. Such a process would allow T cells, after encountering specific antigen in lymph nodes, to be retained transiently to ensure proper activation and clonal expansion (Lo et al. 2005).

In addition regulating lymphocyte exit from secondary lymphoid organs, the S1P1 receptor also facilitates entry into secondary lymphoid organs in a cell- and tissue-specific manner. S1P1 receptor deficient lymphocytes are defective in integrin-dependent homing to lymph nodes via high endothelial venules (Halin et al. 2005).

The S1P1 receptor is also essential for the proper location of marginal zone B cells within the spleen. When the S1P1 receptor is genetically missing, the marginal zone B cells respond to the follicular chemokine CXCL13 and migrate to the follicles. Similarly, activation of marginal zone B cells by antigen or LPS causes down-regulation of the S1P1 receptor and migration into the follicles (Cinamon et al. 2004).   The S1P3 receptor, while not critical for the localization of marginal zone B cells, is important for the organization macrophages and endothelial cells of the marginal sinus (Girkontaite et al. 2004).

The phenotypic similarity of mice treated with FTY720 and mice with genetic deletions of the S1P1 receptor is consistent with FTY720 inhibition of the S1P1 receptor-signaling pathway on lymphocytes, possibly by inactivation caused by constant stimulation with the agonist.

7. Conclusions and perspectives

S1P, as a signaling molecule, is well-suited for regulating lymphocyte trafficking. Through interactions with G-protein coupled receptors, S1P potently regulates activities of lymphocytes (migration) and endothelial cells (barrier enhancement) essential for lymphocyte movement. An S1P gradient exists within the body with high concentrations in blood and lymph nodes, and lower concentrations in tissues, which can induce migration responses as well as differentially regulate functional activity of lymphocytes. In addition, S1P levels can be regulated locally to allow alterations of trafficking during acute conditions such as inflammation.

The S1P1 receptor appears to have the dominant role in the regulation of lymphocyte trafficking. While clear evidence exists that the lymphocyte-expressed receptor is critical for trafficking, the receptor contribution on endothelial cells has not yet been established. Given the importance of the S1P1 receptor for endothelial function, however, it may also play a key role.

The S1P-S1P receptor system has enormous therapeutic importance, as illustrated by the results obtained with FTY720. As our understanding of the system expands, a new generation of compounds that affect receptor function and S1P generation may emerge as agents that can profoundly modify immune responses.

8. References

Allende ML, Dreier JL, Mandala S, Proia RL (2004a) Expression of the sphingosine-1-phosphate receptor, S1P1, on T-cells controls thymic emigration. *J Biol Chem*, **279**, 15396-15401.

Allende ML, Proia RL (2002) Lubricating cell signaling pathways with gangliosides. *Curr Opin Struct Biol*, **12**, 587-592.

Allende ML, Sasaki T, Kawai H, Olivera A, Mi Y, van Echten-Deckert G, Hajdu R, Rosenbach M, Keohane CA, Mandala S, Spiegel S, Proia RL (2004b) Mice deficient in sphingosine kinase 1 are rendered lymphopenic by FTY720. *J Biol Chem*, **279**, 52487-52492.

Allende ML, Yamashita T, Proia RL (2003) G-protein coupled receptor S1P1 acts within endothelial cells to regulate vascular maturation. *Blood*, **102**, 3665-3667.

Billich A, Bornancin F, Devay P, Mechtcheriakova D, Urtz N, Baumruker T (2003) Phosphorylation of the imunomodulatory drug FTY720 by sphingosine kinases. *J Biol Chem*, **278**, 47408-47415.

Brinkmann V, Davis MD, Heise CE, Albert R, Cottens S, Hof R, Bruns C, Prieschl E, Baumruker T, Hiestand P, Foster CA, Zollinger M, Lynch KR (2002) The immune modulator FTY720 targets sphingosine 1-phosphate receptors. *J Biol Chem*, **277**, 21453-21457.

Brinkmann V, Lynch KR (2002) FTY720: targeting G-protein-coupled receptors for sphingosine 1-phosphate in transplantation and autoimmunity. *Curr Opin Immunol*, **14**, 569-575.

Brinkmann V, Pinschewer D, Chiba K, Feng L (2000) FTY720: a novel transplantation drug that modulates lymphocyte traffic rather than activation. *Trends Pharmacol Sci*, **21**, 49-52.

Chae SS, Proia RL, Hla T (2004) Constitutive expression of the S1P1 receptor in adult tissues. Prostaglandins Other Lipid Mediat, 73, 141-150.

Chi H, Flavell RA (2005) Cutting edge: regulation of T cell trafficking and primary immune responses by sphingosine 1-phosphate receptor 1. *J Immunol*, **174**, 2485-2488.

Chiba K, Yanagawa Y, Masubuchi Y, Kataoka H, Kawaguchi T, Ohtsuki M, Hoshino Y (1998) FTY720, a novel immunosuppressant, induces sequestration of circulating mature lymphocytes by acceleration of lymphocyte homing in rats. I. FTY720 selectively decreases the number of circulating mature lymphocytes by acceleration of lymphocyte homing. *J Immunol*, **160**, 5037-5044.

Cinamon G, Matloubian M, Lesneski MJ, Xu Y, Low C, Lu T, Proia RL, Cyster JG (2004) Sphingosine 1-phosphate receptor 1 promotes B cell localization in the splenic marginal zone. *Nat Immunol*, **5**, 713-720.

Clair T, Aoki J, Koh E, Bandle RW, Nam SW, Ptaszynska MM, Mills GB, Schiffmann E, Liotta LA, Stracke ML (2003) Autotaxin hydrolyzes sphingosylphosphorylcholine to produce the regulator of migration, sphingosine-1-phosphate. *Cancer Res*, **63**, 5446-5453.

Cyster JG (2005) Chemokines, sphingosine-1-phosphate, and cell migration in secondary lymphoid organs. *Annu Rev Immunol*, **23**, 127-159.

Czeloth N, Bernhardt G, Hofmann F, Genth H, Forster R (2005) Sphingosine-1-phosphate mediates migration of mature dendritic cells. *J Immunol*, **175**, 2960-2967.

De Ceuster P, Mannaerts GP, Van Veldhoven PP (1995) Identification and subcellular localization of sphinganine-phosphatases in rat liver. *Biochem J*, **311 (Pt 1)**, 139-146.

Degroote S, Wolthoorn J, van Meer G (2004) The cell biology of glycosphingolipids. *Semin Cell Dev Biol*, **15**, 375-387.

Dorsam G, Graeler MH, Seroogy C, Kong Y, Voice JK, Goetzl EJ (2003) Transduction of multiple effects of sphingosine 1-phosphate (S1P) on T cell functions by the S1P1 G protein-coupled receptor. *J Immunol*, **171**, 3500-3507.

Duong CQ, Bared SM, Abu-Khader A, Buechler C, Schmitz A, Schmitz G (2004) Expression of the lysophospholipid receptor family and investigation of lysophospholipid-mediated responses in human macrophages. *Biochim Biophys Acta*, **1682**, 112-119.

Edsall LC, Spiegel S (1999) Enzymatic measurement of sphingosine 1-phosphate. *Anal Biochem*, **272**, 80-86.

El Bawab S, Roddy P, Qian T, Bielawska A, Lemasters JJ, Hannun YA (2000) Molecular cloning and characterization of a human mitochondrial ceramidase. *J Biol Chem*, **275**, 21508-21513.

Futerman AH, Hannun YA (2004) The complex life of simple sphingolipids. *EMBO Rep*, **5**, 777-782.

Girkontaite I, Sakk V, Wagner M, Borggrefe T, Tedford K, Chun J, Fischer KD (2004) The sphingosine-1-phosphate (S1P) lysophospholipid receptor S1P3 regulates MAdCAM-1+ endothelial cells in splenic marginal sinus organization. *J Exp Med*, **200**, 1491-1501.

Goetzl EJ, Graler MH (2004) Sphingosine 1-phosphate and its type 1 G protein-coupled receptor: trophic support and functional regulation of T lymphocytes. *J Leukoc Biol*, **76**, 30-35.

Goetzl EJ, Rosen H (2004) Regulation of immunity by lysosphingolipids and their G protein-coupled receptors. *J Clin Invest*, **114**, 1531-1537.

Graeler M, Goetzl EJ (2002) Activation-regulated expression and chemotactic function of sphingosine 1-phosphate receptors in mouse splenic T cells. *Faseb J*, **16**, 1874-1878.

Graeler M, Shankar G, Goetzl EJ (2002) Cutting edge: suppression of T cell chemotaxis by sphingosine 1-phosphate. *J Immunol*, **169**, 4084-4087.

Graler MH, Bernhardt G, Lipp M (1998) EDG6, a novel G-protein-coupled receptor related to receptors for bioactive lysophospholipids, is specifically expressed in lymphoid tissue. *Genomics*, **53**, 164-169.

Graler MH, Goetzl EJ (2004) The immunosuppressant FTY720 down-regulates sphingosine 1-phosphate G-protein-coupled receptors. *Faseb J*, **18**, 551-553.

Graler MH, Huang MC, Watson S, Goetzl EJ (2005) Immunological effects of transgenic constitutive expression of the type 1 sphingosine 1-phosphate receptor by mouse lymphocytes. *J Immunol*, **174**, 1997-2003.

Hakomori S (2003) Structure, organization, and function of glycosphingolipids in membrane. *Curr Opin Hematol*, **10**, 16-24.

Halin C, Scimone ML, Bonasio R, Gauguet JM, Mempel TR, Quackenbush E, Proia RL, Mandala S, von Andrian UH (2005) The S1P-analog FTY720 differentially modulates T-cell homing via HEV: T-cell-expressed S1P1 amplifies integrin activation in peripheral lymph nodes but not in Peyer patches. *Blood*, **106**, 1314-1322.

Hla T (2004) Physiological and pathological actions of sphingosine 1-phosphate. *Semin Cell Dev Biol*, **15**, 513-520.

Hla T, Lee MJ, Ancellin N, Paik JH, Kluk MJ (2001) Lysophospholipids--receptor revelations. *Science*, **294**, 1875-1878.

Hla T, Maciag T (1990) An abundant transcript induced in differentiating human endothelial cells encodes a polypeptide with structural similarities to G-protein-coupled receptors. *J Biol Chem*, **265**, 9308-9313.

Idzko M, Panther E, Corinti S, Morelli A, Ferrari D, Herouy Y, Dichmann S, Mockenhaupt M, Gebicke-Haerter P, Di Virgilio F, Girolomoni G, Norgauer J (2002) Sphingosine 1-phosphate induces chemotaxis of immature and modulates cytokine-release in mature human dendritic cells for emergence of Th2 immune responses. *Faseb J*, **16**, 625-627.

Im DS, Heise CE, Ancellin N, O'Dowd BF, Shei GJ, Heavens RP, Rigby MR, Hla T, Mandala S, McAllister G, George SR, Lynch KR (2000) Characterization of a novel sphingosine 1-phosphate receptor, Edg-8. *J Biol Chem*, **275**, 14281-14286.

Ishii I, Friedman B, Ye X, Kawamura S, McGiffert C, Contos JJ, Kingsbury MA, Zhang G, Brown JH, Chun J (2001) Selective loss of sphingosine 1-phosphate signaling with no obvious phenotypic abnormality in mice lacking its G protein-coupled receptor, LP(B3)/EDG-3. *J Biol Chem*, **276**, 33697-33704.

Jaillard C, Harrison S, Stankoff B, Aigrot MS, Calver AR, Duddy G, Walsh FS, Pangalos MN, Arimura N, Kaibuchi K, Zalc B, Lubetzki C (2005) Edg8/S1P5: an oligodendroglial receptor with dual function on process retraction and cell survival. *J Neurosci*, **25**, 1459-1469.

Jolly PS, Bektas M, Olivera A, Gonzalez-Espinosa C, Proia RL, Rivera J, Milstien S, Spiegel S (2004) Transactivation of sphingosine-1-phosphate receptors by FcepsilonRI triggering is required for normal mast cell degranulation and chemotaxis. *J Exp Med*, **199**, 959-970.

Kharel Y, Lee S, Snyder AH, Sheasley-O'neill S L, Morris MA, Setiady Y, Zhu R, Zigler MA, Burcin TL, Ley K, Tung KS, Engelhard VH, Macdonald TL, Lynch KR (2005) Sphingosine kinase 2 is required for modulation of lymphocyte traffic by FTY720. *J Biol Chem*, **280**, 36865-36872.

Kim JI, Jo EJ, Lee HY, Cha MS, Min JK, Choi CH, Lee YM, Choi YA, Baek SH, Ryu SH, Lee KS, Kwak JY, Bae YS (2003) Sphingosine 1-phosphate in amniotic fluid modulates cyclooxygenase-2 expression in human amnion-derived WISH cells. *J Biol Chem*, **278**, 31731-31736.

Kiuchi M, Adachi K, Kohara T, Minoguchi M, Hanano T, Aoki Y, Mishina T, Arita M, Nakao N, Ohtsuki M, Hoshino Y, Teshima K, Chiba K, Sasaki S, Fujita T (2000) Synthesis and immunosuppressive activity of 2-substituted 2-aminopropane-1,3-diols and 2-aminoethanols. *J Med Chem*, **43**, 2946-2961.

Kluk MJ, Hla T (2002) Signaling of sphingosine-1-phosphate via the S1P/EDG-family of G-protein-coupled receptors. *Biochim Biophys Acta*, **1582**, 72-80.

Kohama T, Olivera A, Edsall L, Nagiec MM, Dickson R, Spiegel S (1998) Molecular cloning and functional characterization of murine sphingosine kinase. *J Biol Chem*, **273**, 23722-23728.

Kono M, Mi Y, Liu Y, Sasaki T, Allende ML, Wu YP, Yamashita T, Proia RL (2004) The sphingosine-1-phosphate receptors S1P1, S1P2, and S1P3 function coordinately during embryonic angiogenesis. *J Biol Chem*, **279**, 29367-29373.

Kveberg L, Bryceson Y, Inngjerdingen M, Rolstad B, Maghazachi AA (2002) Sphingosine 1 phosphate induces the chemotaxis of human natural killer cells. Role for heterotrimeric G proteins and phosphoinositide 3 kinases. *Eur J Immunol*, **32**, 1856-1864.

Lee H, Liao JJ, Graeler M, Huang MC, Goetzl EJ (2002) Lysophospholipid regulation of mononuclear phagocytes. *Biochim Biophys Acta*, **1582**, 175-177.

Liu CH, Hla T (1997) The mouse gene for the inducible G-protein-coupled receptor edg-1. *Genomics*, **43**, 15-24.

Liu H, Sugiura M, Nava VE, Edsall LC, Kono K, Poulton S, Milstien S, Kohama T, Spiegel S (2000a) Molecular cloning and functional characterization of a novel mammalian sphingosine kinase type 2 isoform. *J Biol Chem*, **275**, 19513-19520.

Liu Y, Wada R, Yamashita T, Mi Y, Deng CX, Hobson JP, Rosenfeldt HM, Nava VE, Chae SS, Lee MJ, Liu CH, Hla T, Spiegel S, Proia RL (2000b) Edg-1, the G protein-coupled receptor for sphingosine-1-phosphate, is essential for vascular maturation. *J Clin Invest*, **106**, 951-961.

Lo CG, Xu Y, Proia RL, Cyster JG (2005) Cyclical modulation of sphingosine-1-phosphate receptor 1 surface expression during lymphocyte recirculation and relationship to lymphoid organ transit. *J Exp Med*, **201**, 291-301.

MacLennan AJ, Browe CS, Gaskin AA, Lado DC, Shaw G (1994) Cloning and characterization of a putative G-protein coupled receptor potentially involved in development. *Mol Cell Neurosci*, **5**, 201-209.

MacLennan AJ, Carney PR, Zhu WJ, Chaves AH, Garcia J, Grimes JR, Anderson KJ, Roper SN, Lee N (2001) An essential role for the H218/AGR16/Edg-5/LP(B2) sphingosine 1-phosphate receptor in neuronal excitability. *Eur J Neurosci*, **14**, 203-209.

Mandala S, Hajdu R, Bergstrom J, Quackenbush E, Xie J, Milligan J, Thornton R, Shei GJ, Card D, Keohane C, Rosenbach M, Hale J, Lynch CL, Rupprecht K, Parsons W, Rosen H (2002) Alteration of lymphocyte trafficking by sphingosine-1-phosphate receptor agonists. *Science*, **296**, 346-349.

Mandala SM, Thornton R, Tu Z, Kurtz MB, Nickels J, Broach J, Menzeleev R, Spiegel S (1998) Sphingoid base 1-phosphate phosphatase: a key regulator of sphingolipid metabolism and stress response. *Proc Natl Acad Sci U S A*, **95**, 150-155.

Mao C, Wadleigh M, Jenkins GM, Hannun YA, Obeid LM (1997) Identification and characterization of Saccharomyces cerevisiae dihydrosphingosine-1-phosphate phosphatase. *J Biol Chem*, **272**, 28690-28694.

Matloubian M, Lo CG, Cinamon G, Lesneski MJ, Xu Y, Brinkmann V, Allende ML, Proia RL, Cyster JG (2004) Lymphocyte egress from thymus and peripheral lymphoid organs is dependent on S1P receptor 1. *Nature*, **427**, 355-360.

McVerry BJ, Garcia JG (2005) In vitro and in vivo modulation of vascular barrier integrity by sphingosine 1-phosphate: mechanistic insights. *Cell Signal*, **17**, 131-139.

Min JK, Yoo HS, Lee EY, Lee WJ, Lee YM (2002) Simultaneous quantitative analysis of sphingoid base 1-phosphates in biological samples by o-phthalaldehyde precolumn derivatization after dephosphorylation with alkaline phosphatase. *Anal Biochem*, **303**, 167-175.

Miyake Y, Kozutsumi Y, Nakamura S, Fujita T, Kawasaki T (1995) Serine palmitoyltransferase is the primary target of a sphingosine-like immunosuppressant, ISP-1/myriocin. *Biochem Biophys Res Commun*, **211**, 396-403.

Muller G, Reiterer P, Hopken UE, Golfier S, Lipp M (2003) Role of homeostatic chemokine and sphingosine-1-phosphate receptors in the organization of lymphoid tissue. *Ann N Y Acad Sci*, **987**, 107-116.

Okajima F (2002) Plasma lipoproteins behave as carriers of extracellular sphingosine 1-phosphate: is this an atherogenic mediator or an anti-atherogenic mediator? *Biochim Biophys Acta*, **1582**, 132-137.

Olivera A, Rivera J (2005) Sphingolipids and the balancing of immune cell function: lessons from the mast cell. *J Immunol*, **174**, 1153-1158.

Olivera A, Spiegel S (2001) Sphingosine kinase: a mediator of vital cellular functions. *Prostaglandins Other Lipid Mediat*, **64**, 123-134.

Paik JH, Skoura A, Chae SS, Cowan AE, Han DK, Proia RL, Hla T (2004) Sphingosine 1-phosphate receptor regulation of N-cadherin mediates vascular stabilization. *Genes Dev*, **18**, 2392-2403.

Paugh SW, Payne SG, Barbour SE, Milstien S, Spiegel S (2003) The immunosuppressant FTY720 is phosphorylated by sphingosine kinase type 2. *FEBS Lett*, **554**, 189-193.

Payne SG, Milstien S, Barbour SE, Spiegel S (2004) Modulation of adaptive immune responses by sphingosine-1-phosphate. *Semin Cell Dev Biol*, **15**, 521-527.

Pettus BJ, Bielawski J, Porcelli AM, Reames DL, Johnson KR, Morrow J, Chalfant CE, Obeid LM, Hannun YA (2003) The sphingosine kinase 1/sphingosine-1-phosphate pathway mediates COX-2 induction and PGE2 production in response to TNF-alpha. *Faseb J*, **17**, 1411-1421.

Prieschl EE, Csonga R, Novotny V, Kikuchi GE, Baumruker T (1999) The balance between sphingosine and sphingosine-1-phosphate is decisive for mast cell activation after Fc epsilon receptor I triggering. *J Exp Med*, **190**, 1-8.

Renkl A, Berod L, Mockenhaupt M, Idzko M, Panther E, Termeer C, Elsner P, Huber M, Norgauer J (2004) Distinct effects of sphingosine-1-phosphate, lysophosphatidic acid and histamine in human and mouse dendritic cells. *Int J Mol Med*, **13**, 203-209.

Rosen H, Alfonso C, Surh CD, McHeyzer-Williams MG (2003) Rapid induction of medullary thymocyte phenotypic maturation and egress inhibition by nanomolar sphingosine 1-phosphate receptor agonist. *Proc Natl Acad Sci U S A*, **100**, 10907-10912.

Rosen H, Goetzl EJ (2005) Sphingosine 1-phosphate and its receptors: an autocrine and paracrine network. *Nat Rev Immunol*, **5**, 560-570.

Roviezzo F, Del Galdo F, Abbate G, Bucci M, D'Agostino B, Antunes E, De Dominicis G, Parente L, Rossi F, Cirino G, De Palma R (2004) Human eosinophil chemotaxis and selective in vivo recruitment by sphingosine 1-phosphate. *Proc Natl Acad Sci U S A*, **101**, 11170-11175.

Ruwisch L, Schafer-Korting M, Kleuser B (2001) An improved high-performance liquid chromatographic method for the determination of sphingosine-1-phosphate in complex biological materials. *Naunyn Schmiedebergs Arch Pharmacol*, **363**, 358-363.

Saba JD, Nara F, Bielawska A, Garrett S, Hannun YA (1997) The BST1 gene of Saccharomyces cerevisiae is the sphingosine-1-phosphate lyase. *J Biol Chem*, **272**, 26087-26090.

Sanchez T, Estrada-Hernandez T, Paik JH, Wu MT, Venkataraman K, Brinkmann V, Claffey K, Hla T (2003) Phosphorylation and action of the immunomodulator FTY720 inhibits vascular endothelial cell growth factor -induced vascular permeability. *J Biol Chem*, **278**, 47281-47290.

Sanchez T, Hla T (2004) Structural and functional characteristics of S1P receptors. *J Cell Biochem*, **92**, 913-922.

Spiegel S, Milstien S (2003) Sphingosine-1-phosphate: an enigmatic signalling lipid. *Nat Rev Mol Cell Biol*, **4**, 397-407.

Taha TA, Argraves KM, Obeid LM (2004) Sphingosine-1-phosphate receptors: receptor specificity versus functional redundancy. *Biochim Biophys Acta*, **1682**, 48-55.

van Meeteren LA, Ruurs P, Christodoulou E, Goding JW, Takakusa H, Kikuchi K, Perrakis A, Nagano T, Moolenaar WH (2005) Inhibition of autotaxin by lysophosphatidic acid and sphingosine 1-phosphate. *J Biol Chem*, **280**, 21155-21161.

van Veldhoven PP, Mannaerts GP (1993) Sphingosine-phosphate lyase. *Adv Lipid Res*, **26**, 69-98.

Vora KA, Nichols E, Porter G, Cui Y, Keohane CA, Hajdu R, Hale J, Neway W, Zaller D, Mandala S (2005) Sphingosine 1-phosphate receptor agonist FTY720-phosphate causes marginal zone B cell displacement. *J Leukoc Biol*, **78**, 471-480.

Wang W, Graeler MH, Goetzl EJ (2005) Type 4 sphingosine 1-phosphate G protein-coupled receptor (S1P4) transduces S1P effects on T cell proliferation and cytokine secretion without signaling migration. *Faseb J*, **19**, 1731-1733.

Xia P, Gamble JR, Rye KA, Wang L, Hii CS, Cockerill P, Khew-Goodall Y, Bert AG, Barter PJ, Vadas MA (1998) Tumor necrosis factor-alpha induces adhesion molecule expression through the sphingosine kinase pathway. *Proc Natl Acad Sci U S A*, **95**, 14196-14201.

Xia P, Wang L, Moretti PA, Albanese N, Chai F, Pitson SM, D'Andrea RJ, Gamble JR, Vadas MA (2002) Sphingosine kinase interacts with TRAF2 and dissects tumor necrosis factor-alpha signaling. *J Biol Chem*, **277**, 7996-8003.

Xie JH, Nomura N, Koprak SL, Quackenbush EJ, Forrest MJ, Rosen H (2003) Sphingosine-1-phosphate receptor agonism impairs the efficiency of the local immune response by altering trafficking of naive and antigen-activated CD4+ T cells. *J Immunol*, **170**, 3662-3670.

Xu Y, Xiao YJ, Zhu K, Baudhuin LM, Lu J, Hong G, Kim KS, Cristina KL, Song L, F SW, Elson P, Markman M, Belinson J (2003) Unfolding the pathophysiological role of bioactive lysophospholipids. *Curr Drug Targets Immune Endocr Metabol Disord*, **3**, 23-32.

Yagi H, Kamba R, Chiba K, Soga H, Yaguchi K, Nakamura M, Itoh T (2000) Immunosuppressant FTY720 inhibits thymocyte emigration. *Eur J Immunol*, **30**, 1435-1444.

Yamaguchi F, Yamaguchi K, Tokuda M, Brenner S (1999) Molecular cloning of EDG-3 and N-Shc genes from the puffer fish, Fugu rubripes, and conservation of synteny with the human genome. *FEBS Lett*, **459**, 105-110.

Yang L, Yatomi Y, Miura Y, Satoh K, Ozaki Y (1999) Metabolism and functional effects of sphingolipids in blood cells. *Br J Haematol*, **107**, 282-293.

Yang Z, Chen M, Fialkow LB, Ellett JD, Wu R, Brinkmann V, Nadler JL, Lynch KR (2003) The immune modulator FYT720 prevents autoimmune diabetes in nonobese diabetic mice. *Clin Immunol*, **107**, 30-35.

Yatomi Y, Ohmori T, Rile G, Kazama F, Okamoto H, Sano T, Satoh K, Kume S, Tigyi G, Igarashi Y, Ozaki Y (2000) Sphingosine 1-phosphate as a major bioactive lysophospholipid that is released from platelets and interacts with endothelial cells. *Blood*, **96**, 3431-3438.

Yatomi Y, Ozaki Y, Ohmori T, Igarashi Y (2001) Sphingosine 1-phosphate: synthesis and release. *Prostaglandins Other Lipid Mediat*, **64**, 107-122.

Yatomi Y, Welch RJ, Igarashi Y (1997a) Distribution of sphingosine 1-phosphate, a bioactive sphingolipid, in rat tissues. *FEBS Lett*, **404**, 173-174.

Yatomi Y, Yamamura S, Ruan F, Igarashi Y (1997b) Sphingosine 1-phosphate induces platelet activation through an extracellular action and shares a platelet surface receptor with lysophosphatidic acid. *J Biol Chem*, **272**, 5291-5297.

Zhang G, Contos JJ, Weiner JA, Fukushima N, Chun J (1999) Comparative analysis of three murine G-protein coupled receptors activated by sphingosine-1-phosphate. *Gene*, **227**, 89-99.

Zhou J, Saba JD (1998) Identification of the first mammalian sphingosine phosphate lyase gene and its functional expression in yeast. *Biochem Biophys Res Commun*, **242**, 502-507.

6-2 Sphingolipids and Lung Vascular Barrier Regulation

Liliana Moreno, Steven M. Dudek, and Joe G.N. Garcia

Department of Medicine, University of Chicago, 5841 South Maryland Avenue, Chicago, Illinois 60637, USA

Summary. Long thought to function primarily as the structural components of lipid membranes, sphingolipids are now also recognized as vitally important signaling mediators regulating a diverse range of functions. We recently described the potent vascular barrier-regulating properties of one of these sphingolipids, the lipid and angiogenic factor sphingosine 1-phosphate (S1P) (Garcia et al, 2001). Since disruption of vascular barrier integrity commonly occurs in highly morbid inflammatory lung conditions, a better understanding of the mechanism of barrier regulation would have important clinical implications. In this chapter, we provide a brief overview of vascular barrier regulation before detailing the mechanisms underlying potent barrier-enhancing effects of S1P in vitro and *in vivo* in models of acute lung injury (ALI) syndromes. The potential ramifications of these findings for the development of specific therapeutic interventions for patients with ALI syndromes are then discussed.

Keywords. permeability, endothelium, sphingosine 1-phosphate, acute lung injury, cytoskeleton

1. Overview of vascular barrier regulation

The vasculature is lined by endothelial cells (EC) which provide a semi-permeable barrier between circulating vascular contents and the surrounding tissues perfused by these vessels. Disruptions in vascular barrier integrity significantly increases permeability to fluid, protein, and circulating cells. This is the central pathophysiological mechanism for many inflammatory disease processes. The lung is particularly sensitive to this type of injury, given the large surface area needed for alveolar gas exchange. Pulmonary EC vascular leak leads to alveolar flooding during inflammatory conditions such as ALI/ARDS (acute lung injury/acute respiratory distress syndrome) and sepsis. This phenomenon contributes significantly to the morbidity of those clinical syndromes. Therefore, we focus here on the pulmonary endothelium as a model for vascular barrier regulation.

EC permeability is primarily determined by the EC cytoskeleton, cell-cell connections, and cell-matrix connections. Accumulated evidence supports a critical role for the endothelial cytoskeleton in dynamic modulation of vascular barrier function. The EC cytoskeleton itself is composed of actin filaments, intermediate filaments, and microtubules (for review, see Dudek & Garcia, 2001). Actin cytoskeleton can rapidly be rearranged into filaments of various shapes and sizes that closely regulate the overall shape, motility, and contractile status of the EC. This dynamic rearrangement is controlled by various actin binding, capping, nucleating, and severing proteins regulating filament size and shape. The primary role of actin filaments in EC permeability was demonstrated in early observations that cytochalasin D, which disrupts the actin cytoskeleton, increased EC permeability while phallacidin, an actin stabilizer, prevented barrier disruption by various agonists (Phillips et al, 1989). Actin filaments interact with myosin to generate EC tensile force that, in turn, drives cell shape changes and barrier regulation. When cellular contraction occurs along the a cell's actin stress fibers, gaps form between adjacent cells and paracellular permeability increases. In addition, the actin microfilament system participates in EC barrier enhancement through focal linkages to multiple membrane adhesive proteins that connect the system to cell-cell (adherens junctions, tight junctions) and cell-matrix (focal adhesion) junctions as well as anchor the endothelium. The functional roles of microtubule and intermediate filament cytoskeletal components in EC barrier regulation are less well defined. However, microtubule disrupting agents, such as nocodazole, induce rapid rearrangement of actin filaments and focal adhesions, cellular contraction, and increase permeability across EC monolayers while microtubule stabilization attenuates these effects (Verin et al, 2001). This suggests that actin filament-microtubule crosstalk is important in EC barrier regulation.

Cell-cell and cell-matrix contacts provide tethering forces essential for EC mechanical stability and barrier maintenance. The primary cell-cell contacts in EC are adherens junctions, consisting of extracellular cadherins that interact through cytoplasmic tails with the catenin family of intracellular proteins to provide direct anchorage to the actin cytoskeleton. As evidence for the critical role of these adherens junctions in EC barrier integrity, infusion of vascular endothelial-cadherin blocking antibody induces lung vascular leak in cultured EC and mice (Corada et al, 1999). Tight junction cell-cell complexes are considered to play a less important role than adherens junctions in EC barrier regulation. The primary cell-extracellular matrix connections are provided by focal adhesion complexes, which utilize transmembrane integrin receptors connected to the EC actin cytoskeleton through multi-protein focal adhesion plaques. These linkages are essential to maintain EC barrier integrity since blocking antibodies to β1 integrin inhibit EC attachment and increase monolayer permeability (Lampugnani, 1991). Therefore, focal adhesion and adherens junction complexes provide essential tethering sites for optimal EC barrier integrity.

2. Sphingolipids and endothelial cell biology

Sphingosine, the backbone for most sphingolipids, is a molecule derived from the degradation of the plasma membrane component sphingomyelin. Sphingosine can then be phosphorylated to form sphingosine 1-phosphate (S1P), a biologically active mediator involved in many diverse cellular functions including vascular barrier regulation. The biochemistry and cellular metabolism of S1P was reviewed in great detail by several authors (Hla, 2003) (Spiegel and Milstien, 2003) (see also the related chapters in this book) Briefly, sphingomyelin is metabolized by sphingomyelinase to produce a variety of lipids mediators including ceramide, which subsequently converts to sphingosine by ceramidase. Ultimately, the phosphorylation of sphingosine by sphingosine kinases generates S1P. Various growth factors, cytokines and hormones stimulate sphingosine kinase activity that increases intracellular levels of S1P (Hannun YA, Luberto C, 2001). S1P can also be reconverted into sphingosine via S1P phosphatases or transformed into hexadecanal and phosphanolamine by S1P lyases. Circulating platelets are rich in sphingosine kinase, which catalyzes sphingosine phosphorylation, but relatively deficient in S1P lyases leading to the accumulation of S1P in platelets and an important cellular source of secreted S1P.. S1P functions extracellularly via receptor ligation (see below) or as an intracellular mediator in different cell types, including EC (Hla 2001).

S1P mediates a wide variety of cellular and biological responses in different cell types through interaction with members of the endothelial differentiation gene (Edg) family of receptors. The Edg receptors are G-protein coupled receptors (Sanchez and Hla 2004), with five known S1P receptors originally identified as Edg-1, Edg-5, Edg-3, Edg-6 and Edg-8, now renamed as $S1P_1$ to $S1P_5$, respectively. These S1P receptors differ in tissue distribution and expression levels. In addition, as G protein coupled receptors, they signal through different G proteins that exerting differential effects intercellularly.

S1P receptors are present in leukocytes, cardiomyocytes, neurons, and endothelial cells. Vascular endothelial cells primarily express two S1P receptors, $S1P_1$ and $S1P_3$ (Hla and Maciag 1990). $S1P_1$ signals through Gi protein and promotes vascular maturation and decreased EC permeability (Garcia, Liu et al. 2001). $S1P_1$ receptor expression is essential for platelet-mediated barrier enhancement as inhibition of $S1P_1$ expression attenuates this effect (Schaphorst et al. 2003) and $S1P_3$ is coupled to Gi and Gq and involved in EC migration. $S1P_2$ was cloned from rat brain and vascular smooth muscle cells (Okazaki et al. 1993). $S1P_4$ is expressed in lymphoid tissues and may play a role in regulating S1P-induced migration (Graler et al. 1998) (Kohno et al. 2003).

Of the five receptors, only the SIP4 receptor, which has a high affinity for the phytosphingoisene-1-phosphate receptor lacks a high affinity for S1P. Non-phosphorylated sphingosine derivatives, such as sphingosine or ceramide, do not bind to S1P receptors. These receptors have 20% homology with cannabinoid receptors and 30% homology with lysophosphatidic acid receptors. Their amino-terminus contains N-linked glycosylation sites, followed by seven transmembrane domains and their loops. The intracellular loops and the C-terminus, containing a Cys residue, have potential sites for phosphorylation by serine/threonine protein kinases.

A structural analog of S1P, FTY720, has become the focus of a significant amount of pharmacological research (Brinkmann, Davis et al. 2002), (Mandala, Hajdu et al. 2002). Once phosphorylated, FTY720 is a potent agonist for the S1P receptors and is in Phase III clinical trials as an immunosuppressive drug following kidney transplantation (Brinkmann V and Lynch KR, 2002). In addition, like S1P, FTY720 appears to be a potent vascular barrier protector agent in endothelial cells in vitro and in vivo (Sanchez and Hla 2004), (Peng, Hassoun et al. 2004).

3. In vitro alterations in the cytoskeleton, monolayer integrity, and angiogenesis

Over the last decade a large body of in vitro data has accumulated that describes a multitude of S1P-mediated effects on EC signaling pathways, structure, barrier function, and migration. S1P induces dramatic cytoskeletal and cell contact rearrangements that drive these physiologic processes by integrating (see McVerry and Garcia, 2005). In this section we will briefly discuss the in vitro alterations responsible for mediating these effects.

The $S1P_1$ receptor (Edg-1) was originally identified as an orphan receptor involved in angiogenesis (Hla and Maciag, 1990). Following its identification as the ligand for this receptor, S1P's impressive stimulatory effects on EC migration and angiogenesis were also identified. We demonstrated that S1P is the most potent endothelial chemoattractant present in serum and is a complete promoter of angiogenesis (English et al, 2000). Angiogenesis is a complex, multi-step process. It begins with the disruption of the existing EC monolayer followed by cell migration directed toward the chemotactic stimulus. Subsequent cell proliferation then leads to the formation of the nascent vessel. Full maturation of the new vessel then requires barrier restoration and enhancement of EC barrier function. S1P is a complete angiogenic factor that can drive this entire process.

Because of its critical role in angiogenesis, we studied, in vitro, S1P's direct effects on EC barrier function. S1P exhibits potent, rapid, and sustained elevations in the barrier function of cultured pulmonary vascular EC as measured by the highly sensitive in vitro method of transendothelial monolayer electrical resistance (Garcia et al, 2001). This barrier-enhancing effect occurs within seconds and lasts for hours. Interestingly, the S1P precursor sphingosine fails to exert EC barrier-enhancing properties in vitro (personal unpublished observation). As described above, secreted S1P is primarily generated through the hydrolysis of membrane lipids in activated platelets that are subsequently released into the bloodstream. Platelets enhance the integrity of the EC barrier by decreasing capillary permeability protein and fluid leakage (Lo et al, 1988) We identified S1P as a major barrier-protective product of platelets (Schaphorst et al, 2003), and therefore, activated platelets are a key source for secreted S1P that mediate EC barrier regulation, migration, and angiogenesis.

The exact mechanisms through which S1P exerts these physiological effects remain an area of intense study, but several critical steps in the signaling cascade have been identified (Fig. 1). To initiate EC barrier enhancement, S1P must first bind to and recruit the $S1P_1$ receptor into membrane lipid rafts. This is followed by Gi-coupled signaling and downstream activation of the small GTPase Rac (Singleton et al, 2005) (Garcia et al, 2001). All of these events are essential for optimal S1P-induced EC barrier enhancement. Rac activation induces rapid translocation of the actin-binding protein, cortactin, to the cell periphery (Dudek et al, 2004) where dramatic cortical-actin ring formation occurs. Cortactin likely participates in promotion of EC barrier integrity in a number of ways including stimulation of peripheral actin polymerization through activation of the Arp2/3 complex, binding to the ZO-1 protein of the tight junction complex, and interaction with the critical barrier regulatory enzyme myosin light chain kinase (Dudek et al, 2004). Combined these cytoskeletal events promote rapid development of a prominent cortical actin ring after S1P. Development of this cortical actin structure is an element in multiple models of EC barrier enhancement shared by diverse stimuli, including shear stress, hepatocyte growth factor, simvastatin, and ATP. All of these stimuli promote actin polymerization in the cell periphery in association with augmentation of vascular barrier integrity (reviewed in McVerry and Garcia, 2005).

In addition, S1P enhances EC barrier function by strengthening of cell-cell and cell-matrix connections (Fig. 1). S1P dramatically increases localization and interaction of vascular endothelial-cadherin and catenin proteins at EC adherens junctions, thereby promoting the formation and stabilization of these critical cell-cell linkages (McVerry and Garcia, 2005). The localization of these proteins occurs within 1 hour of S1P stimulation, indicating the induction of functional complex assembly with increased affinity to the EC cytoskeleton. S1P also augments vascular barrier function by strengthening EC focal adhesion connections to the underlying extracellular matrix. Concomitant with peripheral cortical actin ring formation, S1P induces $p60^{src}$-mediated phosphorylation of focal adhesion kinase (FAK) and redistribution of the focal adhesion proteins paxillin and FAK to the cell periphery (Shikata et al, 2003a) (Shikata et al, 2003b). These adhesion protein rearrangements promote focal adhesion linkages and further enhance EC barrier function. Thus, S1P improves vascular barrier function by altering all three of the primary EC permeability regulatory structural elements: the EC cytoskeleton, cell-cell connections, and cell-matrix connections.

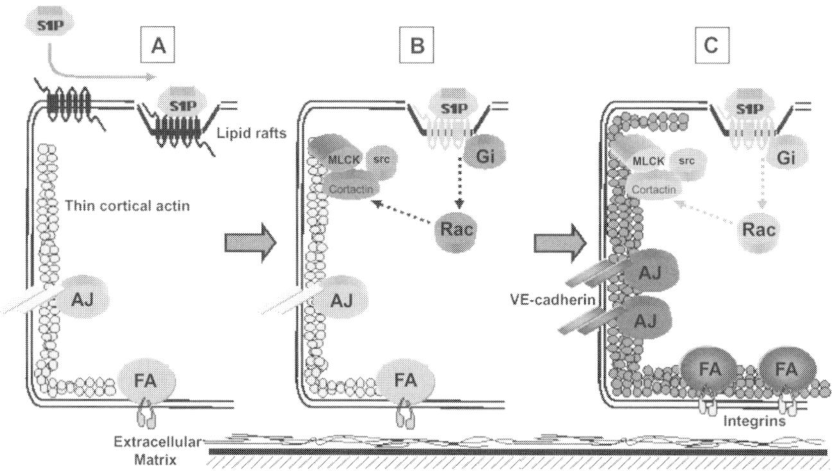

Fig. 1. Mechanisms of S1P-Mediated Endothelial Barrier Enhancement. Under baseline conditions, pulmonary endothelial cells (EC) are characterized by a thin cortical actin ring with associated cell-cell (adherens junctions) and cell-matrix (focal adhesions) connections that provide the structural framework for barrier integrity. A. S1P stimulation results in rapid recruitment of its $S1P_1$ receptor into membrane lipid rafts. B. $S1P_1$ receptor activation is coupled to Gi signaling and subsequent activation of the small GTPase Rac. Rac stimulates rapid translocation of the actin binding protein cortactin to the cell periphery where it interacts with the barrier regulatory enzyme, myosin light chain kinase (MLCK). p60src phosphorylation of either of these proteins increases their interaction. C. The cortical actin ring at the periphery of the EC is markedly enhanced in conjunction with rearrangement and strengthening of adherens junctions and focal adhesions. This sequence of events occurs within minutes of S1P stimulation and results in dramatically increased EC barrier function.

4. In vivo applications of sphingolipids: acute lung injury and ischemia/reperfusion injury

Pulmonary EC vascular leak leads to alveolar flooding during inflammatory conditions such as ALI/ARDS and sepsis. This leakage contributes significantly to morbidity of these clinical syndromes. Despite recent advances in supportive care, these inflammatory lung syndromes remain a significant cause of intensive care unit mortality. The lack of current specific therapies for ALI has led various investigators to search for more effective therapeutic interventions. Since our in vitro data convincingly demonstrate that S1P produces potent EC barrier enhancement through ligation of the S1P receptors, we worked with animal models of ALI to demonstrate in vivo the protective therapeutic potential of S1P and related compounds.

Exciting data shows that S1P has impressive protective effects in a murine model of endotoxin-induced lung injury (Peng et al. 2004). The pulmonary vasculature is particularly susceptible to dynamic alterations in the integrity of the barrier given the extended surface area of the lung microcirculation. Fluid and macromolecules will move into the interstitial and alveolar space when the lung vascular barrier is disrupted, leading to clinically profound lung edema and physiologic dysfunction as seen in syndromes of acute lung injury (ALI) and sepsis. Exposing mice to bacterial lipopolysaccharide (LPS) induces a pulmonary injury that mimics ALI or sepsis syndromes seen in humans. The ALI murine model (Peng et al. 2004) is characterized predominantly by a massive neutrophil infiltration in bronchoalveolar content and, in lung tissue, fluid and protein leak, edema and increases in vascular permeability. The barrier-disrupting phenotype of this model suggested that, based on the known in vitro barrier- protective and anti-inflammatory properties of S1P, it may reverse or prevent ALI. In an isolated, perfused murine lung model, S1P infusion produced a rapid and significant reduction in lung weight (Peng et al., 2004). In a murine model of LPS-induced ALI, intravenously administered S1P (1 hour after LPS) reduced LPS-mediated intravascular extravasation of Evans blue dye (EBD)-linked albumin, reduced bronchoalveolar lavage protein content, and reduced lung tissue leukocyte content (myeloperoxidase activity) at 24 hrs (Peng et al., 2004) (Fig. 2). Consistent with systemic barrier-enhancing properties, S1P significantly decreased vascular leakage and reduced lung tissue leukocyte content in renal tissues from LPS-exposed mice. These studies indicate that S1P significantly decreased pulmonary and renal vascular leakage and inflammation in a murine model of LPS-induced ALI, and may represent a novel therapeutic strategy for alleviating vascular barrier dysfunction.

To further investigate the role of S1P as a vascular barrier-protective compound, we developed a canine model of ALI induced by exposure to both bacterial LPS and mechanical ventilation (McVerry et al. 2004). In this model, LPS, in combination with mechanical ventilation produces hypoxemia, increased shunt, reduced compliance, and lung edema in these animals. When S1P was administered concomitantly with LPS combined with 6 hours of high tidal volume mechanical ventilation, significant attenuation of both alveolar and vascular barrier dysfunction was noted as well as significant reductions in shunt formation. Whole lung computer axial tomographic (CAT) imaging revealed that S1P reduced lung water content after 6h of LPS exposure (Fig. 2). In addition, the protective effects of S1P in ALI models in large animals was demonstrated as well as an essential step taken towards translating these therapies to human application (McVerry and Garcia, 2005).

FTY720, a structural analog of S1P, is a potent agonist for S1P receptors following phosphorylation by sphingosine kinase. This compound has similar EC barrier-promoting effects as S1P. It potently reversed transmonolayer permeability induced by vascular endothelial cell growth factor (VEGF) in vitro as well as VEGF-induced vascular permeability in a mouse ear assay (Sanchez et al, 2003). We employed FTY720 in our murine model of endotoxin-induced lung injury described above and demonstrated that it has similar vascular barrier-protective effects as S1P. A single intra-peritoneal injection of FTY720 significantly attenuated LPS-induced lung injury in mice at 24 hrs (Peng et al, 2004). These observations are also very promising for potential rapid development of a specific therapeutic intervention for patients with ALI syndromes. FTY720 is currently in Phase III clinical trials as an immunosuppressive drug after kidney transplantation. We anticipate conducting a clinical trial with FTY in patients with lung leak in the intensive care unit.

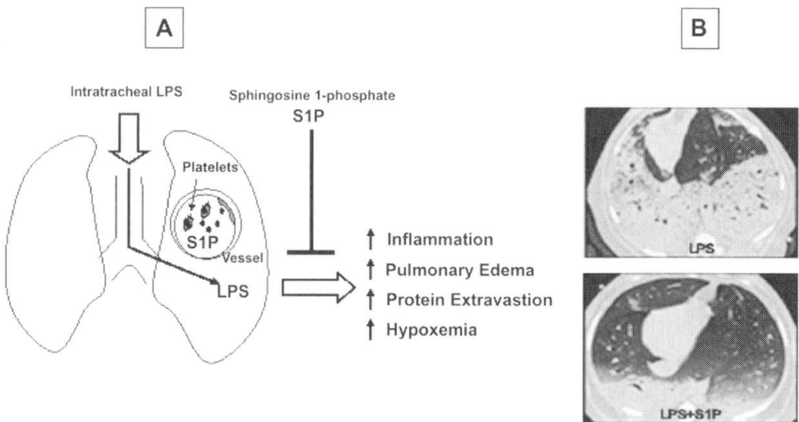

Fig. 2. S1P inhibits LPS-induced acute lung injury (ALI) in vivo. A. This schematic illustrates the murine LPS-induced ALI model utilized by the authors and described in detail in the text. Intratracheally administered LPS causes inflammation, lung edema formation, protein extravasation, and hypoxemia in the exposed mice. S1P, produced endogenously by platelets or given intravenously in this model, acts on the pulmonary endothelium to significantly attenuate the development of these ALI parameters in vivo. B. The top CT scan image shows the fluid-filled lungs of a dog after exposure to intratracheal LPS. The bottom CT image demonstrates significant reduction in this injury in the lungs of a separate LPS-exposed dog after S1P therapy.

Ischemia/reperfusion (I/R) injury of the lung is characterized by alveolar damage, lung edema, and inflammation, and represents a common sequela in patients who have undergone lung transplantation, with I/R injury often occurring in the donor lung within hours with attendant significant morbidity and mortality. Despite improvements in lung preservation techniques, I/R injury remains a significant cause of lung transplant failure and highlights the need for novel therapeutic interventions. We investigated the ability of S1P to enhance pulmonary endothelial cell barrier function and subsequently attenuate I/R lung injury. S1P was administered (i.v.) 15 min before ischemia. Indices of lung vascular permeability and inflammation, including bronchoalveolar lavage cell counts, differentials and albumin concentrations, were taken and later assessed at the end of the I/R time period. Separately, myeloperoxidase (MPO) from leukocytes was measured in left lung homogenates. BAL fluid from animals pre-treated with S1P significant decreased in total number of cells (32.4% decrease), neutrophils (63.3% decrease) and albumin concentration (57.2% decrease) when compared to controls. In addition, S1P administered prior to ischemia decreased lung tissue MPO (63%) (Moreno et al. 2005). Our findings suggest that S1P significantly attenuates vascular permeability and inflammation associated with I/R injury. Ultimately, these findings may have profound clinical implications as S1P may serve as a useful adjunct to current preservation approaches employed during lung transplantation allowing for improved patient outcomes.

In summary, studies using animal models of ALI demonstrate the potent anti-inflammatory and barrier regulatory effects of S1P when administered intravenously to small (mice and rats) and large animals. Continued investigation of the mechanisms that enable S1P to exert its barrier promoting effects will provide insight into signaling pathways and other promising targets for therapeutic agents, such as FTY720. Thus, these findings establish a bridge between in vitro observations and functional properties in vivo and hopefully provide a link to exciting translational studies in patients in the near future.

5. References

Brinkmann V, Davis MD, Heise CE, Albert R, Cottens S, Hof R, Bruns C, Prieschl E, Baumruker T, Hiestand P, Foster CA, Zollinger M, Lynch KR (2002). The immune modulator FTY720 targets sphingosine 1-phosphate receptors. *J Biol Chem*, **277**, 21453-21457.
Corada M, Mariotti M, Thurston G, Smith K, Kunkel R, Brockhaus M, Lampugnani MG, Martin-Padura I, Stoppacciaro A, Ruco L, McDonald DM, Ward

PA, Dejana E (1999) Vascular endothelial-cadherin is an important determinant of microvascular integrity in vivo. *Proc Natl Acad Sci USA*, **96**, 9815–9820.

Dudek SM, Jacobson JR, Chiang ET, Birukov KG, Wang P, Zhan X, Garcia JGN (2004) Pulmonary endothelial cell barrier enhancement by sphingosine 1-phosphate: roles for cortactin and myosin light chain kinase. *J Biol Chem*, **279**, 24692-24700.

Dudek SM, Garcia JG (2001) Cytoskeletal regulation of pulmonary vascular permeability. *J Appl Physiol*, **91**, 1487-1500.

English D, Welch Z, Kovala AT, Harvey K, Volpert OV, Brindley DN, Garcia JG (2000) Sphingosine 1-phosphate released from platelets during clotting accounts for the potent endothelial cell chemotactic activity of blood serum and provides a novel link between hemostasis and angiogenesis. *FASEB J*, **14**, 2255-2265.

Garcia JG, Liu F, Verin AD, Birukova A, Dechert MA, Gerthoffer WT, Bamberg JR, English D (2001) Sphingosine 1-phosphate promotes endothelial cell barrier integrity by Edg-dependent cytoskeletal rearrangement. *J Clin Invest*, **108**, 689-701.

Graler MH, Bernhardt G, Lipp M (1998). EDG6, a novel G-protein-coupled receptor related to receptors for bioactive lysophospholipids, is specifically expressed in lymphoid tissue. *Genomics*, **53**, 164-169.

Hannun YA, Luberto C, Argraves KM (2001) Enzymes of sphingolipid metabolism: from modular to integrative signaling. Biochem, 40, 4893-4903.

Hla T (2003) Signaling and biological actions of sphingosine 1-phosphate. *Pharmacol Res*, **47**, 401-407.

Hla T (2001) Sphingosine 1-phosphate receptors. *Prostaglandins*, **64**, 135-142.

Hla T, Maciag T (1990) An abundant transcript induced in differentiating human endothelial cells encodes a polypeptide with structural similarities to G-protein-coupled receptors. *J Biol Chem*, **265**, 9308-9313.

Kohno T, Matsuyuki H, Inagaki Y, Igarashi Y (2003) Sphingosine 1-phosphate promotes cell migration through the activation of Cdc42 in Edg-6/S1P4-expressing cells. *Genes Cells*, **8**, 685-697.

Lampugnani MG, Resnati M, Dejana E, Marchisio PC (1991) The role of integrins in the maintenance of endothelialmonolayer integrity. *J Cell Biol*, **112**, 479–490.

Lo SK, Burhop KE, Kaplan JE, Malik AB (1988) Role of platelets in maintenance of pulmonary vascular permeability to protein. *Am J Physiol*, **254**, H763-771.

Mandala S, Hajdu R, Bergstrom J, Quackenbush E, Xie J, Milligan J, Thornton R, Shei GJ, Card D, Keohane C, Rosenbach M, Hale J, Lynch CL, Rupprecht K, Parsons W, Rosen H (2002) Alteration of lymphocyte trafficking by sphingosine-1-phosphate receptor agonists. *Science*, **296**, 346-349.

McVerry BJ, Garcia JG (2005) In vitro and in vivo modulation of vascular barrier integrity by sphingosine 1-phosphate: mechanistic insights. *Cell Signal*, **17**, 131-139.

McVerry BJ, Peng X, Hassoun PM, Sammani S, Simon BA, Garcia JG. (2004) Sphingosine 1-phosphate reduces vascular leak in murine and canine models of acute lung injury. *Am J Respir Crit Care Med*, **170**, 987-993.

Moreno L, Bonde P, Jacobson J, Garcia JG (2005) Attenuation of rodent Ischemia-Rperfusion Acute Lung Injury by Sphingosine-1-Phosphate. *Proc Am Thorac Soc*, **2**, A476.

Okazaki H, Ishizaka N, Sakurai T, Kurokawa K, Goto K, Kumada M, Takuwa Y (1993) Molecular cloning of a novel putative G protein-coupled receptor expressed in the cardiovascular system. *Biochem Biophys Res Commun*, **190**, 1104-1109.

Peng X, Hassoun PM, Sammani S, McVerry BJ, Burne MJ, Rabb H, Pearse D, Tuder RM, Garcia JG (2004) Protective effects of sphingosine 1-phosphate in murine endotoxin-induced inflammatory lung injury. *Am J Respir Crit Care Med*, **169**, 1245-1251.

Phillips PG, Lum H, Malik AB, Tsan M (1989) Phallacidin prevents thrombin-induced increases in endothelial permeability to albumin. *Am J Physiol Cell Physiol*, **257**, C562–567.

Sanchez T, Hla T (2004) Structural and functional characteristics of S1P receptors. *J Cell Biochem*, **92**, 913-922.

Sanchez T, Estrada-Hernandez T, Paik JH, Wu MT, Venkataraman K, Brinkmann V, Claffey K, Hla T (2003) Phosphorylation and action of the immunomodulator FTY720 inhibits vascular endothelial cell growth factor-induced vascular permeability. *J Biol Chem*, **278**, 47281-47290.

Schaphorst KL, Chiang E, Jacobs KN, Zaiman A, Natarajan V, Wigley F, Garcia JG (2003) Role of sphingosine-1 phosphate in the enhancement of endothelial barrier integrity by platelet-released products. *Am J Physiol Lung Cell Mol Physiol*, **285**, L258-267.

Shikata Y, Birukov KG, Birukova AA, Verin A, Garcia JG (2003) Involvement of site-specific FAK phosphorylation in sphingosine-1 phosphate- and thrombin-induced focal adhesion remodeling: role of Src and GIT. *FASEB J*, **17**, 2240-2249.

Shikata Y, Birukov KG, Garcia JG (2003) S1P induces FA remodeling in human pulmonary endothelial cells: role of Rac, GIT1, FAK, and paxillin. *J Appl Physiol*, **94**, 1193-1203.

Singleton PA, Dudek SM, Chiang ET Garcia JG (2005) Regulation of sphingosine 1-phosphate-induced endothelial cytoskeletal rearrangement and barrier enhancement by S1P1 receptor, PI3 kinase, Tiam1/Rac1 and alpha-actinin. *FASEB J*, **19**, 1646-1656.

Spiegel S, Milstien S (2003) Sphingosine-1-phosphate: an enigmatic signalling lipid. *Nat Rev Mol Cell Biol*, **4**, 397-407.

Verin AD, Birukova A, Wang P, Birukov K, Garcia JGN (2001) Microtubule disassembly increases endothelial cell barrier dysfunction: role of microfilament crosstalk and myosin light chain phosphorylation. *Am J Physiol Lung Cell Mol Physiol*, **281**, L565–574.

6-3 Signaling Mechanisms for Positive and Negative Regulation of Cell Motility by Sphingosine-1-Phosphate Receptors

Yoh Takuwa[1], Naotoshi Sugimoto[1], Noriko Takuwa[1],
and Yasuyuki Igarashi[2]

[1]Department of Physiology, Kanazawa University Graduate School of Medicine, 2-6-11 Takara-machi, Kanazawa 920-8640; [2]Department of Biomembrane and Bifunctional Chemistry, Hokkaido University Graduate School of Pharmaceutical Sciences, Sapporo 060-0812, Japan

Summary. Sphingosine-1-phosphate (S1P) exerts positive and negative effects on cell migration apparently in a cell-type-dependent manner. Our data suggest that the bimodal actions of S1P on cell migration is due to receptor subtype-specific positive and negative regulation of Rho family GTPase, Rac; $S1P_1$ and $S1P_3$ mediate Rac stimulation and chemotaxis whereas $S1P_2$ mediates Rac inhibition and chemorepulsion. The stimulatory effects of $S1P_1$ and $S1P_3$ on Rac and, subsequently on migration, are mediated by G_i. The inhibitory effect of $S1P_2$ acts on $G_{12/13}$ and Rho. S1P exerts inhibitory effects on some tumor cell migration and invasion via $S1P_2$. $S1P_2$ also mediates the inhibition of hematogenous metastasis. In contrast, exogenously expressed $S1P_1$ has the reverse effect, it stimulates invasion and metastasis. S1P also exerts a similar bimodal action on vascular endothelial cells and, thereby, angiogenesis. The examples suggest that control of S1P receptor activity using a receptor subtype-specific agonist and antagonist may have beneficial effects on disorders, including cancer, and vascular diseases.

Keywords. Sphingosine-1-phosphate, Edg receptor, Cell migration, Rac

1. Introduction

Sphingosine-1-phosphate (S1P) is a lysophospholipid growth factor that circulates in the blood. It can induce a wide variety of biological responses in diverse cell types. S1P is present in the blood at a concentration of 10^{-7} M largely in a form that is bound to plasma proteins, including albumin and lipoproteins, and is released by activated platelets and other cell types (Yatomi et al., 2001). Identification of SIP receptors at the cell surface strongly indicates that many, if not all, of SIP's diverse biological activities of S1P are mediated through these receptors (Hla et al., 2001; Takuwa, 2002; Spiegel and Milstien, 2003; Ishii et al., 2004). Among 5 S1P receptor subtypes, $S1P_1$/Edg (for endothelial differentiation gene)-1, $S1P_2$/Edg5, $S1P_3$/Edg3, are widely expressed in various tissues (Hla et al., 2001; Takuwa, 2002; Spiegel and Milstien, 2003; Ishii et al., 2004).

Cell migration, a critical component of cellular responses, participates in many physiological and pathological processes. These include embryonic morphogenesis, angiogenesis, wound healing, atherogenesis, inflammation and tumor cell dissemination (Lauffenburger and Horwitz, 1996). Cell migration is regulated in positive and negative directions by a variety of extracellular cues called chemoattractants and chemorepellants, respectively. A number of biologically active substances, including chemokines, cytokines and growth factors   act as chemoattractants to induce chemotaxis, i.e. cell migration toward a higher concentration of a chemoattractant. Certain factors including some members of the semaphorine and the ephrine families were shown to directly repel neuronal and vascular cells as chemorepellants. However, our knowledge about chemorepellants or migration-inhibitory substances is rather limited at present compared to what we know about chemoattractants (Lauffenburger and Horwitz, 1996).

More than 10 years ago, Igarashi and his colleagues that S1P exerted inhibitory effects on migration and invasion of certain tumor cells including melanoma cells and osteosarcoma cells (Sadahira et al., 1992). They also showed that S1P inhibited migration of vascular smooth muscle cells (VSMCs) (Bornfeldt et al., 1995) and neutrophils (Kawa et al., 1997) that were directed toward chemoattractants. More recently, it was shown that S1P by itself was capable of stimulating, rather than inhibiting, chemotaxis of vascular endothelial cells (Lee et al., 1999) and mouse embryonic fibroblasts (Liu et al., 2000). S1P appears to be a unique, bimodal regulator of cell migration.

In this chapter, we describe molecular mechanisms underlying S1P bimodal regulation, including S1P receptor subtype-specific, positive and negative regulation of cell motility and Rho family GTPases, and S1P regulation of invasion and metastasis of malignant melanoma cells via endogenous S1P receptor through the control of Rho GTPases.

2. Receptor subtype-specific, bimodal regulation of cell motility and the small GTPase Rac by S1P

After the identification of the S1P-specific Edg family receptors, S1P was found to stimulate chemotaxis in vascular endothelial cells via endogenously expressed $S1P_1$ and $S1P_3$ (Lee et al., 1999; Liu et al., 2000). However, it remained unknown how S1P induced inhibition of cell migration in several other cell types, although an extracellular mode of this inhibitory S1P action was suggested (Yamamura et al., 1997). We showed the activity of each of the S1P receptors on cell migration in Boyden chamber assay by employing Chinese hamster ovary (CHO) cells that stably overexpressed each of the widely expressed subtypes, $S1P_1$, $S1P_2$ and $S1P_3$. It is known that CHO cells overexpressing exogenous chemoattractant receptors can vigorously respond to a concentration gradient of a respective chemoattractant with stimulation of directed cell migration in a trans-well migration assay. S1P stimulated directed cell migration, i.e. chemotaxis, of either $S1P_1$- or $S1P_3$-expressing CHO cells (Okamoto et al., 2000; Sugimoto et al., 2003). These two cell types were quite sensitive to S1P with the threshold concentrations as low as 0.1-1 nM of S1P. It was also reported that S1P stimulated chemotaxis in HEK293 cells overexpressing $S1P_1$ (Wang et al., 1999). In sharp contrast, S1P did not stimulate trans-well migration of $S1P_2$-expressing cells and vector-transfected control cells (Okamoto et al., 2000). However, the addition of a potent chemoattractant insulin-like growth factor (IGF-I) unveiled a strong inhibitory effect of S1P on cell migration via $S1P_2$; IGF-I stimulated chemotaxis in three S1P receptor subtype-expressing cells and vector control cells to comparable degrees. When S1P was present together with IGF-I in the lower well of Boyden chamber, IGF-I-directed chemotaxis was completely abolished only in $S1P_2$-expressing CHO cells. The $S1P_2$–mediated inhibitory effect of S1P on cell migration was potent with the threshold concentration of as low as 1 nM, and was dependent on an S1P concentration gradient. Thus, S1P acts as a chemorepellent.

Chemoattractants receptors activate complex, multiple signaling cascades including protein tyrosine kinases, lipid kinases and the low molecular weight GTPases (Lauffenburger and Horwitz, 1996). These signaling events lead in concert to regulate actin organization and myosin motor function, which constitute essential processes for cell migration.

Among the low molecular weight G proteins, the Rho family GTPases comprise three major members, Rho, Rac and Cdc42, and exert distinct regulatory actions on actin cytoskeletons, thus playing an important role in cell motility (Hall, 1999; Nobes and A.Hall., 1999); Rho mediates formation of stress fibers and focal adhesions, while Rac and Cdc42 direct peripheral actin assembly that results in formation of lamellipodia and filopodia, respectively, especially at the leading edge of a migrating cell. For example, expression of a dominant negative form of Rac, N^{17}-Rac, in a variety of cell types inhibits directed cell migration toward diverse chemoattractants, including PDGF, LPA, epidermal growth factor (EGF), insulin and colony stimulating factor-1 (Hall, 1999; Nobes and A.Hall, 1999; Banyard et al., 2002) indicating that Rac is required for chemotaxis toward these attractants. The expression of dominant negative forms of the Rho family GTPases, N^{17}-Rac and N^{17}-Cdc42, but not N^{19}-RhoA, inhibited chemotaxis toward S1P in $S1P_1$- and $S1P_3$-expressing CHO cell types and IGF-I in all the CHO cell types, indicating that IGF-I- and S1P-directed chemotaxis was dependent on Rac and Cdc42 (Okamoto et al., 2000). S1P as well as IGF-I increased an amount in a GTP-bound, active form of Rac in $S1P_1$- and $S1P_3$-expressing CHO cells and all the CHO cell types, respectively. In $S1P_2$-expressing cells, S1P by itself did not change Rac activity. However, S1P dramatically abolished IGF-I-induced Rac activation and lamellipodia formation with a dose-response relationship similar to that for S1P inhibition of IGF-I-directed chemotaxis (Okamoto et al., 2000). Either S1P or IGF-I did not affect GTP-loading of Cdc42 in any of CHO cell types. Rho regulation by S1P and IGF-I was a little complicated; stimulation of the attractant receptor $S1P_1$ did not change RhoA activity, but both the repellant receptor $S1P_2$ and the attractant $S1P_3$ stimulated RhoA. IGF-I did not affect RhoA or Cdc42 activity (Okamoto et al., 2000). Thus, all these data in CHO cells was consistent with the notion that stimulation of Rac activity, but not of Cdc42 or RhoA, above the resting level was necessary for stimulation of chemotaxis in this cell type. The three Edg family S1P receptors transmit overlapping but distinct sets of signals to the Rho family GTPases (Okamoto et al., 2000; Takuwa, 2002; Takuwa et al.,2002; Sugimoto et al., 2003), thus allowing receptor subtype-specific, distinct regulatory activities on cell migration (Fig. 1).

Chemoattractants that act via G protein-coupled receptors are generally shown to mediate chemotaxis and Rac activation via the heterotrimeric G_i protein. However, signaling mechanisms of chemorepellant receptors were largely unknown. The C-terminal mini-peptide of the heterotrimeric G protein α-subunits (Gα-CTs) serve as inhibitors specific for respective G proteins (Koch et al., 1994). The expression of either $G_{12}\alpha$-CT or $G_{13}\alpha$-CT, but not $G_s\alpha$-CT or $G_q\alpha$-CT or pertussis toxin, abrogated $S1P_2$–mediated inhibition of Rac and cell migration. Moreover, either $S1P_2$-$G_{12\alpha}$ or $S1P_2$-$G_{13\alpha}$ fusion receptor (Seifert et al., 1999), but not $S1P_2$-$G_{q\alpha}$ fusion receptor, mediated S1P inhibition of Rac and cell motility (Sugimoto et al., 2003). Our data indicated that the $G_{12/13}$ family protein coupled $S1P_2$ to inhibition of Rac, cell migration and lamellipodia in CHO cells (Sugimoto et al., 2003). Furthermore we found that downstream of $G_{12/13}$, Rho mediated $S1P_2$ inhibition of Rac and cell migration through a mechanism not involving Rho kinase as a Rho effector (Sugimoto et al., 2003). Very recently, Sanchez et al. (Sanchez et al., 2005) showed that $S1P_2$ mediated inhibition of cell migration through stimulating the 3'-phosphatase of phosphoinositides, PTEN, in a Rho kinase-dependent manner. This may suggest the possibility that more than a single mechanism could contribute to $S1P_2$ mediated inhibition of cell migration and that the mechanism behind $S1P_2$-mediated cell migration inhibition may be different among cell types.

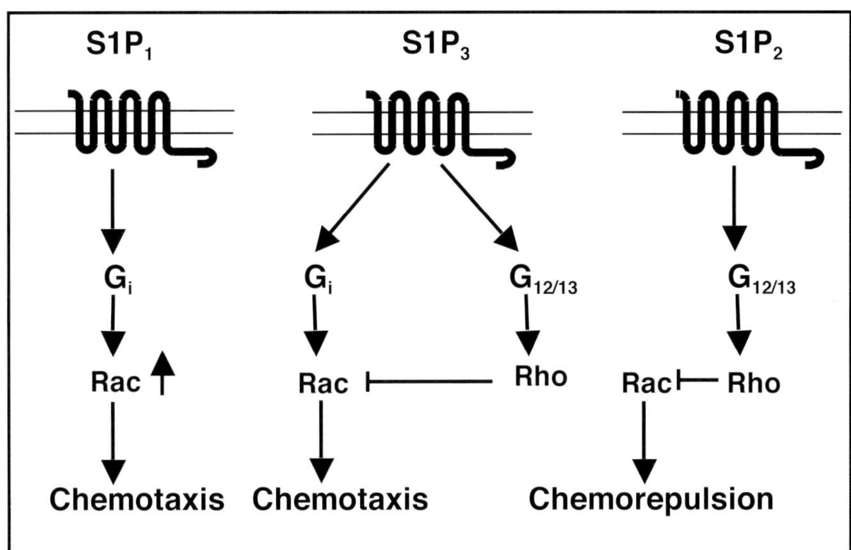

Fig. 1. Bimodal regulation of the Rho family GTPases and cell migration by the attractant receptors $S1P_1$ and $S1P_3$, and the repellent receptor $S1P_2$. $S1P_1$ and $S1P_3$ mediate Rac activation and chemotaxis via Gi. $S1P_2$ mediates Rac inhibition and migration inhibition (chemorepulsion), which is mediated by G12/13 and Rho. $S1P_3$ also stimulates Rho and generates an inhibitory signal for Rac, which is uncovered by inactivation of Gi with pertussis toxin.

$S1P_3$ mediated Rac activatin and chemotaxis despite stimulating Rho (Okamoto et al., 2000). We found in pertussis toxin-pretreated CHO cells expressing $S1P_3$ that S1P induced inhibition of Rac, cell migration and lamellipodia just as in CHO cells expressing $S1P_2$ (Sugimoto et al., 2003). Thus, inactivation of G_i by pertussis toxin treatment converted $S1P_3$ into the $S1P_2$–like repellant receptor. These results indicate that integration of counteracting signals from the G_i- and the $G_{12/13}$-Rho pathways directs either positive or negative regulation of Rac and thus cell migration, upon activation of a single S1P receptor isoform.

3. Stimulation and inhibition of melanoma cell invasion and metastasis by S1P receptors

Igarashi and his colleagues (Sadahira et al., 1992) first reported that S1P inhibits trans-well migration as well as invasion across the Matrigel layer of B16 mouse melanoma cells. They showed that S1P immobilized on glass beads exhibited similar inhibitory action on migration of B16 melanoma cells (Yamamura et al., 1997). They also demonstrated the presence of a single class of specific, binding site for a radio-labelled S1P on B16 melanoma cells. These observations suggested that S1P inhibited melanoma cell migration through an extracellular action by specific binding to a cell surface receptor. On the other hand, other studies suggested S1P-induced migration inhibition through its intracellular action (Wang et al., 1999). Our observation that the heterologous expression of $S1P_2$ mediated cell migration inhibition in CHO cells prompted us to investigate the possibility that $S1P_2$ receptor might mediate migration inhibition of B16 melanoma cells in response to S1P. We found that B16 melanoma cells detectably expressed $S1P_2$, but not other S1P receptor subtypes (Arikawa et al., 2003). The $S1P_2$-selective antagonist JTE013, which specifically blocks S1P binding to $S1P_2$ and prevents Ca^{2+} mobilization and ERK activation mediated by $S1P_2$, completely abolished the inhibitory effect of S1P on B16 cell migration. Thus, B16 melanoma cell is the example of inhibitory regulation of cell migration by endogenously expressed $S1P_2$.

In B16 cells, S1P inhibited and stimulated Rac and RhoA, respectively, as in $S1P_2$-overexpressing CHO cells, and that both of these responses were totally abrogated by JTE013. S1P did not affect Cdc42 activity. Consistent with the observation in $S1P_2$-expressing CHO cells (Okamoto et al., 2000), suppression of endogenous Rac activity by adenovirus-mediated expression of N^{17}Rac resulted in inhibition of migration of B16 cells (Arikawa et al., 2003). On the other hand, the inactivation of cellular Rho by C3 toxin reversed S1P inhibition of migration in B16 cells. In addition, random migration activity of B16 cell in the absence of S1P was also stimulated, rather than inhibited, after C3 toxin pretreatment, suggesting that the basal Rho activity is inhibitory for cell migration. This latter observation is consistent with a previous report in Rat1 fibroblasts (Banyard et al., 2000). Similar to CHO cells, the Rho kinase inhibitors Y27632 or HA1077, at the concentrations that effectively inhibited stress fiber formation, did not reverse S1P inhibition of migration. The results indicate that a RhoA effector other than Rho kinase, and cell responses different from formation of stress fibers or focal adhesions, are responsible for mediating negative regulation of cell motility and invasion by S1P.

S1P$_2$ also mediated inhibition of B 16 cell invasion across the Matrigel matrix, which is Rac-dependent (Arikawa et al., 2003). With regard to the latter point, it is of note that Rac1 is reported to mediate matrix metalloproteinase (MMP)-2 activation in fibrosarcoma cells (Zhuge and Xu, 2001). Requirement of Rac activity for both cell motility and invasion through extracellular matrix has been reported in Rat1 fibroblasts and T lymphoma cells as well (Banyard et al., 2000). Wang et al. (Wang et al., 1999) showing that S1P inhibited migration of MDA-MB-231 breast carcinoma cells which express S1P$_1$ and S1P$_3$ but not S1P$_2$, and postulated that this S1P-inhibitory action of tumor cell migration was due to an intracellular action of S1P. It is possible that the role for Rac in regulation of migration or invasion might be distinct among cells of epithelial and non-epithelial origin through mechanisms involving the regulation of cell adhesion molecules. Invasion of B 16 cells was inhibited by N^{17}Rac, like cell migration. On the other hand, unlike cell migration, invasion in the absence of S1P was inhibited by C3 toxin, suggesting that invasion is dependent upon both Rac and Rho. Inhibition of invasion by the expression of N^{17}RhoA, as well as V^{14}RhoA, was reported previously in fibroblasts (Banyard et al., 2000). Despite this, the inhibitory effect of S1P on B16 cell invasion was partially reversed by C3 toxin pretreatment, underscoring the role for Rho in S1P inhibition of invasion. These composite observations indicate that S1P$_2$-mediated Rho stimulation and Rac inhibition are both involved as mechanisms for S1P-induced inhibition of B16 cell migration and invasion. However, Rho stimulation by itself doe not appear to be sufficient for migration inhibition in the absence of Rac inhibition. This notion is based on the observation obtained in S1P$_3$-overexpressing B16 cells (see below), in which the addition of S1P together with JTE013 stimulated RhoA activity via S1P$_3$, but stimulated, rather than inhibited, Rac activity and migration. Rho may be involved in migration and invasion inhibition through negatively regulating Rac, as in migration inhibition in CHO cells (Sugimoto et al., 2003).

Metastasis is a temporo-spacially architectured multistep phenomenon that takes place in the context of host environment. Rho family GTPases and their extracellular regulators were implicated in transformation and invasive phenotypes (Banyard et al., 2000), however, their roles in metastasis *in vivo* were largely unknown. B16 cells that are injected into mouse tail veins form multiple metastatic nodules in the lung two or three weeks later. We found that daily intraperitoneal injection of S1P reduced the number of metastatic nodules (Yamaguchi et al., 2003). More interestingly, transient treatment of B16 cells with S1P just before the tail vein injection also effectively reduced he number of metastatic nodules, suggesting that S1P might inhibit very early steps of pulmonary metastasis including attachment to the pulmonary capillary endotelium. Adenovirus-mediated transient expression of N^{17}Rac markedly inhibited pulmonary metastasis, suggesting that Rac inhibition might underly S1P inhibition of metastasis. The stable overexpression of $S1P_2$ augmented S1P-induced cell migration inhibition compared to vector control cells, whereas the overexpression of either $S1P_1$ or $S1P_3$ greatly attenuated S1P inhibition of migration or stimulated migration. In $S1P_2$–overexpressing cells S1P-induced metastasis inhibition was augmented. In contrast, that of either $S1P_1$ or $S1P_3$ resulted in aggravation of metastasis in an S1P-dependent manner. These observations suggest that Rac is required at a very early stage of hematogenous metastasis before invasion or proliferation and that G-protein-coupled receptors could participate in regulation of metastasis in receptor subtype-specific ligand-dependent manners.

4. Differential regulation of Rac and cell migration by endogenous Edg receptors in vascular smooth muscle and endothelial cells

As mentioned above, S1P displays unique bimodal activities on two vascular cell types, endothelial cells (Lee et al., 1999; Wang et al., 1999) and VSMCs (Ryu et al., 2002). S1P inhibits VSMC migration, whereas it stimulates endothelial cell migration. Both VSMCs and endothelial cells in culture express multiple S1P receptor subtypes. VSMCs express abundant levels of $S1P_2$ and $S1P_3$, and endothelial cells express $S1P_1$ and $S1P_3$ (Ryu et al., 2002; Usui et al., 2004). The expression of $S1P_1$ in VSMCs and $S1P_2$ in endothelial cells is very faint. The observations, together with the fact that $S1P_1$ and $S1P_3$ are chemotactic whereas $S1P_2$ is inhibitory, suggest that the vascular cell type-specific, distinct expression pattern of the S1P receptor subtypes can account for differential cell motility responses of the two vascular cell types. The S1P receptors exerted bimodal regulation on Rac activity in vascular cells; S1P inhibited PDGF-induced Rac activation in VSMCs although S1P alone did not affect Rac very much, whereas S1P by itself robustly stimulated Rac activity in endothelial cells. The overexpression of $S1P_1$ in VSMCs rendered the cells chemotactic toward S1P with elevation of Rac, whereas that of $S1P_2$ augmented S1P inhibition of PDGF-directed chemotaxis. The overexpression of $S1P_1$ in endothelial cells also made the cells repulsive in response to S1P. These observations are consistent with the notion that that an integration of the S1P receptor subtype-selective, positive and negative signals on cellular Rac activity is a critical determinant for eventual direction of regulation on cell motility by S1P.

5. Future directions

The S1P receptors, especially $S1P_1$, $S1P_2$ and $S1P_3$, are expressed widely in a variety of tissues during the fetal life as well as the adult life. Both attractive and repulsive signals guide migrating cells in tissue remodeling and angiogenesis as well as organogenesis and morphogenesis. It is likely that $S1P_2$, a unique negative regulator of Rac, might be involved in these pathological and physiological processes in concert with $S1P_1$ and $S1P_3$, which are positive regulators of Rac, and presumably also other S1P receptors. It would be particularly interesting to see whether these S1P receptors play any role in pathological processes such as inflammation, tumor cell progression and atherosclerosis. Elucidation of pathological roles for the S1P signaling system is expected to lead to development of novel therapeutic strategies including the use of specific antagonists and agonists for S1P receptor subtypes and inhibitors of synthesis and degradation of S1P, and therefore should deserve intensive investigation.

Acknowledgments. This work was supported by grants from the Ministry of Education, Science and Culture of Japan, and the Hoh-Ansha Foundation.

6. References

Arikawa K, Takuwa N, Yamaguchi H, Sugimoto N, Kitayama J, Nagawa H, Takehara K, Takuwa Y. (2003) Ligand-dependent inhibition of B16 melanoma cell migration and invasion via endogenous S1P2 G protein-coupled receptor. Requirement of inhibition of cellular Rac activity. *J Biol Chem*, **278**, 32841-32851.

Banyard, J., B. Anand-Apte, M. Symons, and B.R.Zetter. 2000. Motility and invasion are differentially modulated by Rho family GTPases. *Oncogene*, **19**, 580-591.

Bornfeldt KE, Graves LM, Raines EW, Igarashi Y, Wayman G, Yamamura S, Yatomi Y, Sidhu JS, Krebs EG, Hakomori S, Ross (1995) Sphingosine-1-phosphate inhibits PDGF-induced chemotaxis of human arterial smooth muscle cells: spatial and temporal modulation of PDGF chemotactic signal transduction. *J Cell Biol*, **130**, 193-206.

Hall, A. 1998. Rho GTPases and the actin cytoskeleton. *Science*, **279**, 509-514

Ishii I, Fukushima N, Ye X, Chun J (2004) Lysophospholipid receptors: signaling and biology. *Annu Rev Biochem*, **73**, 321-354.

Kawa S, Kimura S, Hakomori S, Igarashi Y (1997) Inhibition of chemotactic motility and trans-endothelial migration of human neutrophils by sphingosine 1-phosphate. *FEBS Lett*, **420**, 196-200.

Koch, W.J., B.E. Hawes, J. Inglese, L.M. Luttrell, and R.J. Lefkowitz. 1994. Cellular expression of the carboxyl terminus of a G protein-coupled receptor kinase attenuates G beta gamma-mediated signaling. *J Biol. Chem*, **269**, 6193-6197.

Lauffenburger DA, Horwitz AF (1996) Cell migration: a physically integrated molecular process. *Cell*, **84**, 359-369.

Lee MJ, Thangada S, Claffey KP, Ancellin N, Liu CH, Kluk M, Volpi M, Shaafi RI, Hla T (1999) Vascular endothelial cell adherens junction assembly and morphogenesis induced by sphingosine-1-phosphate. *Cell*, **99**, 301-312.

Liu Y, Wada R, Yamashita T, Mi Y, Deng CX, Hobson JP, Rosenfeldt HM, Nava VE, Chae SS, Lee MJ, Liu CH, Hla T, Spiegel S, Proia RL (2000) Edg-1, the G protein-coupled receptor for sphingosine-1-phosphate, is essential for vascular maturation. *J Clin Invest*, **106**, 951-961.

Nobes, C.D., and A.Hall. 1999. Rho GTPases control polarity, protrusion, and adhesion during cell movement. *J Cell Biol*, **144**, 1235-1244.

Okamoto H, Takuwa N, Yokomizo T, Sugimoto N, Sakurada S, Shigematsu H, Takuwa, Y (2000) Inhibitory regulation of Rac activation, membrane ruffling, and cell migration by the G protein-coupled sphingosine-1-phosphate receptor EDG5 but not EDG1 or EDG3. *Mol Cell Biol*, **20**, 9247-9261.

Ryu, Y., N. Takuwa, N. Sugimoto, S. Sakurada, S. Usui, H. Okamoto, O. Matsui and Y. Takuwa. 2002. Sphingosine-1-Phosphate, a Platelet-Derived Lysophospholipid Mediator, Negatively Regulates Cellular Rac Activity and Cell Migration in Vascular Smooth Muscle Cells. *Circ Res*, **90**, 325-332.

Sadahira Y, Ruan F, Hakomori S, Igarashi Y (1992) Sphingosine 1-phosphate, a specific endogenous signaling molecule controlling cell motility and tumor cell invasiveness. *Proc Natl Acad Sci U S A*, **89**, 9686-9690.

Sanchez T, Thangada S, Wu MT, Kontos CD, Wu D, Wu H, Hla T. (2005) PTEN as an effector in the signaling of antimigratory G protein-coupled receptor. *Proc Natl Acad Sci U S A*, **102**, 4312-4317.

Seifert, R., K. Wenzel-Seifert, and B.K. Kobilka. 1999. GPCR-Galpha fusion proteins: molecular analysis of receptor-G-protein coupling. *Trends Pharmacol Sci*, **20**, 383-389.

Spiegel S, Milstien S (2003) Sphingosine-1-phosphate: an enigmatic signalling lipid. *Nat Rev Mol Cell Biol*, **4,** 397-407.

Sugimoto, N., Takuwa, N., Okamoto, H., Sakurada, S., and Takuwa, Y. (2003) Inhibitory and stimulatory regulation of Rac and cell motility by the G12/13-Rho and Gi pathway integrated downstream of a single G protein-coupled sphingosine-1-phosphate receptor isoform. *Mol Cell Biol*, **23**, 1534-1545.

Takuwa Y (2002) Subtype-specific differential regulation of Rho family G proteins and cell migration by thc Edg family sphingosine-1-phosphate receptors. *Biochim Biophys Acta*, **1582**, 112-120.

Takuwa Y, Takuwa N, Sugimoto N. (2002) The Edg family G protein-coupled receptors for lysophospholipids: their signaling properties and biological activities. *J Biochem*, **131**, 767-771.

S. Usui, N. Sugimoto, N. Takuwa, S. Sakagami, S. Takata, S. Kaneko, Y. Takuwa. (2004) Blood lipid mediator sphingosine-1-phosphate potently stimulates platelet derived growth factor-A and –B chain expression through S1P1-Gi-Ras-MAPK-dependent induction of krüppel-like factor 5. *J Biol Chem*, **279**, 12300-12311.

Wang F, Van Brocklyn JR, Hobson JP, Movafagh S, Zukowska-Grojec Z, Milstine S, Spiegel S. (1999) Sphingosine 1-phosphate stimulates cell migration through a G(i)-coupled cell surface receptor. Potential involvement in angiogenesis. *J. Biol Chem*, **274**, 35343-35350.

F. Wang, J. R. Van Brockly, L. Edsall, V. E. Nava, S. Spiegel. (1999) Sphingosine-1-phosphate inhibits motility of human breast cancer cells independently of cell surface receptors. *Cancer Res*, **59**, 6185-6191.

Yamaguchi H, Kitayama J, Takuwa N, Arikawa K, Inoki I, Takehara K, Nagawa H, Takuwa Y. (2003) Sphingosine-1-phosphate receptor subtype-specific positive and negative regulation of Rac and hematogenous metastasis of melanoma cells. *Biochem J*, **374**, 715-722.

Yamamura S, Yatomi Y, Ruan F, Sweeney EA, Hakomori S, Igarashi Y (1997) Sphingosine 1-phosphate inhibits motility of human breast cancer cells independently of cell surface receptors. *Biochemistry*, **36**, 10751-10759.

Yatomi Y, Ozaki Y, Ohmori T, Igarashi Y (2001) Sphingosine 1-phosphate: synthesis and release. *Prostaglandins Other Lipid Mediat*, **6**, 107-122.

Zhuge Y, Xu J. (2001) Rac1 mediates type I collagen-dependent MMP-2 activation. role in cell invasion across collagen barrier. *J Biol Chem*, **276**, 16248-16256.

6-4 Sphingosine 1-Phosphate-Related Metabolism in the Blood Vessel

Yutaka Yatomi[1], Shinya Aoki[1], and Yasuyuki Igarashi[2]

[1]Department of Laboratory Medicine, Graduate School of Medicine, The University of Tokyo, Tokyo 113-8655, Japan; [2]Department of Biomembrane and Biofunctional Chemistry, Graduate School of Pharmaceutical Sciences, Hokkaido University, Sapporo 060-0812, Japan

Summary. Sphingosine 1-phosphate (S1P) is an important intercellular lipid mediator, especially in vascular biology. Blood platelets store S1P abundantly and release this bioactive lysophospholipid extracellularly upon stimulation. Vascular endothelial and smooth muscle cells respond dramatically to this platelet-derived lipid. Regulating S1P biological activities may be valuable to treat vascular disorders, and, as such, requires a better understanding of those mechanisms that mediated vascular S1P. Various factors may mediate the level and function of plasma S1P in vivo, these include S1P release from platelets and S1P distribution between albumin and lipoproteins as well as S1P receptor (S1P) expression and lipid phosphate phosphatase activity on vascular cells. In atherosclerotic diseases, where the plasma levels of lipids and lipoproteins or endothelial cell functions are altered, modulating effects of S1P may be of pathophysiological importance.

Keywords. Sphingosine 1-phosphate, platelets, vascular endothelial cells, vascular smooth muscle cells, vascular biology, lipid phosphate phosphatase

1. Introduction

Sphingosine 1-phosphate (S1P) is a bioactive sphingolipid capable of inducing a wide spectrum of biological responses, mainly through a family of G protein-coupled cell-surface receptors named $S1P_{1-5}$ (originally EDG-1, 5, 3, 6, and 8, respectively) (Kluk and Hla 2002, Lynch 2002, Spiegel and Milstien 2002, Takuwa 2002, Hla 2003, Spiegel and Milstien 2003). S1P is an important intercellular lipid mediator, especially in vascular biology (Liu et al. 2000, Yatomi et al. 2001, Karliner 2002, Okajima 2002, Panetti 2002, Siess 2002, Ozaki et al. 2003). Blood platelets store S1P abundantly and release this bioactive lysophospholipid extracellularly upon stimulation. Vascular endothelial cells (ECs) and smooth muscle cells (SMCs) respond dramatically to this platelet-derived lipid. Recent reports detail the functional roles of S1P in platelet-vascular cell interactions (Liu et al. 2000, Yatomi et al. 2001, Karliner 2002, Okajima 2002, Panetti 2002, Siess 2002, Ozaki et al. 2003). SIP interactions with Ecs may sustain vascular system integrity and mediate physiological wound healing processes such as vascular repair. This important lipid, however, can become atherogenic and thrombogenic, thereby causing, or aggravating, cardiovascular diseases. Furthermore, S1P interaction with SMCs induces a variety of responses, including vasoconstriction. Hence, it seems likely that the manipulation of S1P would have invaluable therapeutic benefits for treating vascular disorders. To this end, it is essential to fully understand the mechanisms that contribute to the regulation of S1P levels in blood vessels. One such mechanism is S1P-related metabolism in the blood vessel.

2. S1P release from activated platelets: source of plasma S1P

S1P forms intracellularly through the phosphorylation of sphingosine (Sph) that is catalyzed by Sph kinase (Kohama et al. 1998, Olivera et al. 1998). The S1P lyase, which degrades S1P into phosphoethanolamine and fatty aldehyde, seems the most important enzyme for degradation of S1P (van Veldhoven and Mannaerts 1993), but phosphatase activity for S1P should not be neglected. Platelets possess a highly active Sph kinase (Stoffel et al. 1973, Yatomi et al. 1995, Yatomi et al. 1997b) but are devoid of S1P lyase, a nearly ubiquitous enzyme present in almost all tissues van (Veldhoven and Mannaerts 1993, Yatomi et al. 1997b) (Figure 1). As a result, platelets store S1P abundantly (Yatomi et al. 1997a, Yatomi et al.

2001). Furthermore, this stored S1P is released extracellularly in response to stimulation by physiological agonists such as thrombin (Yatomi et al. 1995, Yatomi et al. 1997b, Yatomi et al. 2001). S1P release is closely correlated with activation of protein kinase C (Figure 2), which is also highly expressed in platelets (Kikkawa et al. 1982).

Platelets, the S1P storage sites in the blood vessel, have a unique metabolic activity related to this lipid. This sphingolipid metabolism in activated platelets was reported recently (Yatomi et al. 2004). When Sph-radio-labeled platelet suspensions were exposed to thrombin or protein kinase C activator 12-*O*-tetradecanoylphorbol 13-acetate (TPA), a decrease in S1P formation and an increase in Sph, ceramide (Cer), and sphingomyelin formation were observed (Yatomi et al. 2004). Sph conversion into Cer, and Cer conversion into sphingomyelin were not affected upon activation, suggesting that S1P dephosphorylation may initiate the formation of sphingolipid signaling molecules. Further studies suggested that the S1P released may undergo dephosphorylation extracellularly, possibly due to the ecto-phosphatase activity. The presence of a transmembrane cycling pathway in stimulated platelets that starts with S1P release and dephosphorylation and leads to the formation of other sphingolipid mediators was suggested (Figure 2) (Yatomi et al. 2004).

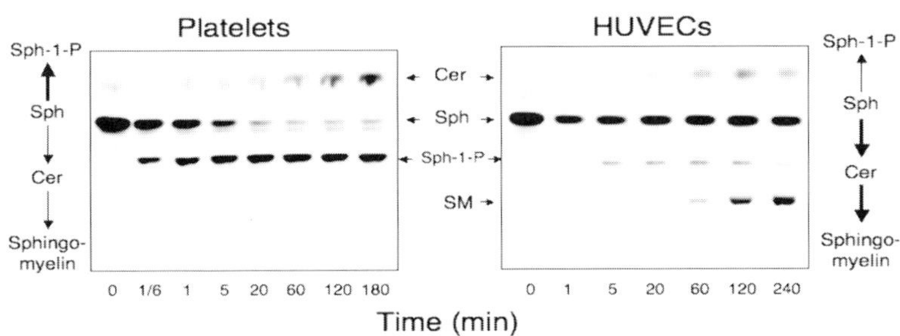

Fig. 1. Metabolism of exogenous [³H]Sph in platelets (left panel) and human umbilical vein ECs (right panel). Lipids were extracted from cells incubated with [³H]Sph for various durations and analyzed for [³H]sphingolipids by TLC autoradiography. Locations of standards lipids are indicated in the middle. SM, sphingomyelin. In platelets (which possess a high Sph kinase activity), the incorporated Sph is rapidly converted into Sph-1-P. The S1P band remains intense even after a long incubation, indicating the stability of S1P in platelets (which lack S1P lyase activity). In contrast, Sph conversion into S1P is weak and transient in human umbilical vein ECs; the formation of sphingomyelin, possibly through Cer, is active and time-dependent.

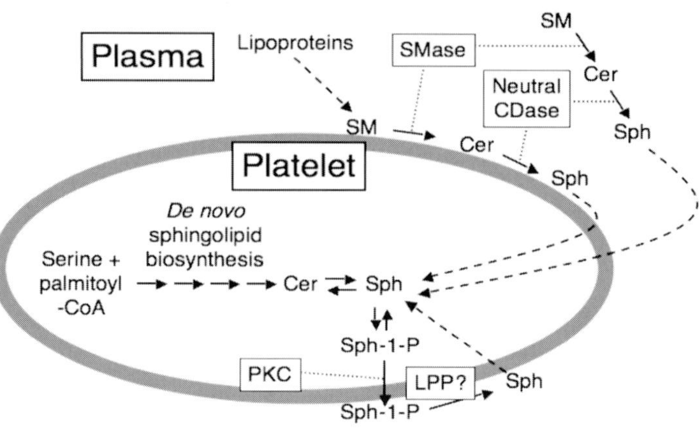

Fig. 2. The S1P-related metabolism in platelets. Platelets, which lack de novo sphingolipid biosynthesis pathway, obtain Sph via at least two pathways: i) plasma generation with subsequent incorporation and ii) generation at the outer leaflet of the plasma membrane, initiated by cell surface sphingomyelin degradation. S1P, formed form Sph and abundantly stored in platelets, is released upon activation, possibly through mediation of protein kinase C (PKC). Part of the released S1P may be dephosphorylated to Sph by the surface LPP and incorporated again into platelets again.

Although the precise mechanism for extracellular release of S1P following platelet activation is not clarified, this must require specific transporter(s) since S1P possesses the polar nature of the head group. Also, extracellular release of S1P into the medium is much higher in the presence of extracellular albumin or lipoproteins than in their absence (Yatomi et al. 2000, Aoki et al. 2005). These protein fractions, especially albumin, may extract S1P from the platelets when intracellular S1P changes upon protein kinase C activation. Since S1P stored in platelets is susceptible to depletion when platelet protein kinase C is activated, this may be the route by which S1P externalization. In this senario, transbilayer movement across the plasma membrane is enhanced as a result of modifications to membrane properties that accompany protein kinase C activation. Phospholipids may be transported by ABC transporters (Higgins 1994, Ruetz and Gros 1994). ABCA7, which is preferentially expressed in platelets and may play an important role there (Sasaki et al. 2003), may be responsible for the S1P release, but this remains to be determined.

Although platelets are an established source for plasma S1P, other source(s) may exist. A small but significant fraction of Sph kinase was found to be secreted extracellularly from human umbilical vein ECs (Ancellin et al. 2002). The secretion was reported to be constitutive and independent of the cell stimulation (Ancellin et al. 2002). It was also shown that the secreted Sph kinase is enzymatically active and produces Sph-1-P; extracellular Sph seems to be a rate-limiting factor (Ancellin et al. 2002). However, the significance of this route for S1P formation in the blood stream remains to be determined since Sph is metabolically stable, *i.e.* S1P is not formed from Sph in plasma, but will converted into S1P in the presence of activated platelets (Aoki et al. 2005).

Sphingosylphosphorylcholine reportedly hydrolysizes into S1P *in vitro*, with the use of phospholipase D of bacterial origin (van Veldhoven et al. 1989). Deacylation of ceramide 1-phosphate into S1P is also theoretically possible. However, these routes have not been reported in mammalian systems. Very recently, autotaxin, which is responsible for serum lysophospholipase D activity to produce lysophosphatidic acid from lyso-glycerophospholipids, was cloned. It hydrolyzes sphingosylphosphorylcholine into S1P (Tokumura et al. 2002, Umezu-Goto et al. 2002). However, the significance of this pathway for S1P production *in vivo* also remains to be clarified.

3. The source of Sph for S1P production in platelets

Since platelets possess high Sph kinase activity and store resultant S1P abundantly, the next important question is how platelets obtain the substrate Sph for S1P production. This remains unanswered because investigations of platelet sphingolipid metabolism primarily tracked labeled Sph incubated with platelets. Platelets themselves lack a *de novo* sphingolipid biosynthesis pathway, which starts from serine and palmitoyl-CoA (Tani et al. 2005). In platelets incubated with radio-labeled serine, only trace levels of labeled sphingomyelin or Cer were detected, although glycerophospholipids were detected under the same conditions. In fact, platelets exhibit little serine palmitoyltransferase activity and a loss in enzyme activity during megakaryocytic differentiation into platelets renders them unable to synthesize sphingolipids *de novo* (Tani et al. 2005). Thus, it is likely that S1P generation in platelets depends on an extracellular supply of sphingolipids.

The generation pathway that supplies Sph to platelets has recent been studied. Platelets specifically incorporated extracellular Sph and rapidly converted it into Sph-1-P; the presence of a novel transporter that transports extracellular Sph to the inside of the cells was predicted (Tani et al. 2005). Furthermore, Sph formed from plasma sphingomyelin by bacterial sphingomyelinase and neutral ceramidase was rapidly incorporated into platelets and converted to Sph-1-P, suggesting that platelets utilize extracellular Sph as a source for S1P (Tani et al. 2005). On the other hand, platelets abundantly express sphingomyelin, possibly supplied from plasma lipoproteins, at the cell surface, supplied by selective transfer, not by endocytosis (Engelmann et al. 1996). Indeed, despite defective *de novo* sphingolipid biosynthesis, platelets are rich in sphingomyelin. Sphingomyelin makes up 21% of a platelet's total phospholipids and 13% of its total lipids, and is localised primarily at the outer leaflet of the plasma membrane (Marcus et al. 1969). Treating platelets with bacterial sphingomyelinase produces Sph at the cell surface, conceivably via membrane-bound neutral ceramidase activity (Tani et al. 2005). Furthermore, the secretory form of acid sphingomyelinase also induces S1P elevation in platelets (Tani et al. 2005). It seems likely that acid sphingomyelinase, secreted from macrophages and vascular ECs as a Zn^{2+}-dependent enzyme (Spence et al.1989, Marathe et al. 1998), constitutively hydrolyzes cell-surface sphingomyelin in platelets and contributes S1P production there. Acid sphingomyelin secreted from activated platelets themselves may stimulate S1P production in an autocrine/paracrine fashion (Simon et al. 1998). Collectively, the supple of Sph in platelets has at least two sources, through the incorporation of plasma Sph generated by secreted enzymes or generation at the outer leaflet of the plasma membrane initiated by cell surface sphingomyelin degradation (Figure 2) (Tani et al. 2005).

That platelets can use Sph at the outer leaflet of the plasma membrane and in the extracellular space implies that neutral ceramidase has an important role. Neutral ceramidase is localized at the plasma membrane as a type II integral membrane protein, and it secreted into the extracellular space after *N*-terminal anchor processing (Riley et al. 2001). In mice, it is released from (mouse) ECs (Romiti et al. 2000) and present in (mouse) serum (Riley et al. 2001). Its activity is also detectable in human plasma and participates in extracellular Sph formation (Tani et al. 2005). Additionally, plasma membrane-bound neutral ceramidase is involved in the hydrolysis of Cer, which is generated at the outer leaflet of the plasma membrane (Tani *et al.*, unpublished result). Moreover, Sph generation reportedly occurs at the cell surface of platelets after treatment with exogenously added sphingomyelinase, and platelets have membrane-bound neutral ceramidase activity (Tani et al. 2005). Importantly, Yoshimura *et al.* recently reported that knocking down the neutral ceramidase gene in zebrafish during embryogenesis results in a defect in blood cell circulation (Yoshimura et al. 2004), how that defect arises, however, is yet to be determined. It is possible that the phenotype may stem from a decrease of plasma S1P levels, since plasma S1P and its receptors S1Ps are important for keeping vascular integrity (Liu et al. 2000). Neutral ceramidase may be very important to S1P production in platelets and hence the regulation of plasma S1P levels (Figure 2).

4. S1P is a normal constituent of human plasma and serum

As might be expected, S1P is a normal constituent of human plasma and serum (Yatomi et al. 1997a, Yatomi et al. 2001). Lipid extraction under alkaline and acidic conditions, followed with acetylation with radioactive acetic anhydride, revealed S1P levels in plasma and serum of about 200 and 500 nM, respectively (Yatomi et al. 1997a). Similar levels of S1P in plasma or serum were obtained using other methods (Okajima 2002), although it should be noted that the reported plasma S1P levels might include S1P released from platelets during *in vitro* manipulation (*e.g.*, centrifugation). Serum S1P levels have been always higher than those in plasma, and it is most likely that the source of discharged S1P during blood clotting is platelets, as they abundantly store S1P, that release the stored S1P extracellularly upon stimulation with thrombin, a product of the coagulation cascade (Yatomi et al. 2001).

As described above, platelets store and release extracellular S1P into the blood. That released S1P was once thought bound to albumin (Yatomi et al. 2000). Okajima *et al.* revealed, however, that when expressed as the per unit amount of protein, S1P is concentrated in lipoprotein fractions with a rank order of high-density lipoprotein (HDL) > low-density lipoprotein (LDL) > very low-density lipoprotein > lipoprotein-deficient plasma (mainly albumin), among plasma and serum components (Murata et al. 2000, Okajima 2002, Kimura et al. 2003). In fact, when the mixture of S1P and plasma was fractionated in a gel-filtration column, the entire S1P co-eluted with protein fractions, which coincide with lipoproteins and albumin by agarose gel electrophoresis (Aoki et al. 2005). When evaluated by polyacrylamide gel electrophoresis, 7.2 ± 3.8%, 53.3 ± 6.4%, and 39.5 ± 7.9% of the radioactivity of S1P in plasma were recovered in LDL, HDL, and albumin fractions, respectively (Aoki et al. 2005). S1P interaction with lipoproteins reportedly reduces its bioactivity and its active concentration level (Murata et al. 2000, Okajima 2002, Kimura et al. 2003). Based on previous results, Kds for S1P receptors are much lower than the concentrations of S1P in the plasma; Kd of S1P for these receptors is ranges from 2 to 30 nM (Lee et al. 1998, Kon et al. 1999, Lee et al. 1999,

Kimura et al. 2003). S1P regulation by lipoproteins of S1P bioactivities might be protective mechanism preventing the full activation of S1P receptors in vascular cells. However, degradation of S1P by ectoenzymes such as lipid phosphate phosphatases (LPPs) or via re-uptake into cells may be blocked when S1P binds to lipoproteins. The half-life of HDL-associated S1P is reportedly about four times longer than that of S1P in the presence of 0.1% bovine serum albumin (but not HDL) when examined using human umbilical vein ECs (Kimura et al. 2001, Okajima 2002, Kimura et al. 2003). Accordingly, HDL-associated S1P may exert a much weaker response in short-term reaction (such as intracellular Ca^{2+} mobilization) compared to free S1P (without HDL), but it may have a more sustained response in the long-term reaction (such as survival and proliferation) (Okajima 2002, Kimura et al. 2003).

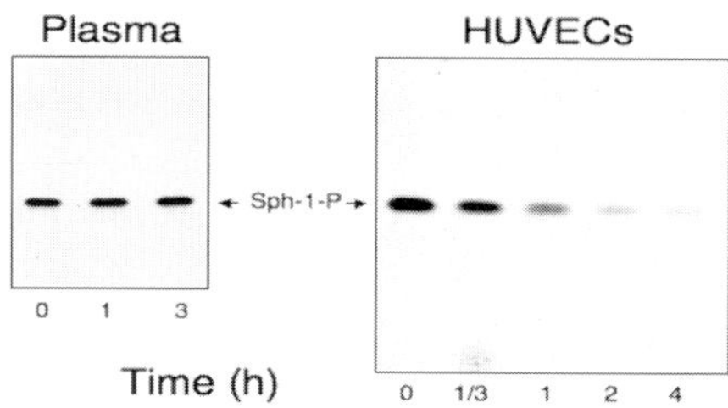

Fig. 3. Metabolic fates of [³H]S1P exogenously added to plasma (left panel) or human umbilical vein ECs (right panel). [³H]S1P was incubated with plasma or the ECs for indicated durations. The lipids were extracted and analyzed by TLC autoradiography for [³H]S1P metabolism.

5. Regulation of S1P plasma levels by LPPs

S1P is stable when incubated with plasma or serum (Yatomi et al. 2001, Aoki et al. 2005), indicating negligible enzymatic activity for S1P degradation (Figure 3, left). However, radioactive S1P was markedly degraded by ectophosphatase activity in the presence of human umbilical vein ECs (Figure 3, right) or whole blood (Aoki et al. 2005). Instead, radioactive Sph, Cer, and sphingomyelin were present (Aoki et al. 2005). It is likely that non-polar Sph, formed from polar S1P via some ectophosphatase activity (Mandala 2001) and is incorporated into ECs before being converted to ceramide (and then to sphingomyelin) intracellularly. Sph (but not S1P) is hydrophobic and easily passes through the lipid bilayer.   Similar results were obtained with the use of platelets (Yatomi et al. 2004) and SMCs (Ohmori et al. 2003).

Recently, several isoenzymes of mammalian LPP, that is, type 2 phosphatidic acid phosphatase, have been cloned. These probably act at the outer leaflet of the cell surface bilayer, accounting for those ecto-lipid phosphate phosphatase activities (toward Sph-1-P, lysophosphatidic acid, or phosphatidic acid) previously described (Mandala 2001, Brindley et al. 2002, Pyne and Pyne 2002, Sciorra and Morris 2002). In fact, ECs (Aoki et al. 2005) and SMCs (Ohmori et al. 2003) express mRNA transcripts for LPP-1, 2, and 3. S1P degradation on the surface of these vascular cells may be important when *in vivo* effects of S1P are evaluated. S1P concentrations in plasma and serum are much higher than the S1P Kd values for its receptors. S1P dephosphorylation at the cell surface may explain why the concentrations of S1P interacting with S1P receptors are lower than the plasma S1P levels. This, then, would suggest another protection mechanism following the full activation of S1Ps. This consistent with a recent *in vitro* transfection study showing that LPPs may limit the bioactivity of S1P by regulating its concentration (interacting with S1P receptors) (Brindley et al. 2002, Pyne and Pyne 2002, Sciorra and Morris 2002).

6. Concluding remarks

In summary, various factors may regulate the levels and bioactivities of plasma S1P *in vivo* including S1P release from platelets, S1P distribution between albumin and lipoproteins, and S1P expression and LPP activity on vascular cells. In various atherosclerotic diseases, where the plasma level of lipids and lipoproteins or EC functions are altered, modulation of S1P effects may provide a novel therapeutic approach to treating or preventing pathophysiological effects.

7. References

Ancellin N, Colmont C, Su J, Li Q, Mittereder N, Chae SS, et al. (2002) Extracellular export of sphingosine kinase-1 enzyme. Sphingosine 1-phosphate generation and the induction of angiogenic vascular maturation. *J Biol Chem,* **277,** 6667-6675.

Aoki S, Yatomi Y, Ohta M, Osada M, Kazama F, Satoh K, Nakahara K, Ozaki Y. (2005) Sphingosine 1-phosphate-related metabolism in the blood vessel. *J Biochem,* **138,** 47-55.

Brindley DN, English D, Pilquil C, Buri K, Ling ZC. (2002) Lipid phosphate phosphatases regulate signal transduction through glycerolipids and sphingolipids. *Biochim Biophys Acta,* **1582,** 33-44.

Engelmann B, Kogl C, Kulschar R, Schaipp B. (1996) Transfer of phosphatidylcholine, phosphatidylethanolamine and sphingomyelin from low- and high-density lipoprotein to human platelets. *Biochem J,* **315,** 781-789.

Higgins CF. (1994) Flip-flop: the transmembrane translocation of lipids. *Cell,* **79,** 393-395.

Hla T. (2003) Signaling and biological actions of sphingosine 1-phosphate. *Pharmacol Res,* **47,** 401-407.

Karliner JS. (2002) Lysophospholipids and the cardiovascular system. *Biochim Biophys Acta,* **1582,** 216-221.

Kikkawa U, Takai Y, Minakuchi R, Inohara S, Nishizuka Y. (1982) Calcium-activated, phospholipid-dependent protein kinase from rat brain. Subcellular distribution, purification, and properties. *J Biol Chem,* **257,** 13341-13348.

Kimura T, Sato K, Kuwabara A, Tomura H, Ishiwara M, Kobayashi I, et al. (2001) Sphingosine 1-phosphate may be a major component of plasma lipoproteins responsible for the cytoprotective actions in human umbilical vein endothelial cells. *J Biol Chem,* **276,** 31780-31785.

Kimura T, Sato K, Malchinkhuu E, Tomura H, Tamama K, Kuwabara A, et al. (2003) High-density lipoprotein stimulates endothelial cell migration and survival through sphingosine 1-phosphate and its receptors. *Arterioscler Thromb Vasc Biol,* **23,** 1283-1288.

Kluk MJ, Hla T. (2002) Signaling of sphingosine-1-phosphate via the S1P/EDG-family of G-protein-coupled receptors. *Biochim Biophys Acta,* **1582,** 72-80.

Kohama T, Olivera A, Edsall L, Nagiec MM, Dickson R, Spiegel S. (1998) Molecular cloning and functional characterization of murine sphingosine kinase. *J Biol Chem,* **273,** 23722-23728.

Kon J, Sato K, Watanabe T, Tomura H, Kuwabara A, Kimura T, et al. (1999) Comparison of intrinsic activities of the putative sphingosine 1-phosphate receptor subtypes to regulate several signaling pathways in their cDNA-transfected Chinese hamster ovary cells. *J Biol Chem,* **274,** 23940-23947.

Lee MJ, van Brocklyn JR, Thangada S, Liu CH, Hand AR, Menzeleev R, et al. (1998) Sphingosine-1-phosphate as a ligand for the G protein-coupled receptor EDG-1. *Science,* **279,** 1552-1555.

Lee MJ, Thangada S, Claffey KP, Ancellin N, Liu CH, Kluk M, et al. (1999) Vascular endothelial cell adherens junction assembly and morphogenesis induced by sphingosine-1-phosphate. *Cell,* **99,** 301-312.

Liu Y, Wada R, Yamashita T, Mi Y, Deng CX, Hobson JP, et al. (2000) Edg-1, the G protein-coupled receptor for sphingosine-1-phosphate, is essential for vascular maturation. *J Clin Invest*, **106**, 951-961.

Lynch KR. (2002) Lysophospholipid receptor nomenclature. *Biochim Biophys Acta*, **1582**, 70-71.

Mandala SM. (2001) Sphingosine-1-phosphate phosphatases. *Prostaglandins Other Lipid Mediat*, **64**, 143-156.

Marathe S, Schissel SL, Yellin MJ, Beatini N, Mintzer R, Williams KJ, Tabas I. (1998) Human vascular endothelial cells are a rich and regulatable source of secretory sphingomyelinase. Implications for early atherogenesis and ceramide-mediated cell signaling. *J Biol Chem*, **273**, 4081-4088.

Marcus AJ, Ullman HL, Safier LB. (1969) Lipid composition of subcellular particles of human blood platelets. *J Lipid Res*, **10**, 108-114.

Murata N, Sato K, Kon J, Tomura H, Yanagita M, Kuwabara A, et al. (2000) Interaction of sphingosine 1-phosphate with plasma components, including lipoproteins, regulates the lipid receptor-mediated actions. *Biochem J*, **352**, 809-815.

Ohmori T, Yatomi Y, Osada M, Kazama F, Takafuta T, Ikeda H, et al. (2003) Sphingosine 1-phosphate induces contraction of coronary artery smooth muscle cells via S1P2. *Cardiovasc Res*, **58**, 170-177.

Okajima F. (2002) Plasma lipoproteins behave as carriers of extracellular sphingosine 1-phosphate: is this an atherogenic mediator or an anti-atherogenic mediator? *Biochim Biophys Acta*, **1582**, 132-137.

Olivera A, Kohama T, Tu Z, Milstien S, Spiegel S. (1998) Purification and characterizaion of rat kidney sphingosine kinase. *J Biol Chem*, **273**, 12576-12583.

Ozaki H, Hla T, Lee MJ. (2003) Sphingosine-1-phosphate signaling in endothelial activation. *J Atheroscler Thromb*, **10**, 125-131.

Panetti TS. (2002) Differential effects of sphingosine 1-phosphate and lysophosphatidic acid on endothelial cells. *Biochim Biophys Acta*, **1582**, 190-196.

Pyne S, Pyne NJ. (2002) Sphingosine 1-phosphate signalling and termination at lipid phosphate receptors. *Biochim Biophys Acta*, **1582**, 121-131.

Riley RT, Enongene E, Voss KA, Norred WP, Meredith FI, Sharma RP, Spitsbergen J, Williams DE, Carlson DB, Merrill AH Jr. (2001) Sphingolipid perturbations as mechanisms for fumonisin carcinogenesis. *Environ Health Perspect*, **109**, 301-308.

Romiti E, Meacci E, Tani M, Nuti F, Farnararo M, Ito M, Bruni P. (2000) Neutral/alkaline and acid ceramidase activities are actively released by murine endothelial cells. *Biochem Biophys Res Commun*, **275**, 746-751.

Ruetz S, Gros P. (1994) Phosphatidylcholine translocase: a physiological role for the mdr2 gene. *Cell*, **77**, 1071-1081.

Sasaki M, Shoji A, Kubo Y, Nada S, Yamaguchi A. (2003) Cloning of rat ABCA7 and its preferential expression in platelets. *Biochem Biophys Res Commun*, **304**, 777-782.

Sciorra VA, Morris AJ. (2002) Roles for lipid phosphate phosphatases in regulation of cellular signaling. *Biochim Biophys Acta*, **1582**, 45-51.

Siess W. (2002) Athero- and thrombogenic actions of lysophosphatidic acid and sphingosine-1-phosphate. *Biochim Biophys Acta*, **1582**, 204-215.

Simon CG Jr, Chatterjee S, Gear AR. (1998) Sphingomyelinase activity in human platelets. *Thromb Res*, **90**, 155-161.

Spence MW, Byers DM, Palmer FB, Cook HW. (1989) A new Zn^{2+}-stimulated sphingomyelinase in fetal bovine serum. *J Biol Chem*, **264**, 5358-5363.

Spiegel S, Milstien S. (2002) Sphingosine 1-phosphate, a key cell signaling molecule. *J Biol Chem*, **277**, 25851-25854.

Spiegel S, Milstien S. (2003) Sphingosine-1-phosphate: an enigmatic signalling lipid. *Nat Rev Mol Cell Biol*, **4**, 397-407.

Stoffel W, Hellenbroich B, Heimann G. (1973) Properties and specificities of sphingosine kinase from blood platelets. *Hoppe-Seyler's Z Physiol Chem*, **354**, 1311-1316.

Takuwa Y. (2002) Subtype-specific differential regulation of Rho family G proteins and cell migration by the Edg family sphingosine-1-phosphate receptors. *Biochim Biophys Acta*, **1582**, 112-120.

Tani M, Sano T, Ito M, Igarashi Y. (2005) Mechanisms of sphingosine and sphingosine 1-phosphate generation in human platelets. *J Lipid Res*, **46**, 2458-2467

Tokumura A, Majima E, Kariya Y, Tominaga K, Kogure K, Yasuda K, et al. (2002) Identification of human plasma lysophospholipase D, a lysophosphatidic acid-producing enzyme, as autotaxin, a multifunctional phosphodiesterase. *J Biol Chem*, **277**, 39436-39442.

Umezu-Goto M, Kishi Y, Taira A, Hama K, Dohmae N, Takio K, et al. (2002) Autotaxin has lysophospholipase D activity leading to tumor cell growth and motility by lysophosphatidic acid production. *J Cell Biol*, **158**, 227-233.

Yatomi Y, Ruan F, Hakomori S, Igarashi Y. (1995) Sphingosine-1-phosphate: a platelet-activating sphingolipid released from agonist-stimulated human platelets. *Blood*, **86**, 193-202.

Yatomi Y, Igarashi Y, Yang L, Hisano N, Qi R, Asazuma N, et al. (1997a) Sphingosine 1-phosphate, a bioactive sphingolipid abundantly stored in platelets, is a normal constituent of human plasma and serum. *J Biochem*, **121**, 969-973.

Yatomi Y, Yamamura S, Ruan F, Igarashi Y. (1997b) Sphingosine 1-phosphate induces platelet activation through an extracellular action and shares a platelet surface receptor with lysophosphatidic acid. *J Biol Chem*, **272**, 5291-5297.

Yatomi Y, Ohmori T, Rile G, Kazama F, Okamoto H, Sano T, et al. (2000) Sphingosine 1-phosphate as a major bioactive lysophospholipid that is released from platelets and interacts with endothelial cells. *Blood*, **96**, 3431-3438.

Yatomi Y, Ozaki Y, Ohmori T, Igarashi Y. (2001) Sphingosine 1-phosphate: synthesis and release. *Prostaglandins Other Lipid Mediat*, **64**, 107-122.

Yatomi Y, Yamamura S, Hisano N, Nakahara K, Igarashi Y, Ozaki Y. (2004) Sphingosine 1-phosphate breakdown in platelets. *J Biochem*, **136**, 495-502.

Yoshimura Y, Tani M, Okino N, Iida H, Ito M. (2004) Molecular cloning and functional analysis of zebrafish neutral ceramidase. *J Biol Chem*, **279**, 44012-44022.

van Veldhoven PP, Foglesong RJ, Bell RM. (1989) A facile enzymatic synthesis of sphingosine-1-phosphate and dihydrosphingosine-1-phosphate. *J Lipid Res*, **30**, 611-616.

van Veldhoven PP, Mannaerts GP. (1993) Sphingosine-phosphate lyase. *Advances Lipid Res*, **26**, 69-98.

Part 7

Advanced Technology in Sphingolipid Biology

7-1 Conventional/Conditional Knockout Mice

Ichiro Miyoshi[1], Tadashi Okamura[2], Noriyuki Kasai[3],
Tetsuyuki Kitmoto[4], Shinichi Ichikawa[5], Soh Osuka[6],
and Yoshio Hirabayashi[6]

[1]Center for Experimental Animal Science, Nagoya City University Graduate School of Medical Sciences, Nagoya 467-8601, Japan, [2]Department of Infectious Diseases, Research Institute, International Medical Center of Japan, Tokyo 162-8655, Japan, [3]Institute for Animal Experimentation, Tohoku University Graduate School of Medicine, Sendai 980-8575, Japan, [4]Department of Neurological Science, Tohoku University Graduate School of Medicine, Sendai 980-8575, Japan, [5]Department of Applied Life Sciences, Niigata University of Pharmacy and Applied Life Sciences, Niigata 956-8603, Japan, [6]CREST and Neuronal Circuit Mechanism Research Group, RIKEN Brain Science Institute, Saitama 351-0198, Japan

Summary. Genetically modified mice, created using transgenic technologies and gene targeting, are essential tools for clinical and fundamental research. In sphinoglipid research this became possible when the genes encoding enzymes in the sphingolipid metabolism pathway were identified and cloned. Using these mouse models, we can study the functions of sphingolipids *in vivo*. However, some of these knockout (KO) strains die in early developmental stages. This indicates that the enzymes and their products, sphingolipids, play a vital role during development. Conditional targeting, using the Cre-loxP system, avoids fetal lethality via a tissue-specific disruption of an allele. Alternatively, introducing tissue-specific transgenes encoding an enzyme with similar activity to that of the disrupted gene prevents embryonic lethality in KO mice. We established gene-modified mice deficient in ceramide glucosyltransferase or serine palmitoyltransferase to gain new insights into the roles of sphingolipids *in vivo*. These are the enzymes that take the first committed steps in sphingolipid synthesis, which includes the synthesis of ceramide, glycosphingolipid, and sphingomyelin. We found that sphingolipids are essential not only for early embryo development, but also for the formation and maintenance of certain organs.

Keywords.   ceramide   glucosyltransferase,   conditional   knockout, gene-modified mice, serine palmitoyltransferase

1. Structure and biosynthesis of sphingolipid

Ceramide is the common structure for all sphingolipids and is synthesized from serine and palmitoyl CoA in four steps. Serine palmitoyltransferase (SPT) consists of two subunits (Sptlc1 and Sptlc2) and synthesizes 3-keto sphinganine, a precursor of ceramide, from serine and palmitoyl CoA (see Appendix). Galactose, glucose, and phosphocholine are transferred to ceramide   via   synthetic   pathways.   Ceramide   glucosyltransferase [UDP-glucose: ceramide ß-1-1' glucosyltransferase (Ugcg)] catalyzes the first glycosylation step in GSL biosynthesis, the transfer of glucose from UDP-glucose to ceramide. Both SPT and Ugcg are key enzymes in GSL biosynthesis, as their respective products, ceramide and glucosylceramide, form the core structure in hundreds of sphingolipids.

2. Study of sphingolipid function using cell lines

In vitro studies using mutant cell lines, such as Chinese hamster ovary cell-derived SPT-defective, temperature-sensitive mutant SPB-1 (Hanada et al. 1992), and B16 melanoma-derived Ugcg-defective GM95 (Ichikawa et al. 1994), were revealing and important. The former could not grow in sphingolipid-deficient medium and this was accompanied by the cessation of *de novo* sphingolipid synthesis. Sphingolipids may also be essential for cell growth and maintenance. Since exogenous sphingosine or sphingo-myelin restored sphingolipid synthesis and cell growth, the cells must have a system for taking up and utilizing sphingolipids from the medium. GM95 was independent of most GSLs: it could grow and maintain its biological ability in the absence of serum in the medium. GSLs may not be indis-pensable for cell growth *in vitro*, when the compensatory up-regulation of sphingomyelin biosynthesis can form a normal microdomain, as in the original cell line.

   Both subunits of SPT and Ugcg are highly conserved in various species, suggesting their essential biological function. Furthermore, there may be more complicated intercellular interaction among heterogeneous cells during development and differentiation than those seen in homogenous cultured cells.

3. Study of sphingolipid function using genetically altered mice

Cloning and characterization of the genes that encode the enzymes involved in the metabolism of sphingolipids (Taniguchi et al. 2001) has led to the development of genetically modified mouse models for studying the role of sphingolipids *in vivo*. It is likely that the diversity in GSL sugar chains is important in biological phenomena. To reveal the functions of GSLs, we established genetically modified mice that could not synthesize the majority of GSLs.

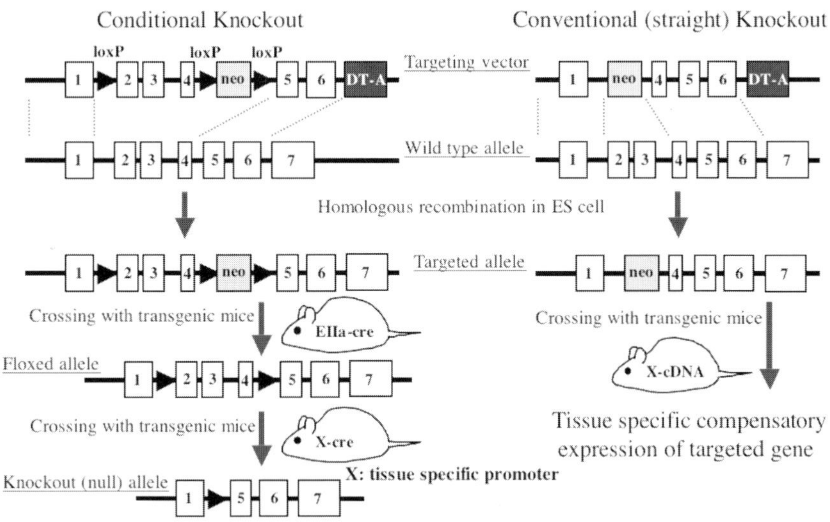

Fig. 1. Strategy for the conventional/conditional disruption of the gene.

3.1 Conditional knockout using the Cre-loxP system

Conventional KO mice have a null mutation introduced into a target allele by homologous recombination. Strains that suffer embryonic lethality are may indicate that the target allele has a vital role development and differentiation. However, embryonically lethal strains cannot undergo further analysis of the gene at different developmental stages. The Cre recombinase (Cre)-loxP system, which was discovered in P1 bacteriophage, was developed to circumvent these hindrances (Raul and Kühn 1997). Using this system, the target allele is flanked by loxP sites (floxed) using homologous recombination. The (floxed) genome sequence between the loxP sites can be deleted only when Cre is expressed and catalyzed under the regulation of its promoter (and enhancer) in a stage- and location-specific, conditional, manner (Fig.1). Cre can be introduced into conditional KO mice by crossing them with transgenic mice harboring a transgene consisting of the specific promoter and the *Cre* gene. The FLP (flipase)-FRT (flipase recombination target) system, which originated from yeast, is also utilized for conditional KO (Dymecki 2000).

By introducing a transgene encoding the corresponding enzyme into a KO mouse, we can establish a simple rescue system that allows for conditional compensation depending on the promoter in the transgene when compared to a conditional knockout. As explained below, either approach will facilitate analyzing the function of GSLs (Fig.1).

3.2 Disruption of the *Ugcg* gene and its rescue by introducing mouse prion promoter-driven human Ugcg

We produced Ugcg-deficient mice to investigate the function and biological roles of GSLs. As Ugcg is the key enzyme directing the first glycosylation step in the biosynthesis of GSLs, deficiency in this enzyme activity results in failure of the biosynthesis of the majority of GSLs. The disruption of exon 6-8 of the *Ugcg* gene resulted in embryonic lethality at around E7.5 during gastrulation, and pathological analyses identified the reproducible phenotype, as seen in KO mice through disruption of exon 7 of the *Ugcg* gene (Yamashita *et al.* 1999). Embryos appeared to die from apoptosis of the ectodermal cells caused by a lack of Ugcg activity.

GSLs are distributed ubiquitously in most animal tissues and are especially abundant in the nervous system. Therefore, we believe that GSLs are also important for the survival of various tissues after birth. Therefore, we attempted to rescue the Ugcg KO mice from embryonic lethality by introducing a transgene composed of human Ugcg cDNA (hUgcg) and the ß-actin/CMV (CAG) (Niwa et al. 1991) or mouse prion (Prp) promoter (Kitamoto *et al.* 2002). The former promoter directs ubiquitous expression as a control, and the latter regulates the strong expression in neurons and cells derived from the neural crest, in addition to most cells at an early developmental stage.

The introduction of the CAG-Ugcg transgene by crossbreeding with transgenic mice rescued the Ugcg KO mice, which appeared normal throughout their life cycle. They became viable, fertile adults (T Okamura unpublished data).

Through crossbreeding with Prph-Ugcg transgenic mice, the compensatory expression of Ugcg from the transgene successfully rescued Ugcg KO mice from early embryonic death. Genotype analysis of newborns from mating between Tg/Tg $ugcg^{+/-}$ parents showed that the rescued $ugcg^{-/-}$ (Tg/Tg $ugcg^{-/-}$) offspring were obtained in the expected Mendelian ratio and appeared normal at birth. However, 51% of the Tg/Tg $ugcg^{-/-}$ neonatal mice died by postnatal day 3 (P3), and all had died by P13. Tg/Tg $ugcg^{-/-}$ mice could be distinguished from Tg/Tg $ugcg^{+/+}$ or Tg/Tg $ugcg^{+/-}$ littermates by their smaller size. An abnormal arrangement of enterocytes and a marked increase in proliferative cells were observed in the prospective crypts of the developing intestine of Tg/Tg $ugcg^{-/-}$ mice. Neither the Ugcg epitope nor GM1 ganglioside was detected in the proliferative cells of Tg/Tg $ugcg^{-/-}$ crypts, using specific antibody or the cholera toxin-B subunit, respectively. These results suggest that the synthesis of GSLs is indispensable not only for embryonic development, but also for the development and maintenance of intestinal organization, although the precise mechanism remains to be elucidated (T Okamura unpublished data). Rescued $Ugcg^{-/-}$ can be used to study the biological roles of GSLs. Interestingly, stem cells in the intestinal crypts showed ceramide synthase-mediated apoptosis in response to certain stresses, such as irradiation (Ch'ang et al. 2005). Therefore, expression of this enzyme may be regulated in a more precisely than previously assumed.

3.3 Conditional knockout by disruption of the Sptlc2 gene

We established Sptlc2 flox mice and then crossbred them with transgenic mice carrying a transgene composed of the *Cre* gene under the regulation of the CAG, lymphocyte-specific protein tyrosine kinase (lck), or keratin 5 (K5) promoters (S Osuka unpublished data). As expected, global disruption of Sptlc2 by CAG-Cre caused embryo lethality at a very early embryonic stage, *i.e.*, before E6.5, suggesting that SPT expression is required at an earlier developmental stage than that of Ugcg.

Specific disruption of floxed Sptlc2 in either thymocytes (T cells) or epidermal cells was produced by crossing Sptlc2 flox mice with lck-Cre or K5-Cre transgenic mice, respectively. Unexpectedly, the cells lacking SPT activity were still viable in either tissue, showing the maintenance of organ structure and some sphingolipids, which may have been obtained from the surroundings, although some malfunctions were observed in each tissue. The exogenous supplementation of small molecules, such as sphingolipids, makes an *in vivo* study complex. However, the involvement of the products and transcripts compensates for this in cells that proliferate at a lower rate. The disruption of *SPT* genes blocks the synthetic pathway downstream and, to some extent, the storage of precursor molecules. Therefore, we should consider the cause of any phenotypic change from both perspectives.

4. Current progress in the study of sphingolipid function using gene-modified mice

As GSLs are abundant and extremely diverse structurally in the brain and nervous tissue, we expected them to play a critical role in neural function. Recently, the function of sphingolipids has been studied using gene-modified mice in which the glycosyltransferase gene is disrupted by targeting. Mice that cannot synthesize complex gangliosides have unusual phenotypes, including abnormal morphologies and functional maintenance of the nervous system (Takamiya et al. 1996), a lower capability for neuroregeneration (Okada et al. 2002), neurological impairment including neurodegeneration (Inoue et al. 2002, Yamashita et al. 2005), and improper spermatogenesis (Takamiya et al. 1998, Sandhoff et al. 2005). Recently, we generated conditional KO mice devoid of glucosylceramide synthesis specifically in nervous tissues by crossing Ugcg flox mice and transgenic mice carrying the nestin promoter-driven Cre transgene. The mice had shortened life spans and cerebellar dysfunction, i.e. loss of Purkinje cells, suggesting that GSLs are not be essential for early brain development, but play a pivotal role in neuron differentiation and brain maturation after birth (Jennemann et al. 2005). A deficiency of GM3 synthase (GM3S) enhanced insulin signaling in the knockout mice, although the mice grew and reproduced normally (Yamashita et al. 2003).

Interestingly, GSLs may do more than simply participate in a synthetic pathway as either a substrate or product. The exogenous addition of glucosylceramide rapidly restores the abnormal morphology of zebra fish, in which Ugcg was inhibited using the knockdown method (see the chapter by M. Ito). Furthermore, glucosylceramide is its own elicitor in some plants (Umemura et al. 2002).

The first report of a disruption in GM3S human homologue that led to disease followed the establishment of GM3S knockout mice. This homozygous mutant of GM3S causes infantile-onset symptomatic epilepsy syndrome in humans, where the knockout mice show enhanced sensitivity to insulin with no obvious seizure activity or a shortened life span (Simpson et al. 2004). Perhaps understanding such differences among species may be a key to unraveling the biological functions of GSLs. A mutation of Sptlc1 was found to be responsible for hereditary sensory neuropathy type 1 (HSN1), although conditional Sptlc2 KO mice with a global disruption allele were embryo-lethal. It is likely that the neurodegeneration that occurs in HSN1 is due to apoptotic cell death associated with increased de novo glucosylceramide synthesis (Bejaoui et al. 2001, Dawkins et al. 2001). Deficiency of the enzyme activity involved in the synthetic pathway enabled us to analyze the roles of GSLs by observing interesting and unexpected phenotypes, in addition to their roles in the breakdown pathway. In order to apply this excellent model to elucidating GSL function, precise profiling of GSL expression at the cellular level is required.

5. References

Bejaoui K, Wu C, Scheffler MD, Haan G, Ashby P, Wu L, de Jong P, Brown RH Jr (2001) SPTLC1 is mutated in hereditary sensory neuropathy, type 1. *Nat Genet*, **27**, 261-262.

Ch'ang HJ, Maj JG, Paris F, Xing HR, Zhang J, Truman JP, Cardon-Cardo C, Haimovitz-Friedman A, Kolesnick R, Fuks Z (2005) ATM regulates target switching to escalating doses of radiation in the intestines. *Nat Med*, **11**, 484-490.

Dawkins JL, Hulme DJ, Brahmbhatt SB, Auer-Grumbach M, Nicholson GA (2001) Mutations in SPTLC1, encoding serine palmitoyltransferase, long chain base subunit-1, cause hereditary sensory neuropathy type I. *Nat Genet*, **27**, 309-312.

Dymecki SM (2000) in *Gene Targeting: A practical approach* (Joyner AL, Ed.), pp37-99, Oxford University Press, Oxford.

Hanada K, Nishijima M, Kiso M, Hasegawa A, Fujita S, Ogawa T, Akamatsu Y (1992) Sphingolipids are essential for the growth of Chinese hamster ovary cells. Restoration of the growth of a mutant defective in sphingoid base biosynthesis by exogenous sphingolipids. *J Biol Chem*, **267**, 23527-23533.

Ichikawa S, Nakajo N, Sakiyama H, Hirabayashi Y (1994) A mouse B16 melanoma mutant deficient in glycolipids. *Proc Natl Acad Sci USA*, **91**, 2703-2707.

Inoue M, Fujii Y, Furukawa K, Okada M, Okumura K, Hayakawa T, Furukawa K, Sugiura Y (2002) Refractory skin injury in complex knock-out mice expressing only the GM3 ganglioside. *J Biol Chem*, **277**, 29881-29888.

Jennemann R, Sandhoff R, Wang S, Kiss E, Gretz N, Zuliani C, Martin-Villalba A, Jager R, Schorle H, Kenzelmann M, Bonrouhi M, Wiegandt H, Grone HJ (2005) Cell-specific deletion of glucosylceramide synthase in brain leads to severe neural defects after birth. *Proc Natl Acad Sci USA*, **102**, 12459-12464.

Kitamoto T, Mohri S, Ironside JW, Miyoshi I, Tanaka T, Kitamoto N, Itohara S, Kasai N, Katsuki M, Higuchi J, Muramoto T, Shin RW (2002) Follicular dendritic cell of the knock-in mouse provides a new bioassay for human prions. *Biochem Biophys Res Commun*, **294**, 280-286.

Niwa H, Yamamura K Miyazaki J (1991) Efficient selection for high-expression transfectants with a novel eukaryotic vector. *Gene*, **108**, 193-199.

Okada M, Itoh Mi M, Haraguchi M, Okajima T, Inoue M, Oishi H, Matsuda Y, Iwamoto T, Kawano T, Fukumoto S, Miyazaki H, Furukawa K, Aizawa S, Furukawa K (2002) b-series Ganglioside deficiency exhibits no definite changes in the neurogenesis and the sensitivity to Fas-mediated apoptosis but impairs regeneration of the lesioned hypoglossal nerve. *J Biol Chem*, **277**, 1633-1636.

Raul MT, Kühn R (1997) in *Laboratory Protocols for Conditional Gene Targeting*, Oxford University Press, Oxford.

Sandhoff R, Geyer R, Jennemann R, Paret C, Kiss E, Yamashita T, Gorgas K, Sijmonsma TP, Iwamori M, Finaz C, Proia RL, Wiegandt H, Grone HJ (2005) Novel class of glycosphingolipids involved in male fertility. *J Biol Chem*, **280**, 27310-27318.

Simpson MA, Cross H, Proukakis C, Priestman DA, Neville DC, Reinkensmeier G, Wang H, Wiznitzer M, Gurtz K, Verganelaki A, Pryde A, Patton MA, Dwek RA, Butters TD, Platt FM, Crosby AH (2004) Infantile-onset symptomatic epilepsy syndrome caused by a homozygous loss-of-function mutation of GM3 synthase. *Nat Genet*, **36**, 1225-1229.

Takamiya K, Yamamoto A, Furukawa K, Yamashiro S, Shin M, Okada M, Fukumoto S, Haraguchi M, Takeda N, Fujimura K, Sakae M, Kishikawa M, Shiku H, Furukawa K, Aizawa S (1996) Mice with disrupted GM2/GD2 synthase gene lack complex gangliosides but exhibit only subtle defects in their nervous system. *Proc Natl Acad Sci USA*, **93**, 10662-10667.

Takamiya K, Yamamoto A, Furukawa K, Zhao J, Fukumoto S, Yamashiro S, Okada M, Haraguchi M, Shin M, Kishikawa M, Shiku H, Aizawa S, Furukawa K (1998) Complex gangliosides are essential in spermatogenesis of mice: possible roles in the transport of testosterone. *Proc Natl Acad Sci USA*, **95**, 12147-12152.

Taniguchi N, Honke K and Fukuda M (2001) in *Handbook of Glycosyltransferase and Related Genes*, Springer Verlag Tokyo.

Umemura K, Ogawa N, Koga J, Iwata M, Usami H (2002) Elicitor activity of cerebroside, a sphingolipid elicitor, in cell suspension cultures of rice. *Plant Cell Physiol*, **43**, 778-784.

Yamashita T, Wada R, Sasaki T, Deng C, Bierfreund U, Sandhoff K, Proia RL (1999) A vital role for glycosphingolipid synthesis during development and differentiation. *Proc Natl Acad Sci USA*, **96**, 9142-9147.

Yamashita T, Hashiramoto A, Haluzik M, Mizukami H, Beck S, Norton A, Kono M, Tsuji S, Daniotti JL, Werth N, Sandhoff R, Sandhoff K, Proia RL (2003) Enhanced insulin sensitivity in mice lacking ganglioside GM3. *Proc Natl Acad Sci USA*, **100**, 3445-3449.

Yamashita T, Wu YP, Sandhoff R, Werth N, Mizukami H, Ellis JM, Dupree JL, Geyer R, Sandhoff K, Proia RL (2005) Interruption of ganglioside synthesis produces central nervous system degeneration and altered axon-glial interactions. *Proc Natl Acad Sci USA*, **102**, 2725-2730.

7-2 Biosynthesis and Function of *Drosophila* Glycosphingolipids

Ayako Koganeya-Kohyama and Yoshio Hirabayashi

Hirabayashi Laboratory Unit, Brain Science Institute, RIKEN, 2-1 Hirosawa, Wako, Saitama 351-0198, Japan

Summary. Glycosphingolipids (GSLs) are primarily located on the outer leaflet of the cellular membrane where they participate in the assembly of signaling molecules, modulation of cell adhesion, and cellular differentiation. Intracellular GSLs are also important for membrane protein trafficking. *Drosophila melanogaster* is emerging as one of the most effective systems for analyzing the fundamental mechanisms underlying development and human diseases. Recently, approaches directed toward studying GSLs in *Drosophila* have been initiated, providing an excellent opportunity to understand the *in vivo* functions of GSLs.This review gives an overview of GSL biosynthesis in *Drosophila* and summarizes the functions of the enzymes that participate in GSL biosynthesis.

Keywords. glucosylceramide, glucosyltransferase, apoptosis, *Drosophila melanogaster*

1. Introduction: *Drosophila* is a powerful model organism for analyzing sphingolipids

1.1 Drosophila melanogaster as an in vivo model

Many of genetic pathways that guide basic developmental processes in *Drosophila* and in vertebrates have remained largely intact throughout evolution. This conservation enabled rapid analysis of the developmental processes in vertebrates, as insights gained from the model genetics organism *Drosophila* can be applied immediately to vertebrates. In many cases, *Drosophila* genes can functionally replace counterparts in other organisms, including vertebrates. Furthermore, *Drosophila* serves as an excellent intermediate model system, filling the niche between unicellular organisms such as yeast and slime mold (*Dichtyoctelium discoideum*), which are ideal for studying cell autonomous eukaryotic functions (e.g., DNA repair and cell division), and vertebrate systems such as mice and zebrafish, which can be used as accurate models for humans.

1.2 *Drosophila* is a highly tractable genetic system

There are five advantages to using *Drosophila* to study developmental processes and human diseases:
  (1) the complete genomic sequence of *Drosophila* is known;
  (2) a vast collection of *Drosophila* gene mutations already exists;
  (3) new mutations can be easily generated;
  (4) convenient techniques exist for specifically inactivating or expressing *Drosophila* genes in dispensable tissues that can be easily examined and measured; and
  (5) second-site modifier genetics can be readily carried out (Huang and Rubin, 2000).
Drosophila genetics remains a powerful tool for broadening our understanding of organism development and genetic pathways.

The fruit of nearly a century's worth of *Drosophila* genetics research includes a huge collection of mutant fly lines and new lines are systematically generated (Rorth *et al.*, 1998; Spradling *et al.*, 1999). In addition, RNA interference (RNAi) technology provides an attractive alternative method for establishing loss-of-function mutants (Nishihara *et al.*, 2004; Gunsalus & Piano, 2005). These mutant collections include various allelic mutants (e.g., hypomorphic, neomorphic), allowing us to understand the phenotypic consequences of reducing the activity of a specific gene. Furthermore, techniques for stably overexpressing or knocking out specific genes in a specific tissue or group of cells in flies are well established. For example, site-specific recombination of *Drosophila* genes can be easily carried out using the yeast transcription factor Gal4/upstream activating sequence (GAL4/UAS) system (Fig. 1) (Brand & Perrimon, 1993) or the yeast recombinase-FLP recombinase target (FLP-FRT) system (Xu & Rubin, 1993). With these methods, researchers have generated many *Drosophila* mutant homologues of vertebrate genes. Finally, transgenic constructs can be readily introduced into flies that cause genes of interest to misexpress along spatial- and temporal-specific patterns.

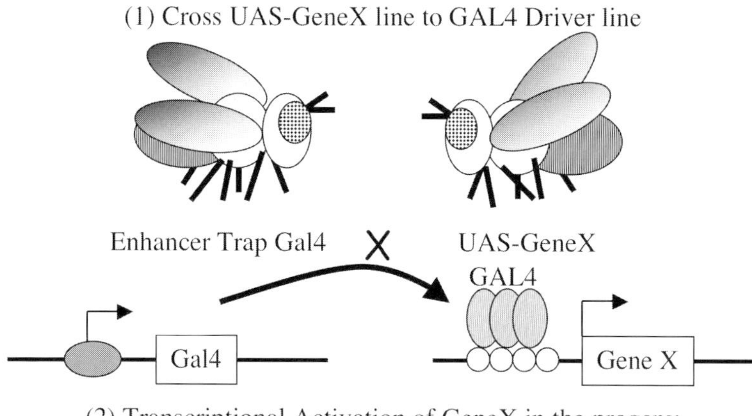

Fig. 1. The UAS/GAL4 expression system. The Gal4-driver fly has a transgene containing the yeast transcription factor Gal4, which is expressed by a tissue-specific promoter. The UAS-cDNA fly has a transgene containing a cDNA that is ligated to the UAS promoter, a target of GAL4. The GAL4-driver and UAS-cDNA transgenic fly lines are mated, and the F1 progeny are used for phenotypic analysis.

2. Glycosphingolipid pathway in *Drosophila*

"GSLs have been characterized in a broad range of animal phyla and have been demonstrated to participate in important cellular and developmental functions" (Hakomori, 1990, 2000; Seppo *et al.,* 2000; Furukawa *et al.,* 2004; Haltiwanger & Lowe, 2004).

As with mammals, *Drosophila* possesses sphingolipids—both simple sphingolipids and complex GSLs. Both zwitterionic and acidic GSLs have been detected in *Drosophila* (Seppo and Tiemeyer, 2000; Seppo *et al.,* 2000). Zwitterionic GSLs contain phosphoethanolamine linked to N-acetylglucosamine (GlcNAc) residues, whereas acidic GSLs contain phosphoethanolamine and glucuronic acid or phosphoethanolamine and hexose and N-acetylhexossamine (Seppo et al., 2000). The most abundant GSLs in *Drosophila* are Glcβ1-ceramide (GlcCer) and Manβ1-4Glcβ1-ceramide (MacCer) (Rietveld *et al.,* 1999).

Although there are considerable differences in the composition of GSLs in *Drosophila* compared to that in vertebrates, the GSLs in both invertebrates and vertebrates share a common element consisting of GlcCer (Fig. 2). *Drosophila* GSLs are derived from a MacCer core, while the majority of vertebrate GSLs are from a Galβ1-4Glcβ1-ceramide (LacCer) core. In *Drosophila,* as well as in other invertebrates, GSL elongation is initiated by the addition of a mannose residue to GlcCer followed by the addition of GlcNAc and additional monosaccharide residues. Two *Drosophila* genes, *egghead* and *brainiac*, encode distinct glycosyltransferases that catalyze these two initial steps of the GSL elongation cascade. This first elongation step is unique to *Drosophila* (invertebrates), since in vertebrates, GSL elongation is initiated by the addition of a galactose residue to form Lac-Cer (Fig. 2)

To date, three genes (*DGlcT-1, egghead,* and *brainiac*) have been analyzed in *Drosophila* that encode key enzymes involved in GSL biosynthesis. These enzymes play important roles in development, differentiation, and apoptosis (Kohyama-Koganeya *et al.*, 2004; Wandall *et al.*, 2005).

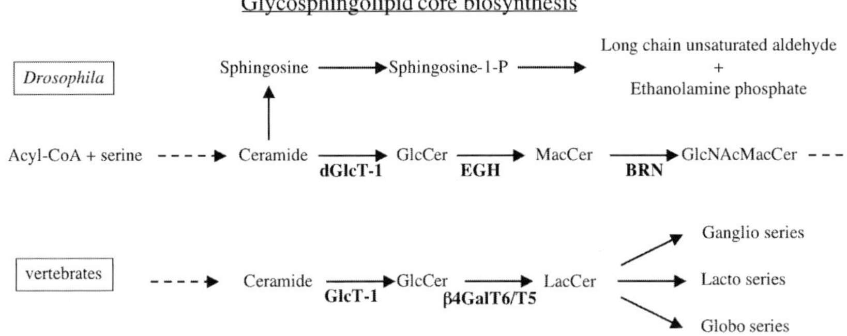

Fig. 2. Comparison of the glycosphingolipid (GSL) core biosynthesis pathway of *Drosophila* and humans. GlcCer is a common precursor of most extended GSLs.

3. Glucosylceramide synthase (GlcT-1/GCS/UGCG)

Glucosylceramide synthase (EC 2.4.1.80) transfers UDP-glucose to ceramide (Cer) to form GlcCer, the core structure of the majority of GSLs (Degroote *et al.,* 2004). Human GlcT-1 was originally cloned from human and mouse cell lines (Ichikawa *et al.*, 1996). To date, GlcT-1 has been cloned from or identified in fish, *Drosophila*, fungi, plants, and bacteria (Leipelt *et al.*, 2001; Kohyama-Koganeya *et al.*, 2004). All species share a common element consisting of GlcCer synthase, suggesting that GlcT-1 and its product, GlcCer, have conserved functions. Indeed, GlcT-1–knockout mice and flies display a similar phenotype: embryonic lethality and excessive cell death in the neuroectoderm (Yamashita *et al.*, 1999; Kohyama-Koganeya *et al.*, 2004).

The *Drosophila* GlcT-1 (DGlcT-1) gene was identified through an exhaustive TBLASTN search of the *Drosophila* genome sequence database. (Kohyama-Koganeya *et al.*, 2004) DGlcT-1 mRNA is expressed ubiquitously throughout development, suggesting that DGlcT-1 is important for development and differentiation. Using RNAi techniques, we showed that a loss of GlcT-1 function enhances apoptotic cell death in the neuroectoderm. This pattern of apoptosis is similar to the pattern we observed in GlcT-1–knockout mice. The enhanced apoptosis in our RNAi model and in the GlcT-1–knockout mice may be due to elevated Cer levels. Whether elevated Cer levels are responsible for apoptosis *in vivo* remains to be determined. GlcCer and GSLs derived from GlcCer may also protect cells from undergoing apoptosis.

Analysis of DGlcT-1 revealed that it functions in major apoptosis signaling pathways. Targeted-expression of GlcT-1 through a UAS/GAL4 system (Fig. 1) partially rescues cell death caused by the pro-apoptotic factors Reaper and Grim (Fig. 3). GlcT-1 has very little effect, however, on cell death caused by Hid, another pro-apoptotic factor whose regulation and function are different from those of Reaper and Grim (Kohyama-Koganeya *et al.*, 2004). These results indicate that DGlcT-1 is an important mediator of apoptosis *in vivo* and provide evidence that the (glyco)sphingolipid biosynthetic pathway genetically interacts with specific cell death pathways.

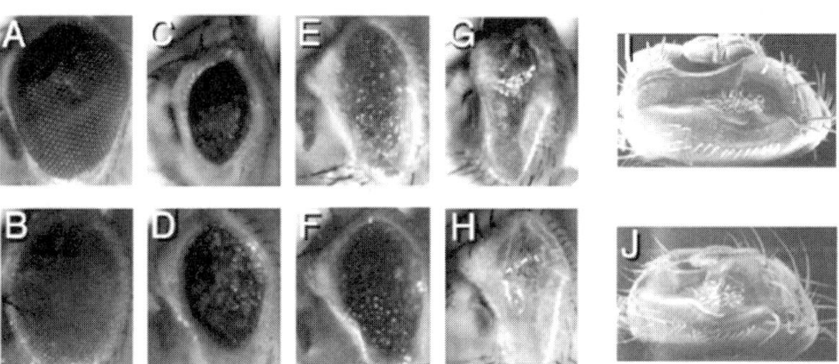

Fig. 3. Suppression of RPR- or GRIM-induced cell death by DGlcT-1. Using the UAS/GAL4expression system, we induced apoptosis with RPR and GRIM. Apoptosis was manifested as extremely rough, small eyes. Co-expression of DGlcT-1with either RPR or GRIM partially suppressed the small-eye phenotype (D and F). DGlcT-1, however, did not rescue HID-induced cell death (G-J).

4. *Egghead* (*egh*) and *brainiac* (*brn*) define lipid glycosyltransferase functions

The *Drosophila* genes *egghead* (*egh*) and *brainiac* (*brn*) were originally identified as neurogenic genes (Goode et al., 1992, 1996a, 1996b). Flies harboring *egh* and *brn* mutations display defects that resemble those produced by loss of Notch function during oogenesis. Consequently, *egh* and *brn* are considered possible modulators of Notch activity. Both *egh* and *brn* mutants display similar, non-additive defects, suggesting that egh and brn proteins act in the same pathway (Goode *et al.*, 1996a).

Recently, *egh* has been shown to encode a GDP-Mannose βGlc β1,4-mannosyltransferase, which is responsible for synthesizing MacCer (Wandall *et al.*, 2003). *brn* encodes a UDP-N-acetlyglucosamine: βMan β1,3-N-acetylglucosaminyltransferase that synthesizes GlcNAcMacCer, the core structure of invertebrate GSLs (Muller *et al.*, 2002; Schwientek *et al.*, 2002). Loss of either gene abolishes GSL biosynthesis. This is consistent with earlier observations by Goode and colleagues (1996a), suggesting that *egh* and *brn* act within the same pathway. Interestingly, *egh* and *brn* mutants lack elongated GSLs and accumulate truncated precursor GSLs (Wandall *et al.*, 2005). Moreover, Wandall and colleagues (2005) also showed that *brn* mutant clones generated by the FRP-FRT system accumulate MacCer.

Despite considerable structural differences between the GSLs of *Drosophila* and vertebrates, homologous glycosyltransferase genes that encode enzymes that catalyze most GSL biosynthetic steps remain conserved (e.g., GlcT-1, GSL-synthesizing enzymes, and glycan-synthesizing and processing enzymes) (Keusch *et al.*, 2000). At present, our understanding is that egghead is the only exception—an *egh* homologue is yet to be found in vertebrates. Interestingly, *Drosophila egh* mutants can be rescued when mammalian LacCer biosynthetic pathway is introduced into the mutant flies using human β4-galactosyltransferase (β4Gal-T6) (Wandall *et al.*, 2005). The observation that brainiac protein can use LacCer as a substrate, while vertebrate orthologs appear to act only on LacCer (Schwientek *et al.*, 2002), suggests that *Drosophila* GSLs do not require a specific core structure in order to function properly during development.

Drosophila that lack zygotic *egh* and *brn* activity die at the pupal stage. *Drosophila* that lack both maternal and zygotic activities of these genes are devoid of elongated GSLs. These flies die during embryonic stages and have severe defectives in neural and epidermal cell type specification. The abnormal phenotype of these mutants demonstrates that elongated GSLs are required for normal development.

5. Conclusion

Drosophila and vertebrate GSLs share a common element consisting of GlcCer. Because of differential glycosyltransferase-mediated modification of GlcCer, *Drosophila* and vertebrate GSLs differ markedly in structure and complexity. Although these structural differences may reflect alternative functions, these are yet to be determined. Phenotypic characterization and genetic epistatic analysis of the enzymes involved in GSL biosynthesis have revealed that *Drosophila* is an excellent model system for the *in vivo* study of GSLs (Fig. 4). Furthermore, we already know that most homologous glycosyltransferases that synthesize GSLs are conserved in *Drosophila*. Recent studies of basic sphingolipid metabolic pathways in *Drosophila* have also demonstrated new and/or conserved functions for sphingolipids (Adachi-Yamada et. al., 1999; Rohrbough et al., 2002; Acharya et al., 2003; Herr et al., 2003, 2004; Acharya and Acharya, 2005). *Drosophila* will provide a foundation for investigating mechanisms that regulate GSL expression and function, thereby enabling us to understand the roles of complex GSLs in several biological processes.

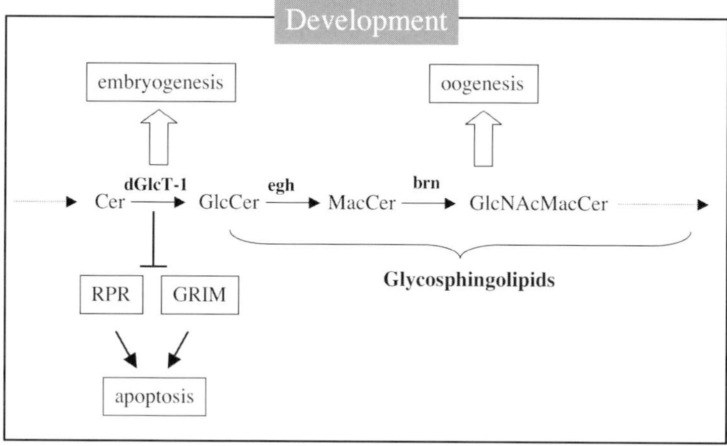

Fig. 4. Schematic diagram summarizing the functions of GSLs in *Drosophila*.

6. References

Acharya, U. and Acharya, J.K. (2005) Enzymes of sphingolipid metabolism in Drosophila melanogaster. Cell Mol Life Sci, 62, 128-142.

Acharya, U., Mowen, M.B., et al. (2004) Ceramidase expression facilitates membrane turnover and endocytosis of rhodopsin in photoreceptors. Proc Natl Acad Sci U S A, 101, 1922-1926. Epub 2004 Feb 1929.

Acharya, U., Patel, S., et al. (2003) Modulating sphingolipid biosynthetic pathway rescues photoreceptor degeneration. Science, 299, 1740-1743.

Adachi-Yamada, T., Gotoh, T., et al. (1999) De novo synthesis of sphingolipids is required for cell survival by down-regulating c-Jun N-terminal kinase in Drosophila imaginal discs. Mol Cell Biol, 19, 7276-7286.

Brand, A.H. and Perrimon, N. (1993) Targeted gene expression as a means of altering cell fates and generating dominant phenotypes. Development, 118, 401-415.

Degroote, S., Wolthoorn, J., et al. (2004) The cell biology of glycosphingolipids. Semin Cell Dev Biol, 15, 375-387.

Furukawa, K., Tokuda, N., et al. (2004) Glycosphingolipids in engineered mice: insights into function. Semin Cell Dev Biol, 15, 389-396.

Goode, S., Melnick, M., et al. (1996a) The neurogenic genes egghead and brainiac define a novel signaling pathway essential for epithelial morphogenesis during Drosophila oogenesis. Development, 122, 3863-3879.

Goode, S., Morgan, M., et al. (1996b) Brainiac encodes a novel, putative secreted protein that cooperates with Grk TGF alpha in the genesis of the follicular epithelium. Dev Biol, 178, 35-50.

Goode, S., Wright, D., et al. (1992) The neurogenic locus brainiac cooperates with the Drosophila EGF receptor to establish the ovarian follicle and to determine its dorsal-ventral polarity. Development, 116, 177-192.

Gunsalus, K.C. and Piano, F. (2005) RNAi as a tool to study cell biology: building the genome-phenome bridge. Curr Opin Cell Biol, 17, 3-8.

Hakomori, S. (1990) Bifunctional role of glycosphingolipids. Modulators for transmembrane signaling and mediators for cellular interactions. J Biol Chem, 265, 18713-18716.

Hakomori, S. (2000) Traveling for the glycosphingolipid path. Glycoconj J, 17, 627-647.

Haltiwanger, R.S. and Lowe, J.B. (2004) Role of glycosylation in development. Annu Rev Biochem, 73, 491-537.

Herr, D.R., Fyrst, H., et al. (2004) Characterization of the Drosophila sphingosine kinases and requirement for Sk2 in normal reproductive function. J Biol Chem, 279, 12685-12694. Epub 12004 Jan 12613.

Herr, D.R., Fyrst, H., et al. (2003) Sply regulation of sphingolipid signaling molecules is essential for Drosophila development. Development, 130, 2443-2453.

Huang, AM. and Rubin, GM. (2000) A misexpression screen identifies genes that can modulate RAS1 pathway signaling in Drosophila melanogaster. Genetics. 2000 Nov;156(3):1219-30.

Ichikawa, S., Sakiyama, H., et al. (1996) Expression cloning of a cDNA for human ceramide glucosyltransferase that catalyzes the first glycosylation step of glycosphingolipid synthesis. Proc Natl Acad Sci U S A, 93, 4638-4643.

Keusch, J.J., Manzella, S.M., et al. (2000) Cloning of Gb3 synthase, the key enzyme in globo-series glycosphingolipid synthesis, predicts a family of alpha 1, 4-glycosyltransferases conserved in plants, insects, and mammals. J Biol Chem, 275, 25315-25321.

Kohyama-Koganeya, A., Sasamura, T., et al. (2004) Drosophila glucosylceramide synthase: a negative regulator of cell death mediated by proapoptotic factors. J Biol Chem, 279, 35995-36002. Epub 32004 Jun 35921.

Leipelt, M., Warnecke, D., et al. (2001) Glucosylceramide synthases, a gene family responsible for the biosynthesis of glucosphingolipids in animals, plants, and fungi. J Biol Chem, 276, 33621-33629. Epub 32001 Jul 33626.

Muller, R., Altmann, F., et al. (2002) The Drosophila melanogaster brainiac protein is a glycolipid-specific beta 1,3N-acetylglucosaminyltransferase. J Biol Chem, 277, 32417-32420. Epub 32002 Jul 32418.

Nishihara, S., Ueda, R., et al. (2004) Approach for functional analysis of glycan using RNA interference. Glycoconj J, 21, 63-68.

Rietveld, A., Neutz, S., et al. (1999) Association of sterol- and glycosylphosphatidylinositol-linked proteins with Drosophila raft lipid microdomains. J Biol Chem, 274, 12049-12054.

Rohrbough, J., Rushton, E., et al. (2004) Ceramidase regulates synaptic vesicle exocytosis and trafficking. J Neurosci, 24, 7789-7803.

Rorth, P., Szabo, K., et al. (1998) Systematic gain-of-function genetics in Drosophila. Development, 125, 1049-1057.

Schwientek, T., Keck, B., et al. (2002) The Drosophila gene brainiac encodes a glycosyltransferase putatively involved in glycosphingolipid synthesis. J Biol Chem, 277, 32421-32429. Epub 32002 Jul 32418.

Seppo, A., Moreland, M., et al. (2000) Zwitterionic and acidic glycosphingolipids of the Drosophila melanogaster embryo. Eur J Biochem, 267, 3549-3558.

Seppo, A. and Tiemeyer, M. (2000) Function and structure of Drosophila glycans. Glycobiology, 10, 751-760.

Spradling, A.C., Stern, D., et al. (1999) The Berkeley Drosophila Genome Project gene disruption project: Single P-element insertions mutating 25% of vital Drosophila genes. Genetics, 153, 135-177.

Wandall, H.H., Pedersen, J.W., et al. (2003) Drosophila egghead encodes a beta 1,4-mannosyltransferase predicted to form the immediate precursor glycosphingolipid substrate for brainiac. J Biol Chem, 278, 1411-1414. Epub 2002 Nov 1425.

Wandall, H.H., Pizette, S., et al. (2005) Egghead and brainiac are essential for glycosphingolipid biosynthesis in vivo. J Biol Chem, 280, 4858-4863. Epub 2004 Dec 4815.

Xu, T. and Rubin, G.M. (1993) Analysis of genetic mosaics in developing and adult Drosophila tissues. Development, 117, 1223-1237.

Yamashita, T., Wada, R., et al. (1999) A vital role for glycosphingolipid synthesis during development and differentiation. Proc Natl Acad Sci U S A, 96, 9142-9147.

7-3 Characterization of Genes Conferring Resistance Against ISP-1/Myriocin-Induced Sphingolipid Depletion in Yeast

Hiromu Takematsu [1,2] and Yasunori Kozutsumi [1,2,3]

[1]Laboratory of Membrane Biochemistry and Biophysics, Graduate School of Biostudies, Kyoto University, Kyoto 606-8501, Japan, [2]CREST, JST, Kawaguchi, Saitama, Japan, and [3]Supra-biomolecular System Research Group, Frontier Research System, RIKEN, Saitama 351-0198, Japan

Summary. ISP-1 is a naturally occurring immunosuppressant with a structure resembling that of sphingosine. ISP-1 inhibits serine palmitoyl-transferase, the primary enzyme of sphingolipid biosynthesis and reduces intracellular sphingolipid, leading to cell death in mammalian cells and in yeast cells. To determine the mechanism underlying cell death induced by sphingolipid deprivation, ISP-1 resistant genes were isolated and characterized in yeast. Two genes were found to be tightly associated with sphingolipid signaling.

Keywords. ISP-1, serine palmitoyltransferase, myriocin, Ypk1, Pkh1, Mss4, Sli1, sphingolipid signaling

1. Introduction

Fungi that infect insects are widely used in Chinese traditional herbal medicine. A caterpillar fungus that is parasitic on moth larva, in particular, has long been used in tonics. ISP-1 was isolated from one such fungi, *Isaria sinclairii* (ATCC 24400), that is parasitic on cicada larva. Structural studies revealed that ISP-1 is identical to myriocin and thermozymocidin ((2S, 3R 4R)-(E)-2-amino-3,4-dihydro-2-hydroxy-methyl-14-oxoeicos-6-snoic acid), which were previously isolated as antibiotics (Fujita et al.,

1994)(Fig.1). ISP-1 was originally isolated as an immunosuppressant. It appears to inhibit the mouse allogenic mixed lymphocyte reaction and allo-reactive cytotoxic T lymphocyte generation *in vivo* with a potency 10- to 100-fold greater than that of cyclosporine A, the most widely used immunosuppressant (Fujita et al., 1994). Unlike cyclosporine A and FK506 (another widely used immunosuppressant), ISP-1 did not suppress IL-2 production in the mixed lymphocyte reaction, but did suppress the IL-2-dependent growth of a mouse cytotoxic T cell line, CTLL-2. The structure of ISP-1 is similar to that of sphingosine (Fig.1). Thus, ISP-1 is also called 'sphingosine-like immunosuppressant'. Indeed, the primary target of ISP-1 was identified as serine palmitoyltransferase (SPT) (Miyake et al., 1995), which catalyzes the first step of sphingolipid biosynthesis, i.e. the condensation of serine and palmitoyl-CoA into ketodihydrosphingosine. The interaction of SPT with ISP-1 suggested that ISP-1 would strongly and competitively inhibit the SPT enzyme activity by forming an external aldimine in the active site of the enzyme (Ikushiro et al., 2004). IL-2-dependent growth of CTLL-2 cells was probably suppressed when SPT was inhibited, thereby triggering a reduction in intracellular levels of sphingolipids (Miyake et al., 1995). Growth suppression ultimately produced apoptosis of the cells (Nakamura et al., 1996), which was caspase-3 independent (Yamaji et al., 2001). In yeast (*Saccharomyces cerevisiae*), ISP-1 also inhibited cell growth by inhibiting SPT (Sun et al., 2000). Yeast cells were arrested primarily with a late S or G2-like morphology, and died. Since ISP-1 is a good inhibitor of SPT, ISP-1 is widely used to reduce sphingolipid levels in mammals and yeast to investigate sphingolipid function. Despite this extensive use of ISP-1, , the mechanism by which ISP-1 induces cell death has not been clarified. To yield a better understanding this mechanism, characterization of those yeast genes responsible for ISP-1 resistance is summarized here. Yeast has species of long chain base (LCB) similar to those in mammalian cells, and studies of LCB metabolic pathways in yeast have contributed greatly to current understanding of the biosynthetic pathway of LCB in mammalian cells (Dickson, 1998).

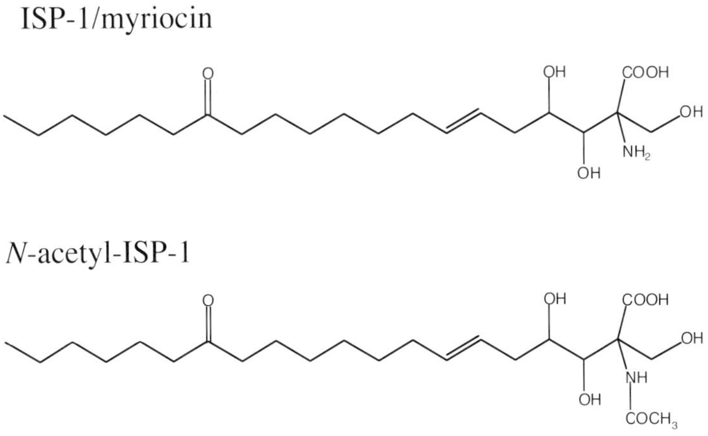

ISP-1/myriocin

N-acetyl-ISP-1

Fig. 1. Structures of ISP-1 and *N*-acetyl-ISP-1.

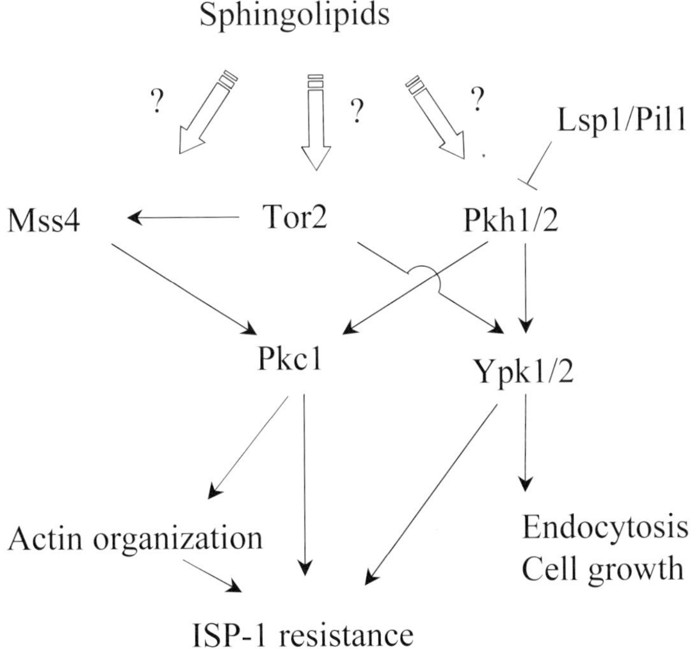

Fig. 2. Tentative pathway of sphingolipid signaling in yeast.

2. Action of ISP-1 and isolation of ISP-1 resistance genes in yeast

ISP-1-induced yeast growth inhibition was triggered by reductions in the intracellular level of sphingolipids due to the inhibition of SPT, as shown by the fact that exogenously added phytosphingosine (PHS), a metabolite of yeast sphingolipids, abolished the growth inhibition (Sun et al., 2000). This was quite similar to the case of CTLL-2 cells, in which sphingosine, a counterpart of PHS in mammals, also suppressed the cell death caused by ISP-1 (Miyake et al., 1995, Nakamura et al., 1996). In contrast to the apoptosis in CTLL-2, yeast cells died with an average DNA content of 4C in the presence of ISP-1, which is consistent with a mismatch between cell division and DNA synthesis in yeast due to starvation for LCB (Pinto et al., 1992). These findings may be related to the inhibition of cytokinesis by psychosine, a metabolite of sphingolipids in mammals (Kanazawa et al., 2000).

ISP-1 resistance genes were isolated as multicopy suppressor genes for ISP-1-induced sphingolipid depletion and were referred to as *SLI* genes (*SLI* denotes sphingosine-like immunosuppressant resistance gene) (Sun et al., 2000). Of the 8 ISP-1 resistance genes that have been isolated in yeast, 3 have been characterized. The features of these 3 are summarized below.

2.1 *SLI2* gene (*YKL126W* ; GenBank accession number CAA81967.1)

SLI2 was a firstly characterized as an ISP-1 resistance gene (Sun et al., 2000). *SLI2* is a homologue of *YPK1* (Chen et al., 1993), which was isolated by hybridization to bovine cyclic AMP-dependent protein kinase cDNA. *YPK1* encodes a serine/threonine protein kinase, which is an ortholog of the mammalian serum- and glucocorticoid-inducible kinase SGK (Casamayor et al., 1999). Yeast contains a gene closely related to *YPK1*, called *YPK2*. Overexpression of *YPK2* also produced ISP-1 resistance. The kinase activity of Ypk1 was shown to be indispensable for resistance to ISP-1. Overexpression of Ypk1 had no effect on sphingolipid metabolism, in sharp contrast to the case of *SLI1*, which increased sphingolipid biosynthesis (see below). The phosphorylated protein band(s) of Ypk1 were increased when the intracellular sphingolipid concentration was decreased by treatment with ISP-1. In contrast, the band(s) were elevated by the addition of PHS, suggesting that the phosphorylation of Ypk1 is regulated by the intracellular sphingolipid concentration. Pkh1 and its closely related protein Pkh2 are upstream kinases of Ypk1/2. Overexpression of *PKH1* caused ISP-1 resistance in yeast (Sun et al., 2000). Furthermore, nanomolar concentrations of LCBs, including PHS, induced the activity of Pkh1/2 *in vitro* (Friant et al., 2001). These data suggest that the Pkh1/2-Ypk1/2 pathway may be one of the sphingolipid-mediated signaling pathways and PHS may be a regulatory molecule for the pathway. Pkh1/2 phosphorylates and activates several protein kinases, including Ypk1/2 and Pkc1 (Casamayor et al., 1999, Inagaki et al., 1999). Indeed overexpression of Pkc1 caused ISP-1 resistance (Kobayashi et al., 2005, Roelants et al., 2004). Thus, Pkh1/2 may play key roles in sphingolipid-mediated signaling pathways in yeast (Fig.2). Two closely related proteins, Lsp1 and Pil1, are candidates as negative regulators of the Pkh1/2-Ypk1/2 pathway. These proteins were initially found to form complexes with Pkh1/2 (Ito et al., 2001). Pkh1/2 phosphorylated these proteins *in vitro* (Zhang et al., 2004). LCBs, including PHS, inhibited the phosphorylation of Pil1 but stimulated the phosphorylation of Lsp1. Such inhibition and stimulation by LCBs was predominantly detected for the phosphorylation by Pkh2 rather than Pkh1. Even when the opposite effects of LCB on the phosphorylation of Pil1 and Lsp1 were produced, these proteins down-regulated the Pkc1-MAP and Ypk1 pathways during heat stress, under which the biosynthesis of LCB was stimulated. While accumulating evidence supports sphingolipid regulation of the Pkh1/2-Ypk1/2 pathway, several discrepancies exist. First, overexpression of Pkh1 does not increase the phosphorylated band(s) of Ypk1, whose production is regulated by sphingolipids, even though Ypk1 activity increased (Roelants et al., 2002). T504 in the activation loop of the kinase is known to be a phosphorylation site of Ypk1 by Pkh1. However, the phosphorylated band(s) of the T504A mutant of Ypk1 were still observed (H.T. and Y.K. unpublished results). Identification of the molecular species of sphingolipids involved in the phosphorylation of Ypk1 *in vivo* will facilitate resolution of these discrepancies.

In mammals, 3-phosphoinositide-dependent kinase-1 (PDK1), which is an orthologue of Pkh1/2, was activated by sphingosine *in vitro*. This activation increased the phosphorylation of several PDK1 substrates (King et al., 2000) and suggests that the PDK1-SGK pathway may be sphingolipid-dependent. However, the relatively high concentration of sphingosine (about 100 μ M for the maximum induction of the PDK1 activity) used in this experiment may indicate that a different characteristic of sphingosine created this effect not its function as a signaling molecule affecting PDK1. PDK1 has a pleckstrin homology domain which binds to phosphatidylinositol (PI) 3,4,5-triphosphate or PI 3,4-bisphosphate (Stokoe et al., 1997). However, Pkh1/2 lacks a pleckstrin homology domain and yeast lacks the ability to generate PI 3,4,5 triphosphate or PI 3,4-bisphosphate (Dove et al., 1997, Hawkins et al., 1993). These observations suggest that yeast uses sphingolipids instead of phosphoinositides as lipid activators.

FKHRL1, a forkhead transcription factor, and Nedd4-2, a ubiquitin-protein ligase, are known downstream targets of SGK in mammals (Brunet et al., 2001, Snyder et al., 2002, Zhou and Snyder, 2005). Although the targets of Ypk1/2 have not been identified yet in yeast, Ypk1 is known to be involved in several downstream signaling events as well as the sphingolipid signaling pathway. A *ypk1* null strain showed a slow-growth phenotype, while a *ypk2* null strain did not show any phenotype (Chen et al., 1993). A mutation, G490 to R in Ypk1, was identified in a yeast mutant defective in endocytosis and the *ypk1* null strain is also defective in endocytosis (deHart et al., 2002). Among these downstream events of Ypk1, resistance against ISP-1-induced sphingolipid depletion required the phosphorylation of T662 (Roelants et al., 2004, Tanoue et al., 2005). However, cell growth- and endocytosis- defects were rescued by a phosphorylation-deficient mutant of T662. Referred to as the PDK2 site, T662 is located in a hydrophobic motif close to the C-terminus of Ypk1, and T504 of the activation loop is called the PDKI site, which is phosphorylated by Pkh1/2, the yeast orthologue of PDK1. In contrast to the PDK2 site, PDK1 phosphorylation is indispensable for all of three phenotypes. Hence, the phosphorylation of the PDK2 site, not the PDK1 site, may specifically involve sphingolipid signaling. This seems inconsistent with the Pkh1/2-Ypk1/2 hypothesis of sphingolipid signaling described above, because it is the PDK1 site that is likely to be phosphorylated by Pkh1/2 and regulated by PHS. PDK2 is believed to be the protein kinase responsible for phosphorylation of the PDK2 site. Several attempts have been made to identify PDK2 using the AGC kinases, which consist of a great number of serine/threonine kinases both in mammals and in yeast (Balendran et al., 1999, Delcommenne et al., 1998, Rane et al., 2001). Recently, the PDK2 site of Ypk2 was reported to be phosphorylated *in vitro* by Tor2, which is one of the targets of rapamycin protein kinases, Tor1/2 (Kamada et al., 2005). Tor1 and Tor2 redundantly regulate growth in a rapamycin-sensitive manner. Tor2 but not Tor1 is involved in the regulation of the actin cytoskeleton in a rapamycin-insensitive manner through the TORC2 complex. Mutants of Ypk1/2 suppressed the lethality of the *tor2* strain but not the *tor1* strain. However, the wild-type Ypk1/2 did not complement the defect of the *tor2* strain (Kamada et al., 2005). In conjunction with the results obtained below (*SLI6* gene), these data suggest that the Tor2-Ypk1/2 pathway is one of the candidates for the sphingolipid signaling pathway in yeast, although the different functions between the mutants and the wild-types of Ypk1/2 in the complementation of the *tor1* strain are not completely understood.

2.2 *SLI6* gene (*YDR208W* ; GenBank accession number CAA92347.1)

The *SLI6* gene is a synonym of *MSS4,* which was originally isolated as a multicopy suppressor of an *stt4* mutation that has a defect in type III PI 4-kinase (Yoshida et al., 1994). *MSS4* encodes type 1 PI 4-phosphate 5-kinase (PI4P5K), which synthesizes PI 4,5-bisphosphate in yeast. The catalytic domain of Mss4 shares high homology to those of mammalian PI4P5Ks. The kinase activity of Mss4 is indispensable for the resistance to ISP-1 (Kobayashi et al., 2005). Like overexpression of Ypk1, overexpression of Mss4 did not increase sphingolipid biosynthesis, which was reduced by ISP-1 treatment. ISP-1 treatment causes defects in both the activity and subcellular localization of Mss4. Mss4 may be involved in the Tor2-Pkc1 signaling pathway, which controls actin organization (Loewith et al., 2002) (Fig.2). In the downstream pathway of Mss4, PI 4,5-bisphosphate mediates the recruitment of Rom2 to the membrane through the pleckstrin-homology domain of Rom2, which is the crucial step for the activation of Rho1/2, a GDP/GTP exchanging factor, and for the subsequent activation of Pkc1. Indeed, ISP-1 induced abnormal localization of Rom2 as well as Mss4. Furthermore, failure of the proper localization of both Rom2 and Mss4 was also observed in *csg2*-deleted cells, which have reduced amounts of mannosylated inositolphosphorylceramide, a yeast sphingolipid (Kobayashi et al., 2005). Furthermore, the calcium-sensitive phenotype of the *csg2* strain was suppressed by *mss4* mutation (Beeler et al., 1998). These data suggest that the Tor2-Pkc1 pathway, including Mss4, participates in sphingolipid signaling.

The real target of sphingolipid would be Mss4 itself or a molecule upstream of Mss4. *MSS4/SLI6* and *YPK1/SLI2* are the ISP-1 resistance genes noted above. Mss4 and Ypk1 are associated with each other through Pkc1, which is a downstream kinase of Mss4 and is activated by Pkh1/2, upstream kinases of Ypk1. Indeed, the slow growth phenotype of a *ypk1*-deleted mutant was rescued by Mss4 overexpression, possibly due to the Pkc1 activation (Kobayashi et al., 2005). The lethal phenotype of *ypk1/2* double mutation was overcome by overexpression of constitutively activated Pkc1 (Roelants et al., 2002, Schmelzle et al., 2002). Further studies on Ypk1 and Mss4 and on signaling molecules related to these proteins would advance the overall understanding of sphingolipid signaling in yeast.

2.3 *SLI1* gene (*YGR212W*; GenBank accession number CAA97239.1)

The *SLI1* gene encodes a polypeptide of 468 amino acids with a molecular mass of 54.5 kDa. Sli1 is not related to any known protein. This protein is a major contributor to ISP-1 resistance in yeast, as indicated by the fact that cell growth of a *sli1*-null mutant was markedly reduced even with low concentrations of ISP-1 and that IC50 of ISP-1 for the growth was decreased to approximately 1 %, that of the parental strain (Momoi et al., 2004). In contrast to *SLI2* and *SLI6*, overexpression of *SLI1* overcame reductions in sphingolipid biosynthesis induced by ISP-1, but did not increase sphingolipid biosynthesis in the absence of ISP-1. Although Sli1 has slight similarity to Atf1 and Atf2, which are yeast *O*-acetyltransferases, the protein has an *N*-acetyltransferase activity, which converts ISP-1 to *N*-acetyl-ISP-1 *in vitro* (Fig.1). *N*-acetyl-ISP-1 lacks the inhibitory activity toward cell growth. The SPT-inhibitory activity of *N*-acetyl-ISP-1 is much lower than that of ISP-1 (Momoi et al., 2004). These data suggest that Sli1 inactivates ISP-1 due to its *N*-acetyltransferase activity toward ISP-1 *in vivo*. The mechanism of action of Sli1 may partly explain why IC50 of ISP-1 for SPT *in vitro* is much lower than that for yeast cell growth *in vivo* (Sun et al., 2000). Sli1 simply decreases the intracellular concentration of ISP-1 by detoxification. Thus, Sli1 seems not to be directly related to the sphingolipid signaling pathway. However, a genetic interaction between *SLI1* and *YPK1* may suggest the involvement of Sli1 in Ypk1-mediated sphingolipid signaling (Momoi et al., 2004).

3. Conclusion

Several lines of evidence suggest that Ypk1/Sli2 and Mss4/Sli6 are involved in sphingolipid signaling; however, key pieces of evidence remain missing. First, what kinase(s) is involved in the sphingolipid-dependent phosphorylation of Ypk1? Pkh1/2 and Tor2 are candidates, but no direct evidence for their role has been reported. Second, what molecular species of sphingolipids is involved in the sphingolipid-dependent recruitment of Mss4 and Rom2? Mannosylated inositolphosphorylceramide is a candi-

date, but, again, a direct demonstration of its function remains unsubstanti-
ated. Further studies are required to address these key questions.

Acknowledgements. This work was supported in part by Grants-in-Aid for Scientific Research (13470488 to Y.K. and 13877373 to Y.K.) from the Japan Society for the Promotion of Science, and Grants-in-Aid for Scientific Research on Priority Areas (12140202 to Y.K.) from the Ministry of Education, Culture, Sports and Technology, CREST and RIKEN.

Abbreviations. The abbreviations used are: SPT, serine palmitoyltransferase; PHS, phytosphingosine; LCB, long chain base; PI, phosphatidylinositol; PI4P5K, phosphatidylinositol 4-phosphate 5-kinase.

4. References

Balendran A, Casamayor A, Deak M, Paterson A, Gaffney P, Currie R, Downes CP and Alessi DR (1999) PDK1 acquires PDK2 activity in the presence of a synthetic peptide derived from the carboxyl terminus of PRK2. *Curr Biol,* **9**, 393-404.

B eeler T, Bacikova D, Gable K, Hopkins L, Johnson C, Slife H and Dunn T (1998) The Saccharomyces cerevisiae TSC10/YBR265w gene encoding 3-ketosphinganine reductase is identified in a screen for temperature-sensitive suppressors of the Ca2+-sensitive csg2Delta mutant. *J Biol Chem,* **273**, 30688-30694.

Brunet A, Park J, Tran H, Hu LS, Hemmings BA and Greenberg ME (2001) Protein kinase SGK mediates survival signals by phosphorylating the forkhead transcription factor FKHRL1 (FOXO3a). *Mol Cell Biol,* **21**, 952-965.

Casamayor A, Torrance PD, Kobayashi T, Thorner J and Alessi DR (1999) Functional counterparts of mammalian protein kinases PDK1 and SGK in budding yeast. *Curr Biol,* **9**, 186-197.

Chen P, Lee KS and Levin DE (1993) A pair of putative protein kinase genes (YPK1 and YPK2) is required for cell growth in Saccharomyces cerevisiae. *Mol Gen Genet,* **236**, 443-447.

deHart AK, Schnell JD, Allen DA and Hicke L (2002) The conserved Pkh-Ypk kinase cascade is required for endocytosis in yeast. *J Cell Biol,* **156**, 241-248.

Delcommenne M, Tan C, Gray V, Rue L, Woodgett J and Dedhar S (1998) Phosphoinositide-3-OH kinase-dependent regulation of glycogen synthase kinase 3 and protein kinase B/AKT by the integrin-linked kinase. *Proc Natl Acad Sci U S A,* **95**, 11211-11216.

Dickson RC (1998) Sphingolipid functions in Saccharomyces cerevisiae: comparison to mammals. *Annu Rev Biochem,* **67**, 27-48.

Dove SK, Cooke FT, Douglas MR, Sayers LG, Parker PJ and Michell RH (1997) Osmotic stress activates phosphatidylinositol-3,5-bisphosphate synthesis. *Nature,* **390**, 187-192.

Friant S, Lombardi R, Schmelzle T, Hall MN and Riezman H (2001) Sphingoid base signaling via Pkh kinases is required for endocytosis in yeast. *Embo J,* **20**, 6783-6792.

Fujita T, Inoue K, Yamamoto S, Ikumoto T, Sasaki S, Toyama R, Chiba K, Hoshino Y and Okumoto T (1994) Fungal metabolites. Part 11. A potent immunosuppressive activity found in Isaria sinclairii metabolite. *J Antibiot (Tokyo),* **47**, 208-215.

Hawkins PT, Stephens LR and Piggott JR (1993) Analysis of inositol metabolites produced by Saccharomyces cerevisiae in response to glucose stimulation. *J Biol Chem,* **268**, 3374-3383.

Ikushiro H, Hayashi H and Kagamiyama H (2004) Reactions of serine palmitoyltransferase with serine and molecular mechanisms of the actions of serine derivatives as inhibitors. *Biochemistry,* **43**, 1082-1092.

Inagaki M, Schmelzle T, Yamaguchi K, Irie K, Hall MN and Matsumoto K (1999) PDK1 homologs activate the Pkc1-mitogen-activated protein kinase pathway in yeast. *Mol Cell Biol,* **19**, 8344-8352.

Ito T, Chiba T, Ozawa R, Yoshida M, Hattori M and Sakaki Y (2001) A comprehensive two-hybrid analysis to explore the yeast protein interactome. *Proc Natl Acad Sci U S A,* **98**, 4569-4574.

Kamada Y, Fujioka Y, Suzuki NN, Inagaki F, Wullschleger S, Loewith R, Hall MN and Ohsumi Y (2005) Tor2 directly phosphorylates the AGC kinase Ypk2 to regulate actin polarization. *Mol Cell Biol,* **25**, 7239-7248.

Kanazawa T, Nakamura S, Momoi M, Yamaji T, Takematsu H, Yano H, Sabe H, Yamamoto A, Kawasaki T and Kozutsumi Y (2000) Inhibition of cytokinesis by a lipid metabolite, psychosine. *J Cell Biol,* **149**, 943-950.

King CC, Zenke FT, Dawson PE, Dutil EM, Newton AC, Hemmings BA and Bokoch GM (2000) Sphingosine is a novel activator of 3-phosphoinositide-dependent kinase 1. *J Biol Chem,* **275**, 18108-18113.

Kobayashi T, Takematsu H, Yamaji T, Hiramoto S and Kozutsumi Y (2005) Disturbance of Sphingolipid Biosynthesis Abrogates the Signaling of Mss4, Phosphatidylinositol-4-phosphate 5-Kinase, in Yeast. *J Biol Chem,* **280**, 18087-18094.

Loewith R, Jacinto E, Wullschleger S, Lorberg A, Crespo JL, Bonenfant D, Oppliger W, Jenoe P and Hall MN (2002) Two TOR complexes, only one of which is rapamycin sensitive, have distinct roles in cell growth control. *Mol Cell,* **10**, 457-468.

Miyake Y, Kozutsumi Y, Nakamura S, Fujita T and Kawasaki T (1995) Serine palmitoyltransferase is the primary target of a sphingosine-like immunosuppressant, ISP-1/myriocin. *Biochem Biophys Res Commun,* **211**, 396-403.

Momoi M, Tanoue D, Sun Y, Takematsu H, Suzuki Y, Suzuki M, Suzuki A, Fujita T and Kozutsumi Y (2004) SLI1 (YGR212W) is a major gene conferring resistance to the sphingolipid biosynthesis inhibitor ISP-1, and encodes an ISP-1 N-acetyltransferase in yeast. *Biochem J,* **381**, 321-328.

Nakamura S, Kozutsumi Y, Sun Y, Miyake Y, Fujita T and Kawasaki T (1996) Dual roles of sphingolipids in signaling of the escape from and onset of apoptosis in a mouse cytotoxic T-cell line, CTLL-2. *J Biol Chem,* **271**, 1255-1257.

Pinto WJ, Srinivasan B, Shepherd S, Schmidt A, Dickson RC and Lester RL (1992) Sphingolipid long-chain-base auxotrophs of Saccharomyces cerevisiae: genetics, physiology, and a method for their selection. *J Bacteriol,* **174**, 2565-2574.

Rane MJ, Coxon PY, Powell DW, Webster R, Klein JB, Pierce W, Ping P and McLeish KR (2001) p38 Kinase-dependent MAPKAPK-2 activation functions as 3-phosphoinositide-dependent kinase-2 for Akt in human neutrophils. *J Biol Chem,* **276**, 3517-3523.

Roelants FM, Torrance PD, Bezman N and Thorner J (2002) Pkh1 and pkh2 differentially phosphorylate and activate ypk1 and ykr2 and define protein kinase modules required for maintenance of cell wall integrity. *Mol Biol Cell,* **13**, 3005-3028.

Roelants FM, Torrance PD and Thorner J (2004) Differential roles of PDK1- and PDK2-phosphorylation sites in the yeast AGC kinases Ypk1, Pkc1 and Sch9. *Microbiology,* **150**, 3289-3304.

Schmelzle T, Helliwell SB and Hall MN (2002) Yeast protein kinases and the RHO1 exchange factor TUS1 are novel components of the cell integrity pathway in yeast. *Mol Cell Biol,* **22**, 1329-1339.

Snyder PM, Olson DR and Thomas BC (2002) Serum and glucocorticoid-regulated kinase modulates Nedd4-2-mediated inhibition of the epithelial Na+ channel. *J Biol Chem,* **277**, 5-8.

Stokoe D, Stephens LR, Copeland T, Gaffney PR, Reese CB, Painter GF, Holmes AB, McCormick F and Hawkins PT (1997) Dual role of phosphatidylinositol-3,4,5-trisphosphate in the activation of protein kinase B. *Science,* **277**, 567-570.

Sun Y, Taniguchi R, Tanoue D, Yamaji T, Takematsu H, Mori K, Fujita T, Kawasaki T and Kozutsumi Y (2000) Sli2 (Ypk1), a homologue of mammalian protein kinase SGK, is a downstream kinase in the sphingolipid-mediated signaling pathway of yeast. *Mol Cell Biol,* **20**, 4411-4419.

Tanoue D, Kobayashi T, Sun Y, Fujita T, Takematsu H and Kozutsumi Y (2005) The requirement for the hydrophobic motif phosphorylation of Ypk1 in yeast differs depending on the downstream events, including endocytosis, cell growth, and resistance to a sphingolipid biosynthesis inhibitor, ISP-1. *Arch Biochem Biophys,* **437**, 29-41.

Yamaji T, Nakamura S, Takematsu H, Kawasaki T and Kozutsumi Y (2001) Apoptosis of CTLL-2 cells induced by an immunosuppressant, ISP-I, is caspase-3-like protease-independent. *J Biochem (Tokyo),* **129**, 521-527.

Yoshida S, Ohya Y, Nakano A and Anraku Y (1994) Genetic interactions among genes involved in the STT4-PKC1 pathway of Saccharomyces cerevisiae. *Mol Gen Genet,* **242**, 631-640.

Zhang X, Lester RL and Dickson RC (2004) Pil1p and Lsp1p negatively regulate the 3-phosphoinositide-dependent protein kinase-like kinase Pkh1p and downstream signaling pathways Pkc1p and Ypk1p. *J Biol Chem,* **279**, 22030-22038.
Zhou R and Snyder PM (2005) Nedd4-2 phosphorylation induces serum and glucocorticoid-regulated kinase (SGK) ubiquitination and degradation. *J Biol Chem,* **280**, 4518-4523.

7-4 Lysenin: A New Probe for Sphingomyelin

Toshihide Kobayashi[1,2,3] and Akiko Yamaji-Hasegawa[4]

[1]Lipid Biology Laboratory, RIKEN (Institute of Physical and Chemical Research) Discovery Research Institute, 2-1, Hirosawa, Wako, Saitama 351-0198, Japan, [2]Supra-Biomolecular System Research Group, RIKEN Frontier Research System, Saitama 351-0198, Japan, [3] INSERM U585, INSA (Institut National des Sciences Appliquees)-Lyon, 20 Ave A. Einstein, 69621 Villeurbanne, France, [4] Laboratory of Cellular Biochemistry, RIKEN Discovery Research Institute, Saitama 351-0198, Japan

Summary. Lysenin is a pore-forming toxin that binds to sphingomyelin in a distribution-dependent manner. Studies of this interaction revealed the heterogeneous organization of sphingomyelin in biomembranes while investigations with non-toxic lysenin helped elucidate the spatial and functional heterogeneity of lipid rafts. This chapter summarizes the characterization of lysenin and discusses the possible applications and limitations of this newly developed sphingomyelin probe.

Keywords. Lipid raft, Lipid probe, Pore-forming toxin, Membrane domain

1. Introduction

Antibodies and toxins that specifically bind to sphingolipids are useful probes to study the distribution and dynamics of these lipids. β-subunits of the cholera and Shiga toxins are effective non-toxic glycosphingolipid specific probes (Pina and Johannes, 2005). Anti-glycosphingolipid antibodies have also been reported. However, because most of these are IgM and are substantial in size, using them as lipid probes is difficult. Several proteins that interact with sphingomyelin have been identified (Bernheimer and Avigad, 1976; Zitzer et al., 2000; Valcarcel et al., 2001; Tomita et al.,

2004). Among these, lysenin is highly specific to sphingomyelin (Yamaji et al., 1998; Ishitsuka and Kobayashi, 2004; Ishitsuka et al., 2005). Recent results indicate that lysenin recognizes sphingomyelin only when the lipid forms clusters (Ishitsuka et al., 2004).

2. Lysenin recognizes sphingomyelin in distribution-dependent manner

Lysenin is a 297-amino-acid protein toxin obtained from the earthworm *Eisenia foetida* (Sekizawa et al., 1997). Lysenin is unique; it uses sphingomyelin as the cell surface receptor though the binding is dependent on the surface distribution of the lipid (Ishitsuka et al., 2004). Thus, the apical membrane of epithelial Madin-Darby canine kidney (MDCK) cells is much more resistant to lysenin than the basolateral membrane even though both contain sphingomyelin. Since the apical membrane is highly enriched with glycosphingolipids, glycosphingolipids might affect the interaction between lysenin and sphingomyelin. Cultured melanoma cells, which are enriched with glycosphingolipids, are resistant to lysenin, further supporting that hypothesis. In contrast, glycosphingolipid-deficient melanoma mutants are susceptible to the toxin. Model membrane experiments showed that lysenin binds to sphingomyelin only when the lipid forms clusters. When sphingomyelin co-exists with glycerophospholipid containing unsaturated fatty acid, the two lipids are phase-separated and sphingomyelin forms clusters. On the other hand, when glycosphingolipids are mixed with sphingomyelin the local density of sphingomyelin decreases thereby inhibiting the formation of sphingomyelin clusters (Fig. 1). Measurements of binding stoichiometry by isothermal calorimetry indicate that one lysenin molecule binds 5-6 sphingomyelin molecules.

These results indicate that the binding of lysenin to sphingomyelin is dependent on the organization of lipids in the membrane. In addition, they show that a lack of lysenin binding does not necessarily indicate a deficiency of sphingomyelin (Ishitsuka et al., 2004). Since lysenin binds to most mammalian cells in culture, sphingomyelin probably forms clusters of at least five molecules in most membranes. Moreover, the distribution of sphingomyelin varies between cell types (e.g. melanoma vs fibroblast) as well as between different membrane domains of the same cell (apical vs basolateral) (Ishitsuka et al., 2004).

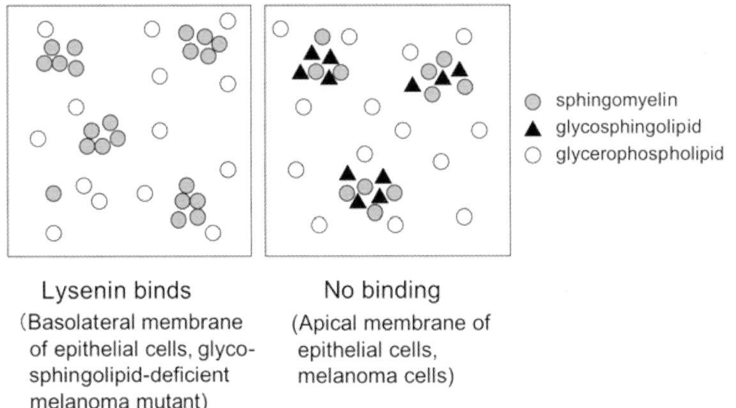

Fig. 1. Lysenin binding depends on sphingomyelin organisation. Lysenin binds to the membrane when sphingomyelin molecules cluster. Lysenin-binding then significantly decreases in the presence of glycosphingolipids because the local density of sphingomyelin is low.

3. Lysenin is a pore-forming toxin

The binding of lysenin to sphingomyelin-rich membrane domains is associated with the its oligomerization (Yamaji-Hasegawa et al., 2003). Lysenin oligomerization is accompanied by pore formation. A negative staining electron micrograph detected oligomers of 10-12 nm in diameter with a 3-5 nm pore. The size of the pore is consistent with biochemical data measuring the inhibition of lysenin-induced hemolysis by carbohydrates and polymers of various sizes. The oligomer is SDS- and heat-resistant. Antibody scanning analysis suggests that the C-terminal region of lysenin is exposed and the N-terminal is hidden in the isolated oligomer complex. This oligomerization, not lysenin binding, is affected by membrane fluidity, as a fluid membrane can accelerate the oligomerization of the protein. Tryptophan fluorescence and differential scanning calorimetry indicate that the lysenin oligomer is inserted into the hydrophobic region of the membrane (Yamaji-Hasegawa et al., 2003). The precise structure of either the lysenin oligomer or monomer is unknown.

Lysenin is part of a family of proteins that includes lysenin-related protein 1 (LRP1, lysenin 2) and LRP2 (lysenin 3) (Kiyokawa et al., 2004). LRP2 is also known as fetidin. The amino acid sequence of LRP1 is 76% identical with and 88% similar to that of lysenin, whereas the sequence of LRP-2 has an 89% identity and 94 % similarity with the lysenin sequence (Fig. 2). Both lysenin and LRP2 have 30 sites in common with the aromatic amino acids. Only one position among these sites is substituted in LRP1, isoleucine for phenylalanine 209. Like lysenin, both LRP1 and LRP2 specifically bind to sphingomyelin and induce hemolysis. However, LRP1 is ten times less active than lysenin and LRP2. Replacing isoleucine 210 with phenylalanine increased LRP1 activity to the levels of lysenin and LRP2. This indicates that aromatic amino acids are important in lysenin binding activity and its toxicity. This was confirmed by systematic mutation of tryptophan to alanine in lysenin. Lysenin contains six tryptophan residues, five are conserved in LRP-1 and LRP-2. The conserved tryptophans were shown to be necessary for the recognition of sphingomyelin and for the hemolytic activity of lysenin, whereas the non-conserved tryptophan was not (Kiyokawa et al., 2004).

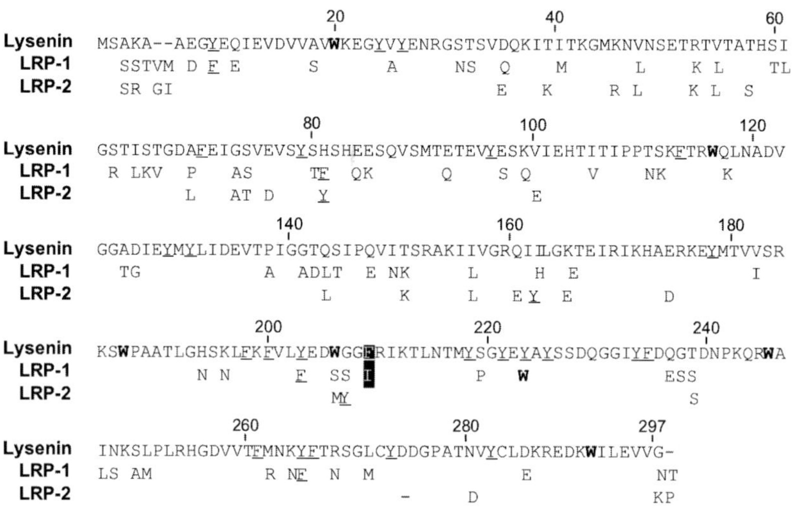

Fig. 2. Sequence alignment of the lysenin family. Accession numbers of the GenBank database are D85846 (lysenin), D85847 (lysenin-related protein (LRP)-1) and D85848 (LRP-2). Protein names are on the left. Sequence numbers for lysenin are shown. Tryptophan residues are in bold. Other aromatic amino acids are underlined. Black bar marks phenylalanine 209 of lysenin, which is substituted for isoleucine 210 in LRP-1.

4. Non-toxic lysenin reveals spatial and functional heterogeneity of sphingolipid-containing membrane domains

Somatic cell mutants defective in sphingomyelin synthesis and ceramide transport can be isolated by exploiting the toxic nature of lysenin (see the chapter by K. Hanada, this volume). Lysenin can also identify sphingomyelin synthase (see the chapter by T. Okazaki, this volume) (Miyaji et al., 2005). However, a non-toxic form of lysenin is required to follow the distribution and dynamics of sphingomyelin in intact cells. A series of truncated lysenins revealed that 137 amino acids of the C-terminus of the protein are necessary and sufficient to bind sphingomyelin (Kiyokawa et al., 2005). This truncated lysenin is a non-toxic, sphingomyelin probe. This peptide does not oligomerize after binding to sphingomyelin-containing membranes, suggesting that oligomerization is required for toxicity. The truncated lysenin and native lysenin showed comparable on-rate of binding to sphingomyelin when measured by surface plasmon resonance. In contrast, dissociation of the truncated lysenin was 100 times faster than that of the native one. This gives a 36-fold difference of overall K_D (5.3 x 10^{-9} M for native lysenin vs 1.9 x 10^{-7} M for truncated lysenin). These results suggest that the oligomerization stabilizes the binding of lysenin to sphingomyelin-containing membranes. The lower affinity of truncated lysenin can be overcome if relatively high concentrations of the peptide are used. Using truncated lysenin and the non-toxic cholera toxin β-subunit, the distribution of sphingomyelin-rich and GM1-rich domains were compared in Jurkat T cells. The results indicate that sphingomyelin and GM1 form domains with a diameter of 100-150 nm. However, they do not co-cluster on the plasma membrane (Fig. 3). Thus, the sphingomyelin-rich and GM1-rich domains are spatially distinct. The cross-linking of truncated lysenin after binding to the membrane induces calcium influx and phosphorylation of ERK kinase. Although the cross-linking of GM1 also induces calcium influx and ERK phosphorylation, the underlying mechanisms appear to be different. Cholera toxin-dependent stimulation is accompanied by protein tyrosine phosphorylation whereas lysenin does not induce significant tyrosine phosphorylation (Kiyokawa et al., 2005). These results thus suggest that the sphingomyelin-rich domain provides a functional signal cascade platform that is distinct from those provided by GM1-rich domains. These findings reveal spatial and functional heterogeneities in sphingolipid-rich membrane domains.

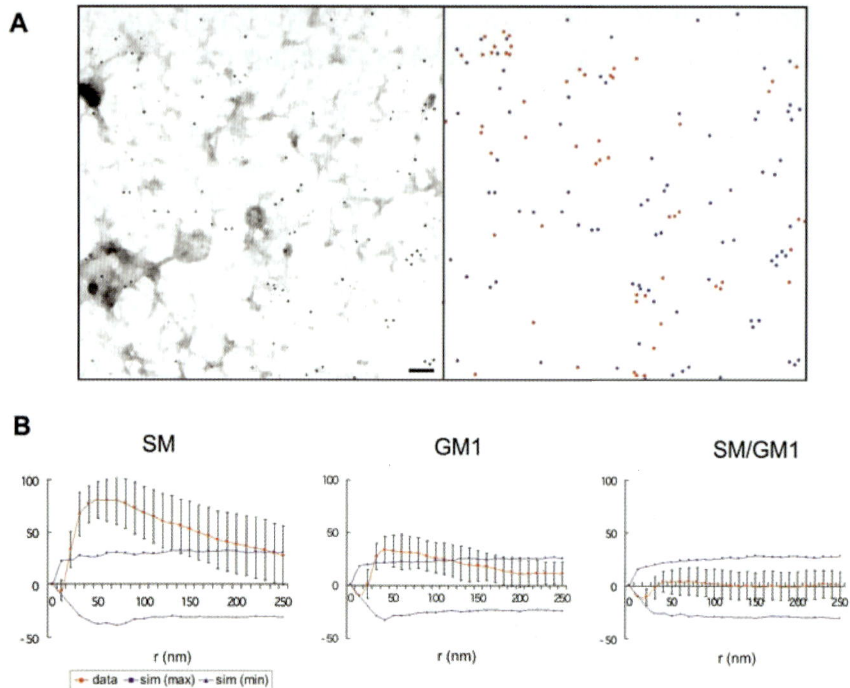

Fig. 3. **Distribution of sphingomyelin-rich and GM1-rich domains on two-dimensional sheets of plasma membrane from Jurkat cells.** A, cells were labeled with GFP-tagged truncated lysenin and biotinylated cholera toxin β-fragment at 4 °C. After fixation, the cells were further labeled with anti-GFP antibody followed by the incubation with anti-IgG-5 nm gold and anti-biotin-10 nm gold. The distribution of gold particles on the plasma membrane was examined under electron microscope after ripping off the membrane. Bar, 100 nm. Right panel, distribution of sphingomyelin (5 nm) indicated in red, whereas the distribution of GM1 (10 nm gold) is in blue. B, analysis of the distribution of sphingomyelin-rich domains and GM1-rich domains using Ripley's K-function. Both sphingomyelin-rich and GM1-rich domains form clusters. Pairwise values for sphingomyelin and GM1 fall within blue lines that represent the range of values expected for pairs of different particles whose distribution are random.

5. Perspectives

Recent advances in biophysical techniques such as FRET (fluorescence resonance energy transfer) and single molecular tracking (see the chapter by A. Kusumi, this volume) make it possible to measure the behavior of membrane components at the nanometer level. Using lysenin as a probe, these techniques may provide detailed information on the distribution and dynamics of sphingomyelin-rich domains in real time. Sphingolipid probes that can follow the fate of lipids in living cells are still limited, and further developments are required for future advances.

6. References

Bernheimer AW, Avigad LS (1976) Properties of a toxin from the sea anemone Stoichacis helianthus, including specific binding to sphingomyelin. *Proc Natl Acad Sci U S A*, **73**, 467-471.

Ishitsuka R, Kobayashi T (2004) Lysenin: A new tool for investigating membrane lipid organization. *Anat Sci Int*, **79**, 184-190.

Ishitsuka R, Sato SB, Kobayashi T (2005) Imaging lipid rafts. *J Biochem* (Tokyo), **137**, 249-254.

Ishitsuka R, Yamaji-Hasegawa A, Makino A, Hirabayashi Y, Kobayashi T (2004) A lipid-specific toxin reveals heterogeneity of sphingomyelin-containing membranes. *Biophys J*, **86**, 296-307.

Kiyokawa E, Baba T, Otsuka N, Makino M, Ohno S, Kobayashi T (2005) Spatial and functional heterogeneity of sphingolipid-rich membrane domains. *J Biol Chem*, **280**, 24072-24084.

Kiyokawa E, Makino A, Ishii K, Otsuka N, Yamaji-Hasegawa A, Kobayashi T (2004) Recognition of sphingomyelin by lysenin and lysenin-related proteins. *Biochemistry*, **43**, 9766-9773.

Lange S, Nussler F, Kauschke E, Lutsch G, Cooper EL, Herrmann A (1997) Interaction of earthworm hemolysin with lipid membranes requires sphingolipids. *J Biol Chem*, **272**, 20884-20892.

Miyaji M, Jin ZX, Yamaoka S, Amakawa R, Fukuhara S, Sato SB, Kobayashi T, Domae N, Mimori T, Bloom ET, Okazaki T, Umehara H (2005) Role of membrane sphingomyelin and ceramide in platform formation for Fas-mediated apoptosis. *J Exp Med*, **202**, 249-259.

Pina DG, Johannes L (2005) Cholera and Shiga toxin B-subunits: thermodynamic and structural considerations for function and biomedical applications. *Toxicon*, **45**, 389-393.

Sekizawa Y, Kubo T, Kobayashi H, Nakajima T, Natori S (1997) Molecular cloning of cDNA for lysenin, a novel protein in the earthworm Eisenia foetida that causes contraction of rat vascular smooth muscle. *Gene*, **191**, 97-102.

Tomita T, Noguchi K, Mimuro H, Ukaji F, Ito K, Sugawara-Tomita N, Hashimoto Y (2004) Pleurotolysin, a novel sphingomyelin-specific two-component cytolysin from edible mushroom Pleurotus ostreatus, assembles into a transmembrane pore complex. *J Biol Chem*, **279**, 26975-26982.

Valcarcel CA, Dalla Serra M, Potrich C, Bernhart I, Tejuca M, Martinez D, Pazos F, Lanio ME, Menestrina G (2001) Effects of lipid composition on membrane permeabilization by sticholysin I and II, two cytolysins of the sea anemone Stichodactyla helianthus. *Biophys J*, **80**, 2761-2774.

Yamaji-Hasegawa A, Makino A, Baba T, Senoh Y, Kimura-Suda H, Sato SB, Terada N, Ohno S, Kiyokawa E, Umeda M, Kobayashi T (2003) Oligomerization and pore formation of a sphingomyelin-specific toxin, lysenin. *J Biol Chem*, **278**, 22762-22770.

Yamaji A, Sekizawa Y, Emoto K, Sakuraba H, Inoue K, Kobayashi H, Umeda M (1998) Lysenin, a novel sphingomyelin-specific binding protein. *J Biol Chem*, **273**, 5300-5306.

Zitzer A, Harris JR, Kemminer SE, Zitzer O, Bhakdi S, Muething J, Palmer M (2000) Vibrio cholerae cytolysin: assembly and membrane insertion of the oligomeric pore are tightly linked and are not detectably restricted by membrane fluidity. *Biochim Biophys Acta*, **1509**, 264-274.

7-5 Structural Biology of Sphingolipid Synthesis

Hiroko Ikushiro[1], Akihiro Okamoto[2], and Hideyuki Hayashi[1]

[1]Department of Biochemistry, Osaka Medical College, Takatsuki, Osaka 569-8686, Japan, [2]Department of Biological Science & Technology, School of High-Technology for Human Welfare, Tokai University, Numazu, Shizuoka 410-0395, Japan

Summary. As most eukaryotic enzymes involved in sphingolipid synthesis are membrane-bound proteins analyses of their biochemical and structural are difficult, but sphingolipid-containing bacteria are useful alternatives for enzyme sources. We studied serine palmitoyltransferase (SPT), the first enzyme of the biosynthetic pathway, using this method. To study recombinant SPT's enzymatic properties, we cloned bacterial SPT genes and then over-expressed them in *Escherichia coli* and purified the water-soluble enzymes. One yielded crystals good enough for X-ray crystallographic analysis, and the structure of SPT in a complex with the amino acid substrate, L-serine, was successfully determined at 2.3 Å resolution. SPT is a homodimer with two pyridoxal 5'-phosphate (PLP) molecules located at the dimer interface. Both subunits contribute side chains to the active sites. The electron density map indicates that a Schiff base is formed between L-serine and PLP in the crystal, so any reaction would stop at the external aldimine intermediate if the co-substrate, palmitoyl-CoA, were absent. Highly conserved amino acids among bacterial and eukaryotic SPTs are also located in the three dimensional structure of this enzyme, and their possible roles in the function of SPT are discussed.

Keywords. bacterial prototype, crystal structure, reaction mechanism, serine palmitoyltransferase, sphingolipid-containing bacteria

1. Introduction

Reverse genetic approaches using yeast or mammalian cell systems successfully identified all the structural genes of eukaryotic enzymes involved in sphingolipid biosynthesis (Nagiec et al. 1996; Dickson 1998; Hanada and Nishijima 2000; Hanada et al. 1997; Yamaoka et al 2004). Detailed enzymological studies, however, remain to be done because most of the biosynthetic enzymes are bound to organelles, including the endoplasmic reticulum, Golgi apparatus and cell membranes (Mandon et al. 1992; Ikeda et al. 2004). Additionally, the cellular contents of native enzymes are too low for purification. Therefore, it is very difficult to obtain sufficient quantities of enzymes from eukaryotic sources to conduct protein crystallization and basic analysis of the catalyses.

For an alternative source of enzymes, we turned to bacteria containing sphingolipids. The outer membranes of bacteria such as *Sphingomonas* or *Sphingobacterium* contain large amounts of complex sphingolipids (Yabuuchi et al. 1979; Yabuuchi et al. 1983; Kawasaki et al. 1994; Naka et al. 2003) that can rapidly provide complex sphingolipids during their growth phase via their biosynthetic systems. Therefore, these organisms should contain all counterparts of eukaryotic sphingolipid biosynthetic enzymes. Serine palmitoyltransferase (SPT), the key enzyme, catalyzes the pyridoxal 5'-phosphate (PLP)-dependent condensation reaction of L-serine with palmitoyl-CoA to generate 3-ketodihydrosphingosine (KDS) (Hanada 2004). All the sphingolipid-containing bacteria examined so far demonstrated significant SPT activity. Several SPT genes were isolated from these bacteria and overexpressed in *Escherichia coli* (Ikushiro et al. 2001). We used protein crystallization in addition to biochemical characterizations using the purified recombinant bacterial enzymes. One of the recombinant SPTs yielded crystals good enough for X-ray crystallographic analysis.

2. Cloning, overproduction, purification and characterization of bacterial SPTs

2.1 Bacterial SPTs function as water-soluble homodimers

The first successful protein purification and subsequent gene cloning of a bacterial SPT was of *Sphingomonas paucimobilis* (Ikushiro et al. 2001). Based on the amino acid sequence similarity between *Sphingomonas* and eukaryotic SPTs, SPT genes of other sphingolipid-containing bacteria such as *Sphingobacterium* or *Bdellovibrio* were also isolated. Importantly, all of bacterial SPTs examined are water-soluble homodimers unlike eukaryotic SPTs, which are heterodimers composed of two tightly membrane-bound subunits named LCB1 and LCB2 (Nagiec et al. 1994; Nagiec et al. 1996; Hanada et al. 1998). Overall the sequence similarity among bacterial and eukaryotic proteins is consistant except for the N-terminal transmembrane regions. This might explain why bacterial SPTs are soluble. *Sphingobacterium multivorum* SPT has 25% identity and 40% similarity with human LCB1 and 31% identity and 40% similarity with human LCB2 (Fig. 2). Strong similarities between the bacterial enzymes exist and the conserved amino acids are distributed throughout the entire polypeptide sequences with ~ 40% identity (Fig. 2). 86 amino acid residues, including the SPT specific PLP-binding motif (GTFSKSXXXXGG), are completely conserved among all bacterial SPTs (21.6% of total amino acid residues of *S. multivorum* SPT).

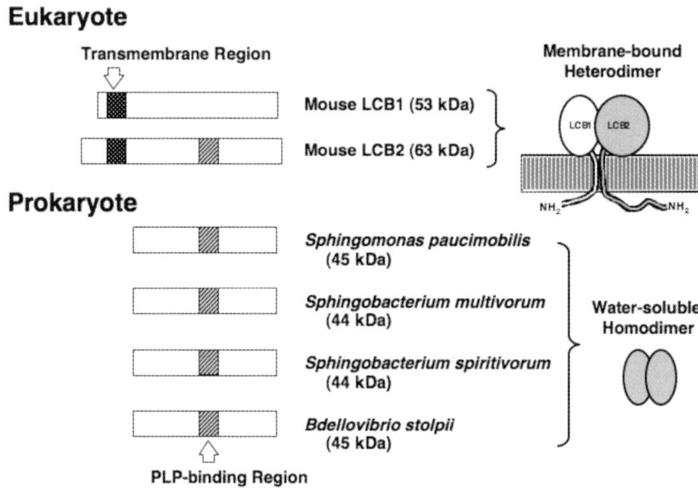

Fig. 1. Schematic views of SPTs. Subunits of eukaryotic SPT, LCB1 and LCB2, form heterodimers and are enriched in the entoplasmic reticulum. These subunits have mutual similarity. While LCB2 carries a lysine residue that was expected to form a Schiff base with PLP, LCB1 does not have such PLP-binding motif.

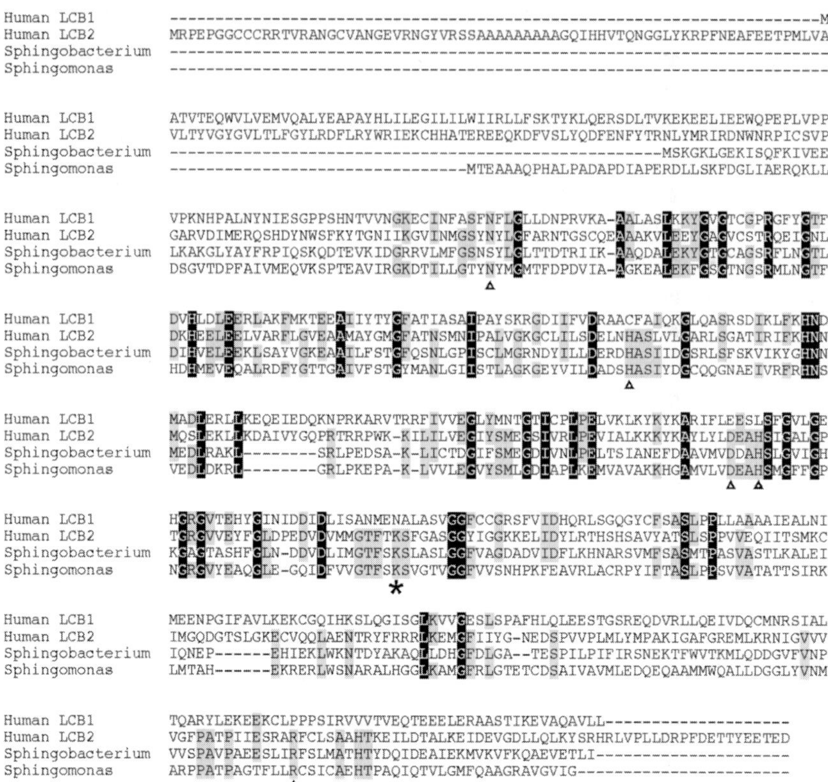

Fig. 2. Comparison of SPT protein sequences. Deduced amino acid sequences of SPT proteins from *Sphingobacterium multivorum* and *Sphingomonas paucimobilis* are compared with those of LCB1 and LCB2 proteins from human. Alignment analysis was performed with GENETYX SVRC. Residues conserved among all proteins are shaded with black, and those conserved between the three of these members with grey. An *asterisk* marks the lysine residue predicted to bind PLP. *Open triangles* mark the residues corresponding to those in the active site of other members of β-oxamine synthase family.

SPT belongs to the β-oxamine synthase family of PLP-dependent enzymes. These enzymes include eukaryotic SPT subunits, 5-aminolevulinic acid synthase (ALAS) in heme biosynthesis, 8-amino-7-oxononanoate synthase (AONS) in biotin biosynthesis, and 2-amino-3-ketobutyrate CoA ligase (KBL) in the threonine utilization pathway (Alexander et al. 1994). Bacterial SPTs have about 30% similarity with the prokaryotic members of this family (Ikushiro et al. 2001; Ikushiro et al. 2003). The structures of the substrate analogue complexes of ALAS, AONS and KBL have been determined (Alexeev et al. 1998; Webster et al. 2000; Schmidt et al. 2001;

Astner et al. 2005). Active site residues responsible for catalytic activity in these family enzymes are completely conserved at the corresponding positions in all bacterial SPTs; Lys244 (of *S. multivorum* SPT) forms a Schiff base linkage with PLP. His138, Asp210, and His213 (of *S. multivorum* SPT) interact directly with PLP (Fig. 2, *asterisk* and *triangles*). Both Asn52 and Arg367 (of *S. multivorum* SPT) might act as potential hydrogen-bonding partners of the carboxylate group of L-serine in the external aldimine complex of SPT.

2.2 Spectroscopic characters of bacterial SPTs

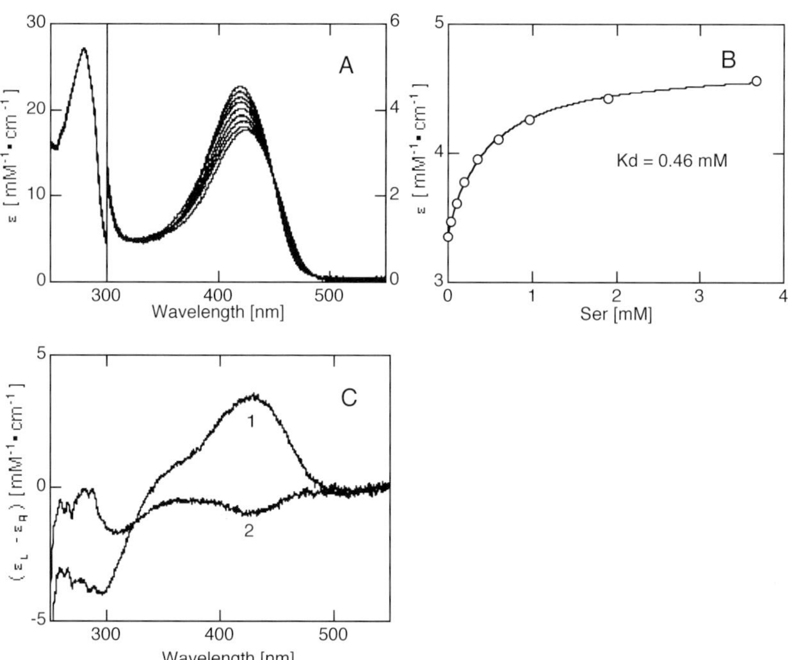

Fig. 3. Absorption and CD spectra of *S. multivorum* SPT. (A) Absorption spectra in the presence of 0, 0.04, 0.1, 0.2, 0.4, 0.6, 1.0, 1.9 and 3.7 mM of L-serine (lines from bottom to top, respectively). (B) Titration curve of the molar extinction coefficient at 422 nm. (C) CD spectra in the absence (line 1) and presence (line 2) of 4.0 mM L-serine.

The absorption spectra of purified SPTs a peak at 426 nm (with an additional peak at 338 nm in some cases), besides a protein absorption peak at 278 nm (Fig. 3, and Ikushiro et al. 2004). Absorption peaks over 300 nm come from the PLP molecule bound to the β-amino group of a lysine residue in the active site (Kallen et al. 1985; Ikushiro et al. 1998). The ratio of the peak height of the PLP-derived absorption band(s) to the protein-derived band indicates that SPTs bind two PLP molecules per dimer. Adding L-serine to the enzymes strenghtened the absorption peak at 426 nm and caused inversion of the CD at the corresponding wavelength. These spectral changes reflect the external aldimine complex formation. Like eukaryotic enzymes, bacterial SPTs metabolize only L-serine among the natural amino acids examined. Spectroscopic analysis of the binding of substrate analogues to SPT showed that the carboxyl group of L-serine is important for recognition by SPT (Ikushiro et al. 2004).

3. Structure of a bacterial SPT

3.1 Overall structure of SPT

We determined the crystal structure of *Sphingobacterium multivorum* SPT at 2.3 Å resolution using the molecular replacement method with the coordinates of KBL, another member of the β-oxamine synthase family. The overall structure of SPT–L-serine complex is shown in Fig. 4A. SPT molecule takes a homodimer structure formed by a head-to-tail association of two monomers. SPT was assigned to Fold Type I of PLP-dependent enzymes. Each monomer consists of three domains, all of which participate in dimerization: the N-terminal domain (residues 2–23), the large (central catalytic) domain (residues 60–294) and the small domain (residues 24–59 and 295–393). The N-terminal domain is formed by two helices, which extend away from the rest of the monomer. The large domain consists of a seven-stranded, twisted β-sheet and nine β-helices in repeated β/β motifs. The small domain is made of two nearly perpendicular β-sheets and three β-helices, cover the external, solvent-directed surface of the sheets.

The SPT molecule has two symmetrical active-site cavities around a molecular 2-fold axis. Each cavity is located at the domain interface of one subunit and at the subunit interface. The active site contains residues from both subunits and is fully formed only when the dimer is assembled. PLP is located at the bottom of the active-site cavity and is covalently bound as a Schiff base to Lys244 in the substrate-free enzyme. Side chains from both subunits contact serine and the PLP moiety; however, the putative catalytic residues come from only one of the two subunits.

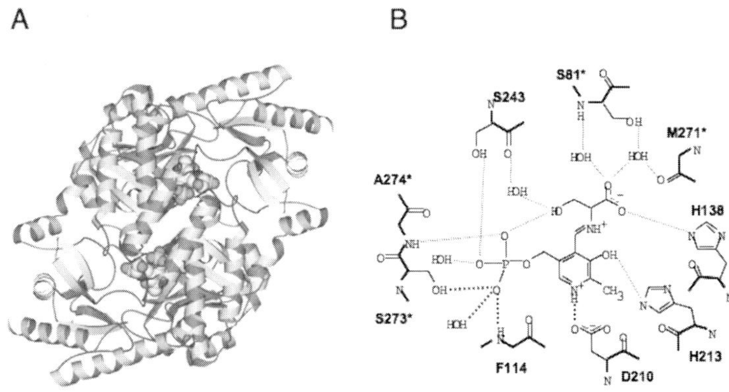

Fig. 4. Crystal structure of *S. multivorum* **SPT.** (A) Overall structure of SPT. (B) Schematic diagram showing hydrogen-bond and salt-bridge interactions of the active-site residues. Putative interactions are shown by *dotted lines*. Ala212 which sandwiches the pyridine ring of PLP with His138 is omitted for clarity.

3.2 Active-site residues and their contribution to catalysis

Fig. 4B shows the schematic diagram of the active-site residues of the SPT–L-serine complex. L-serine is covalently bound to PLP through a Schiff base linkage, forming the so-called 'external aldimine'. The hydroxyl group of L-serine forms hydrogen bonds with the phosphate group of PLP and a water molecule. The carboxyl group forms hydrogen bonds with the side chain of His138 and two water molecules. The pyridine ring of PLP is sandwiched between the side chain of Ala212 and the imidazole ring of His138. The phosphate group of PLP forms six hydrogen bonds with the side-chains of Ser273* and Ser243, the main chain amide nitrogens of Ala274* and Phe114 (where *asterisk* indicates a residue from another subunit) and two water molecules and acts as an anchor to fix the cofactor to the active site. His213 interacts with the phenolic oxygen O3 of PLP. Asp210 donates a proton to the nitrogen atom of the pyridine ring of PLP. The later interaction, found in most PLP-dependent enzymes, increases the electron-withdrawing potential of the pyridine ring. The tight binding allows PLP to remain bound during catalysis, in which PLP temporally forms a Schiff base with L-serine and loses its covalent bond to Lys244.

The entire reaction of SPT follows an Ordered Bi-Bi mechanism, in which L-serine binds first, followed by the binding of palmitoyl-CoA. The crystal structure shows that the β-proton of L-serine is not perpendicular to the plane of the conjugated system of the PLP–L-serine Schiff base. Therefore, the β-deprotonation does not occur at this stage. The binding of palmitoyl-CoA would promote deprotonation, probably by switching the hydrogen bond acceptor of the substrate hydroxyl group from His138 to Asn52 and Arg367 and by making the conformation of the Schiff base favorable for deprotonation. This is considered to be a mechanism to minimize undesired side reactions by suppressing the formation of the highly-reactive carbanion until the entrance of the second substrate. Palmitoyl-CoA would be then attacked by the β-carbanion to form a transitional acylated intermediate. Deprotonation and reprotonation at the β-carbon would yield KDS, which is released from PLP and the PLP–Lys244 internal Schiff base is regenerated.

The reaction mechanism of SPT can only be clarified by interpreting the results of the biochemical and physicochemical analyses based on the three-dimensional structure of SPT. The structural information of the bacterial enzymes involved in sphigolipid synthesis would provide us with a better understanding of the more complex eukaryotic homologues and the molecular basis of the disorders in which these enzymes are involved.

Acknowledgements. The authors thank Takato Yano for a careful reading of the manuscript. This work was supported by Grants-in-Aid 16770103 (to H. I.) and 16570125 (to H.H.) from the Ministry of Education, Culture, Sports, Science, and Technology of Japan.

4. References

Alexander FW, Sandmeier E, Mehta PK, and Christen P (1994) Evolutionary relationships among pyridoxal-5'-phosphate-dependent enzymes. Regio-specific alpha, beta and gamma families. *Eur J Biochem*, **219**, 953-960.

Alexeev D, Alexeeva M, Baxter RL, Campopiano DJ, Webster SP and Sawyer L (1998) The crystal structure of 8-amino-7-oxononanoate synthase: a bacterial PLP-dependent, acyl-CoA-condensing enzyme. *J Mol Biol*, **284**, 401-419.

Astner I, Schulze JO, van den Heuvel J, Jahn D, Schubert WD, and Heinz DW (2005) Crystal structure of 5-aminolevulinate synthase, the first enzyme of heme biosynthesis, and its link to XLSA in humans. *EMBO J*, **24**, 3166-3177.

Dickson RC (1998) Sphingolipid functions in Saccharomyces cerevisiae: comparison to mammals. *Annu Rev Biochem*, **67**, 27-48.

Hanada K (2004) Serine palmitoyltransferase, a key enzyme of sphingolipid metabolism. Biochim *Biophys Acta*, **1632**, 16-30.

Hanada K, and Nishijima M (2000) Selection of mammalian cell mutants in sphingolipid biosynthesis. *Methods Enzymol*, **312**, 304-317.

Hanada K, Hara T, Fukasawa M, Yamaji A, Umeda M, Nishijima M (1998) Mammalian cell mutants resistant to a sphingomyelin-directed cytolysin. Genetic and biochemical evidence for complex formation of the LCB1 protein with the LCB2 protein for serine palmitoyltransferase. *J Biol Chem*, **273**, 33787-33794.

Hanada K, Hara T, Nishijima M, Kuge O, Dickson RC, Nagiec MM (1997) A mammalian homolog of the yeast LCB1 encodes a component of serine palmitoyltransferase, the enzyme catalyzing the first step in sphingolipid synthesis. *J Bio Chem*, **272**, 32108-3211.

Ikeda M, Kihara A, and Igarashi Y (2004) Sphingosine-1-phosphate lyase SPL is an endoplasmic reticulum-resident, integral membrane protein with the pyridoxal 5'-phosphate binding domain exposed to the cytosol. *Biochem Biophys Res Commun*, **325**, 338-343.

Ikushiro H, Hayashi H, and Kagamiyama H (2004) Reactions of serine palmitoyltransferase with serine and molecular mechanisms of the actions of serine derivatives as inhibitors. *Biochemistry*, **43**, 1082-1092.

Ikushiro H, Hayashi H, and Kagamiyama H (2003) Bacterial serine palmitoyltransferase: a water-soluble homodimeric prototype of the eukaryotic enzyme. *Biochim Biophys Acta*, **1647**, 116-120.

Ikushiro H, Hayashi H, and Kagamiyama H (2001) A water-soluble homodimeric serine palmitoyltransferase from Sphingomonas paucimobilis EY2395^T strain. Purification, characterization, cloning, and overproduction. *J Biol Chem*, **276**, 18249-18256.

Ikushiro H, Hayashi H, Kawata Y, and Kagamiyama H (1998) Analysis of the pH- and ligand-induced spectral transitions of tryptophanase: activation of the coenzyme at the early steps of the catalytic cycle. *Biochemistry*, **37**, 3043-3052

Kallen RG, Korpela T, Martell AE, Matsushima Y, Metzler CM, Metzler DE, Morozov YV, Ralston IM, Savin FA, Torchinsky YM, and Ueno H (1985) Chemical and spectroscopic properties of pyridoxal and pyridoxamine phosphate. in *Transaminases* (Christen P, and Metzler DE, Eds.), pp 37-108, John Wiley & Sons, New York.

Kawasaki S, Moriguchi R, Sekiya K, Nakai T, Ono E, Kume K, and Kawahara K (1994) The cell envelope structure of the lipopolysaccharide-lacking gram-negative bacterium *Sphingomonas paucimobilis*. *J Bacteriol*, **176**, 284-290.

Mandon EC, Ehses I, Rother J, van Echten G, and Sandhoff K (1992) Subcellular localization and membrane topology of serine palmitoyltransferase, 3-dehydrosphinganine reductase, and sphinganine N-acyltransferase in mouse liver. *J Biol Chem*, **267**, 11144-11148.

Nagiec MM, Lester RL and Dickson RC (1996) Sphingolipid synthesis: identification and characterization of mammalian cDNAs encoding the Lcb2 subunit of serine palmitoyltransferase. *Gene*, **177**, 237-241.

Nagiec MM, Baltisberger JA, Wells GB, Lester RL, and Dickson RC (1994) The LCB2 gene of Saccharomyces and the related LCB1 gene encode subunits of serine palmitoyltransferase, the initial enzyme in sphingolipid synthesis. *Proc Natl Acad Sci USA*, **91**, 7899-7902.

Naka T, Fujiwara N, Yano I, Maeda S, Doe M, Minamino M, Ikeda N, Kato Y, Watabe K, Kumazawa Y, Tomiyasu I, and Kobayashi K (2003) Structural analysis of sphingophospholipids derived from *Sphingobacterium spiritivorum*, the type species of genus Sphingobacterium. *Biochim Biophys Acta*, **1635**, 83-92.

Schmidt A, Sivaraman J, Li Y, Larocque R, Barbosa JA, Smith C, Matte A, Schrag JD, and Cygler M (2001) Three-dimensional structure of 2-amino-3-ketobutyrate CoA ligase from Escherichia coli complexed with a PLP-substrate intermediate: inferred reaction mechanism. *Biochemistry*, **40**, 5151-5160.

Webster SP, Alexeev D, Campopiano DJ, Watt RM, Alexeeva M, Sawyer L, and Baxter RL (2000) Mechanism of 8-amino-7-oxononanoate synthase: spectroscopic, kinetic, and crystallographic studies. *Biochemistry*, **39**, 516-528.

Yabuuchi E, Tanimura E, Kosako Y, Ohyama A, Yano I and Yamamoto A (1979) Flavobacterium devorans ATCC10829: A strain of *Pseudomonas paucimobilis*. *J Gen Appl Microbiol*, **25**, 95-107.

Yabuuchi E, Kaneko T, Yano I, Moss CW, and Miyoshi N (1983) *Sphingobacterium* gen. nov., *Sphingobacterium spiritivorum* comb. nov., *Sphingobacterium multivorum* comb. nov., *Sphingobacterium mizutae* sp. Nov., and *Flavobacterium indologenes* sp. nov.: glucose-nonfermenting Gram-negative rods in CDC groups IIK-2 and IIb. *Int J Syst Bacteriol*, **33**, 580–598.

Yamaoka S, Miyaji M, Kitano T, Umehara H, and Okazaki T (2004) Expression cloning of a human cDNA restoring sphingomyelin synthesis and cell growth in sphingomyelin synthase-defective lymphoid cells. *J Biol Chem*, **279**, 18688-18693.

7-6 A Computer Visualization Model for the De Novo Sphingolipid Biosynthetic Pathway

Jin Young Hong[1], Jeremy C. Allegood[2], Samuel Kelly[2], Elaine Wang[2], Alfred H. Merrill Jr.[2], and May Dongmei Wang[1]

[1]The Wallace H. Coulter Department of Biomedical Engineering, [2]School of Biology, and [1,2]the Petit Institute of Bioengineering and Bioscience, Georgia Institute of Technology, Atlanta, GA 30332, USA

Summary. The *de novo* biosynthesis pathway for sphingolipids has thousands of individual components. This creates difficulties for scientists who conduct "sphingolipidomic" analysis of cells and must deal with a very large dataset. This article reviews various visualization techniques for large datasets, and introduces a new visualization system for sphingolipid biosynthesis, SphingoViz.

Keywords. sphingolipidomic, SphingoViz, pathway visualization

1.1 Introduction

Sphingolipids are major components of membranes and regulate diverse aspects of cell structure and function. The *de novo* biosynthesis pathway for sphingolipids has thousands of individual components not only due to diversity in the head groups (Merrill, 2002; Suzuki, 2002), but also due to the branching that occurs at the dihydroceramide-ceramide steps, which has a multiplier effect on the number of downstream molecular species (Fig. 1) (for overviews see Merrill and Sandhoff, 2002; www.sphingomap.org). Relatively high-throughput methods, such as mass spectrometry (Sullards and Merrill, 2001; Merrill et al., 2005), are available to analyze large numbers of these compounds, but this produces data sets that are so complex that it is difficult for scientists to interpret all of the information. Various visualization techniques, especially those for large datasets and pathways, are reviewed here and a new visualization system, SphingoViz, to study the sphingolipid biosynthetic pathway is introduced.

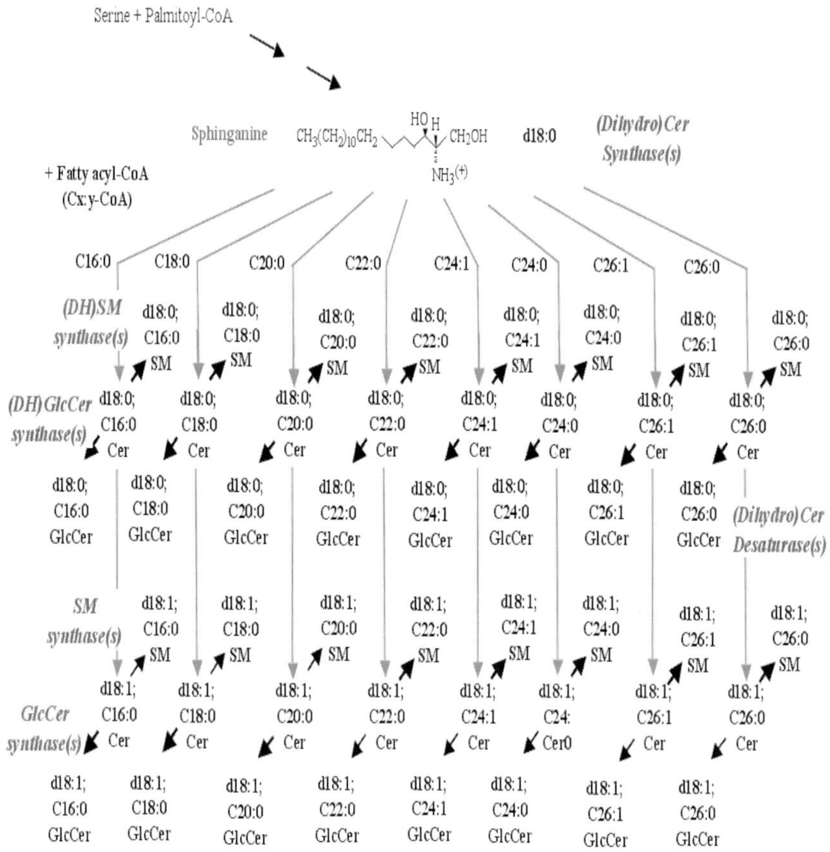

Fig. 1. Depiction of the "multiplier effect" of acylation of sphinganine (d18:0) in the *de novo* sphingolipid biosynthetic pathway using an abbreviated nomenclature (Fahy et al., 2005). Acylation of sphinganine (d18:0) with fatty acids with the carbon number and double bonds shown is followed by desaturation to ceramide (d18:1 followed by the fatty acid) as well as addition of headgroups for sphingomeylin (SM) or glucosylceramide (GlcCer). The diagram is actually a simplification because it does not show the branches to galactosylceramide and ceramide 1-phosphate, nor the additional types of backbones, such as 4-hydroxysphinganine (phytosphingosine), that are made by some cell types.

1.2 Related Work on Pathway Visualization

Interpreting complex, large data sets has always challenging, hence information visualization is a viable technique to assist the interpretation. Researchers developed techniques such as overview-and-detail to handle large and complex data structures, where "Overview" provides the overall pattern of data, and "detail" displays precise information. It uses, among others, a 'focus+context' technique (Sarker and Brown, 1992; Lamping et al., 1995; Jankun-Kelly and Ma, 2003), which shows the region-of-interest in detail while the remaining area has less detail. This way, the area of interest can be illustrated within the overall structure together.

In pathway visualization, most pathways are depicted graphically, where nodes (metabolites) are interconnected with edges. There are several visualization techniques to graph, or network, data. i) The fisheye lens technique to explore graphs (Sarker and Brown, 1992) depends on a focal point and degree of interest (DOI) function where the region-of-interest can blow up for display with high resolution and large fonts, while the rest of the graph remains in low resolution and small fonts. Users can change focus area visualization at will, but may find the distortion annoying. ii) A focus and context approach can display large hierarchical structures using the tree-like structure within a hyperbolic plane that is mapped to a circular display region (Lamping et al., 1995). In the hyperbolic browser, the user selects a node and, with animation, that area is brought to the center. iii) A rooted tree using a circular layout can be used to explore large data sets (Melancon and Herman, 1998). Focal points are determine in this visualization by assigning a scaling factor to each node, however it is easy to lose track of the underlying tree-structure with a large number of levels because any node can be a root. iv) A parent and children family of nodes can be arranged in a circular layout similar to the previous method, (Teoh and Ma, 2002) were the parent/child node can moved between the inner and circumference regions of the circle. In this way, the limited display space is used more efficiently to visualize large hierarchies. Potential overlapping between children in the inner and outer circles may complicated efforts to differentiate levels of a hierarchical structure. v) GScope also uses the fisheye view (Sarker and Brown, 1992) to visualize a complicated biomolecular network (Toyoda et al., 2003). With GScope the selected node is displayed in the center of the screen with all details, while others are represent in less detail. Views change using a smooth animation and alternative network display options are available to visualize the whole network. This option allows users to understand the overall structure of the network. GScope displays the list of metabolites in the network, and it allows interaction between this list and the visualization.

Nonetheless, the distorted view could annoy users attempting to understand the overall structure when a change of focus occurs

For these overview-and-detail methods, two or more viewing panels may be used. One panel could show overall information while the other displays detailed information of a part of the data. User interaction, such as zooming and panning (Bederson et al., 1996), is important for effective exploration of the complex data set. For example, PathViewer (Sirava et al., 2002) visualizes metabolic pathways with an automatic graph layout algorithm. This tool can display the overall structure of the pathway in an overview viewing panel, and can display level-of-detail content information of the pathway in the detail viewing panel through zooming in and out. PathViewer allows the user to select the desired pathway, but animation is not always smooth and may distract users. In addition, if the graph structure is complex, the display appears cluttered complicating interactive efforts to thoroughly explore the data.

PathwayAssist   (http://www.ariadnegenomics.com/products/pathway. html) provides a magnifier window for a detailed view of regions within the general pathway. Zooming in and out of the pathway using a menu button is also possible. This tool uses NLP (Natural Language Processing) algorithm (Friedman et al., 2001), which can read the literature from the internet and create the pathway automatically. But againif the pathway structure is complicated, such as the sphingolipid pathway, the overall view will be hard to understand due to limited interactivity.

Cytoscape (Shannon et al., 2003) visualizes molecular networks using both overview and detail viewing panels. It has various automated network layout algorithms, such as spring embedded (Eades, 1984), circular, hierarchical layout to allow the user more control over how the target area of interest is presented. In addition, Cytoscape provides tools to study the pathway, for example, by zooming in and out, connecting to other database, and editing functionality. However, when the pathway becomes large and complicated, the system slows down and interaction with the visualization viewer and the network structure is harder.

In addition to laying out the graph structure in 2D or with double viewing panels, there have also been efforts to visualize the structure in 3D. Robertson et al. (1991) lay out a tree structure into a 3D cone shape with each node placed at the apex of a cone and its children at the base of the cone. This layout algorithm was later improved (Carriere et al., 1995) by traversing the tree structure to compute the approximate size of the base from bottom to top. This prevents node occlusion and supports the imaging of a large tree structure in 3D. Munzner (1998) visualizes a large graph structure by laying it out in 3D hyperbolic space and the additional space in this 3D hyperbolic browser can support large graphs, dynamic exploration, and interactive browsing, but the projection of the 3D layout onto a 2D display screen causes occlusion. Brandes et al. (2004) visualize metabolic pathways by stacking up each pathway in 3D for the comparative analysis of pathway across species. However, this visualization does not provide much user interaction for further exploration, and it does not fit to the large graph structure.

2. Design and development of visualization system

Similar to gene ontologies (Ashburner et al., 2000; Zeeberg et al., 2003), which organize genes as hierarchical categories based on biological process, molecular functions and sub-cellular localization, the sphingolipid pathway can be represented as a hierarchical structure. We have developed a visualization system, SphingoViz (Sphingolipid Visualization), to represent the sphingolipid pathway using a 'focus+context' technique.

2.1 Layout algorithm

The sphingolipid pathway can be viewed as a complex hierarchical structure where a substrate is a parent and its product is a child. The substrate can associate with different products, and a parent can have several children depending on which enzyme is involved in the reaction. Nodes in the hierarchy represent metabolites, so SphingoViz displays the sphingolipid pathway's hierarchical structure in a circular layout as illustrated in Fig. 2. On the right, the parent node A (a substrate) is placed in the center of the large circle, and the children (products) of this node, B, C, D and E, are evenly positioned at the circumference of this circle. B and D are also parents of other nodes (nodes F and G plus H, respectively). Node D is placed at the center of a small circle, and its children, G and H, are placed at the circumference. Similarly, nodes B and F are placed at a small circle as illustrated in Fig. 2. Through this recursive process, SphingoViz creates a circular layout of sphingolipid pathway hierarchy.

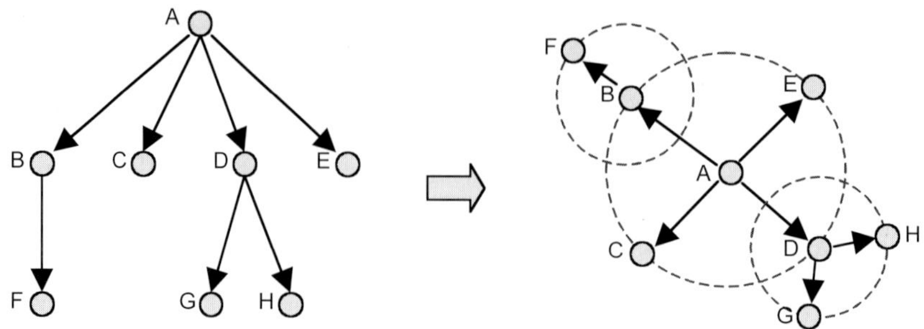

Fig. 2. Conversion of a top-down pathway into a circular representation.

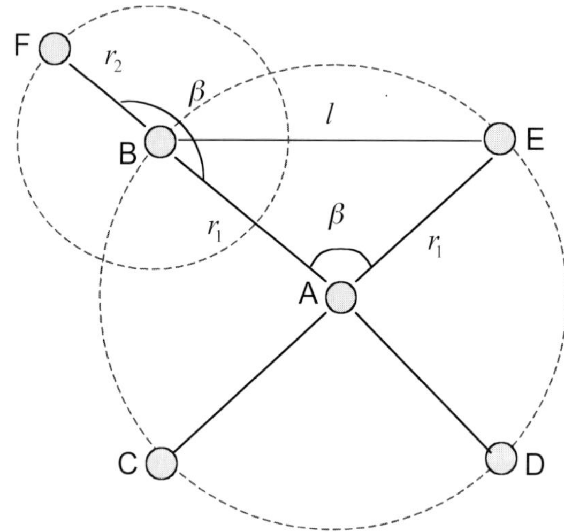

Fig. 3. Computation of radius and angle to place nodes at each level.

When the size of pathway is large or as users progress from beginning nodes to intermediate ones in a pathway, space becomes congested and representations overlap. To prevent overlapping, SphingoViz decreases the radius of the circle around each node as its distance to the center of the focus node increases. A node with a miniscule radial length is not displayed. Radial size depends on the number of nodes. For example, in Figure 2, the number of children of node A determines the radius of small circles, which contains node B and D. This is illustrated in more detail in Figure 3. The radius of the first circle r_1 in Figure 3 is defined initially and then nodes B, C, D, and E are evenly placed by an angle α defined as follows

$$\pi = \frac{2\pi}{n}$$

where n is the number of children of a node A. In this case, $n = 4$ (nodes B, C, D, and E). The next step is to compute the length l between the nodes B and E as:

$$l^2 = r_1^2 + r_1^2 \, \alpha \, 2r_1^2 \cos\alpha = 2r_1^2(1 \, \alpha \, \cos\alpha).$$

The radius of next level r_2 is less than half of the length l to ensure that the nodes at the next level do not overlap.

$$r_2 < \frac{l}{2}$$

In addition, r_2 is computed to prevent overlap between nodes at subsequent levels. Hence, the radius decreases as the number of children increases. We define the following formula for this requirement.

$$r_2 = \frac{l}{1.8n^{1/3}} \qquad \text{if } n < 6$$

$$r_2 = \frac{l}{2.6n^{1/6}} \qquad \text{otherwise}$$

The angle β in Figure 3 to locate node F is computed as follows

$$\pi = \frac{2\pi}{n_k + 1}$$

where n_k is the number of children of the current node. In this case, $n_k = 1$ and the current node is B. After the radius and angle are defined, the nodes are placed evenly on the circle. This process is repeated until the radius r_k is very small. For example, the radius $r_3, r_4, r_5, ..$ are computed by

$$r_k = \frac{l}{2} \qquad \text{if } n_k = 1$$

$$r_k = \frac{l}{2.3n_k^{1/5}} \qquad \text{if } n_k < 9$$

$$r_k = \frac{l}{3.5} \qquad \text{otherwise.}$$

Applying this algorithm to a balanced 5-level hierarchy data set, with each node having the same number of children, the SphingoViz creates a great visual representation (Fig. 4). The focused region is the root, which is displayed in detail. As the children nodes get farther away from this root node, they are represented in less detail.

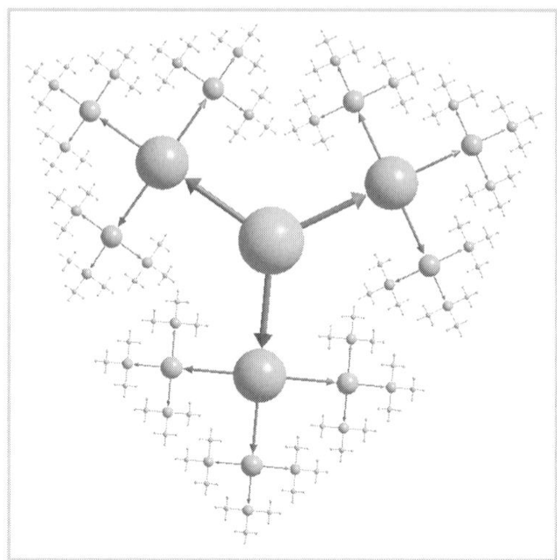

Fig. 4. Visualization of balanced tree with depth size of 5, and 3 children per node using a circular layout algorithm.

2.2 Interactive visualization using focus+context paradigm

When a user clicks on a region of interest in SphingoViz, that node will be brought into the central focused area.   When the change-of-focus occurs, Smooth animation prevents users from being distracted by sudden visual changes and sustains the whole pathway visually.   When other 'focus+context' techniques, such as the hyperbolic tree and circular tree discussed above, (Lamping et al., 1995; Melancon and Herman, 1998; Munzner, 1998) visualizations of the embedded data structure are less clear when change-of-focus occurs.   That is, as the area of interest moves to the center of screen, the root-children relationship is not intuitive or visible anymore. To avoid this problem, SphingoViz highlights the path from the starting metabolite (root) to the currently clicked metabolite (green node) in yellow (Fig. 5).

The remaining nodes around this highlighted path are distributed by the layout algorithm described in the previous section.   This way, a user can see the overall data structure of a pathway with the region-of-interest in more detail.   The entire path from the root to the currently clicked node is highlighted, so when there are many nodes in the highlighted path, the path can be larger than the screen size.   To prevent this problem, the length of the highlighted path is adjusted so that the main path is always visualized inside the screen space.

Fig. 5 illustrates how the focused area changes using a glycosphingolipid pathway. In this example, the starting metabolite is ceramide, which is the focused region in the first panel, and when a user clicks the node "Galβ3 GalNAcβ4 Galβ4Glcβ-Cer" (shown as GA1), the node is represented in more detail as shown in the last panel in Fig. 5.

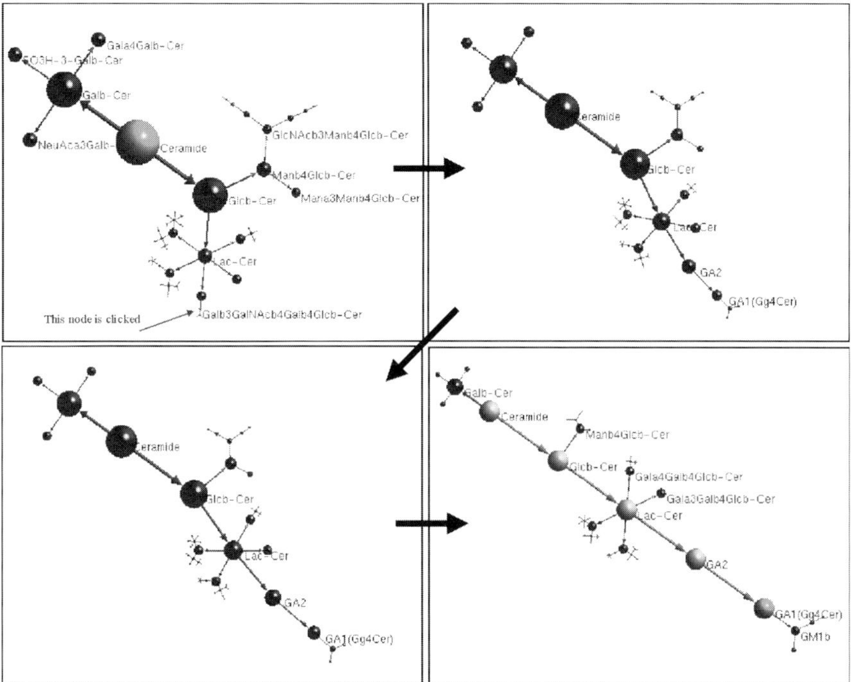

Fig. 5. Smooth animation of the glycosphingolipid pathway when node Galβ3 GalNAcβ4 Galβ4Glcβ-Cer (GA1) is selected. The full path from root to selected node is targeted with the side structure along this path represented in less detail.

3. SphingoViz functionalities

SphingoViz provides interactive visualization to help the researcher interpret complex data structure quickly. To this end, searching, editing, connecting to databases and facilitating differential comparative studies are part of its standard functionalities.

3.1 Comparative study of differential experimental datasets

As shown in Fig. 6, SphingoViz uses a color scale (blue to red) to represent the amount of each compound in the pathway, which is also displayed by a number in the parenthesis next to the compound. Changes in the amounts of each compound with time are seen by clicking through multiple screens, which is illustrated in Fig. 6 by showing three timepoints as a series (i.e., time zero at the top and after one, middle, and two hours, bottom). SphingoViz also allows side-by-side views of two data sets where one represents a control group and the other an experimental group such as cells that have either been altered genetically, treated with a drug, or other perturbation. Thus, one can readily recognize changes with time and differences between experimental and control groups by observing the colors.

In the case of the results shown in Fig. 6, the experiment was to compare *de novo* sphingolipid biosynthesis in two cell types: Hek293 cells (left) and Hek293 cells that have been transfected with cDNA for both subunits of the first enzyme of the pathway, serine palmitoyltransferase (SPT1/SPT2) (right), which results in a >2 fold increase in SPT activity (unpublished results). To follow the appearance of newly made molecules, the cells have been incubated with [U-^{13}C]palmitic acid and the individual sphingolipid species were analyzed by tandem mass spectrometry (Merrill et al. 2005). SphingoViz screens that were selected for Fig. 6 show [^{13}C]sphinganine (the large node shown at the hub) and downstream [^{13}C]-labelled ceramides with different fatty acyl chain lengths (the nodes that radiate from sphinganine). This labeling protocol produces four versions of each compound, all of which can be quantified by mass spectrometry: molecules that have been present since the beginning of the treatment (i.e., with [^{12}C] in both the sphingoid base backbone and fatty acid sidechain); molecules made *de novo* from [^{13}C]palmitate and contain [^{13}C] in the sphingoid base backbone or in both the sphingoid base backbone and amide-linked fatty acid (i.e., newly made sphingolipids are the sum of these species); and molecules that have been made from a pre-existing sphingoid base (i.e., that is [^{12}C]-labeled) and a newly added [^{13}C]fatty acid. Here, the sum of backbone-labeled and backbone-plus-fatty-acid-labeled species are shown since this represents the compounds synthesized *de novo* over this time course. Thus, the relative amounts of the newly made, [^{13}C]-labeled sphingolipids change from dark blue (zero) to brighter colors (red for the highest amounts) over time.

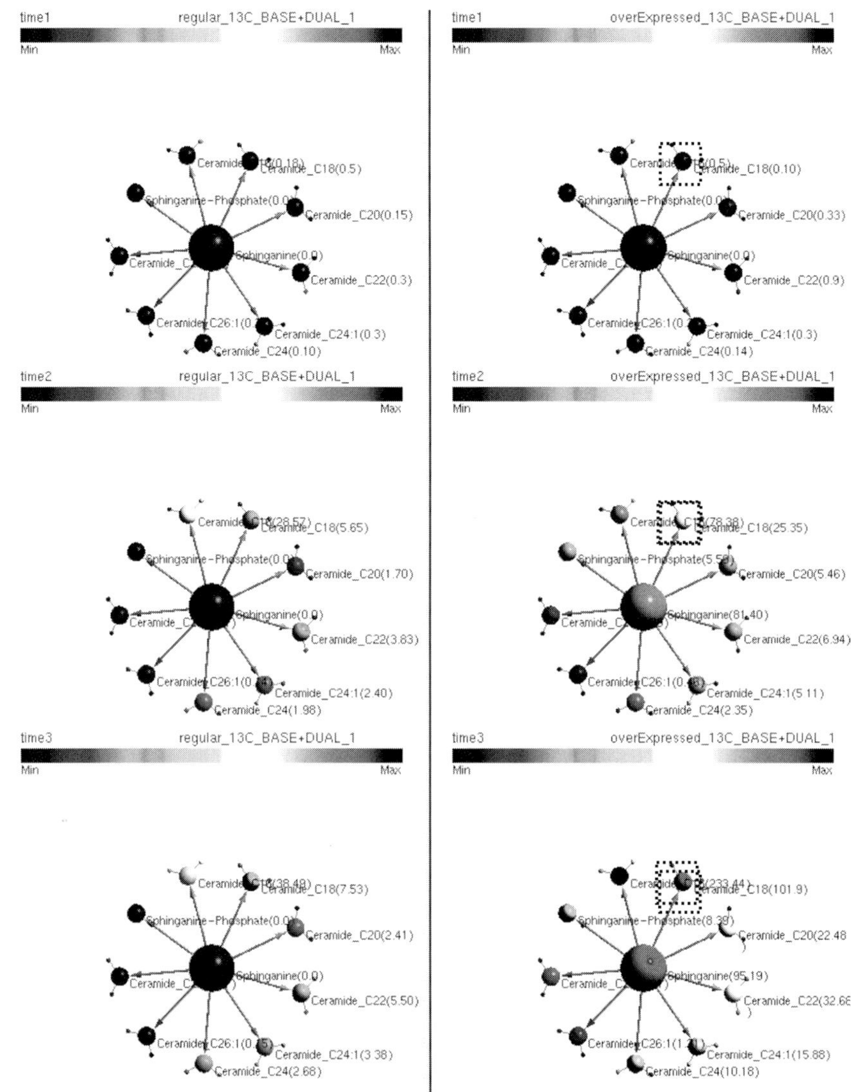

Fig. 6. Visualization of differences in ceramide biosynthesis by control (left) and SPT overexpressing (right) panels, as described in the text.

Comparison of the different chain length ceramides that are made (the nodes that encircle sphinganine at the hub in Fig. 6), reveals not only that the amounts of some species increase more rapidly over time than others (i.e., C16 >> C18 = C22 > C20 = C24 = C24:1 for the Hek cells, left), but also, that there are differences in the amounts and relative distribution for the cells overexpressing serine palmitoyltransferase (i.e., comparing the right versus the left panels).   This comparison clearly shows that the SPT overexpressing cells have a disproportionate increase in C18-ceramides (i.e., note that in Hek cells, the colors for C18 and C22 are essentially the same shade of light blue whereas in the SPT overexpressing cells, C18-species—which has been highlighted by a dashed box--is clearly more red-shifted than C22-subspecies is).   Hence, the use of such color schemes allow rapid perusal of the results of an experiment to see differences that might not be as noticeable if the results were shown in large tables or complicated bar graphs.

This unexpected finding, i.e., that overexpression of the first enzyme of *de novo* sphingolipid biosynthesis is accompanied induction of the ceramide synthase that specifically makes C18-ceramides (LASS1), has subsequently been confirmed by analysis of the amounts of LASS1 mRNA (by QRT-PCR) and protein (by Western blotting) (Ying Liu, unpublished). Hence, SphingoViz can easily visualize changes over time and enable comparisons between groups.   This type of data can, of course, also be displayed in table or more conventional graph; however, the differences are more easily seen in this new way of depicting the data.

3.2 Connection with other information database or software

When researchers study pathways, they are often interested in the chemical structure of each compound.   In SphingoViz, each node in the gly-cosphingolipid pathway is connected to its 2D chemical structure database, and can be displayed in the structure viewer at the bottom of SphingoViz. Fig. 7 shows a portion of the glycosphingolipid pathway after ganglioside GM3 is clicked.   The pathway viewer highlights the path from the root, ceramide, to ganglioside GM3, and a 2D structure of GM3 is displayed in the structure viewer.   In addition to connecting to chemical structural data themselves, these structural data are also linked to 'ChemDraw', which is a software tool to create, edit, and analyze chemical structures.   When the 2D structure in structure viewer is clicked, ChemDraw software is auto-matically activated.

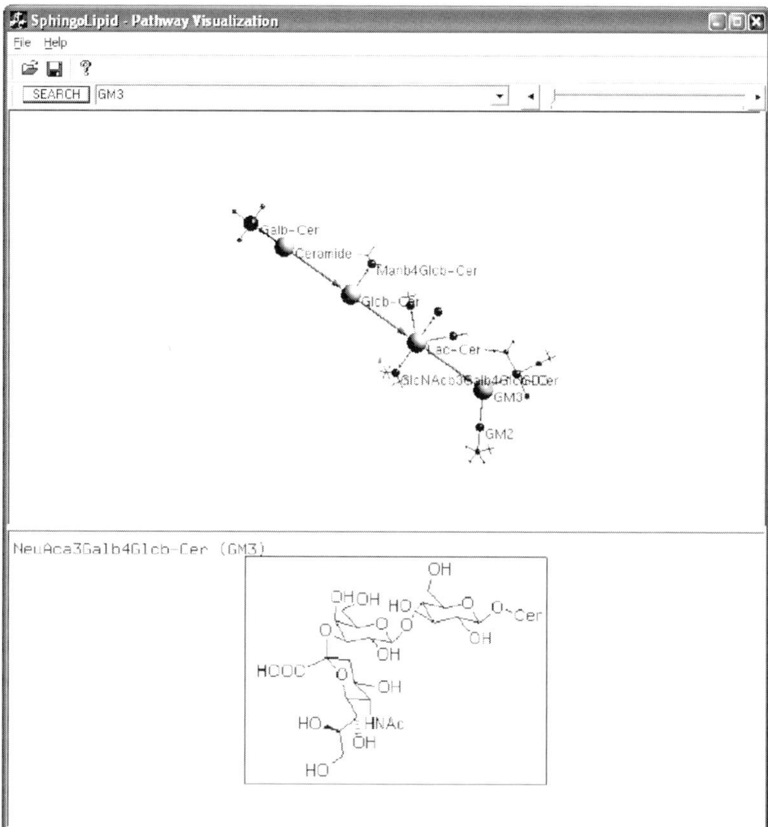

Fig. 7.Glycosphingolipid pathway visualization when ganglioside GM3 is clicked to display its structure in the structure viewer (bottom).

3.3 Multiple searching

SphingoViz also provides searching functionality, which is critical to mine huge data sets effectively. SphingoViz provides two methods to search on one node or several nodes in the pathway: a node can be directly found either by typing the name of a compound into the text entry in the dialog bar, or by using a slide bar, which contains all of the names of compounds in the pathway. Taking one of two actions will result in displaying the whole path from the root to the searched node with smooth animation.

In addition to searching on the exact name of a node, SphingoViz has multiple-search functionality. For example, it will search all the compounds that contain ganglioside GM1 in their structures when the user types GM1 into the text entry. If the user selects one of the items from the list, SphingoViz will display the whole path from the root (ceramide) to the clicked item.

3.4 Annotation

SphingoViz can be annotated so that a user can add or delete a node in the pathway as needed, and to edit the node information. After a user changes the information of the pathway, the user can save the new pathway and can retrieve it for future use.

4. Conclusion

Using a 'focus+context' technique, SphingoViz visualizes many features of complicated pathways in shingolipid biological processes effectively, is interactive, and facilitates comparative analysis. As such, it is a useful for to help researchers observe changes in the amounts of many compounds by observing color distribution and then connect to other information data sources. As SphingoViz develops the entire "sphingolipidome" will be visiualised (http://www.sphingomap.org) as well as other pathways within the lipidome (see LipidMap, http://www.lipidmaps.org). SphingoViz will be soon available from Bio-MIBLab (http://www.miblab.gatech.edu) and sphinGOMAP (http://www.sphingomap.org/).

Acknowledgements. This research has been supported by Georgia Cancer Coalition Distinguished Cancer Scholar Award, a Seed Grant from Institute of Bioengineering and Bioscience of Georgia Tech, and NIH Bioresearch Partnership Grant (NIH R01CA108468).

5. References

Bederson B, Hollan J, Perlin K, Meyer J, Bacon D, Furnas G (1996) Pad++: A Zoomable Graphical Sketchpad for Exploring Alternate Interface Physics. *Journal of Visual Languages and Computing*, **7**, 3-31.

Brandes U, Dwyer T, Schreiber F (2004) Visualizing Related Metabolic Pathways in Two and a Half Dimensions. Proc. 11th Intl. Symp. Graph Drawing (GD '03) © Springer-Verlag 111-122.

Carriere J, Kazman R (1995) Interacting with Huge Hierarchies: Beyond Cone Trees. *Proc. of the IEEE Conference on Information Visualization*, 74-81.

Eades P (1984) A heuristic for graph drawing. *Congressus Numerantium*, 142-160.

Fahy E, Subramaniam S, Brown H, Glass C, Merrill A, Murphy R, Raetz C, Russell D, Seyama Y, Shaw W, Shimizu T, Spener F, van Meer G, VanNieuwenhze M, White W, Witztum J, Dennis E (2005) A comprehensive classification system for lipids. *J Lipid Res*, **46**, 839-862.

Friedman C, Kra P, Yu H, Krauthammer M, Rzhetsky A (2001) GENIES: a natural-language processing system for the extraction of molecular pathways from journal articles. *Bioinformatics*, **17**, S74-S82.

http://www.ariadnegenomics.com/products/pathway.html, PathwayAssist

http://www.sphingomap.org

Jankun-Kelly T, Ma K (2003) MoireGraphs: Radial Focus+context Visualization and Interaction for Graphcs with Visual Nodes. *Proceedings of IEEE 2003 Symposium on Information Visualization*, 59-66.

Kroesen BJ, Jacobs S, Pettus BJ, Sietsma H, Kok JW, Hannun YA, de Leij LF (2003) BcR-induced apoptosis involves differential regulation of C16 and C24-ceramide formation and sphingolipid-dependent activation of the proteasome. *J Biol Chem*, **278**, 14723-14731.

Lamping J, Rao R, Pirolli P (1995) A focus+context technique based on hyperbolic geometry for visualizing large hierarchies. *SIGCHI*, 401-408.

M. Ashburner M, Ball C, Blake J, Botstein D, Butler H, Cherry J, Davis A, Dolinski K, Dwight S, Eppig J, Harris M, Hill D, Issel-Tarver L, Kasarskis A, Lewis S, Matese J, Richardson J, Ringwald M, Rubin G, Sherlock G (2000) Gene ontology: tool for the unification of biology. The Gene Ontology Consortium. *Nat Genet*, **25**, 25-29.

Melancon G, Herman I (1998) Circular Drawings of Rooted Trees. *Reports of the Center for Mathematics and Computer Sciences*, Report number INS-9817.

Merrill A (2002) Minireview: De Novo Sphingolipid Biosynthesis: A Necessary, but Dangerous, Pathway. *J Biol Chem*, **277**, 25843-25846.

Merrill A and Sandhoff K (2002) Sphingolipids: metabolism and cell signaling. *Biochmistry of Lipids, Lipoproteins and Membranes*, **4**, 373-407.

Merrill A, Sullards MC, Allegood J, Kelly S, Wang E (2005) Sphingolipidomics: high-throughput, structure-specific, and quantitative analysis of sphingolipids by liquid chromatography tandem mass spectrometry. *Methods*, **36**, 207-224

Munzner T (1998) Exploring Large Graphs in 3D Hyperbolic Space. *IEEE Computer Graphics & Applications*, **18**, 18-23.

Robertson G, Mackinlay J, Card S (1991) Cone trees: Animated 3D visualizations of hierarchical information. *In Proc. of ACM SIGCHI conference on Human Factors in Computing Systems,* 189-194.

Sarkar M, Brown M (1992) Graphical Fisheye Views of Graphs. *Conference on Human Factors in Computing Systems*, 83-91.

Shannon P, Markiel A, Ozier O, Baliga N, Wang J, Ramage D, Anim N, Schwikowsky B, Ideker T (2003) Cytoscape: A software environment for Integrated Models of Biomolecular Interaction Networks. *Genome Research*, **13**, 2498-2504.

Sirava M, Schäfer T, Eiglsperger M, Kaufmann M, Kohlbacher O, Bornberg-Bauer E, Lenhof H (2002) BioMiner-modeling, analyzing, and visualizing biochemical pathways and networks. *Bioinformatics*, 219-230.

Sullards MC, Merrill A (2001) Analysis of Sphingosine 1-Phosphate, Ceramides, and Other Bioactive Sphingolipids by High-Performance Liquid Chromatography-Tandem Mass Spectrometry. *Science's SIKE*, **67**, PL1.

Suzuki A (2002) Map3: Biosynthetic Pathways of Glycosphingolipids. *Handbook of glycosyltransferases and related genes*, 636-646.

Teoh S, Ma K (2002) RINGS: A Technique for Visualizing Large Hierarchies. *Graph Drawing*, 268-275.

Toyoda T, Mochizuki Y, Konagaya A (2003) GSCope: a clipped fisheye viewer effective for highly complicated biomolecular network graphs. *Bioinformatics (Application note)*, **19**, 437-438.

Zeeberg B, Feng W, Wang G, Wang MD, Fojo A, Sunshine M, Narasimhan S, Kane D, Reinhold W, Lababidi S, Bussey K, Riss J, Barrett J, Weinstein J (2003) GoMiner: a resource for biological interpretation of genomic and proteomic data. *Genome Biology*, **4**, R28.

Appendix

I. Inhibitors of Sphingolipid Biosynthesis

Jin-ichi Inokuchi

The experimental approach to deplete cellular (glyco)sphingolipids with the specific inhibitors for their biosynthetic cascade has been proved to be useful to identify functions of endogenous (glyco)sphingolipids. The site of action of each inhibitor is summarized on the biosynthesis and metabolism map Fig. 1. The detailed information for the inhibitors of sphingolipid synthesis and glycosphingolipid synthesis (glucosylceramide synthase) are listed in Table I and II, respectively.

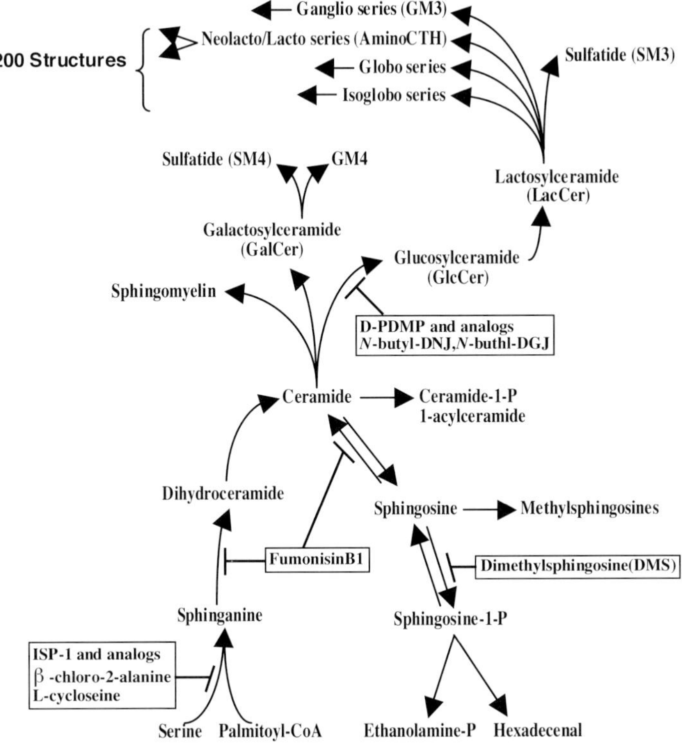

Fig. 1. Inhibitors of (glyco)sphingolipid biosynthesis.

Table 1. Inhibitors for sphingolipid synthesis.

Structure	IC_{50}	Conc. to be used in cell culture
Serine palmitoyltransferase inhibitors		
Myriocin(ISP-1) [a),b)]	0.3nM	47nM
SphingofunginB [c),d)]	3.2nM	ND
ViridiofunjinA [d)]	4.7nM	ND
β - chloro-L-alanine [e)] $\quad$ $ClCH_2CH(NH_2)COOH$	ND	5~25mM
L-cycloserine [f)]	ND	2mM
Ceramide synthase inhibitor		
FumonisinB1 [g),h)] $\quad$ $R=COCH_2CH(COOH)CH_2COOH$	0.1μM	1~50μM

a) Miyake Y, Kozutsumi Y, Nakamura S, Fujita T, Kawasaki T. *Biochem. Biophys. Res. Commun.*,**211**,396(1995)

b) Nakamura S, Kozutsumi Y, Sun Y, Miyake Y, Fujita T, Kawasaki T. *J. Biol. Chem.*.**271**,1255(1996)

c) Zweerink MM, Edison AM, Wells GB, Pinto W, Lester RL. *J. Biol. Chem.* **267**,25032(1992)

d) Mandala SM, Thornton RA, Frommer BR, Dreikorn S, Kurtz MB. *J. Antibiot.*, **50**,339(1997)

e) Gillard BK, Harrell RG, Marcus DM. *Glycobiology*, **6**,33(1996)

f) Wakita H, Nishimura K, Tokura Y, Furukawa F, Takigawa M. *J. Invest. Dermatol.*,**107**,336(1996)

g) Wang E, Norred WP, Bacon CW, Riley RT, Merrill AH Jr. *J. Biol. Chem.*, **266**,14486(1991)

h) Merrill AH Jr, Wang E, Gilchrist DG, Riley RT. *Adv. Lipid Res.*, **26**,215(1993)

Table 2. Glucosylceramide synthase inhibitors.

Structure	IC_{50} [μM]	Conc. to be used in cell culture [μM]
PDMP analogs		
D-*threo*-PDMP [a),b)] 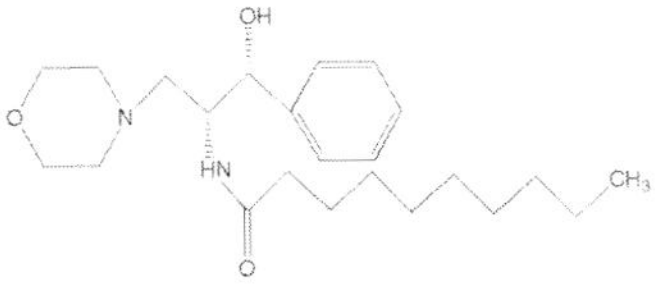	5	5~20
D-*threo*-PPPP(P4) [c),d)] 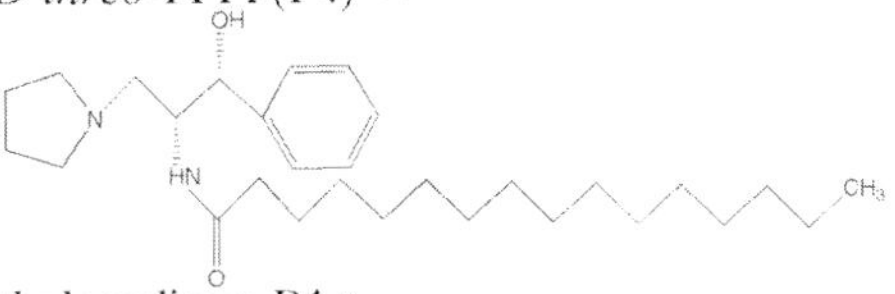	0.5	5~20
Ethylenedioxy-P4 [d)] 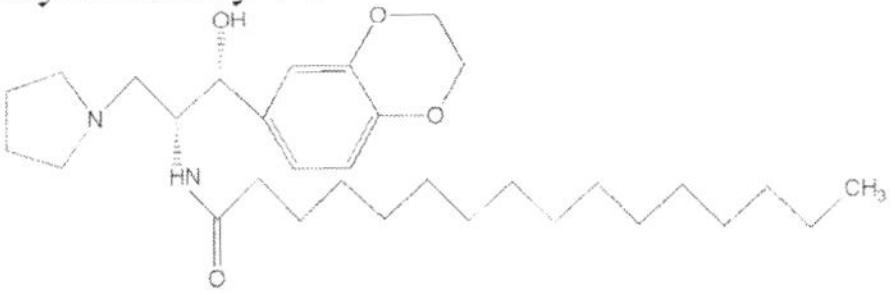	0.1	0.01~0.1
D-*threo*-PBPP [e)] 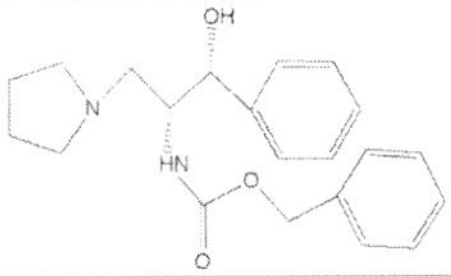	0.3	5~160
N-alkylimino sugars		
N-butyl-DNJ [f)] 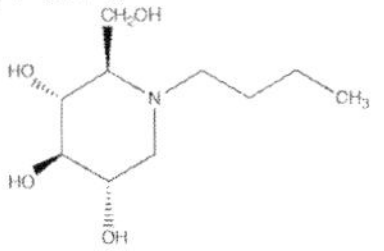	20	5~500
N-butyl-DGJ [g)] 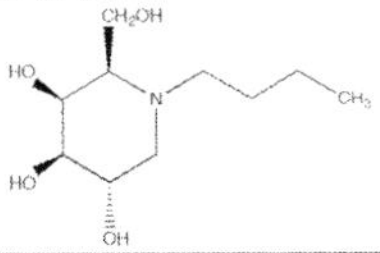	41	5~500

a) Inokuchi J, Radin NS. *J. Lipid Res.*, **28**,565(1987)

b) Radin NS, Shayman JA, Inokuchi J. *Adv. Lipid Res.*,**26**,183(1993)

c) Abe A, Radin NS, Shayman JA, Wotring LL, Zipkin RE, Sivakumar R, Ruggieri JM, Carson KG, Ganem B. *J. Lipid Res.*,**36**, 611(1995)

d) Abe A, Gregory S, Lee L, Killen PD, Brady RO, Kulkarni A, Shayman JA. *J. Clin. Invest.*,**105**, 1563(2000)

e) Jimbo M, Yamagishi K, Yamaki T, Nunomura K, Kabayama K, Igarashi Y, Inokuchi J. *J. Biochem.*, (Tokyo),**127**,485(2000)

f) Platt FM, Neises GR, Dwek RA, Butters TD. *J. Biol. Chem.*,269,8362(1994)

g) Platt FM, Neises GR, Karlsson GB, Dwek RA, Butters TD. *J. Biol. Chem.*.**269**,27108(1994)

II. The Genes in Yeast and Mammals

Yasuyuki Igarashi, Akio Kihara, and Yoshio Hirabayashi

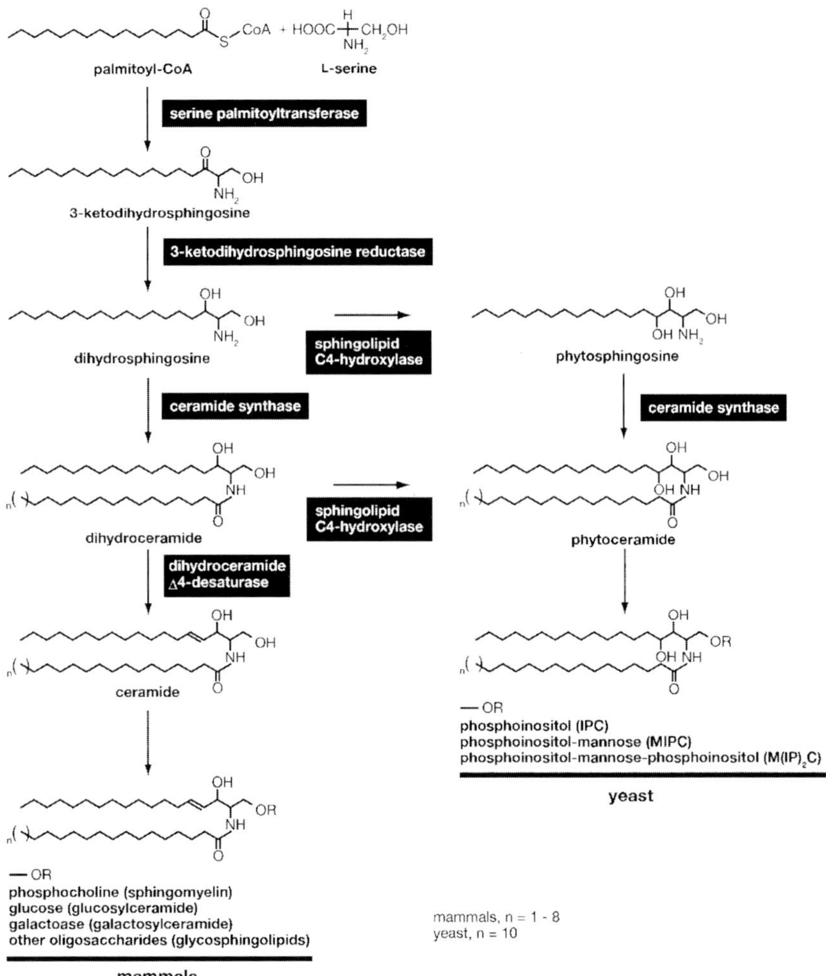

Fig. 1. Pathways of sphingolipid biosynthesis in yeast and mammals.

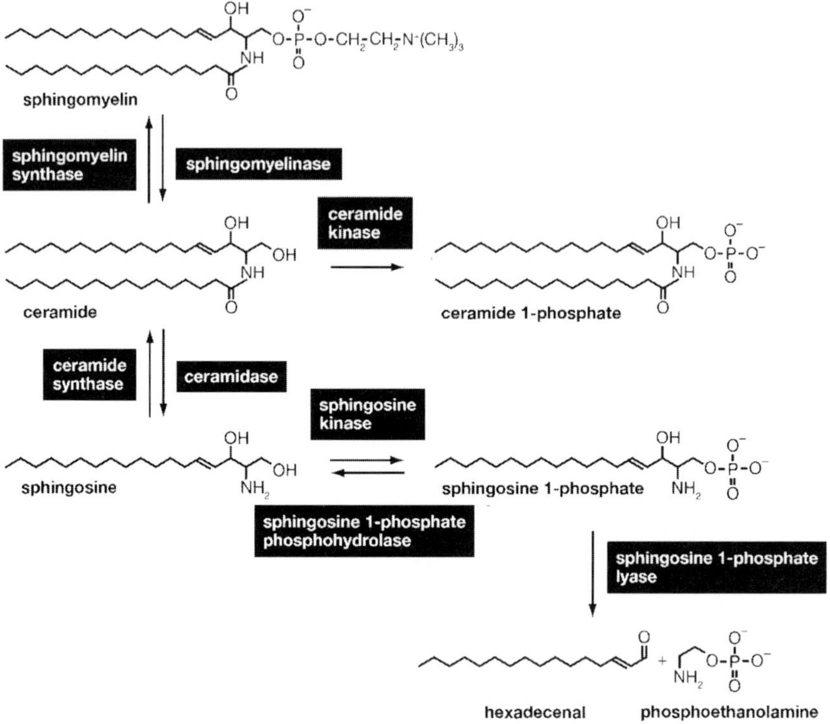

Fig. 2. Sphingomyelin metabolic pathways.

Table 1. Genes involved in de novo sphingolipid synthesis and sphingomyelin metabolism.

Gene	*S. cerevisiae*	*Homo sapiens*
Serine palmitolytransferase	*LCB1/TSC2/END8/YMR296c* (M63674) *LCB2/TSC1/SCS1/YDR062w* (L33931) *TSC3/YBR085c-a* (Z35955)	*SPTLC1/LCB1* (Y08685) *SPTLC2/LCB2* (Y08686)
3-Ketodihydrosphingosine reductase	*TSC10/YBR265w* (Z36134)	*FVT1* (X63657)
Ceramide synthase	*LAG1/YHL003c* (AY558514) *LAC1/YKL008c* (Z28008) *LIP1/YMR298w* (X80836)	*LASS1/UOG1/LAG1* (AF105005) *LASS2/TRH3* (BC010032) *LASS3* (NM_178842) [a] *LASS4/TRH1* (NM_024552) *LASS5/TRH4* (BC032565) *LASS6* (NM_203463)
Sphingolipid C-4 hydroxylase/ Δ4-desaturase	*SUR2/SYR2/YDR297w* (U07171) [b]	*DES1* (AF002668) [c] *DES2* (NM_206918) [d]
Sphingomyelin synthase/ IPC synthase	*AUR1/YKL004w* (U49090)	*SMS1* (AB154421) *SMS2* (NM_152621)
Sphingomyelinase/Inositol phosphosphingolipid phospholipase C	*ISC1/YER019w* (U18778)	*SMPD1/ASM* (M59916) *SMPD2/nSMase* (AJ222801) *SMPD3/nSMase 2* (AJ250460) *alk-SMase* (AY230663)
Ceramidase	*YPC1/YBR183w* (AF191745) *YDC1/YPL087w* (AF214455)	*ASAH1/ASAH/AC* (U70063) *ASAH2/NCDase/Mito-CDase* (AF449759, AF250847) [e] *ASAH3/ACER1* (AF347024) *PHCA/APHC* (AF214454)
Sphingosine kinase	*LCB4/YOR171c* (Z75078) *LCB5/YLR260w* (U17244)	*SPHK1* (BC030553, NM_021972, AF200328) [e] *SPHK2* (BC006161, AF245447) [e]
Sphingosine 1-phosphate phoshohydrolase	*LCB3/YSR2/LBP1/YJL134w* (Z49410) *YSR3/LBP2/YKR053c* (Z28278)	*SPP1/SGPP1* (AJ293294) *SPP2/SGPP2* (AF542512)
Sphingosine 1-phosphate lyase	*DPL1/BST1/YDR294c* (U51031)	*SPL/SGPL* (AI128825)
Ceramide kinase	not found	*CERK* (AB079066)

[a] There is no report describing the enzyme activity at present.

[b] Sphinsoshine C-4 hydroxylase, which prefers sphinganine to dihydroceramide.

[c] Dihydroceramide Δ4-desaturase.

[d] Dihydroceramide C-4 hydroxylase/Δ4-desaturase, a bifunctional enzyme.

[e] Variants that differ in their N-terminal lengths exist.

III. Structure of Glycosphingolipids and Metabolic Pathway

Akemi Suzuki

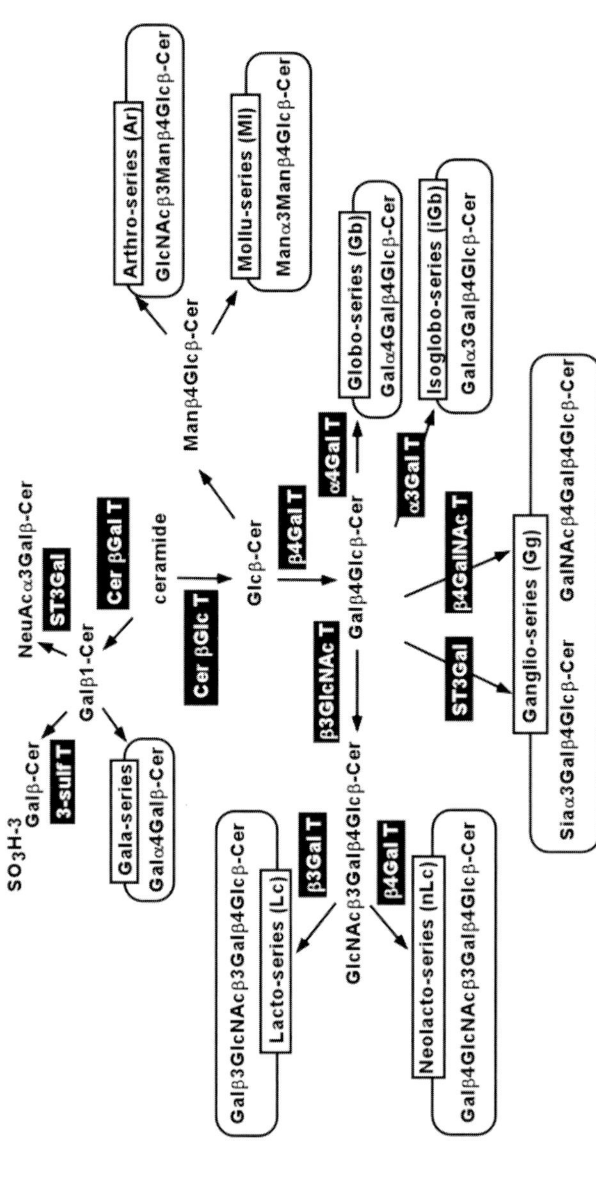

Fig. 1. Map of Glycosphingolipid Biosynthesis.

Data base for biosynthetic pathways of sphingolipids and glycospingolipids is available at the web site: www.sphingomap.org. Arthro-series is found in arthropods and Mollu-series in mollusks. These maps are revised versions of those published in Handbook of Glycosyltransferases and Related Genes, Springer-Verlag Tokyo, 2002.

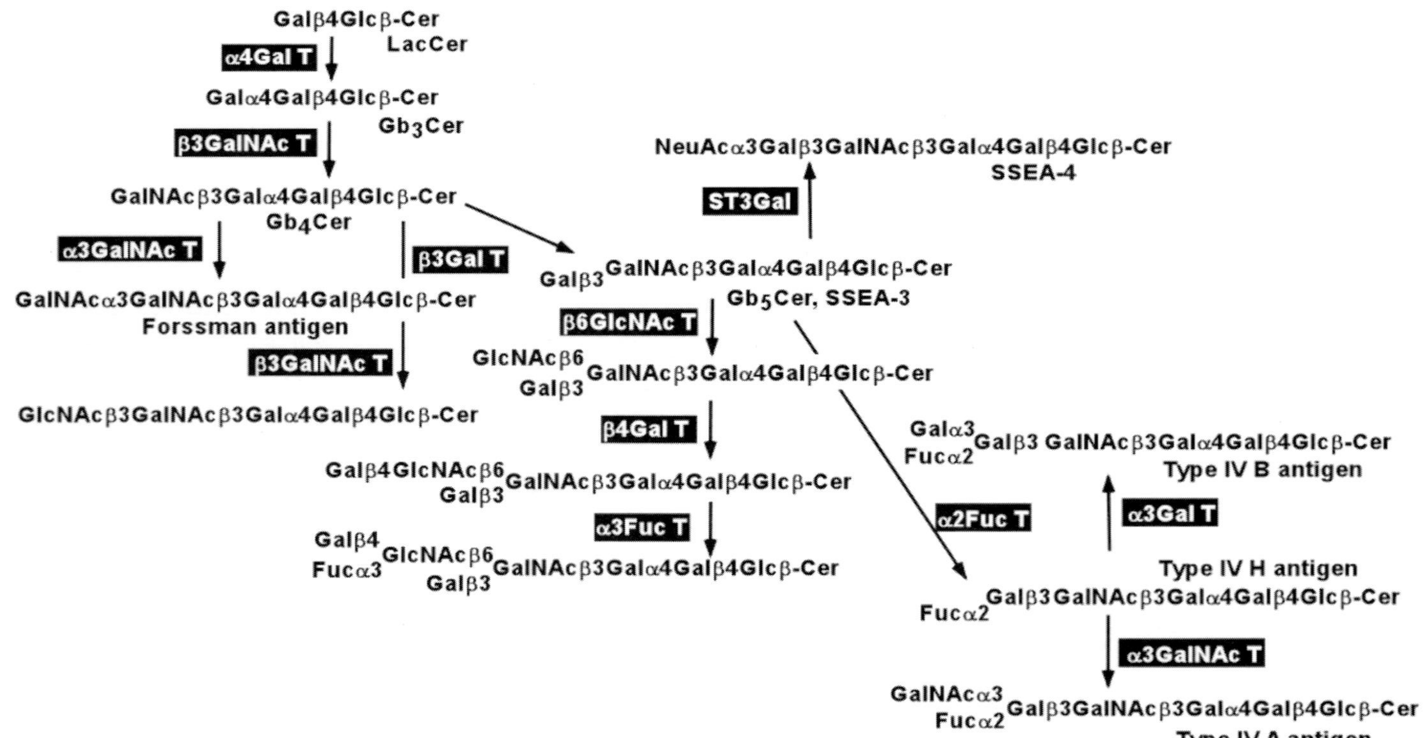

Fig. 2. Map of Globo-series Glycosphingolipids.

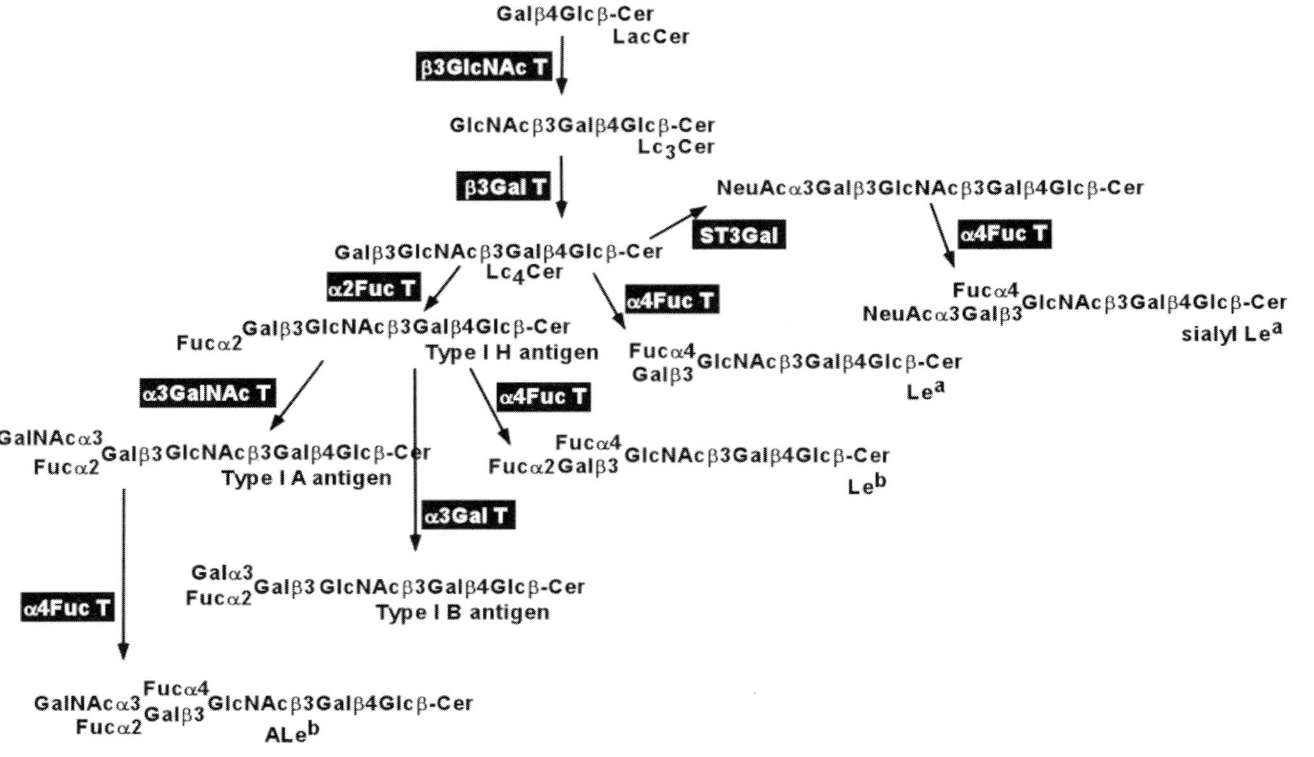

Fig. 3. Map of Lacto-series Glycosphingolipids.

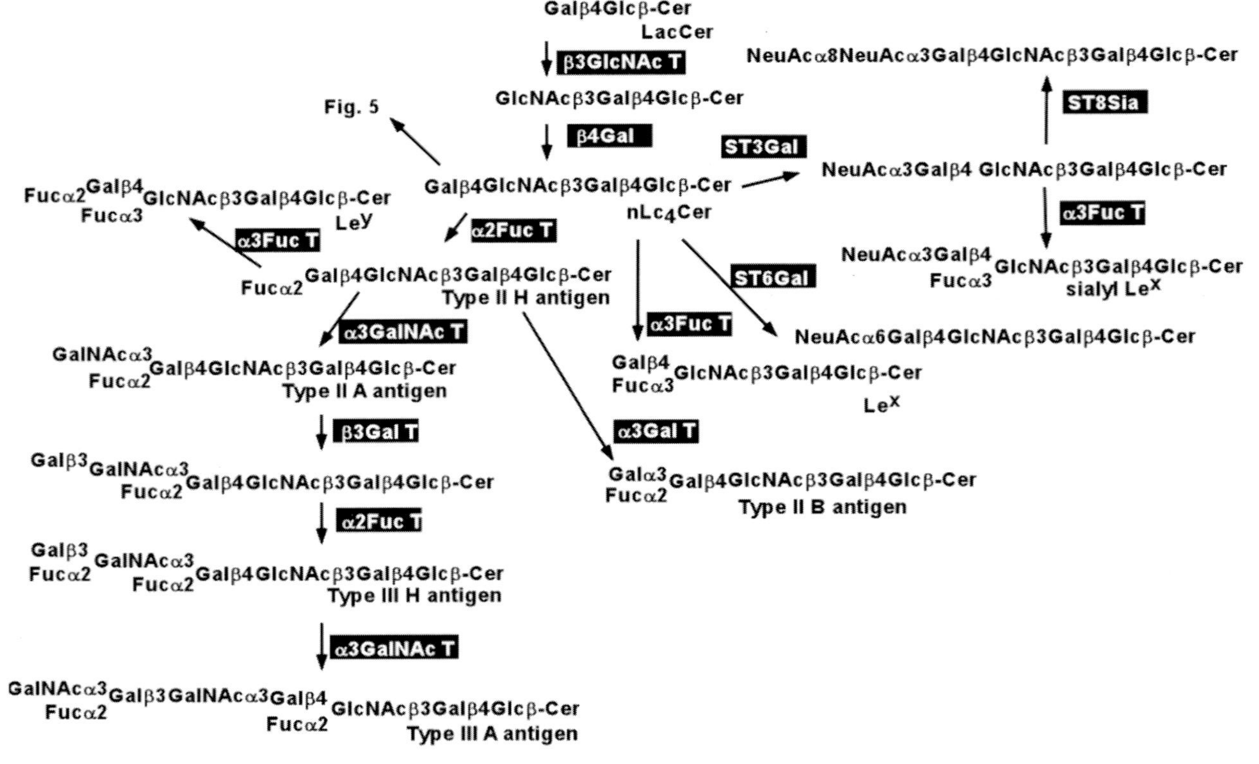

Fig. 4. Map of Neolacto-series Glycosphingolipids (1)

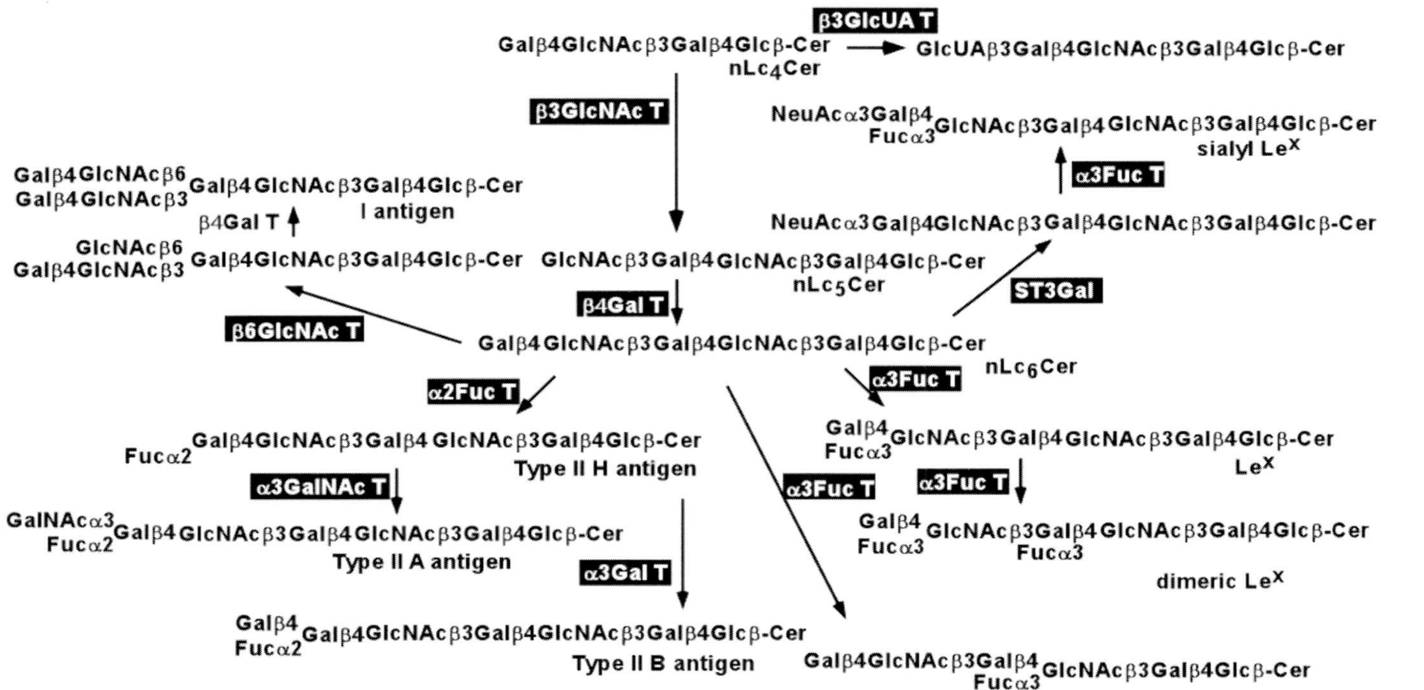

Fig. 5. Map of Neolacto-series Glycosphingolipids (2)

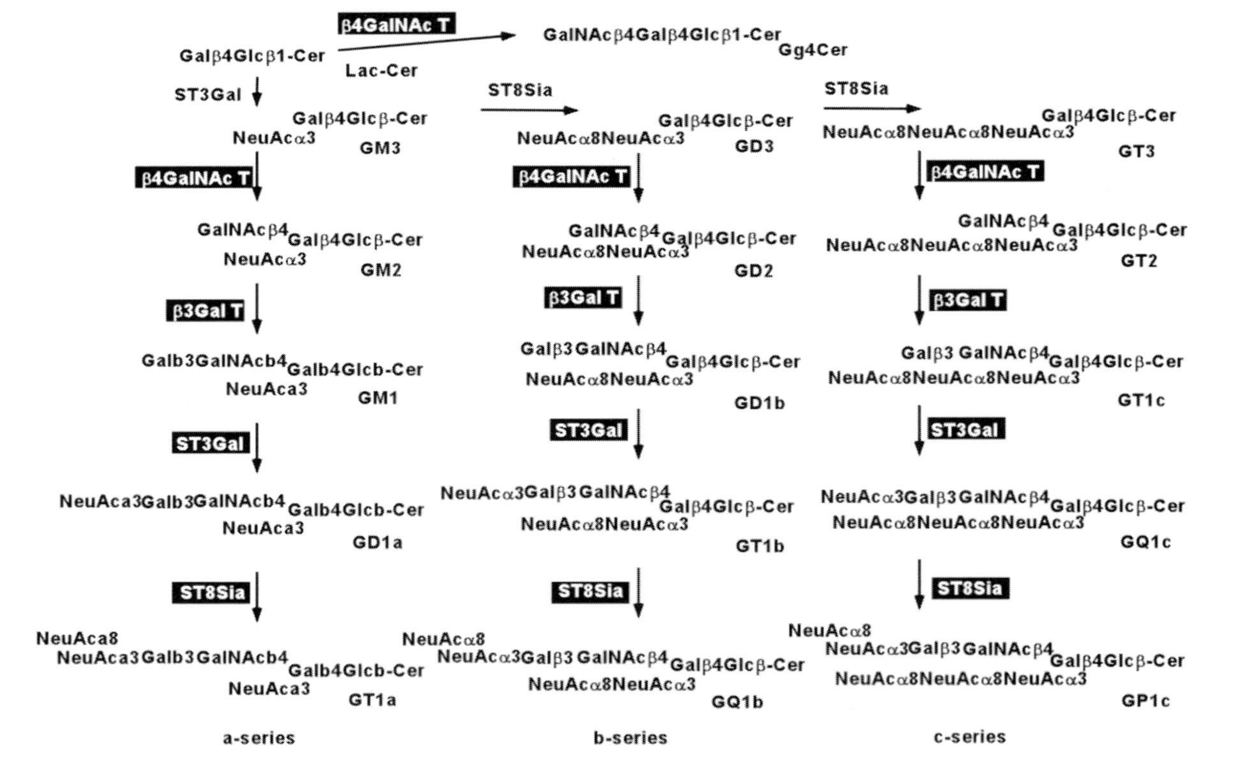

Fig. 6. Map of Gangliosides (1)

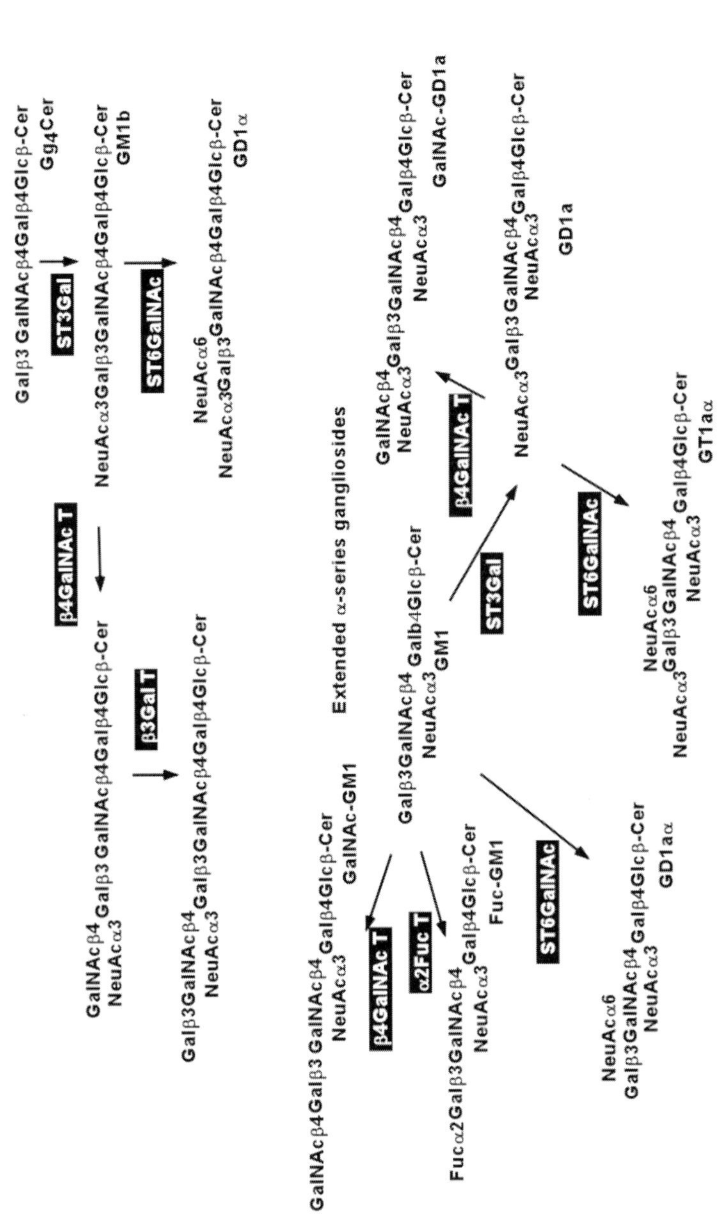

Fig. 7. Map of Gangliosides (2)

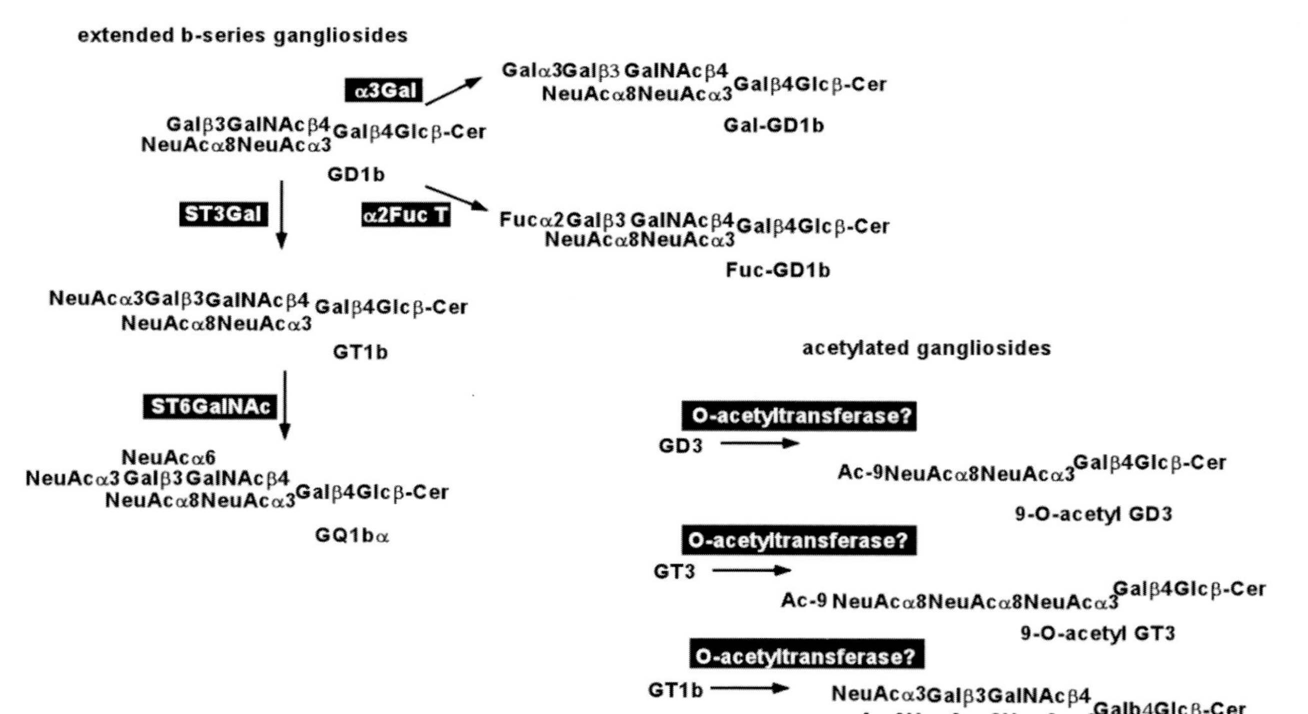

Fig. 8. Map of Gangliosides (3)

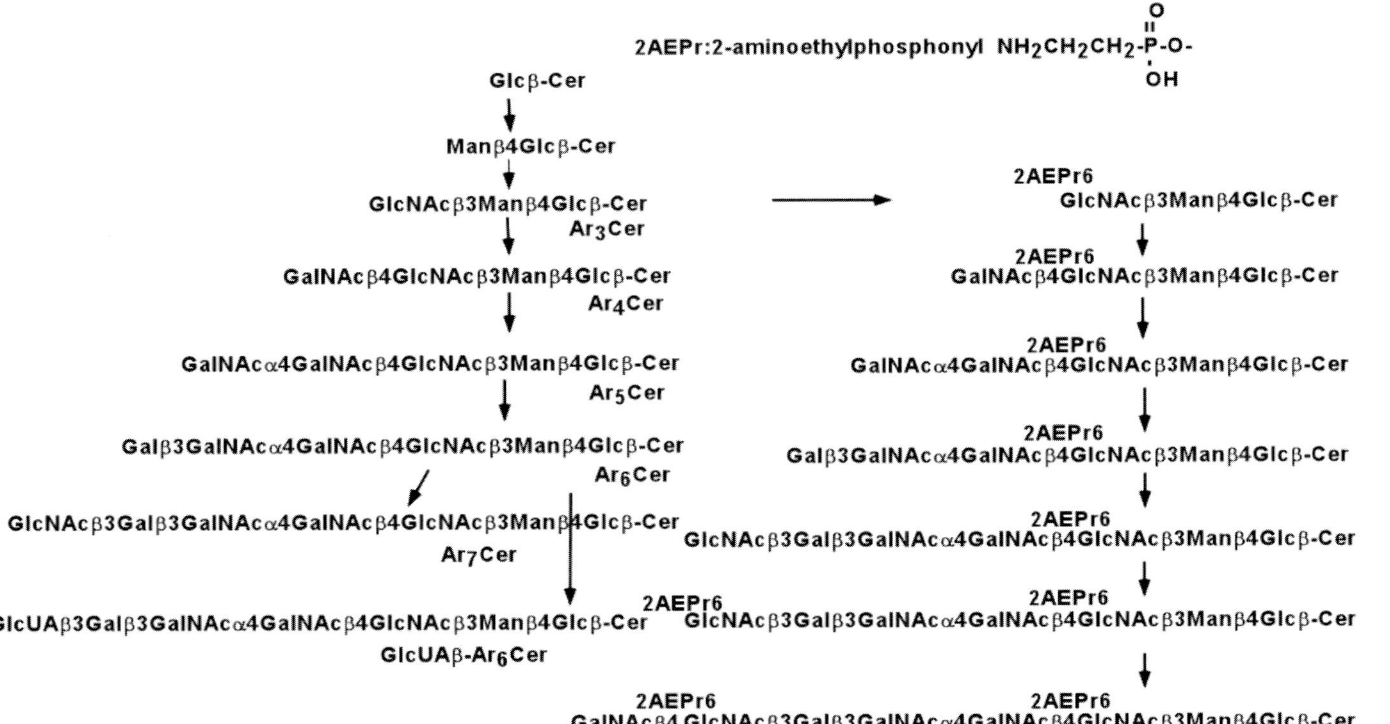

Fig. 9. Map of Arthro-series.

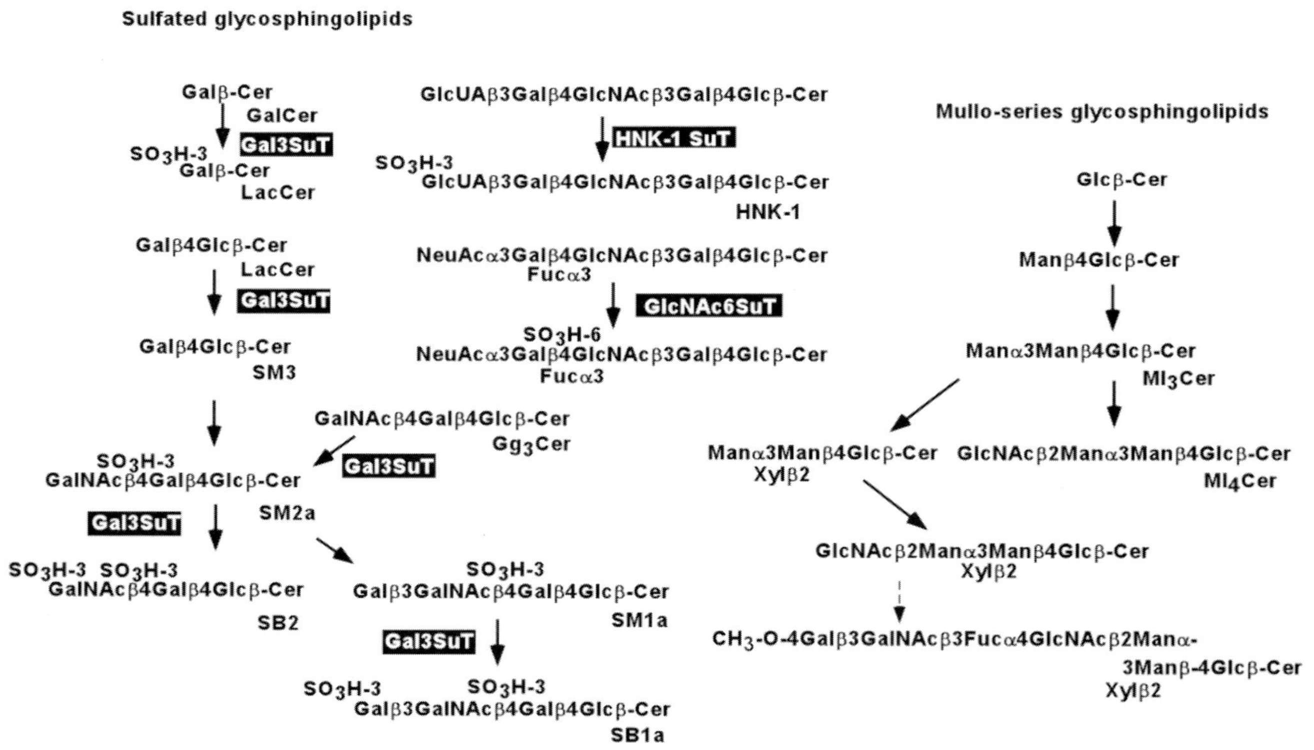

Fig. 10. Map of Sulfated Glycosphingolipids and Mullo-series.

Keyword Index

A

acid sphingomyelinase   233
acute lung injury (ALI)   403
adhesion   253
aggregation   329
Alzheimer's disease   309, 319
amyloid beta protein   309, 319
angiogenesis   183
apoptosis   69, 167, 285, 337, 453
ATP binding cassette transporter   263
axon   253
axonal degeneration   329

B

B cell receptor   245
B cell survival   245
bacterial prototype   483

C

C4-hydroxylase   57
calcium   285
cancer   363
carcinogenesis   363
caveolae   233
caveolin 1,2   233
CD38   245

cell migration   415
ceramide   49, 57, 69, 153, 167, 183, 337
ceramide glucosyltransferase   263, 443, 453
ceramide kinase   207, 337
ceramide synthase   153, 345
ceramide1-phosphate   207
CERKL   337
CERT   107
conditional knockout   443
crystal structure   483
cytoskeleton   403

D

delta4-desaturase   57
detergent-insoluble microdomain   233, 273
detergent-resistant mincrodomain (DRM)   233, 273
dihydroceramide synthase   49
disease   3, 25
Drosophila melanogaster   453

E

Edg receptor   385, 415
endothelium   403

F

FAN (factor associated with
         N-SMase activation)   167
FFAT   107
flip-flop   95
folate   345
fumonisin   345

G

galactosylceramide   83
ganglioside   309
ganglisoide GM3   273
Gaucher disease   83
gene-midified mice   443
glucosylceramide   83, 453
glucosylceramide synthase
         263, 443, 453
glucosyltransferase   263, 443,
         453
glycosphingolipidoses   285
glycosphingolipids   295, 319
G-protein coupled receptors
         (GPCRs)   385, 415
guidance   253

H

hereditary sensory neuropathy
         (HSN)   329

I

imino sugars   83
immunity   385
inflammation   285
insulin receptor   273
insulin resistance   273
intestinal epithelial cells   57
ISP-1   463

L

LASS gene   49
lateral segregation   123
lipid phosphate phosphatase
         427
lipid probe   475
lipid raft   233, 253, 273, 475
long-chain base   95, 219
lysenin   475

M

membrane domain(s)   245, 475
microdomain   309
Mss4   463
multidrug resistance   263
myriocin   463

N

neural tube defects   345
neuron   253
neutral ceramidase   183
neutral sphingomyelinase   233
Niemann Pick disease (NPD)
         167

O

organelle   123

P

pathway visualization   493
PDMP   83
permeability   403
phospholipids   285
photoreceptors   337
phytoceramide   57
phytosphingosine   95

Pkh1   141, 463
Pkh2   141
plasma membrane   233
platelets   427
pleckstrin homology (PH) domain   107, 207
pore-forming toxin   475
protein transport   319, 329
purification   49
pyridoxal 5'-phosphate   219

R

rab GTPases   295
Rac   415
raft   233, 253, 273, 475
reaction mechanism   483
regulation   25
Retinitis Pigmentosa   337

S

S1P lyase (SPL)   219
S1P receptor   183
Saccharomyces cerevisiae   141
secretase   319
seed   309
serine palmitoyltransferase   25, 153, 443, 463, 483
Sgp1l   219
Sli1   463
sphingolipid(s)   123, 263, 385
sphingolipid bases   153
sphingolipid biological functions   3
sphingolipid biosynthesis   25
sphingolipid metabolism   3
sphingolipid metabolites   363
sphingolipid signaling   463
sphingolipid storage diseases   295

sphingolipid structure   3
sphingolipid synthesis   153
sphingolipid-containing bacteria   483
sphingolipidomic   493
sphingomyelin   69, 107
sphingomyelin synthase   69
sphingomyelinase   167
sphingosine   95, 141, 197
sphingosine kinase   197
sphingosine-1-phosphate (S1P)   183, 197, 219, 385, 403, 415, 427
SphingoViz   493
SPTLC1   329
START   107
stress   25

T

tissue-specific distribution   57
trafficking   295
translocation   123
transport and sorting   123

V

vascular biology   427
vascular endothelial cells   427
vascular smooth muscle cells   427

Y

Ypk1   463

Z

zebrafish early development   183